THETA FUNCTIONS AND KNOTS

THETA FUNCTIONS AND KNOTS

Răzvan Gelca
Texas Tech University, USA

World Scientific

NEW JERSEY • LONDON • SINGAPORE • BEIJING • SHANGHAI • HONG KONG • TAIPEI • CHENNAI

Published by

World Scientific Publishing Co. Pte. Ltd.
5 Toh Tuck Link, Singapore 596224
USA office: 27 Warren Street, Suite 401-402, Hackensack, NJ 07601
UK office: 57 Shelton Street, Covent Garden, London WC2H 9HE

Library of Congress Cataloging-in-Publication Data
Gelca, Răzvan, 1967– author.
Theta functions and knots / by Răzvan Gelca (Texas Tech University, USA).
pages cm
Includes bibliographical references and index.
ISBN 978-9814520577 (hard cover : alk. paper)
1. Functions, Theta. 2. Knot theory. I. Title.
QA345.G35 2014
515'.984--dc23
2014005501

British Library Cataloguing-in-Publication Data
A catalogue record for this book is available from the British Library.

In-house Editor: Elizabeth Lie

To Alejandro Uribe and Charles Frohman, from whom I learned to love mathematical physics and topology.

Preface

Theta Functions and Knots combines two ways of mathematical reasoning, one analytical, based on computations and formulas, one geometric, based on spacial and combinatorial intuition. As the title discloses, the book is a discussion about how several areas of mathematics interact, rather than just an intensive study of one particular area. Quantum mechanics, complex analysis, low dimensional topology, and representation theory are brought together by theta functions. There are other fields of mathematics such as algebraic geometry, number theory, differential equations, also related to theta functions, that the book does not discuss because so many other sources do. Instead it focuses on some ideas of the author and his collaborators that allow representation theory to connect theta functions with low dimensional topology.

The book can be read in two different perspectives: as a text on theta functions that emphasizes their quantum mechanical, representation theoretical, and topological aspects, or as an introduction to Edward Witten's Chern-Simons theory through its simplest case, sometimes called the "trivial" case, abelian Chern-Simons theory. It is the unifying force of Chern-Simons theory that is responsible for the many facets of mathematics that come together in this book. The physical nature of Chern-Simons theory motivated the author to introduce Riemann's theta functions through quantum mechanics, in the spirit of Yuri Manin, although a purely algebraic geometric approach is also possible, and is actually embedded in the quantum mechanical approach.

Theta Functions and Knots addresses some developments in the theory of theta functions that occurred during the two decades preceeding the writing of the book, thus filling a gap in the literature. The book is an introduction to both the theory of theta functions and the related concepts from other fields of mathematics. It is accessible to all those curious about the interplay between theta functions and knots, as well as to those who want to become initiated in Chern-Simons theory.

The main concepts and results are covered in depth, to offer a self-contained presentation and to make the reader aware of all difficulties and subtleties. This means that the parts of complex analysis, low dimensional topology, quantum mechanics, and representation theory that lie outside the background of a mathematician without a particular interest in these areas are discussed in detail. Historical and bibliographical information is given whenever possible.

The narrative is focused on the theta functions associated to a Riemann surface, on the action of the finite Heisenberg group on theta functions discovered by André Weil, and on the action of the modular group on theta functions, which is the product of nineteenth century mathematics dating back to the works of Jacobi. They are discussed from various points of view. The central role is played by the representation theory of the Heisenberg group from which the entire abelian Chern-Simons theory is recovered.

The author would like to thank Alastair Hamilton, Jozef Przytycki, Zhenghan Wang, Ernesto Lupercio, Mara Neusel, Roger Barnard, and John McCleary for their suggestions and their help in the completion of this book. The author would also like to thank the many contributors to Wikipedia for offering guidance through the labyrinth of contemporary mathematics.

An active errata will be kept on author's web page. Any corrections, additions, or comments can be sent to *rgelca@gmail.com*.

Contents

Chapter 1

Prologue

The main actors of this book are **classical theta functions** and the **Gaussian linking number**. They were brought together by Edward Witten using a quantum field theory whose Lagrangian is the Chern-Simons functional. In this chapter we offer the reader a first encounter with theta functions, the linking number, and Witten's Chern-Simons theory. The discussion is less formal, less detailed, and less rigorous then the rest of the book, and should be read like a historical summary. We will revisit these notions later, with more rigor and detail.

1.1 The history of theta functions

In the development of mathematics, theta functions appeared in early 19th century, as tools for studying elliptic integrals and elliptic functions. The reader should be aware that a great body of mathematics was devoted to elliptic integrals and functions. Below we only sketch a few of the most important contributions; for more details we recommend [Houzel (1978)].

1.1.1 *Elliptic integrals and theta functions*

Elliptic integrals appeared in the 17th and 18th century in computations of arc-lengths of curves and in models of classical mechanics such as pendulums and springs. The first example was the integral computing the arc-length of an **ellipse**, which gave the name of the entire class. For the ellipse

$$x = a\sin\phi, \quad y = b\cos\phi, \quad \phi \in [0, 2\pi]$$

the arc-length is computed by the integral

$$\int_P^Q ds = a\int_{\phi_P}^{\phi_Q} \sqrt{1-k^2\sin^2\phi}d\phi$$

where $k^2 = 1-b^2/a^2$ is the eccentricity. The integrand is not the derivative of an elementary function.

An example from mechanics is the integral

$$y = \int \frac{x^2dx}{\sqrt{1-x^4}},$$

which Jacob Bernoulli found to describe the equilibrium position of an elastic rod with fixed extremities, subject to a force. Neither is this integral computable through elementary functions.

Lagrange was the first to consider the general case. For him elliptic integrals are of the form

$$\int R(x,y)dx$$

where R is a rational function, and $y = \sqrt{P(x)}$ with P a polynomial of degree 3 or 4 without multiple roots. Legendre reduced elliptic integrals to three types:

$$\int \frac{d\phi}{\sqrt{1-k^2\sin^2\phi}}, \int\sqrt{1-k^2\sin^2\phi}d\phi, \int\frac{d\phi}{(1+n\sin^2\phi)\sqrt{1-k^2\sin^2\phi}},$$

where $x = \sin\phi$.

Since these integrals cannot be reduced to known functions, an intense effort was devoted to their study and their approximate computation. An important role was played by **addition formulas**, first discovered by Euler in the case of the arc-length of the lemniscate, which found their apogee in the works of Abel and Jacobi.

To explain these formulas, let us take a look at the simpler situation of the trigonometric functions $\sin u$ and $\cos u$. We are familiar with them because mathematics started with geometry. But say if mathematics had started with polynomials, then the natural introduction of sine and cosine would have been through their inverse functions

$$\arcsin u = \int_0^u \frac{dx}{\sqrt{1-x^2}}, \quad \arccos u = -\int_0^u \frac{dx}{\sqrt{1-x^2}}.$$

Then $\sin u$ and $\cos u$ would be the inverses of these integrals.

One has the addition formula

$$\sin(u+v) = \sin u\cos v + \cos u \sin v.$$

Moreover, $\sin u$ and $\cos u$ are periodic functions of period 2π. This period is 4 times the "complete integral"

$$\arcsin(1) - \arcsin(0) = \int_0^1 \frac{dx}{\sqrt{1-x^2}}.$$

Abel had the idea of considering the inverse functions of elliptic integrals. These are what we now call **elliptic functions**. Jacobi, inspired by Abel, defined the inverse $\operatorname{am} u$ of the elliptic integral of the first kind

$$u = \int \frac{d\phi}{\sqrt{1-k^2\sin^2\phi}}.$$

Gudermann later introduced the notation

$$\operatorname{sn} u = \sin \operatorname{am} u, \quad \operatorname{cn} u = \cos \operatorname{am} u, \quad \operatorname{dn} u = \sqrt{1-k^2\sin^2 \operatorname{am} u}.$$

The functions sn and cn are periodic with period $4K$ and dn is periodic with period $2K$, where K is the complete integral

$$u(\pi/2) - u(0) = \int_0^{\frac{\pi}{2}} \frac{d\phi}{\sqrt{1-k^2\sin^2\phi}}.$$

Abel showed that elliptic functions satisfy addition theorems similar to those for trigonometric functions, for example

$$\operatorname{sn}(u+v) = \frac{\operatorname{sn} u \operatorname{cn} v \operatorname{dn} v + \operatorname{sn} v \operatorname{cn} u \operatorname{dn} u}{1-k^2\operatorname{sn}^2 u \operatorname{sn}^2 v}.$$

Jacobi was able to extend these elliptic functions to a complex variable. In this context they differ from the trigonometric functions sine and cosine by two significant properties: they are meromorphic and they are doubly periodic. The periods of sn, cn, dn are respectively $(4K, 2iK')$, $(4K, 2(K+iK'))$, $(2K, 4iK')$, where

$$K' = \int_0^{\pi/2} \frac{d\phi}{\sqrt{1-k'^2\sin^2\phi}}, \text{ with } k' = \sqrt{1-k^2}.$$

Jacobi gave these three elliptic functions global analytical representations as quotients of holomorphic functions

$$\operatorname{sn} u = \frac{1}{\sqrt{k}}\frac{H(u)}{\Theta(u)}, \quad \operatorname{cn} u = \sqrt{\frac{k'}{k}}\frac{H_1(u)}{\Theta(u)}, \quad \operatorname{dn} u = \sqrt{k'}\frac{\Theta_1(u)}{\Theta(u)},$$

by introducing the four **theta functions**

$$\Theta\left(\frac{2Kx}{\pi}\right) = \vartheta(x;q) = \sum_{n\in\mathbb{Z}}(-1)^n q^{n^2} e^{2nix},$$
$$H\left(\frac{2Kx}{\pi}\right) = \vartheta_1(x;q) = \sum_{n\in\mathbb{Z}} i^{2n+1} q^{(n+1/2)^2} e^{(2n+1)ix},$$
$$H_1\left(\frac{2Kx}{\pi}\right) = \vartheta_2(x;q) = \sum_{n\in\mathbb{Z}} q^{(n+1/2)^2} e^{(2n+1)ix}$$
$$\Theta_1\left(\frac{2Kx}{\pi}\right) = \vartheta_3(x;q) = \sum_{n\in\mathbb{Z}} q^{n^2} e^{2nix},$$

where $q = e^{-\pi K'/K}$. Θ_1 and H_1 are computed in terms of Θ and H as $\Theta_1(u) = \Theta(K-u)$ and $H_1(u) = H(K-u)$.

The functions Θ and H are not doubly periodic, as Liouville's theorem would prohibit that, nevertheless they are periodic with periods $2K$ respectively $4K$ and they are as close as being doubly periodic as possible with

$$\Theta(u + 2iK') = -e^{-i\pi u} q^{-1} \Theta(u),$$
$$H(u + 2iK') = -e^{-\pi K'/K + 2\pi K' - i\pi u} q^{-1} H(u).$$

Nowadays it is customary to replace x by πz, and to use the parameter τ instead of q, with $q = \exp(\pi i\tau)$, allowing τ to range in the upper half-plane of the complex plane (so $K'/K = -i\tau$). Also, following Riemann, one emphasizes the theta function

$$\theta(z;\tau) = \theta_{00}(z;\tau) = \vartheta_3(x;q) = \sum_{n\in\mathbb{Z}} e^{\pi i\tau n^2 + 2\pi inz},$$

and relates the others to it by

$$\theta_{01}(z;\tau) = \vartheta(x;q) = \theta\left(z + \frac{1}{2};\tau\right),$$
$$\theta_{10}(z;\tau) = \vartheta_2(x;q) = e^{\frac{1}{4}\pi i\tau + \pi iz}\theta\left(z + \frac{1}{2}\tau;\tau\right)$$
$$\theta_{11}(z;\tau) = -\vartheta_1(x;q) = e^{\frac{1}{4}\pi i\tau + \pi i(z+\frac{1}{2})}\theta\left(z + \frac{1}{2}\tau + \frac{1}{2};\tau\right).$$

The periodicity condition reads

$$\theta(z + m + n\tau;\tau) = e^{-\pi in^2\tau - 2\pi inz}\theta^\tau(z;\tau).$$

Changes of coordinates were employed in order to improve the numerical computations of elliptic integrals. They led to the study of the behavior of elliptic functions and of theta functions under coordinate transformations. Jacobi discovered the identity

$$\theta\left(\frac{z}{\tau};-\frac{1}{\tau}\right)=(-i\tau)^{1/2}\exp\left(\frac{\pi}{\tau}iz^2\right)\theta(z;\tau).$$

This has led to the discovery of the action of the entire modular group

$$PSL(2,\mathbb{Z})=\left\{z\mapsto\frac{az+b}{cz+d}\,|\,ad-bc-1\right\}$$

on theta functions.

Theta functions were also used previously by Gauss for the same purpose as Jacobi, in his work on the arithmetic-geometric mean which he related to the elliptic integral of the first kind. They were used by Jacob Bernoulli and Euler in number theory, and by Poisson and Fourier for solving the heat equation. For the latter purpose one restricts the variable z to a real number x and the parameter τ to it with t real. Then

$$\frac{\partial}{\partial t}\theta(x;it)=\frac{1}{4\pi}\frac{\partial^2}{\partial x^2}\theta(x;it),$$

with the initial condition $\theta(x;0)=\sum_{n\in\mathbb{Z}}\delta(x-n)$, where δ is Dirac's delta function.

1.1.2 *The work of Riemann*

Riemann introduced a completely new point of view in the theory of elliptic integrals and theta functions.

To study a function $w(z)$ defined by a polynomial equation $F(z,w)=0$ of degree m in z and n in w, Riemann introduced a complex surface Σ (a Riemann surface), which is an n-sheeted covering of the plane, and on which w is a well defined holomorphic function. Adding the point at infinity turns Σ into a compact, orientable surface. Riemann's programme was to study the integrals of rational functions $R(z,w)$ along paths in Σ.

The Riemann surface associated to $w(z)$ can have the genus g equal to 0 (when w can be solved explicitly in terms of z, which case is totally uninteresting), 1 (the case studied extensively by Abel and Jacobi), or higher. The function

$$u(x)=\int_a^x R(z(t),w(t))dt$$

is univalent in the complement of a finite collection of curves that form a basis for the first homology group of Σ. The $2g$ integrals of $R(z,w)$ along these curves define the **periods** of the inverse function of u. The periods define a lattice in $\mathbb{C}^g$. The quotient of $\mathbb{C}$ by this lattice is the **Jacobian variety**. The function u defined by integrating $R(z,w)$ along paths in Σ is well defined from Σ to the Jacobian variety.

To study the elliptic functions on the Jacobian variety, Riemann introduced his multivariable theta function

$$\theta(\mathbf{z};\Pi) = \sum_{\mathbf{n}\in\mathbb{Z}^g} e^{\pi i \mathbf{n}^T \Pi \mathbf{n} + 2\pi i \mathbf{n}^T \mathbf{z}},$$

where Π is a $g \times g$ matrix with positive-definite imaginary part. The matrix Π depends on the complex structure of the Riemann surface Σ. In genus 1, Π is the complex number τ encountered above.

Example 1.1. These ideas are best understood when applied to the classical example of the Weierstrass elliptic integral

$$u(z) = \int_z^\infty \frac{dt}{\sqrt{t(t-1)(t-\lambda)}}, \quad 0 < \lambda < 1.$$

Here the Riemann surface is associated to the function w defined by $w^2 = \sqrt{z(z-1)(z-\lambda)}$. The function

$$w : \mathbb{C} \to \mathbb{C}, \quad w = \sqrt{z(z-1)(z-\lambda)}$$

is univalent on $\mathbb{C}\backslash([0,\lambda] \cup [1,\infty))$, so the Riemann surface associated to it is obtained by cutting open the complex sphere $\mathbb{C}P^1 = \mathbb{C} \cup \infty$ along two segments, $[0,\lambda]$ and $[1,\infty]$, taking two copies of the result, and gluing them along their boundaries as shown in Figure 1.1. This construction identifies the torus $S^1 \times S^1$ with the algebraic variety $z_1^2 z_2 = z_3(z_3 - z_2)(z_3 - \lambda z_2)$ in the complex projective space $\mathbb{C}P^2$ (with $w = z_1/z_2$, and $z = z_3$).

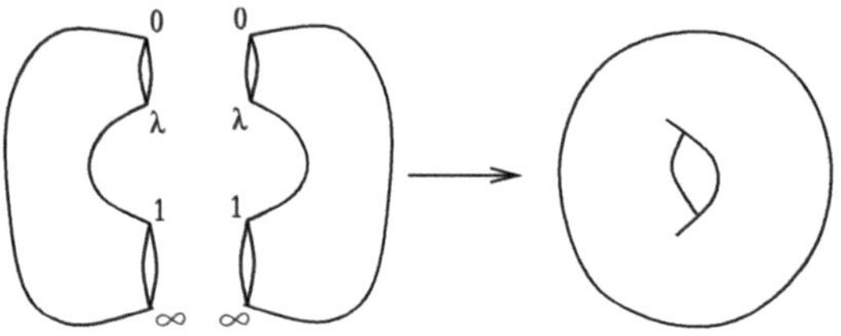

Fig. 1.1 The Riemann surface of $w = \sqrt{z(z-1)(z-\lambda)}$

The integral $u(z)$ is well defined only up to multiples of the values of two integrals. One of them is the integral along either of the branch cuts, the other is the integral along a curve that crosses each of the two branch cuts once. These are **Riemann's periods** ω_1 and ω_2. Then u has well-defined values in the 2-dimensional torus $\mathbb{C}/(\mathbb{Z}\omega_1 + \mathbb{Z}\omega_2)$. This torus is the Jacobian variety associated to the elliptic integral.

The inverse of u is the **Weierstrass elliptic function** $\wp$, which can be interpreted either as a meromorphic function on the Jacobian variety, or as the doubly periodic meromorphic function on $\mathbb{C}$,

$$\wp(z;\omega_1,\omega_2) = \frac{1}{z^2} + \sum_{(m,n)\neq(0,0)} \left[\frac{1}{(z+m\omega_1+n\omega_2)^2} - \frac{1}{(m\omega_1+n\omega_2)^2}\right].$$

One can always arrange, by a change of coordinates, for ω_1 to be 1, in which case it is standard to write $\tau = \omega_2$. In terms of theta functions

$$\wp(z;\tau) = \pi^2\theta^2(0;\tau)\theta_{10}^2(0;\tau)\frac{\theta_{01}^2(z;\tau)}{\theta_{11}(z;\tau)} - \frac{1}{3}\pi^2(\theta^4(0;\tau) + \theta_{10}^4(0;\tau)).$$

In Chapter 4 we will develop the theory of Riemann's theta functions. For this we will employ the tools of quantum mechanics, since theta functions are yet another example of pure mathematics into which quantum theory offers new insights. We are motivated to take this approach by Witten's abelian Chern-Simons theory which is physical in its nature. The functions considered there are of a slightly finer structure. They include a parameter, which is an integer N that plays the role of Planck's constant; in Riemann's situation $N = 1$. They are called **theta functions with characteristics** in [Mumford (1983)].

The reader will notice that in Chapter 4 and the subsequent chapters we write $\theta^\tau(z)$ and $\theta^\Pi(\mathbf{z})$ instead of $\theta(z;\tau)$ and $\theta(\mathbf{z};\Pi)$. This is done so as to simplify formulas and computations, and to emphasize that z is the actual variable while τ and Π are (fixed) parameters depending on the complex structure of the Riemann surface.

1.2 The linking number

1.2.1 *The definition of the linking number*

In mathematics, a **knot** is an embedding of a circle in $\mathbb{R}^3$. The embedding of a disjoint union of finitely many circles in $\mathbb{R}^3$ is called a **link**; each circle is called a **link component**. Two knots or links are the same if one can be deformed continuously into the another without crossing itself.

Knots (and similarly links) are represented by knot diagrams, which are projections of the knot onto a plane, so that all multiple points are double points with the strands crossing transversally. One specifies which strand crosses over. We are only concerned with smooth embeddings, in which case knots and links have finitely many crossings. Two examples are shown in Figure 1.2, the figure-eight knot and the Hopf link. It is customary to

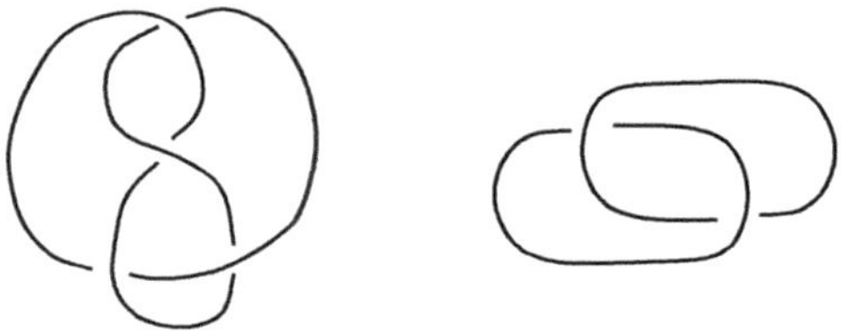

Fig. 1.2 The figure-eight knot and the Hopf link

add the point at infinity and therefore consider knots and links in the 3-dimensional sphere S^3. The theory is the same as for $\mathbb{R}^3$, but S^3 has the advantage of being a compact manifold.

In 1833, while computing the work done on a magnetic pole moving along a closed curve in the presence of a loop of current, Gauss discovered the linking number of two non-intersecting curves. He started with the Biot-Savart Law, which computes the magnetic field $\mathbf{B}$ at a given point $\mathbf{r}$ produced by an electric field of a steady current I in a thin closed wire modeled by a curve γ_1 in the space. The Biot-Savart Law states that

$$\mathbf{B}(\mathbf{r}) = \frac{\mu_0}{4\pi}\int_{\gamma_1} \frac{I}{\|\mathbf{r}-\mathbf{r}_1\|^3}\frac{d\mathbf{r}_1}{ds} \times (\mathbf{r}-\mathbf{r}_1)ds,$$

where μ_0 is the magnetic permeability of the vacuum and $\mathbf{r}_1(s)$ is the parametrization of γ_1. Evaluating the work of the magnetic field along a second curve γ_2 parametrized by $\mathbf{r}_2(t)$ one obtains

$$\begin{aligned}\int_{\gamma_2} \mathbf{B}\cdot\frac{d\mathbf{r}_2}{dt}dt &= \frac{\mu_0}{4\pi}\int_{\gamma_2}\int_{\gamma_1} \frac{I}{\|\mathbf{r}_2-\mathbf{r}_1\|^3}\left(\frac{d\mathbf{r}_1}{ds}\times(\mathbf{r}_2-\mathbf{r}_1)\right)\cdot\frac{d\mathbf{r}_2}{dt}dsdt \\ &= \frac{\mu_0}{4\pi}\int_{\gamma_1}\int_{\gamma_2} I\frac{\mathbf{r}_1-\mathbf{r}_2}{\|\mathbf{r}_1-\mathbf{r}_2\|^3}\cdot\left(\frac{d\mathbf{r}_1}{ds}\times\frac{d\mathbf{r}_2}{dt}\right)dtds.\end{aligned}$$

Here we used the formula $(\mathbf{a}\times\mathbf{b})\cdot\mathbf{c} = -\mathbf{b}\cdot(\mathbf{a}\times\mathbf{c})$. Setting $\mu_0 = I = 1$, Gauss produced the following formula for the linking number of γ_1 and γ_2:

$$\text{lk}(\gamma_1,\gamma_2) = \frac{1}{4\pi}\int_{\gamma_1}\int_{\gamma_2}\frac{\mathbf{r}_1-\mathbf{r}_2}{\|\mathbf{r}_1-\mathbf{r}_2\|^3}\cdot\left(\frac{d\mathbf{r}_1}{ds}\times\frac{d\mathbf{r}_2}{dt}\right)dtds,$$

where γ_1, γ_2 are the two curves parameterized by $\mathbf{r}_1$ respectively $\mathbf{r}_2$. Note that $\text{lk}(\gamma_1,\gamma_2) = \text{lk}(\gamma_2,\gamma_1)$, in other words, we can switch γ_1 and γ_2 and obtain the same value of the work.

As we will see below, the linking number counts the number of times one of the two curves winds around the other. The linking number will play an important role in our discussion, and is an instance of the interplay between analysis and combinatorial topology.

The value of the integral is invariant under isotopy, meaning that we are allowed to deform the curves in the 3-dimensional space as long as they don't cross each other. Much more is true, namely we have the following result.

Theorem 1.1. *If γ_1 and γ_1' are two curves such that $\gamma_1 \cup (-\gamma_1')$ bounds a surface Σ, then $lk(\gamma_1, \gamma_2) = lk(\gamma_1', \gamma_2)$.*

Proof. The proof is based on the Ampére Law, which states that the magnetic field around the boundary of a surface is proportional to the total current passing through the surface. This is a particular case of Stokes' Theorem, and our task is to show that the current passing through the surface is zero. We consider the situation where the electric current passes through γ_2.

Consider the parametrizations $\mathbf{r}_1(s) = (x(s), y(s), z_1(s))$, $\mathbf{r}_2(t) = (\tilde{x}(t), \tilde{y}(t), \tilde{z}(t))$ of γ_1 respectively γ_2 and set

$$P(x,y,z) = \int_{\gamma_2} \frac{-(\tilde{y}-y)d\tilde{z} + (\tilde{z}-z)d\tilde{y}}{((\tilde{x}-x)^2 + (\tilde{y}-y)^2 + (\tilde{z}-z)^2)^{3/2}}$$

$$Q(x,y,z) = \int_{\gamma_2} \frac{(\tilde{x}-x)d\tilde{z} - (\tilde{z}-z)d\tilde{x}}{((\tilde{x}-x)^2 + (\tilde{y}-y)^2 + (\tilde{z}-z)^2)^{3/2}}$$

$$R(x,y,z) = \int_{\gamma_2} \frac{-(\tilde{x}-x)d\tilde{y} + (\tilde{y}-y)d\tilde{x}}{((\tilde{x}-x)^2 + (\tilde{y}-y)^2 + (\tilde{z}-z)^2)^{3/2}}.$$

Set also

$$\alpha_{\gamma_2} = \frac{1}{4\pi}(Pdx + Qdy + Rdz). \tag{1.1}$$

Then

$$\text{lk}(\gamma_1, \gamma_2) = \int_{\gamma_1} \alpha_{\gamma_2} = \frac{1}{4\pi}\int_{\gamma_1} Pdx + Qdy + Rdz.$$

Let $\gamma = \gamma_1' \cup (-\gamma_1)$. Stokes' theorem for γ and Σ reads

$$\int_\gamma \alpha_{\gamma_2} = \iint_\Sigma d\alpha_{\gamma_2},$$

or explicitly

$$\int_\gamma Pdx + Qdy + Rdz = \iint_\Sigma \left(\frac{\partial Q}{\partial x} - \frac{\partial P}{\partial y}\right) dxdy + \left(\frac{\partial R}{\partial y} - \frac{\partial Q}{\partial z}\right) dydz + \left(\frac{\partial P}{\partial z} - \frac{\partial R}{\partial x}\right) dzdx.$$

We will show that the form α_{γ_2} is closed, namely that

$$\frac{\partial Q}{\partial x}-\frac{\partial P}{\partial y}=\frac{\partial R}{\partial y}-\frac{\partial Q}{\partial z}=\frac{\partial P}{\partial z}-\frac{\partial R}{\partial x}=0.$$

We only verify $\frac{\partial Q}{\partial x}-\frac{\partial P}{\partial y}=0$, the other equalities being similar. The part of this expression containing $d\tilde{z}$ is equal to

$$\begin{aligned}
&\int_{\gamma_2} -2((\tilde{x}-x)^2+(\tilde{y}-y)^2+(\tilde{z}-z)^2)^{-3/2}\\
&\quad +3(\tilde{x}-x)^2((\tilde{x}-x)^2+(\tilde{y}-y)^2+(\tilde{z}-z)^2)^{-5/2}\\
&\quad +3(\tilde{y}-y)^2((\tilde{x}-x)^2+(\tilde{y}-y)^2+(\tilde{z}-z)^2)^{-5/2}d\tilde{z}\\
&=\int_{\gamma_2} ((\tilde{x}-x)^2+(\tilde{y}-y)^2+(\tilde{z}-z)^2)^{-3/2}\\
&\quad +3(\tilde{z}-z)^2((\tilde{x}-x)^2+(\tilde{y}-y)^2+(\tilde{z}-z)^2)^{-5/2}d\tilde{z}\\
&=\int_{\gamma_2} \frac{\partial}{\partial \tilde{z}}((\tilde{x}-x)^2+(\tilde{y}-y)^2+(\tilde{z}-z)^2)^{-3/2}d\tilde{z}=0.
\end{aligned}$$

On the other hand, only $\frac{\partial Q}{\partial x}$ has a $d\tilde{x}$ in it, so the part containing $d\tilde{x}$ in $\frac{\partial Q}{\partial x}-\frac{\partial P}{\partial y}$ is equal to

$$\begin{aligned}
&3\int_{\gamma_2} ((x-\tilde{x})^2+(y-\tilde{y})^2+(z-\tilde{z})^2)^{-5/2}(x-\tilde{x})(z-\tilde{z})d\tilde{x}=\\
&\int_{\gamma_2} \frac{\partial}{\partial \tilde{x}}\frac{z-\tilde{z}}{((x-\tilde{x})^2+(y-\tilde{y})^2+(z-\tilde{z})^2)^{3/2}}d\tilde{x}=0.
\end{aligned}$$

The term involving dy' is treated similarly. The conclusion follows. □

Remark 1.1. Σ can be the surface traced by γ_1 while being deformed into γ_1' but it could be any surface bounded by γ_1 and γ_1', showing that the linking number is a homological invariant. In fact $\mathrm{lk}(\gamma_1,\gamma_2)$ is the homology class of γ_1 in $H_1(S^3\backslash\gamma_2,\mathbb{Z})=\mathbb{Z}$.

As such the linking number is a link invariant for oriented links with two components (i.e. for links whose components are oriented). But more is true, namely the linking number is an integer and it can be computed combinatorially from a link diagram. To see why this is so, deform γ_1 so that it consists of several arcs connected to circles that link γ_2, such as in Figure 1.3. As the arcs are traveled back and forth, it suffices to replace γ_1 with the union of the circles. We can compute the linking number separately for each circle, and then add up.

Next, we can deform γ_2 so that everything besides the arc that links with the circle γ_1 is pushed towards infinity, and that arc and the circle

Fig. 1.3 The computation of the linking number

look like the z-axis and the circle $x^2 + y^2 = 1$ in $\mathbb{R}^3$. Parametrizing γ_2 by $\mathbf{r}_2 = (0, 0, t)$, there are two possibilities for the parametrization of γ_1:

$$\mathbf{r}_1 = (\cos s, \sin s, 0) \text{ or } \mathbf{r}_2 = (\cos(-s), \sin(-s), 0).$$

The first case corresponds to the situation from Figure 1.3, and we compute

$$\begin{aligned}\operatorname{lk}(\gamma_1, \gamma_2) &= \frac{1}{4\pi}\int_{-\infty}^{\infty}\int_0^{2\pi} \frac{(\cos s, \sin s, -t)}{(1+t^2)^{3/2}} \cdot [(-\sin s, \cos s, 0) \times (0,0,1)]dsdt \\ &= \frac{1}{4\pi}\int_{-\infty}^{\infty}\int_0^{2\pi} \frac{1}{(1+t^2)^{3/2}} dsdt = \frac{1}{2}\left.\frac{t}{\sqrt{1+t^2}}\right|_{-\infty}^{\infty} = 1.\end{aligned}$$

In the other case the value of the integral is -1.

This allows a combinatorial computation of the linking number from a link diagram. Define the sign of a crossing using Figure 1.4, with the crossing on the left being positive and the one on the right negative. In a link diagram, each crossing is of one of these two types. Consider only the crossings where γ_1 crosses over γ_2, and let n_1 and n_2 be the number of positive, respectively negative crossings. Then

$$\operatorname{lk}(\gamma_1, \gamma_2) = n_1 - n_2.$$

If we perform the computation using both the over and the under crossings, we obtain twice the linking number.

Remark 1.2. The above computation shows that if D is an oriented disk that intersects the closed curve γ at one point, and if ∂D is oriented so that D is on the left, then $\operatorname{lk}(\partial D, \gamma) = \pm 1$. The sign is $+1$ if the 3-dimensional frame obtained by adjoining to an orientation frame of D the tangent vector of γ is positively oriented and -1 if this frame is negatively oriented.

Example 1.2. Depending on the orientation of the link components, the linking number of the Hopf link (Figure 1.2) is either 1 or -1.

Fig. 1.4 A positive and a negative crossing

If we are given just one knotted oriented curve, γ, and consider a knot diagram of it, then we can use the same rules for the crossings in this diagram. The difference between the number of positive crossings and the number of negative crossings is called the **writhe** of the knot diagram. The writhe is not a knot invariant, since it changes if we introduce twists (kinks) like those shown in Figure 1.5.

Fig. 1.5 Positive and negative twists

If we are given a link with two components γ_1 and γ_2, then we can compute $\mathrm{lk}(\gamma_1, \gamma_2)$ from a link diagram, by taking the difference between the total number of positive crossings and that of negative crossings, then subtracting the sum of the writhes of the projections of γ_1 and γ_2, and then dividing the answer by 2.

1.2.2 *The Jones polynomial*

In 1984, Vaughan F.R. Jones discovered a polynomial invariant of knots, which can be computed recursively [Jones (1985)]. For a knot K, the Jones polynomial is a one-variable polynomial, $V_K(t)$, that is equal to 1 for the unknot, and can be computed for any other knot by orienting it and then applying repeatedly the relation

$$t^{-1}V_{K_+}(t) - tV_{K_-}(t) = (t^{1/2} - t^{-1/2})V_{K_0}(t). \tag{1.2}$$

The relation (1.2) is called a **skein relation**. In this expression, K_+, K_-, K_0 denote three oriented knots or links which have the same diagram except for a crossing, and that crossing is positive for K_+, negative for K_-, while for K_0 the crossing is erased and the strands are connected so that the orientations agree.

Definition 1.1. A skein relation for a knot and link invariant is a linear relation among the invariants of the knots and links obtained by modifying one crossing according to some rules.

In the case of the Jones polynomial, the crossing is modified by changing it from an undercrossing to an overcrossing or vice-versa, and by cutting it open and joining the ends so that the crossing disappears.

Example 1.3. The computation of the Jones polynomial for the right-handed trefoil knot is shown in Figure 1.6.

One obtains

$$\begin{aligned} V_{\text{trefoil}}(t) &= t^2 + (t^{3/2} - t^{1/2})V_{\text{Hopf}}(t) \\ &= t^2 + (t^{3/2} - t^{1/2})[t^2(-t^{1/2} - t^{-1/2}) + (t^{3/2} - t^{1/2})] \\ &= t + t^3 - t^4. \end{aligned}$$

Jones discovered his polynomial while using combinatorial methods for understanding how an algebra of quantum observables lies inside another. No intrinsic topological definition of the Jones polynomial is known at the date of publication of this book. That the Jones polynomial is a topological invariant, namely that it is independent of which projection of the knot is considered, follows by checking its invariance under **Reidemeister moves**. These are the following:

I. the elimination or addition of a twist,
II. the separation or overlapping of two strands,
III. the passing of a strand over/under/between two crossing strands.

Figure 1.7 depicts one Reidemeister move of each type. Reidemeister's theorem [Reidemeister (1926)] states that any two diagrams of the same knot and link can be changed into one another by a finite sequence of such moves.

Any quantity that is associated to knots and links and is invariant under the Reidemeister moves is a knot and link invariant. In particular, we could have defined the linking number by counting the positive and negative crossings, as in §1.2.1, and then check invariance under Reidemester moves in order to prove that it is a topological invariant.

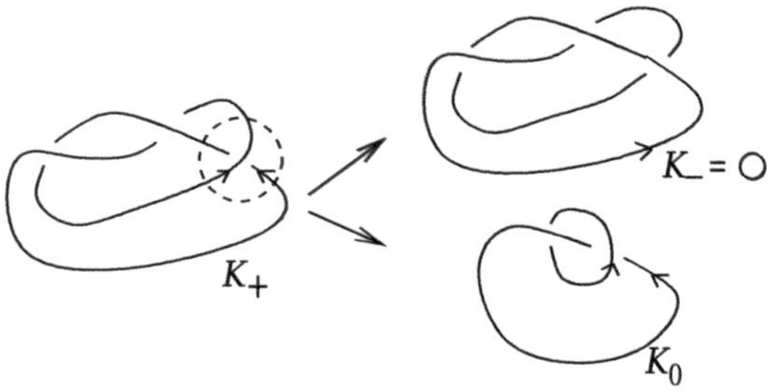

$$V_{K_+}(t) = t^2 + (t^{3/2} - t^{1/2})V_{K_0}(t)$$

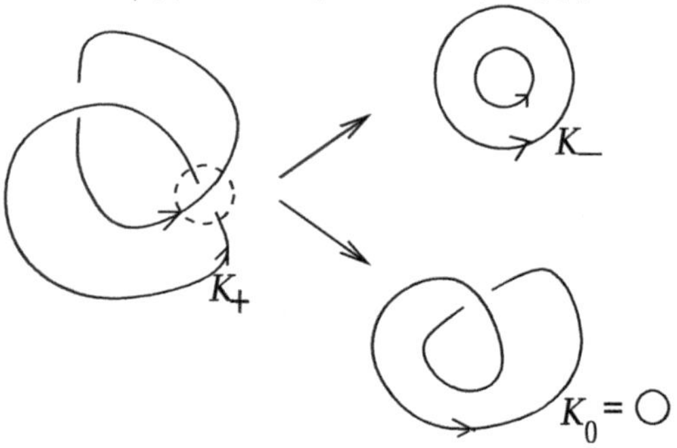

$$V_{K_+}(t) = t^2 V_{K_-}(t) + (t^{3/2} - t^{1/2})$$

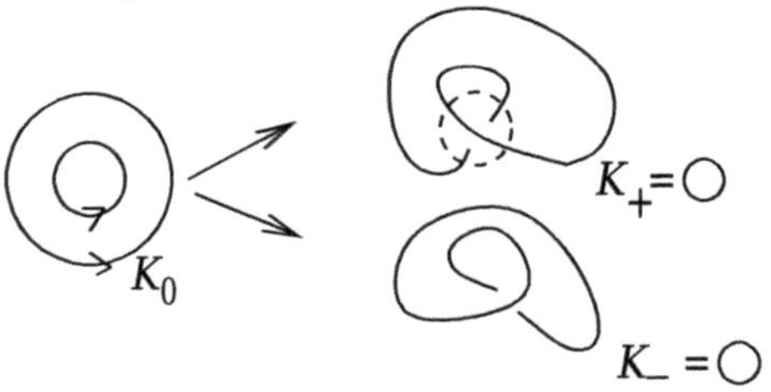

$$V_{K_0}(t) = -t^{1/2} - t^{-1/2}.$$

Fig. 1.6 Computation of the Jones polynomial of the trefoil knot

1.2.3 *Computing the linking number from skein relations*

Returning to the linking number, by analogy with the Jones polynomial we can also compute it from skein relations in a link diagram. For this we introduce the notion of framed oriented links.

Definition 1.2. A **framing** of a smooth knot $\gamma : S^1 \to \mathbb{R}^3$ is a choice of a smooth vector valued function $f : S^1 \to \mathbb{R}^3\backslash\{0\}$ such that for each $\tau \in S^1$, $f(\tau)$ is orthogonal to the tangent vector $\gamma'(\tau)$. A knot endowed with a framing is called a **framed knot**.

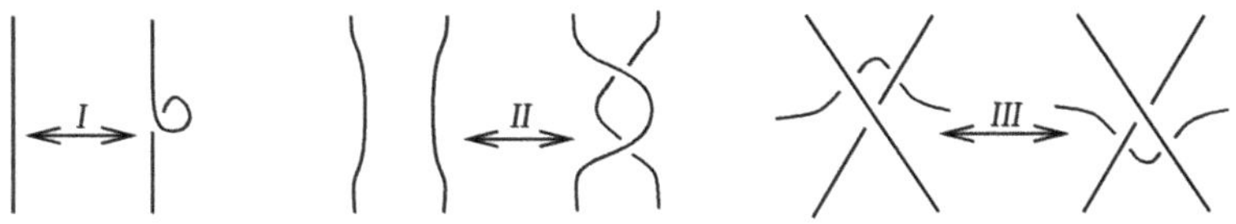

Fig. 1.7 The Reidemeister moves

In a more intuitive language, a framed knot is an embedded ribbon; just view the vector $f(\tau)$ as a little arrow attached to the point $\gamma(\tau)$, $\tau \in S^1$, and then the ribbon is determined by these vectors. This is not entirely correct, since $-f$ defines the same ribbon, but the two can be changed one into the other by spinning the framing 180° about the knot. Note that physical knots made out of rope are framed by nature, the rope plays the same role as the ribbon in keeping track of the twistings of the knot around itself.

Definition 1.3. A **framed link** is a link whose components are framed knots.

Given a framed link and a plane, the link can always be deformed continuously without crossing itself so that the framing becomes parallel to the plane. In this case if we are given a link diagram we can recover the framing from the diagram, by taking a regular neighborhood of the projection in the plane. We say that the link defined by the diagram has the **blackboard framing**.

Reidemeister's Theorem can be adapted to show that two diagrams represent the same framed link if they can be transformed into one another by Reidemeister II and III moves. However, Reidemeister I move changes the framing.

We agree that a link diagram consisting of only unknotted disjoint circles is associated the constant polynomial equal to 1.

Fig. 1.8 Skein relations for the linking number

Given a link diagram of an oriented link L with blackboard framing, apply the **skein relations** from Figure 1.8 until the diagram consists only of unknotted disjoint circles, then replace these circles by the constant polynomial equal to 1. The result is the polynomial $t^{\mathrm{lk}(L)}$ for some integer $\mathrm{lk}(L)$. There is an integral formula for computing the exponent $\mathrm{lk}(L)$.

To explicate it, define the link $L_{\rightarrow}$ to be the parallel copy of L in the direction of the framing. Alternatively, we can think that L and $L_{\rightarrow}$ contain respectively the boundary components of the annuli that comprise the framed link. Recall the form α_γ defined in (1.1). Define the 1-form $\alpha_L = \sum \alpha_\gamma$, where γ ranges over the components of L. Then

$$\mathrm{lk}(L) = \int_{L_{\rightarrow}} \alpha_L = \sum \int_{\gamma_{\rightarrow}} \alpha_L,$$

the sum being taken over all components of $L_{\rightarrow}$. Stokes' Theorem implies that $\mathrm{lk}(L)$ is a framed oriented link invariant.

If L consists of two components, γ_1 and γ_2, then

$$t^{\mathrm{lk}(L)}/(t^{\mathrm{lk}(\gamma_1)}t^{\mathrm{lk}(\gamma_2)}) = t^{2\mathrm{lk}(\gamma_1,\gamma_2)}.$$

For this reason we will call the skein relations from Figure 1.8 the skein relations of the linking number.

1.3 Witten's Chern-Simons theory

A few years after Jones made his discovery, Edward Witten explained the Jones polynomial using quantum field theory [Witten (1989)]. This was done in order to give an intrinsic definition to the Jones polynomial independent of knot diagrams. Let us briefly explain Witten's work, so that the reader will understand some ideas that led to the writing of this book. We will return with a few more details in Chapter 9.

Let G be a compact Lie group with Lie algebra $\mathfrak{g}$. This is the gauge group of the theory. Consider the fields with symmetry group G on a 3-dimensional manifold M without boundary. The presence of such a field is determined by its action on the phase of a particle moving through M. The phase is described by a vector, which is rotated by an element of G.

Each field has a potential, which is a $\mathfrak{g}$-connection A in the trivial principal bundle $M \times G$. As such, A is defined by a $\mathfrak{g}$-valued 1-form, denoted also by A. If we integrate the connection along a loop, we obtain an element of G, by which the phase of the particle is rotated when it travels along that loop. This element of G is called the **holonomy** of the connection along the loop. In Witten's model, the classical observables are the **traces of the holonomies** of such connections in a certain representation of the

Lie group. So, from the point of view of classical physics, the measurable quantities are the functions of fields

$$A \mapsto \mathrm{tr}_V \mathrm{hol}_\gamma(A)$$

where γ is a loop and V is a representation of G. These are called **Wilson lines**.

This theory has a Lagrangian, the **Chern-Simons functional**

$$L(A) = \frac{1}{4\pi} \int_M tr(A \wedge dA + \frac{2}{3} A \wedge A \wedge A),$$

which was introduced by S.S. Chern and J. Simons in [Chern and Simons (1974)].

Witten outlined a way of producing a quantum theory with this Lagrangian, using Feynmann integrals. Planck's constant is chosen to be the reciprocal of an integer: $h = \frac{1}{N}$. To the classical observable that is the Wilson line defined by the curve γ and the representation V one associates the Feynmann integral

$$\int e^{\frac{i}{h}L(A)} \mathrm{tr}_V \mathrm{hol}_\gamma(A) \mathcal{D}A. \tag{1.3}$$

This integral is taken over the infinite dimensional space of all connections. As such, it is not well defined mathematically. It should be thought of as an average of the quantities $\mathrm{tr}_V \mathrm{hol}_\gamma(A)$ over all connections A, where the average is weighted by $e^{i/hhL(A)}$.

Witten claimed that for $G = SU(2)$, the manifold M equal to S^3, γ a knot K, and V the standard 2-dimensional representation of $SU(2)$, the value of the integral (1.3) equals the Jones polynomial evaluated at a root of unity. That is

$$\int e^{iNL(A)} \mathrm{tr}_V \mathrm{hol}_\gamma(A) \mathcal{D}A = V_K\left(e^{\frac{\pi i}{N}}\right).$$

While this fact cannot be established rigorously, certain properties of Feynmann integrals suggest that this is indeed so. These properties allow the localization of the computation, by cutting the manifold M into pieces, computing the integral on each piece, and then combining the results according to certain rules. A crossing can then be placed inside a ball. The Feynmann integral on a ball with two strands inside would be a 2-dimensional vector. Hence the three vectors associated to the ball of the crossing in K_+, K_- respectively K_0 are linearly dependent. The linear dependence is the skein relation of the Jones polynomial.

Of interest to us, in this book, is the case

$$G = U(1) = \{z \in \mathbb{C} \,|\, |z| = 1\}.$$

This case is known as the **abelian Chern-Simons theory**, and was first studied by Albert Schwarz in [Schwarz (1978)]. For similar reasons this case would be related to the linking number. Here, the vectors associated to a crossing are 1-dimensional, the linear dependence of two such vectors giving rise to the skein relations from Figure 1.9.

Fig. 1.9 Skein relations for $U(1)$-knot invariants

While Witten's constructs lacked rigorous foundation, they gave rise to rich mathematics. First, everything depends only on the topology of the manifold M and of the embedded loop γ, so we are in the presence of a **topological quantum field theory** (TQFT). Such theories have been formalized by M.F. Atiyah [Atiyah (1988)], [Atiyah (1990)]. In particular,

$$Z(M) = \int e^{\frac{i}{h}L(A)}\mathcal{D}A$$

is a topological invariant of the manifold M.

Then, the invariance of Witten's integrals under isotopies of the curve γ means invariance under Reidemeister moves. The third of these moves relates to the Yang-Baxter equation in statistical dynamics, and hence to quantum groups.

Both the Chern-Simons Lagrangian, and the Wilson line, are invariant under the changes of coordinates of the field. These changes of coordinates are called gauge transformations. They are defined by smooth functions $g : M \to G$, and change the potential by

$$A \mapsto g^{-1}Ag + g^{-1}dg.$$

Witten's integrals are therefore taken over equivalence classes of connections.

The possibility of computing Witten's integrals by decomposing the manifold into pieces gives rise to an extension of these integrals to manifolds with boundary. To a manifold with boundary corresponds a vector in a finite dimensional vector space that is associated to the boundary surface. These vector spaces arise from the quantization of the space of $\mathfrak{g}$-connections on the boundary surface. Using the symmetries of the system and symplectic reduction, one reduces the quantization problem to the quantization of the space of **flat $\mathfrak{g}$-connections (those with curvature zero) on a surface, modulo gauge transformations**. This moduli space is now a finite dimensional space, which is endowed with the structure of a finite dimensional algebraic variety. The smooth part of this variety has all the good properties of the phase space of a classical mechanical system. The problem now becomes quantum mechanical, and the standard quantization procedure is to replace the moduli space of connections by a Hilbert space, and the Wilson lines of curves on the surface by linear operators on this Hilbert space.

In the case $G = U(1)$, if we endow the surface that is the boundary of the manifold with a complex structure, then the moduli space of flat connections becomes a complex torus, which turns out to be the Jacobian variety of the surface. The Hilbert space of the quantum mechanical system is the space of Riemann's theta functions, and this is how theta functions enter the picture. Witten's considerations therefore show that there is a close relationship between theta functions and the linking number of knots. The aim of this book is to put this relationship in a perspective different from Witten's, using representation theory, and as such, to avoid the heuristical considerations of quantum field theory.

Throughout the book, whenever we refer to Chern-Simons theory, we mean the constructs of Edward Witten mentioned above and the mathematics they gave rise to.

Chapter 2

A quantum mechanical prototype

The development of physics in the second half of the 20th century gave rise to quantum physical interpretations of several mathematical theories. The theory of theta functions is such an example. To help understand how theta functions arise from quantum mechanics, we first present a fundamental example, that of the quantization of a finite number of free one-dimensional particles. We do this also because this example is prototypical for more general cases of Witten's Chern-Simons theory (cf. [Gelca and Uribe (2010)]).

The Stone-von Neumann theorem states that up to a unitary equivalence the quantization model is unique, so our task is apparently simple. But the model can be realized in many ways, each of which offering a different perspective on the subject. The purpose of this chapter is to guide the reader through a variety of models for the quantization of finitely many one-dimensional particles, in order to build the necessary background for subsequent chapters.

2.1 The quantization of a system of finitely many free one-dimensional particles

2.1.1 *The classical mechanics of finitely many free particles in a one-dimensional space*

Let n be a positive integer. Consider a system of n particles, each having the mass equal to 1, in a 1-dimensional space. We assume that there are no constraints, so the 1-dimensional space is $\mathbb{R}$. In classical mechanics, the future evolution of the system is determined by the force fields and the initial positions and momenta of the particles. Hence the coordinates of interest are the positions ξ_j and momenta η_j, $j = 1, 2, \ldots, n$, which are coordinates in the **phase space** $\mathbb{R}^n \times \mathbb{R}^n$ of the system. The observable quantities are smooth functions of $\boldsymbol{\xi} = (\xi_j)$ and $\boldsymbol{\eta} = (\eta_j)$. For example, the kinetic energy is equal to $K(\boldsymbol{\xi}, \boldsymbol{\eta}) = \frac{1}{2}\sum_j \eta_j^2$.

There is an observable that plays a special role, the Hamiltonian H, which is the total energy of the system. (As an example, a system of n harmonic oscillators has the Hamiltonian $H(\boldsymbol{\xi}, \boldsymbol{\eta}) = \frac{1}{2}\sum_j(\xi_j^2 + \eta_j^2)$.)

The phase space is endowed with a **symplectic form**, namely a closed non-degenerate 2-form, given by

$$\omega = -d\boldsymbol{\xi}^T \wedge d\boldsymbol{\eta} = -\sum_j d\xi_j \wedge d\eta_j.$$

As such, $(\mathbb{R}^n \times \mathbb{R}^n, \omega)$ is a **symplectic manifold**. The symplectic form associates to each $f \in C^\infty(\mathbb{R}^n \times \mathbb{R}^n, \mathbb{R})$ a Hamiltonian vector field X_f on $\mathbb{R}^n \times \mathbb{R}^n$ by

$$-df(\cdot) = \omega(X_f, \cdot).$$

Concretely

$$X_f = \left(\frac{\partial f}{\partial \boldsymbol{\eta}}\right)^T \frac{\partial}{\partial \boldsymbol{\xi}} - \left(\frac{\partial f}{\partial \boldsymbol{\xi}}\right)^T \frac{\partial}{\partial \boldsymbol{\eta}}.$$

This vector field defines a Hamiltonian flow on $\mathbb{R}^n \times \mathbb{R}^n$ which preserves the form ω.

Example 2.1. For the coordinate functions ξ_j and η_j,

$$X_{\xi_j} = -\frac{\partial}{\partial \eta_j}, \quad X_{\eta_j} = \frac{\partial}{\partial \xi_j}.$$

Given two functions $f, g \in C^\infty(\mathbb{R}^n \times \mathbb{R}^n, \mathbb{R})$, the smooth function $\{f, g\} = \omega(X_f, X_g)$ is called the **Poisson bracket** of f and g. In coordinates it is given by

$$\{f, g\} = \left(\frac{\partial f}{\partial \boldsymbol{\xi}}\right)^T \frac{\partial g}{\partial \boldsymbol{\eta}} - \left(\frac{\partial f}{\partial \boldsymbol{\eta}}\right)^T \frac{\partial g}{\partial \boldsymbol{\xi}} = \sum_j \frac{\partial f}{\partial \xi_j}\frac{\partial g}{\partial \eta_j} - \frac{\partial f}{\partial \eta_j}\frac{\partial g}{\partial \xi_j}.$$

The Poisson bracket can be extended by linearity to complex valued functions as well.

We can also think about the Poisson bracket infinitesimally, as a Lie bracket of Hamiltonian vector fields, by the formula

$$[X_f, X_g] = X_{\{f,g\}}.$$

Identifying constant vector fields (which are Hamiltonian) with vectors in the tangent space to the phase space, we obtain a Lie bracket in the tangent space.

The Poisson bracket is used for describing the evolution of the system. In this formalism, Hamilton's equations

$$\frac{d\xi_j}{dt} = \frac{\partial H}{\partial \eta_j}, \quad \frac{d\eta_j}{dt} = -\frac{\partial H}{\partial \xi_j}$$

are written as

$$\frac{d\xi_j}{dt} = \{\xi_j, H\}, \quad \frac{d\eta_j}{dt} = \{\eta_j, H\},$$

and in general, an observable quantity evolves according to the equation

$$\frac{df}{dt} = \{f, H\}. \tag{2.1}$$

This equation should be interpreted as saying that, as the particle moves along the trajectory $(\boldsymbol{\xi}(t), \boldsymbol{\eta}(t))$, the instantaneous rate of change of the observable f at a time t is the function $\{f, H\}$ evaluated at t.

Because of the identities

$$\{f, g\} = -\{g, f\}, \quad \{f, \{g, h\}\} + \{g, \{h, f\}\} + \{h, \{f, g\}\} = 0,$$

the space of smooth functions, $C^\infty(\mathbb{R}^n \times \mathbb{R}^n, \mathbb{R})$, endowed with the Poisson bracket is a Lie algebra. This algebra contains a finite-dimensional Lie algebra generated by the position and momentum functions, which has the structure equations

$$\{\xi_j, \xi_k\} = \{\eta_j, \eta_k\} = 0, \quad \{\eta_j, \xi_k\} = \delta_{jk},$$

where δ_{jk} is the Kronecker symbol. This is the **Heisenberg Lie algebra** $\mathfrak{h}(\mathbb{R}^n)$. As a vector space, it is $\mathbb{R}^{2n+1}$, with the last coordinate being the constant functions.

The changes of coordinates are those that preserve the symplectic form ω. As such, they preserve the Poisson bracket and hence the mechanics of the system. Among them we distinguish the linear ones, which form the symplectic group

$$Sp(2n, \mathbb{R}) = \left\{ h \in GL(2n, \mathbb{R}) \;\middle|\; h^T \begin{bmatrix} 0 & I_n \\ -I_n & 0 \end{bmatrix} h = \begin{bmatrix} 0 & I_n \\ -I_n & 0 \end{bmatrix} \right\}.$$

In a more general situation, classical mechanics describes systems of several particles subject to constraints. The constraints confine the positions of particles to a manifold N. The momenta are covectors in the cotangent bundle to N, and hence the phase space of the system is T^*N. Classical observables are smooth functions on T^*N. The cotangent bundle comes equipped with a symplectic form, which gives rise to a Poisson bracket in the same way as above, and the same Hamiltonian formalism models the evolution of observables.

If we impose constraints on momenta as well (somewhat artificially), we obtain a phase space that is a manifold M still endowed with a symplectic form ω (a symplectic manifold). As we will see in §4.3, theta functions come from a system of particles with periodic positions and momenta, in which case the phase space is a torus.

2.1.2 *The Schrödinger representation*

What is quantization?

Quantization is a procedure for passing from classical to quantum mechanics, in which the concepts of classical physics are replaced by quantum physical counterparts. We will describe this procedure in the Heisenberg formalism, the so called matrix mechanics.

First, we introduce **Planck's constant** h, with the usual convention that $\hbar = \frac{h}{2\pi}$ is the **reduced Planck's constant.**

In the general setting, quantization replaces

- the phase space M of a classical mechanical system by a **Hilbert space $\mathcal{H}$ of states** (or wave functions);
- the real-valued classical observables, which are smooth real-valued functions f on M, by self-adjoint operators $\mathrm{op}(f)$ on $\mathcal{H}$, called **quantum observables.**

The Poisson bracket $\{f, g\}$ becomes a multiple of the commutator of operators, more precisely $\frac{\hbar}{i}[\mathrm{op}(f), \mathrm{op}(g)]$, where $[\cdot,\cdot]$ is the commutator. Quantization must be done respecting Dirac's conditions:

1. $\mathrm{op}(1) = I$, where I is the identity operator;
2. if $f = c_1 f_1 + c_2 f_2$ then $\mathrm{op}(f) = c_1\mathrm{op}(f_1) + c_2\mathrm{op}(f_2)$;
3. (the correspondence principle) $\mathrm{op}(\{f, g\}) = \frac{i}{\hbar}[\mathrm{op}(f), \mathrm{op}(g)] + O(\hbar)$;
4. the representation of the quantum observables on the Hilbert space is irreducible.

The first condition states that the constant functions become multiples of the identity operator, and the second that quantization is a linear transformation from the space of functions to that of linear operators. The second condition allows us to extend quantization, by linearity, to complex-valued observables. The fourth condition requires that quantization has no redundancies (such as being the direct sum of two copies of the same model).

The third condition means that quantization is "almost" a Lie algebra homomorphism. The impossibility to obtain a true homomorphism is demonstrated by the Groenewold - van Hove theorem (see [Hall (2000)] for more details). Nevertheless, the restriction of the quantization procedure to the Heisenberg Lie algebra must be a Lie algebra homomorphism. This is one of the formulations of Heisenberg's uncertainty principle. If we set $P_j = \mathrm{op}(\eta_j)$, $Q_j = \mathrm{op}(\xi_j)$ then

$$[P_j, Q_k] = -\delta_{jk} i\hbar I = -\delta_{jk} \frac{ih}{2\pi} I, \quad [P_j, P_k] = [Q_j, Q_k] = 0. \tag{2.2}$$

These are called the **canonical commutation relations**. The first of them implies that the product of the standard deviations in measuring the position and the momentum of a given particle is at least $\hbar/2$ (the **Heisenberg uncertainty principle**), while the second implies that the momenta of two particles, or alternatively their positions, could be measured simultaneously with any desired accuracy. The canonical commutation relations imply that quantization yields a representation of the Heisenberg Lie algebra $\mathfrak{H}(\mathbb{R}^n)$ as a Lie algebra of linear operators on $L^2(\mathbb{R}^n)$ with the Lie bracket

$$\{A, B\} = \frac{i}{\hbar}(AB - BA).$$

Hamilton's equation (2.1) for the time evolution of an observable becomes Schrödinger's equation in the Heisenberg formalism

$$\frac{d\,\mathrm{op}(f)}{dt} = \{\mathrm{op}(f), \mathrm{op}(H)\} = \frac{i}{\hbar}[\mathrm{op}(f), \mathrm{op}(H)],$$

describing the time evolution of a quantum observable.

Quantization makes no provisions about changes of coordinates, except when they come from Hamiltonian flows. In that case, the Hamiltonian flow becomes a semigroup of unitary transformations of the Hilbert space.

The standard Schrödinger representation

Let us return to our model of n particles in a 1-dimensional space. It is standard to let the Hilbert space of the corresponding quantum system be $L^2(\mathbb{R}^n, d\boldsymbol{\xi})$. The canonical commutation relations (2.2) are satisfied by the operators

$$P_j \psi(\boldsymbol{\xi}) = -i\hbar \frac{\partial \psi}{\partial \xi_j}(\boldsymbol{\xi}) = -i \frac{h}{2\pi} \frac{\partial \psi}{\partial \xi_j}(\boldsymbol{\xi}), \quad Q_j \psi(\boldsymbol{\xi}) = \xi_j \psi(\boldsymbol{\xi}). \tag{2.3}$$

The Stone-von Neumann theorem shows that any other choice is equivalent to this one. This representation of the Heisenberg Lie algebra $\mathfrak{H}(\mathbb{R}^n)$ on $L^2(\mathbb{R}^n, d\boldsymbol{\xi})$ is called the **Schrödinger representation**. We convene to identify $\mathfrak{H}(\mathbb{R}^n)$ with the Lie algebra of operators $\oplus_j(\mathbb{R}Q_j \oplus \mathbb{R}P_j) \oplus \mathbb{R}I$.

The self-adjoint operators defining the quantum positions and momenta are necessarily unbounded, and there are some issues about their domains in stating the Stone-von Neumann theorem. One can avoid these issues by exponentiating the operators. For that, we multiply the elements of the Heisenberg Lie algebra by $2\pi i$ and then exponentiate, to obtain a Lie group consisting of unitary operators. This is the Heisenberg group with real entries.

Definition 2.1. The **Heisenberg group with real entries** $\mathbf{H}(\mathbb{R}^n)$ is $\mathbb{R}^n \times \mathbb{R}^n \times U(1)$ with multiplication

$$\left(\mathbf{p}, \mathbf{q}, e^{2\pi i t}\right)\left(\mathbf{p}', \mathbf{q}', e^{2\pi i t'}\right) = \left(\mathbf{p}+\mathbf{p}', \mathbf{q}+\mathbf{q}', e^{2\pi i\left(t+t'+\frac{h}{2}(\mathbf{p}^T\mathbf{q}'-\mathbf{q}'^T\mathbf{p})\right)}\right).$$

The Heisenberg group is a $U(1)$-extension of $\mathbb{R}^n \times \mathbb{R}^n$, defined by the cocycle

$$c((\mathbf{p}, \mathbf{q}), (\mathbf{p}', \mathbf{q}')) = e^{\pi i h(\mathbf{p}^T\mathbf{q}' - \mathbf{q}^T\mathbf{p}')}.$$

Concretely, it is realized by exponentiation as

$$\left(\mathbf{p}, \mathbf{q}, e^{2\pi i t}\right)\psi(\boldsymbol{\xi}) = \exp\left(2\pi i\left(-i\hbar\mathbf{p}^T\frac{\partial}{\partial\boldsymbol{\xi}} + \mathbf{q}^T\boldsymbol{\xi} + tI\right)\right)\psi(\boldsymbol{\xi}).$$

For convenience, we will ignore the factor of $2\pi i$ and denote

$$\exp(\mathbf{p}^T\mathbf{P} + \mathbf{q}^T\mathbf{Q} + tI) = \left(\mathbf{p}, \mathbf{q}, e^{2\pi i t}\right),$$

where $\mathbf{P} = (P_1, P_2, \ldots, P_n)$ and $\mathbf{Q} = (Q_1, Q_2, \ldots, Q_n)$.

Proposition 2.1. *The action of the Heisenberg group with real entries on* $L^2(\mathbb{R}^n, d\boldsymbol{\xi})$ *is given by*

$$\exp(\mathbf{p}^T\mathbf{P} + \mathbf{q}^T\mathbf{Q} + tI)\psi(\boldsymbol{\xi}) = e^{2\pi i\mathbf{q}^T\cdot\boldsymbol{\xi} + \pi i h\mathbf{p}^T\cdot\mathbf{q} + 2\pi i t}\psi(\boldsymbol{\xi} + h\mathbf{p}).$$

Proof. The exponential of a differentiation operator is a translation and the exponential of the multiplication by a function is the multiplication by the exponential of that function. Hence

$$\begin{aligned}
\exp(p_jP_j)\psi(\xi_1, \xi_2, \ldots, \xi_n) &= \exp\left(2\pi i\left(-i\frac{h}{2\pi}p_j\frac{\partial}{\partial\xi_j}\right)\right)\psi(\xi_1, \xi_2, \ldots, \xi_n) \\
&= \psi(\xi_1, \xi_2, \ldots, \xi_j + hp_j, \ldots, \xi_n), \\
\exp(q_jQ_j)\psi(\xi_1, \xi_2, \ldots, \xi_n) &= e^{2\pi i q_j\xi_j}\psi(\xi_1, \xi_2, \ldots, \xi_n), \\
\exp(2\pi i tI)\psi(\xi_1, \xi_2, \ldots, \xi_n) &= e^{2\pi i t}\psi(\xi_1, \xi_2, \ldots, \xi_n).
\end{aligned}$$

The exponential of the momentum acts as a translation, while the exponential of the position acts as the multiplication by a character (i.e. by a $U(1)$-valued function). Since the operators P_j commute with each other, and the operators Q_j also commute with each other, the formula from the statement works whenever either $\mathbf{p}$ or $\mathbf{q}$ is zero.

The general case follows by applying the particular form of the Baker-Campbell-Hausdorff formula

$$e^X e^Y = e^{X+Y+\frac{1}{2}[X,Y]},$$

which holds whenever the second commutators are zero, namely when the commutator of X and Y is a central element. In our case

$$\exp(\mathbf{p}^T\mathbf{P})\exp(\mathbf{q}^T\mathbf{Q}) = \exp(\mathbf{q}^T\mathbf{P} + \mathbf{q}^T\mathbf{Q} + \frac{1}{2}[\mathbf{p}^T\mathbf{P}, \mathbf{q}^T\mathbf{Q}]),$$

and the conclusion follows. □

From the formula derived in Proposition 2.1 we can deduce that this representation of the abstractly defined $\mathbf{H}(\mathbb{R}^n)$ is faithful, so $\mathbf{H}(\mathbb{R}^n)$ can indeed be identified with a group of unitary operators acting on $L^2(\mathbb{R}^n, d\boldsymbol{\xi})$. This action of the Heisenberg group on $L^2(\mathbb{R}^n, d\boldsymbol{\xi})$ is called the **Schrödinger representation of the Heisenberg group**.

The canonical commutation relations (2.2) take the exponential form

$$\begin{aligned} \exp P_j \exp Q_k &= e^{2\pi i h \delta_{jk}} \exp Q_k \exp P_j, \\ \exp P_j \exp P_k &= \exp P_k \exp P_j, \\ \exp Q_j \exp Q_k &= \exp Q_k \exp Q_j. \end{aligned}$$

Theorem 2.1. *(Stone-von Neumann) The Schrödinger representation of the Heisenberg group* $\mathbf{H}(\mathbb{R}^n)$ *is the unique irreducible unitary representation of this group such that* $\exp(tI)$ *acts as* $e^{2\pi i t}I$ *for all* $t \in \mathbb{R}$.

In other words, if R is another irreducible unitary representation of the group $\mathbf{H}(\mathbb{R}^n)$ onto some Hilbert space $\mathcal{H}$, then there is a unitary homomorphism

$$W : L^2(\mathbb{R}^n, d\boldsymbol{\xi}) \to \mathcal{H}$$

such that

$$W^{-1}R(\mathbf{p}, \mathbf{q}, t)W = \exp(\mathbf{p}^T\mathbf{P} + \mathbf{q}^T\mathbf{Q} + tI).$$

A proof of this classical result can be found in [Hall (2013)], [Lion and Vergne (1980)], or [Folland (1989)].

The Schrödinger representation in the momentum representation

There is actually another irreducible unitary representation of the Heisenberg group on $L^2(\mathbb{R}^n, d\boldsymbol{\eta})$, the L^2-space of functions depending on the momenta $\eta_1, \eta_2, \ldots, \eta_n$, given by

$$(\mathbf{p}, \mathbf{q}, t)\psi(\boldsymbol{\eta}) = \exp\left(2\pi i\left(\mathbf{p}^T\boldsymbol{\eta} + i\frac{h}{2\pi}\mathbf{q}^T\frac{\partial}{\partial\boldsymbol{\eta}} + tI\right)\right)\psi(\boldsymbol{\eta}). \tag{2.4}$$

This representation arises by choosing the momentum and position operators as

$$P_j\psi(\eta_j) = \eta_j\psi(\eta_j), \quad Q_j\psi(\eta_j) = i\hbar\frac{\partial\psi}{\partial\eta_j}.$$

Because the variable of the state function is the momentum, this is referred to as the **quantization in the momentum representation**, as opposed to the one before, which is referred to as **quantization in the position representation**.

Proposition 2.2. *The Schrödinger representation in the momentum representation is given by*

$$\exp(\mathbf{p}^T\mathbf{P} + \mathbf{q}^T\mathbf{Q} + tI)\psi(\boldsymbol{\eta}) = e^{2\pi i\mathbf{p}^T\boldsymbol{\eta} - \pi i h\mathbf{p}^T\mathbf{q} + 2\pi i t}\psi(\boldsymbol{\eta} - h\mathbf{q}).$$

By the Stone-von Neumann Theorem, the two representations are equivalent. The unitary equivalence is the Fourier transform

$$(\mathcal{F}_\hbar\psi)(\boldsymbol{\eta}) = h^{n/2}\int_{\mathbb{R}^n}\psi(\boldsymbol{\xi})e^{-2\pi i h\boldsymbol{\xi}^T\boldsymbol{\eta}}d\boldsymbol{\xi}. \tag{2.5}$$

This is because of the fundamental property of the Fourier transform that it transforms differentiation into multiplication by the variable and vice-versa.

2.1.3 *Weyl quantization*

We have seen in the previous section a rule for quantizing the positions and momenta of n particles, as multiplication and differentiation operators. We then exponentiated these operators, obtaining the Schrödinger representation of the Heisenberg group. One can think of the Schrödinger representation as giving a rule for quantizing exponential functions by the correspondence

$$\exp(2\pi i(\mathbf{p}^T\boldsymbol{\eta} + \mathbf{q}^T\boldsymbol{\xi})) \mapsto \exp(\mathbf{p}^T\mathbf{P} + \mathbf{q}^T\mathbf{Q}). \tag{2.6}$$

This quantization rule can be further extended to all classical observables, as we will explain below. It is the oldest quantization procedure, and was introduced by Hermann Weyl in 1931. One should point out that there are infinitely many distinct quantization procedures of a system of n one-dimensional particles that satisfy Dirac's conditions. We focus on Weyl's, since it plays an important role in Chern-Simons theory (see [Gelca and Uribe (2003)]). The method is discussed in more detail in many publications, of which we recommend [Folland (1989)] and [Faddeev and Yakubovskii (2009)].

The quantization of the exponentials of positions and momenta defined by (2.6) is extended by linearity to all trigonometric polynomials. The Schrödinger representation is thus extended to the group algebra of the Heisenberg group. Weyl's method further extends this quantization method to arbitrary functions by decomposing them into "integrals of exponentials" by using the Fourier transform and then quantizing the exponentials under the integral.

Concretely, for $f \in C^\infty(\mathbb{R}^n \times \mathbb{R}^n, \mathbb{C})$ let

$$\hat{f}(\tilde{\boldsymbol{\xi}}, \tilde{\boldsymbol{\eta}}) = \iint f(\boldsymbol{\xi}, \boldsymbol{\eta}) \exp(-2\pi i(\boldsymbol{\xi}^T \tilde{\boldsymbol{\xi}} + \boldsymbol{\eta}^T \tilde{\boldsymbol{\eta}})) d\boldsymbol{\xi} d\boldsymbol{\eta}$$

be its Fourier transform. Then

$$f(\boldsymbol{\xi}, \boldsymbol{\eta}) = \iint \hat{f}(\tilde{\boldsymbol{\xi}}, \tilde{\boldsymbol{\eta}}) \exp(2\pi i(\tilde{\boldsymbol{\xi}}^T \boldsymbol{\xi} + \tilde{\boldsymbol{\eta}}^T \boldsymbol{\eta})) d\boldsymbol{\xi} d\boldsymbol{\eta}.$$

The Weyl quantization of f is the operator

$$\mathrm{op}(f) = \iint \hat{f}(\tilde{\boldsymbol{\xi}}, \tilde{\boldsymbol{\eta}}) \exp(2\pi i(\tilde{\boldsymbol{\eta}}^T \mathbf{P} + \tilde{\boldsymbol{\xi}}^T \mathbf{Q})) d\tilde{\boldsymbol{\xi}} d\tilde{\boldsymbol{\eta}},$$

where the operator $\exp(2\pi i(\tilde{\boldsymbol{\eta}}^T \mathbf{P} + \tilde{\boldsymbol{\xi}}^T \mathbf{Q}))$ is defined by the Schrödinger representation.

These operators act on states by the formula

$$\mathrm{op}(f)\psi(\boldsymbol{\xi}) = \iint f\left(\mathbf{u}, \frac{1}{2}(\boldsymbol{\xi} + \boldsymbol{\xi}')\right) e^{(i/\hbar)(\boldsymbol{\xi} - \boldsymbol{\xi}')\mathbf{u}} \psi(\boldsymbol{\xi}') d\boldsymbol{\xi}' d\mathbf{u}.$$

The function f is called the **Weyl symbol** of the operator $\mathrm{op}(f)$.

Of course, not every smooth function has a Fourier transform, but at least we know that functions that define tempered distributions do, so at least for these functions the quantization procedure is well defined (see [Folland (1989)]). For two such functions f and g, the product $\mathrm{op}(f)\mathrm{op}(g)$ is again an operator that arises through Weyl quantization. The Weyl symbol of this operator is unique, and is given by

$$(f * g)(\boldsymbol{\xi}, \boldsymbol{\eta}) = \left(\frac{2}{h}\right)^{2n} \int_{\mathbb{R}^{4n}} e^{4\pi i h(\boldsymbol{\xi}'^T \boldsymbol{\eta}'' - \boldsymbol{\eta}'' \boldsymbol{\xi}')}$$
$$\times f(\boldsymbol{\xi} + \boldsymbol{\xi}', \boldsymbol{\eta} + \boldsymbol{\eta}') g(\boldsymbol{\xi} + \boldsymbol{\xi}'', \boldsymbol{\eta} + \boldsymbol{\eta}'') d\boldsymbol{\xi}' d\boldsymbol{\eta}' d\boldsymbol{\xi}'' d\boldsymbol{\eta}''.$$

In the realm of deformation quantization (see [Bayen *et al.* (1978a)] [Bayen *et al.* (1978b)]), by treating Planck's constant in this formula as a variable, we obtain a $*$-product for functions in the plane. This is called the Moyal product.

2.2 The quantization of finitely many free one-dimensional particles via holomorphic functions

Because this book is about theta functions, and they are holomorphic functions, we translate the quantization method from §2.1 in the language of complex analysis. The alternative quantization model was discovered by Bargmann and Segal [Bargmann (1961)], [Segal (1962)], [Segal (1963)]. Below, we give a detailed overview of the notions and results that are relevant to this book. The presentation parallels that from [Hall (2000)], with some input from [Folland (1989)].

2.2.1 *The Segal-Bargmann quantization model*

In this alternative model, the wave functions are holomorphic functions on $\mathbb{C}^n$, where the phase space $\mathbb{R}^n \times \mathbb{R}^n$ is identified with $\mathbb{C}^n$ by letting the position coordinates be the real parts and the momentum coordinates be the imaginary parts. This means that $\mathbf{z} = \mathbf{x} + i\mathbf{y}$ with $\mathbf{x} = \boldsymbol{\xi}$ and $\mathbf{y} = \boldsymbol{\eta}$. Then $\bar{\mathbf{z}} = \mathbf{x} - i\mathbf{y}$ and

$$\frac{\partial}{\partial z_j} = \frac{1}{2}\left(\frac{\partial}{\partial x_j} - i\frac{\partial}{\partial y_j}\right), \quad \frac{\partial}{\partial \bar{z}_j} = \frac{1}{2}\left(\frac{\partial}{\partial x_j} + i\frac{\partial}{\partial y_j}\right), \quad j = 1, 2, \ldots, n.$$

The origin of this quantization model lies in the observation, due to Fock, that for a holomorphic function $\phi(z)$,

$$\frac{\partial}{\partial z}(z\phi(z)) - z\frac{\partial}{\partial z}\phi(z) = \phi(z).$$

This means that if we let, for $j = 1, 2, \ldots, n$, $a_j = \hbar\frac{\partial}{\partial z_j}$ and $a_j^\dagger = M_{z_j}$ (the operator of multiplication by z_j) then

$$[a_j, a_k^\dagger] = \delta_{jk}\hbar I, \quad [a_j, a_k] = [a_j^\dagger, a_k^\dagger] = 0, \quad j, k = 1, 2, \ldots, n. \qquad (2.7)$$

Under the identifications

$$a_j = \frac{1}{\sqrt{2}}(Q_j + iP_j) \text{ and } a_j^\dagger = \frac{1}{\sqrt{2}}(Q_j - iP_j), \quad j = 1, 2, \ldots, n,$$

the equations (2.7) are equivalent to the canonical commutation relations (2.2). This is one of the most common choices. The operators $a_j^\dagger$ and a_j,

$j = 1, 2, \ldots, n$, are called creation and annihilation operators (because of how they act on polynomials, which in this case are states of the system).

We could also set $a_j = iP_j$ and $a_j^\dagger = Q_j$, and the canonical commutation relations would be satisfied. However, quantization requires that P_j and Q_j be self-adjoint operators. So the choice of the Hilbert space restricts the definition of the creation and annihilation operators and the way one relates them to P_j and Q_j.

To bring ourselves closer to the theory of theta functions, we choose the creation and annihilation operators to be

$$\hat{a}_j = M_{z_j} + 2\hbar\frac{\partial}{\partial z_j} \text{ and } \hat{a}_j^\dagger = M_{z_j}, \tag{2.8}$$

and set

$$\hat{a}_j = Q_j + iP_j \text{ and } \hat{a}_j^\dagger = Q_j - iP_j. \tag{2.9}$$

Then

$$\left[M_{z_j} + 2\hbar\frac{\partial}{\partial z_j}, M_{z_k}\right]\phi = \left(M_{z_j} + 2\hbar\frac{\partial}{\partial z_j}\right)(z_k\phi) - M_{z_k}\left(z_j\phi + 2\hbar\frac{\partial\phi}{\partial z_j}\right)$$
$$= z_jz_k\phi + 2\hbar\delta_{jk}\phi + 2\hbar z_k\frac{\partial\phi}{\partial z_j} - z_kz_j\phi - 2\hbar z_k\frac{\partial\phi}{\partial z_j} = 2\hbar\delta_{jk}\phi.$$

The equality

$$\left[M_{z_j} + 2\hbar\frac{\partial}{\partial z_j}, M_{z_k}\right] = 2\hbar\delta_{jk}I$$

yields

$$[Q_j + iP_j, Q_k - iP_k] = 2\hbar\delta_{jk}I,$$

which is equivalent to $[P_j, Q_k] = -\delta_{jk}i\hbar I$, and so the canonical commutation relations are again satisfied. Below, we will choose the Hilbert space so that P_j and Q_j defined by (2.9) are self-adjoint.

The Segal-Bargmann space

The fact that the operators P_j and Q_j are self-adjoint is equivalent to a_j and $a_j^\dagger$ being one the adjoint of the other. The operators defined in (2.8) have this property in the case where the Hilbert space is the Segal-Bargmann space.

Definition 2.2. The **Segal-Bargmann space**, $\mathcal{H}L^2(\mathbb{C}^n, d\mu_\hbar)$, is the space of holomorphic functions on $\mathbb{C}^n$ that are L^2-integrable with respect to the measure

$$d\mu_\hbar = (\pi\hbar)^{-n/2}e^{-\|\operatorname{Im}\mathbf{z}\|^2/\hbar}d\mathbf{x}d\mathbf{y}.$$

The inner product on the Segal-Bargmann space is

$$\langle \phi_1(\mathbf{z}), \phi_2(\mathbf{z})\rangle = (\pi\hbar)^{-n/2}\int_{\mathbb{C}} \phi_1(\mathbf{z})\overline{\phi_2(\mathbf{z})}e^{-\|\operatorname{Im}\mathbf{z}\|^2/\hbar}\mathbf{dxdy}.$$

The next proposition shows that this is indeed a correct choice.

Proposition 2.3. *On the Segal-Bargmann space $\mathcal{H}L^2(\mathbb{C}^n, d\mu_\hbar)$, the operators*

$$\hat{a}_j = M_{z_j} + 2\hbar\frac{\partial}{\partial z_j}, \quad \hat{a}_j^\dagger = M_{z_j},$$

where M_j is multiplication by z_j, are one the adjoint of the other.

Proof. The condition $\langle \hat{a}_j\phi_1, \phi_2\rangle = \left\langle \phi_1, \hat{a}_j^\dagger\phi_2\right\rangle$ is equivalent to

$$\left\langle z_j\phi_1 + 2\hbar\frac{\partial\phi_1}{\partial z_j}, \phi_2\right\rangle = \left\langle \phi_1, \frac{z_j}{\hbar}\phi_2\right\rangle.$$

We thus have to prove

$$\int_{\mathbb{C}^n}\left(z_j\phi(\mathbf{z}) + 2\hbar\frac{\partial\phi_1}{\partial z_j}(\mathbf{z})\right)\overline{\phi_2(\mathbf{z})}e^{-\|\operatorname{Im}\mathbf{z}\|^2/\hbar}\mathbf{dxdy}$$
$$= \int_{\mathbb{C}^n}\phi_1(\mathbf{z})\overline{\frac{z_j}{\hbar}\phi_2(\mathbf{z})}e^{-\|\operatorname{Im}\mathbf{z}\|^2/\hbar}\mathbf{dxdy}.$$

Let us point out that the canonical commutation relations cannot be satisfied by bounded operators, and hence neither the position and momentum, nor the creation and annihilation operators can be bounded. Consequently they are not defined for all functions in the Hilbert space, so we must restrict these operators to a (dense) subspace. In order to check the desired property about adjoints, we restrict ourselves to functions ϕ_1 and ϕ_2 that grow slowly enough at infinity so that we can use integration by parts in which the boundary terms are zero. We have

$$\int_{\mathbb{C}^n}\left(z_j\phi(\mathbf{z}) + 2\hbar\frac{\partial\phi_1}{\partial z_j}(\mathbf{z})\right)\overline{\phi_2(\mathbf{z})}e^{-\|\operatorname{Im}\mathbf{z}\|^2/\hbar}\mathbf{dxdy}$$
$$= \int_{\mathbb{C}^n} z_j\phi(\mathbf{z})\overline{\phi_2(\mathbf{z})}e^{-\|\operatorname{Im}\mathbf{z}\|^2/\hbar}\mathbf{dxdy} + \int_{\mathbb{C}^n} 2\hbar\frac{\partial\phi_1}{\partial z_j}(\mathbf{z})\overline{\phi_2(\mathbf{z})}e^{-\|\operatorname{Im}\mathbf{z}\|^2/\hbar}\mathbf{dxdy}.$$

The second integral is equal to

$$2\hbar\int_{\mathbb{C}^n}\left[\left(\frac{\partial}{\partial x_j} - i\frac{\partial}{\partial y_j}\right)\phi_1(\mathbf{z})\right]\overline{\phi_2(\mathbf{z})}e^{-\|\operatorname{Im}\mathbf{z}\|^2/\hbar}\mathbf{dxdy}$$
$$= -2\hbar\int_{\mathbb{C}^n}\phi_1(\mathbf{z})\left[\left(\frac{\partial}{\partial x_j} - i\frac{\partial}{\partial y_j}\right)\overline{\phi_2(\mathbf{z})}e^{-\|\operatorname{Im}\mathbf{z}\|^2/\hbar}\right]\mathbf{dxdy}$$
$$= -2\hbar\int_{\mathbb{C}^n}\phi_1(\mathbf{z})\left[\frac{\partial\overline{\phi_2}}{\partial z_j}(\mathbf{z}) + \overline{\phi_2(\mathbf{z})}\frac{\partial}{\partial z_j}e^{-\|\operatorname{Im}\mathbf{z}\|^2/\hbar}\right]\mathbf{dxdy}.$$

In this expression, the integral is equal to

$$-\int_{\mathbb{C}^n}\phi_1(\mathbf{z})\overline{\frac{\partial\phi_2}{\partial\bar{z}_j}(\mathbf{z})}e^{-\|\operatorname{Im}\mathbf{z}\|^2/\hbar}d\mathbf{x}d\mathbf{y}-\int_{\mathbb{C}^n}\phi_1(\mathbf{z})\overline{\phi_2(\mathbf{z})}\frac{\partial}{\partial z_j}e^{-\|\operatorname{Im}\mathbf{z}\|^2/\hbar}d\mathbf{x}d\mathbf{y}.$$

The first integral is zero, since $\phi_2(\mathbf{z})$ is holomorphic. For the second, note that

$$\frac{\partial}{\partial z_j}e^{-\|\operatorname{Im}\mathbf{z}\|^2/\hbar}=\frac{\partial}{\partial z_j}e^{(\mathbf{z}-\bar{\mathbf{z}})^T(\mathbf{z}-\bar{\mathbf{z}})/4\hbar}=\frac{z_j-\bar{z}_j}{2\hbar}e^{-\|\operatorname{Im}\mathbf{z}\|^2/\hbar}.$$

Substituting we obtain the desired equality. □

By looking at power functions it is not hard to determine an orthonormal basis for the Segal-Bargmann space.

Proposition 2.4. *An orthonormal basis for $\mathcal{H}L^2(\mathbb{C}^n, d\mu_\hbar)$ consists of the functions*

$$\Phi_{k_1k_2\ldots k_n}(\mathbf{z})=\frac{z_1^{k_1}z_2^{k_2}\cdots z_n^{k_n}}{\sqrt{(2\hbar)^{k_1+k_2+\cdots+k_n}k_1!k_2!\cdots k_n!}}e^{-\mathbf{z}^T\mathbf{z}/4\hbar},\ (k_1,k_2,\ldots,k_n)\in\mathbb{Z}^n.$$

Proof. First let us show that the functions $\Phi_{k_1k_2\ldots k_n}$ form a dense set in the Segal-Bargmann space. It is easy to check that they are elements of this space.

Consider a function $\phi(\mathbf{z})\in\mathcal{H}L^2(\mathbb{C}^n, d\mu_\hbar)$ and expand $e^{\mathbf{z}^T\mathbf{z}/4\hbar}$ as a power series. Then we can write

$$\phi(\mathbf{z})=e^{-\mathbf{z}^T\mathbf{z}/4\hbar}\sum c_{m_1m_2\ldots m_n}z_1^{m_1}z_2^{m_2}\cdots z_n^{m_n}.$$

Note that

$$|e^{\mathbf{z}^T\mathbf{z}/4\hbar}|^2=e^{\mathbf{z}^T\mathbf{z}/2\hbar}|=e^{(\|\operatorname{Re}\mathbf{z}\|^2-\|\operatorname{Im}\mathbf{z}\|^2)/2\hbar},$$

and so

$$|e^{-\mathbf{z}^T\mathbf{z}/2\hbar}|e^{-\|\operatorname{Im}z\|^2/\hbar}=e^{-\|\mathbf{z}\|^2/2\hbar}.$$

Switching to polar coordinates we compute

$$(\pi\hbar)^{n/2}\left\langle\phi(\mathbf{z}),z_1^{k_1}z_2^{k_2}\cdots z_n^{k_n}e^{-\mathbf{z}^T\mathbf{z}/4\hbar}\right\rangle$$

$$\int_{(0,\ldots,0)}^{(\infty,\ldots,\infty)}\int_{(0,\ldots,0)}^{(2\pi,\ldots,2\pi)}(r_1e^{-i\theta_1})^{k_1+1}\cdots(r_ne^{-i\theta_n})^{k_n+1}e^{-\|\mathbf{r}\|^2/2\hbar}$$

$$\times\sum c_{m_1m_2\cdots m_n}(r_1e^{i\theta_1})^{m_1}\cdots(r_ne^{i\theta_n})^{m_n}d\boldsymbol{\theta}d\mathbf{r}.$$

Using the Lebesgue dominated convergence theorem we can write this as the limit of the integrals taken over the polydisks $[0,R]^n$, with $R\to\infty$. On the polydisk $[0,R]^n$ the power series of ϕ is uniformly convergent, so this is further equal to

$$\lim_{R\to\infty}\sum c_{m_1m_2\ldots m_n}\prod_{j=1}^n\int_0^R e^{-r_j^2/2\hbar}r_j^{m_j+k_j+1}\int_0^{2\pi}e^{i(m_j-k_j)\theta_j}d\theta_j dr_j.$$

The integral in θ_j is zero, unless $m_j = k_j$ for all $j = 1,2,\ldots,n$. So the above expression is equal to

$$(2\pi)^n c_{k_1k_2\cdots k_n}\prod_{j=1}^n\int_0^\infty r^{2k_j+1}e^{-r_j^2/2\hbar}dr_j.$$

An iterated integration by parts computes

$$\int_0^\infty r^{2k+1}e^{-r^2/2\hbar}dr = \frac{1}{2}k!(2\hbar)^k.$$

We conclude that

$$\left\langle \phi(\mathbf{z}), z_1^{k_1}z_2^{k_2}\cdots z_n^{k_n}e^{\mathbf{z}^T\mathbf{z}/2\hbar}\right\rangle = c_{k_1k_2\ldots k_n}(2\hbar)^{k_1+k_2+\cdots+k_n}k_1!k_2!\cdots k_n!.$$

This relation implies on the one hand that the functions from the statement span a dense subset of the Segal-Bargmann space, because a function orthogonal to all of them has the power series expansion identically equal to zero. But this also implies that these functions have norm one and are pairwise orthogonal. This completes the proof. □

The Segal-Bargmann transform

By the Stone-von Neumann Theorem, there should be a unitary operator

$$L^2(\mathbb{R}^n, d\mathbf{x}) \to \mathcal{H}L^2(\mathbb{C}^n, d\mu_\hbar),$$

that interpolates between the representation of the Heisenberg Lie algebra described in §2.1 and the one on the Segal-Bargmann space defined by (2.8) and (2.9). This unitary operator is the Segal-Bargmann transform.

Definition 2.3. The **Segal-Bargmann transform** is

$$(\mathcal{T}_{SB}\psi)(\mathbf{z}) = (2\pi\hbar)^{-n/2}\int_{\mathbb{R}^n} e^{-(\mathbf{z}-\boldsymbol{\xi})^T(\mathbf{z}-\boldsymbol{\xi})/2\hbar}\psi(\boldsymbol{\xi})d\boldsymbol{\xi}. \qquad (2.10)$$

Theorem 2.2. *The Segal-Bargmann transform (2.10) has the following properties:*

(i) For all $\psi \in L^2(\mathbb{R}^n, d\mathbf{x})$, the integral is convergent and its value is a function in $\mathcal{H}L^2(\mathbb{C}^n, d\mu_\hbar)$.

(ii) The map $\mathcal{T}_{SB}$ *is a unitary operator from* $L^2(\mathbb{R}^n, d\mathbf{x})$ *onto* $\mathcal{H}L^2(\mathbb{C}^n, d\mu_\hbar)$.

(iii) For $j = 1, 2, \dots, n$, *and* $\phi \in \mathcal{H}L^2(\mathbb{C}^n, d\mu_\hbar)$,

$$\mathcal{T}_{SB}(Q_j + iP_j)\mathcal{T}_{SB}^{-1}\phi = z_j\phi + 2\hbar\frac{\partial\phi}{\partial z_j}$$

$$\mathcal{T}_{SB}(Q_j - iP_j)\mathcal{T}_{SB}^{-1}\phi = z_j\phi.$$

Proof. (i) The function under the integral is square integrable, so the integral converges, and hence the Segal-Bargmann transform is well defined. Because for each $\boldsymbol{\xi}$, the integrand is a holomorphic function of $\mathbf{z}$, and because integration is obtained through a limiting process from summation, we can apply Morera's Theorem and conclude that the Segal-Bargmann transform yields a holomorphic function.

(iii) Because we do not know yet that $\mathcal{T}_{SB}$ has an inverse, we prove instead

$$\mathcal{T}_{SB}(Q_j + iP_j)\psi = \left(z_j + 2\hbar\frac{\partial}{\partial z_j}\right)\mathcal{T}_{SB}\psi \tag{2.11}$$

$$\mathcal{T}_{SB}(Q_j - iP_j)\psi = z_j\mathcal{T}_{SB}\psi. \tag{2.12}$$

Again we work only with smooth functions that decay fast at infinity.

We have

$$\begin{aligned}
\mathcal{T}_{SB}P_j\psi &= -i\hbar(2\pi\hbar)^{-n/2}\int_{\mathbb{R}^n} e^{-(\mathbf{z}-\boldsymbol{\xi})^T(\mathbf{z}-\boldsymbol{\xi})/2\hbar}\frac{\partial\psi}{\partial\xi_j}d\boldsymbol{\xi} \\
&= i\hbar(2\pi\hbar)^{-n/2}\int_{\mathbb{R}^n} \psi(\boldsymbol{\xi})\frac{\partial}{\partial\xi_j}e^{-(\mathbf{z}-\boldsymbol{\xi})^T(\mathbf{z}-\boldsymbol{\xi})/2\hbar}d\boldsymbol{\xi} \\
&= i\hbar(2\pi\hbar)^{-n/2}\int_{\mathbb{R}^n} \frac{(z_j-\xi_j)}{\hbar}e^{-(\mathbf{z}-\boldsymbol{\xi})^T(\mathbf{z}-\boldsymbol{\xi})/2\hbar}\psi(\boldsymbol{\xi})d\boldsymbol{\xi} \\
&= iz_j(2\pi\hbar)^{-n/2}\int_{\mathbb{R}^n} e^{-(\mathbf{z}-\boldsymbol{\xi})^T(\mathbf{z}-\boldsymbol{\xi})/2\hbar}\psi(\boldsymbol{\xi})d\boldsymbol{\xi} \\
&\quad - i(2\pi\hbar)^{-n/2}\int_{\mathbb{R}^n} e^{-(\mathbf{z}-\boldsymbol{\xi})^T(\mathbf{z}-\boldsymbol{\xi})/2\hbar}\xi_j\psi(\boldsymbol{\xi})d\boldsymbol{\xi}.
\end{aligned}$$

This means that

$$\mathcal{T}_{SB}P_j = iz_j\mathcal{T}_{SB} - i\mathcal{T}_{SB}Q_j.$$

Solving for $z_j\mathcal{T}_{SB}$ we obtain (2.12).

To prove (2.11), we compute

$$\begin{aligned}
\hbar\frac{\partial}{\partial z_j}\mathcal{T}_{SB}\psi &= (2\pi\hbar)^{-n/2}\int_{\mathbb{R}^n} \hbar\frac{\partial}{\partial z_j}e^{-(\mathbf{z}-\boldsymbol{\xi})^T(\mathbf{z}-\boldsymbol{\xi})/2\hbar}\psi(\boldsymbol{\xi})d\boldsymbol{\xi} \\
&= (2\pi\hbar)^{-n/2}\int_{\mathbb{R}^n} (-z_j+\xi_j)e^{-(\mathbf{z}-\boldsymbol{\xi})^T(\mathbf{z}-\boldsymbol{\xi})/2\hbar}\psi(\boldsymbol{\xi})d\boldsymbol{\xi} \\
&= -z_j\mathcal{T}_{SB}\psi(\mathbf{z}) + \mathcal{T}_{SB}(\xi_j\psi)(\mathbf{z}).
\end{aligned}$$

Thus

$$2\hbar\frac{\partial}{\partial z_j}\mathcal{T}_{SB} = -2z_j\mathcal{T}_{SB} + 2\mathcal{T}_{SB}Q_j.$$

Adding to this equality the equation (2.12) we obtain (2.11).

(ii) We will prove that the operator $\psi \to (\pi\hbar)^{-n/4}e^{-\|\operatorname{Im}\mathbf{z}\|^2/2\hbar}\mathcal{T}_{SB}(\psi)$ is an isometry from $L^2(\mathbb{R}^n, d\mathbf{x})$ into $L^2(\mathbb{C}^n, d\mathbf{x}d\mathbf{y})$. We rewrite the function $(\pi\hbar)^{-n/4}e^{-\|\operatorname{Im}\mathbf{z}\|^2/2\hbar}\mathcal{T}_{SB}(\psi)$ as

$$(2\pi\hbar)^{-n/2}(\pi\hbar)^{-n/4}\int_{\mathbb{R}^n} e^{-\frac{1}{2\hbar}(\|\mathbf{x}\|^2-\|\mathbf{y}\|^2+2i\mathbf{y}^T\mathbf{x}-2\mathbf{x}^T\boldsymbol{\xi}-2i\mathbf{y}^T\boldsymbol{\xi}+\|\xi\|^2+\|\mathbf{y}\|^2)}\psi(\boldsymbol{\xi})d\boldsymbol{\xi}$$

$$= (2\pi\hbar)^{-n/2}(\pi\hbar)^{-n/4}\int_{\mathbb{R}^n} e^{-\frac{i}{\hbar}\mathbf{y}^T(\mathbf{x}-\boldsymbol{\xi})}e^{-\frac{1}{2\hbar}\|\mathbf{x}-\boldsymbol{\xi}\|^2}\psi(\boldsymbol{\xi})d\boldsymbol{\xi}$$

$$= (2\pi\hbar)^{-n/2}(\pi\hbar)^{-n/4}\int_{\mathbb{R}^n} e^{-\frac{i}{\hbar}\mathbf{y}^T\tilde{\boldsymbol{\xi}}}e^{-\frac{1}{2\hbar}\|\tilde{\boldsymbol{\xi}}\|^2}\psi(\mathbf{x}-\tilde{\boldsymbol{\xi}})d\tilde{\boldsymbol{\xi}}.$$

Consider the identification of $L^2(\mathbb{R}^{2n}, d\mathbf{x}d\boldsymbol{\xi})$ with the closure of $L^2(\mathbb{R}^n, d\mathbf{x}) \otimes L^2(\mathbb{R}^n, d\boldsymbol{\xi})$ in the L^2-norm, and let $\psi_0(\boldsymbol{\xi}) = e^{-\|\boldsymbol{\xi}\|^2/2\hbar}$. The integral is obtained by applying to the function $\psi(\mathbf{x})\otimes\psi_0(\tilde{\boldsymbol{\xi}})$ the change of variable

$$(\mathbf{x}, \tilde{\boldsymbol{\xi}}) \mapsto (\mathbf{x}-\tilde{\boldsymbol{\xi}}, \tilde{\boldsymbol{\xi}})$$

followed by the (scaled) Fourier transform in the first variable

$$F(\tilde{\boldsymbol{\xi}}, \mathbf{x}) \mapsto \int_{\mathbb{R}^n} e^{-\frac{i}{\hbar}\mathbf{y}^T\tilde{\boldsymbol{\xi}}}F(\tilde{\boldsymbol{\xi}}, \mathbf{x})d\tilde{\boldsymbol{\xi}}.$$

The first transformation is unitary, while the second is a unitary multiplied by $(2\pi\hbar)^{n/2}$. Hence the value of the integral is equal to

$$(2\pi\hbar)^{n/2}\|\psi\otimes\psi_0\|_2 = (2\pi\hbar)^{n/2}(\pi\hbar)^{n/4}\|\psi\|_2,$$

where we used the fact that

$$\|\psi_0\|_2 = \left(\int_{\mathbb{R}^n} e^{-\|\boldsymbol{\xi}\|^2/\hbar}d\boldsymbol{\xi}\right)^{1/2} = (\pi\hbar)^{n/4}.$$

This proves our claim. It follows that the Segal-Bargmann transform is an isometry.

Let us show that it is onto. We compute

$$\mathcal{T}_{SB}((\pi\hbar)^{-n/4}\psi_0)(\mathbf{z}) = (\pi\hbar)^{-n/2}e^{-\mathbf{z}^T\mathbf{z}/4\hbar}\int_{\mathbb{R}^n} e^{-(\mathbf{z}/\sqrt{2}-\sqrt{2}\xi)^T(\mathbf{z}/\sqrt{2}-\sqrt{2}\xi)/2\hbar}d\boldsymbol{\xi}$$

$$= e^{-\mathbf{z}^T\mathbf{z}/4\hbar}.$$

Using (2.12) we obtain

$$\mathcal{T}_{SB}((\pi\hbar)^{-n/4}(a_1^\dagger)^{k_1}(a_2^\dagger)^{k_2}\cdots(a_n^\dagger)^{k_n}\psi_0)(\mathbf{z}) = z_1^{k_1}z_2^{k_2}\cdots z_n^{k_n}e^{-\mathbf{z}^T\mathbf{z}/4\hbar},$$

with the convention that $a_j^\dagger = Q_j - iP_j$. It is not hard to see that the functions $(\pi\hbar)^{-n/2}(a_1^\dagger)^{k_1}(a_2^\dagger)^{k_2}\cdots(a_n^\dagger)^{k_n}\psi_0)$ are L^2-integrable[1]. It follows from Proposition 2.4 that the Segal-Bargmann transform is onto, and we are done. □

2.2.2 *The Schrödinger representation and the Weyl quantization in the holomorphic setting*

The Schrödinger representation in the holomorphic setting

Through exponentiation we obtain the Schrödinger representation of the Heisenberg group $\mathbf{H}(\mathbb{R}^n)$ on the Segal-Bargmann space.

Theorem 2.3. *The Heisenberg group $\mathbf{H}(\mathbb{R}^n)$ acts on the Segal-Bargmann space by*

$$\exp(\mathbf{p}^T\mathbf{P} + \mathbf{q}^T\mathbf{Q} + tI)\phi(\mathbf{z}) = e^{\pi i h\mathbf{q}^T(\mathbf{p}+i\mathbf{q})+2\pi i\mathbf{q}^T\mathbf{z}+2\pi it}\phi\left(\mathbf{z} + h(\mathbf{p}+i\mathbf{q})\right).$$

Proof. Solving for P_j and Q_j in (2.9) we obtain

$$P_j = -i\hbar\frac{\partial}{\partial z_j} \text{ and } Q_j = M_{z_j} + \hbar\frac{\partial}{\partial z_j}.$$

Note that if α and β are constants, then

$$e^{\alpha M_{z_j}}\phi(\mathbf{z}) = e^{\alpha z_j}\phi(\mathbf{z}) \text{ and } e^{\beta\frac{\partial}{\partial z_j}}\phi(\mathbf{z}) = \phi(z_1,\ldots,z_j+\beta,\ldots,z_n).$$

Hence

$$\begin{aligned}\exp(p_jP_j)\phi(\mathbf{z}) &= e^{2\pi i p_j P_j}\phi(\mathbf{z}) = e^{2\pi\hbar p_j\frac{\partial}{\partial z_j}}\phi(\mathbf{z})\\ &= \phi(z_1,\ldots,z_j+hp_j,\ldots,z_n).\end{aligned}$$

Applying the Baker-Campbell-Hausdorff formula, we obtain

$$\begin{aligned}\exp(q_jQ_j)\phi(\mathbf{z}) &= e^{2\pi i q_j Q_j}\phi(\mathbf{z}) = e^{2\pi i q_j(M_z+\hbar\frac{\partial}{\partial z_j})}\phi(\mathbf{z})\\ &= e^{-\frac{1}{2}[2\pi i q_j\hbar\frac{\partial}{\partial z_j},2\pi i q_j M_{z_j}]}e^{2\pi i q_j\hbar\frac{\partial}{\partial z_j}}e^{2\pi i q_j M_{z_j}}\phi(\mathbf{z})\\ &= e^{\pi h q_j^2}e^{2\pi i q_j\hbar\frac{\partial}{\partial z_j}}e^{2\pi i q_j z_j}\phi(\mathbf{z})\\ &= e^{-\pi h q_j^2+2\pi i q_j z_j}\phi\left(z_1,\ldots,z_j+ihq_j,\ldots,z_n\right).\end{aligned}$$

The general formula from the statement follows by applying once more the Baker-Campbell-Hausdorff formula as in the proof of Proposition 2.1. □

[1] They are the Hermite functions.

Weyl quantization via Toeplitz operators

Weyl quantization in the holomorphic setting has a concise description using Toeplitz operators. Let us explain it.

The Segal-Bargmann space is a closed subspace of $L^2(\mathbb{C}^n, d\mu_\hbar)$. Let

$$\Pi_{SB} : L^2(\mathbb{C}^n, d\mu_\hbar) \to \mathcal{H}L^2(\mathbb{C}^n, d\mu_\hbar),$$

be the orthogonal projection.

Consider a function $f \in C^\infty(\mathbb{R}^n \times \mathbb{R}^n, \mathbb{C})$ that grows sufficiently slowly at infinity so that for every $\phi \in \mathcal{H}L^2(\mathbb{C}^n, d\mu_\hbar)$, the product $f\phi$ is in $L^2(\mathbb{C}^n, d\mu_\hbar)$.

Definition 2.4. The **Toeplitz operator** with symbol f is the operator $T_f : \mathcal{H}L^2(\mathbb{C}^n, d\mu_\hbar) \to \mathcal{H}L^2(\mathbb{C}^n, d\mu_\hbar)$ given by

$$T_f(\phi) = \Pi_{SB}(f\phi).$$

In order to understand Toeplitz operators, we need to understand the projection operator Π_{SB}. The formula for the projection

$$\Pi_{SB} f = \sum_{k_1, k_2, \ldots, k_n} \langle f, \Phi_{k_1 k_2 \ldots k_n} \rangle \, \Phi_{k_1 k_2 \ldots k_n}$$

translates into

$$\Pi_{SB} f(\mathbf{z}) = \int_{\mathbb{C}^n} K_{SB}(\mathbf{z}, \mathbf{w}) \phi(\mathbf{w}) e^{-\|\operatorname{Im} \mathbf{w}\|^2/\hbar} d\mathbf{x}_\mathbf{w} d\mathbf{y}_\mathbf{w}$$

where

$$K_{SB}(\mathbf{z}, \mathbf{w}) = \sum_{k_1, k_2, \ldots, k_n} \Phi_{k_1, k_2, \ldots, k_n}(\mathbf{z}) \Phi_{k_1, k_2, \ldots, k_n}(\overline{\mathbf{w}}), \tag{2.13}$$

and $\mathbf{w} = \mathbf{x}_\mathbf{w} + i\mathbf{y}_\mathbf{w}$. This means that Π_{SB} is an **integral operator** with **kernel** $K_{SB}(\mathbf{z}, \mathbf{w})$ given by (2.13).

Lemma 2.1. *The kernel of the projection operator* Π_{SB} *is given by*

$$K_{SB}(\mathbf{z}, \mathbf{w}) = e^{-(\mathbf{z}-\bar{\mathbf{w}})^T(\mathbf{z}-\bar{\mathbf{w}})/4\hbar}.$$

Proof. We compute

$$\begin{aligned} K_{SB}(\mathbf{z}, \mathbf{w}) &= e^{\mathbf{z}^T\mathbf{z}/4\hbar} e^{\overline{\mathbf{w}}^T\overline{\mathbf{w}}/4\hbar} \sum_{k_1, k_2, \ldots, k_n} \frac{(z_1\overline{w_1})^{k_1}(z_2\overline{w_2})^{k_2}\cdots(z_n\overline{w_n})^{k_n}}{(2\hbar)^{k_1 k_2 \cdots k_n} k_1! k_2! \cdots k_n!} \\ &= e^{\mathbf{z}^T\mathbf{z}/4\hbar} e^{\overline{\mathbf{w}}^T\overline{\mathbf{w}}/4\hbar} e^{\mathbf{z}^T\overline{\mathbf{w}}/2\hbar} = e^{-(\mathbf{z}-\bar{\mathbf{w}})^T(\mathbf{z}-\bar{\mathbf{w}})/4\hbar}. \end{aligned}$$

□

Theorem 2.4. *The operator associated by Weyl quantization to the smooth function* $f(\mathbf{x},\mathbf{y})$ *on the phase space is the Toeplitz operator on the Segal-Bargmann space with symbol* $e^{-\hbar\Delta/4}f(\mathbf{x},-\mathbf{y})$, *where* Δ *is the Laplace operator*

$$\Delta = \sum_j \left(\frac{\partial}{\partial x_j^2} + \frac{\partial}{\partial y_j^2}\right).$$

Before we proceed with the proof of this theorem, let us explain the statement. The exponential of the negative of the Laplacian, referred to as the **inverse heat kernel**, can be defined using the power series expansion of the exponential function. Or we can phrase this result as saying that the operator associated to the function $e^{\hbar\Delta/4}f$ is the same as the Toeplitz operator with symbol f, and here $e^{\hbar\Delta/4}$ is the usual heat kernel (which describes the propagation of heat if its initial distribution was defined by the function f).

Proof. We will only do the proof in the case where the symbol is an exponential function, because we are only interested in the Schrödinger representation. Recall that $\mathbf{x} = \boldsymbol{\xi}$ and $\mathbf{y} = \boldsymbol{\eta}$, so we are looking at exponential functions of the form

$$f(\mathbf{x},\mathbf{y}) = e^{2\pi i(\mathbf{p}^T\mathbf{y}+\mathbf{q}^T\mathbf{x})}.$$

We compute

$$\Delta f(\mathbf{x},\mathbf{y}) = \Delta e^{2\pi i(\mathbf{p}^T\mathbf{y}+\mathbf{q}^T\mathbf{x})} = -4\pi^2(\mathbf{p}^T\mathbf{p}+\mathbf{q}^T\mathbf{q})e^{2\pi i(\mathbf{p}^T\mathbf{y}+\mathbf{q}^T\mathbf{x})},$$

hence

$$e^{-\frac{\hbar\Delta}{4}}f(\mathbf{x},-\mathbf{y}) = e^{\frac{\pi h}{2}(\mathbf{p}^T\mathbf{p}+\mathbf{q}^T\mathbf{q})}e^{2\pi i(-\mathbf{p}^T\mathbf{y}+\mathbf{q}^T\mathbf{x})}.$$

We see that the Toeplitz operator with symbol $e^{\hbar\Delta/4}e^{2\pi i(-\mathbf{p}^T\mathbf{y}+\mathbf{q}^T\mathbf{x})}$ is a scalar multiple of the Toeplitz operator with symbol $e^{2\pi i(-\mathbf{p}^T\mathbf{y}+\mathbf{q}^T\mathbf{x})}$. Let us compute that action of the latter on the elements of the orthonormal basis. Because the integrals are computed separately in each variable, it suffices to consider the one-dimensional case. So let $z = x+iy$ be the complex variable. We switch to the Planck's constant h (instead of the reduced Planck's constant $\hbar$), so that the kernel of the projection is

$$K_{SB}(z,w) = e^{-\frac{(z-\bar{w})^2\pi}{2h}}$$

and the measure of integration is

$$e^{-\frac{2\pi y^2}{h}}dxdy.$$

We consider the function

$$(h/\pi)^k k!\Phi_k(z) = z^n e^{-\frac{z^2\pi}{2h}},$$

and compute the action of the Toeplitz operator with symbol $e^{2\pi i(-py+qx)}$ on it. We have

$$\begin{aligned}
&\int_{\mathbb{R}^2} e^{2\pi i(-px+qy)} e^{-\frac{(z-x+iy)^2\pi}{2h}} e^{-\frac{(x+iy)^2\pi}{2h}} e^{-\frac{2\pi y^2}{h}} (x+iy)^k dxdy \\
&\quad = e^{-\frac{z^2\pi}{2h}} \int_{\mathbb{R}^2} e^{-\frac{\pi}{h}[x^2-x(z+2ihq)]} e^{-\frac{\pi}{h}[y^2+iy(z+2hp)]} (x+iy)^k dxdy \\
&\quad = e^{-\frac{z^2\pi}{2h}} e^{\frac{\pi}{4h}[(z+2ihq)^2-(z+2hp)^2]} \\
&\quad \times \int_{\mathbb{R}^2} e^{-\frac{\pi}{h}(x-\frac{z-2ihq}{2})^2} e^{(y+\frac{iz+2ihp}{2})^2} (x+iy)^2 dxdy.
\end{aligned}$$

By changing the variables of integration to $u = x - \frac{z-2ihq}{2}$ and $v = y + i\frac{z+2hp}{2}$, the above expression becomes

$$\begin{aligned}
&e^{-\frac{z^2\pi}{2h}} e^{izq\pi - zp\pi + \pi h(p^2-q^2)} \int_{\mathbb{R}^2} e^{-\frac{\pi}{h}u^2} e^{-\frac{\pi}{h}v^2} [u+iv+z+h(p+iq)]^k dxdy \\
&\quad = e^{-\frac{\pi h}{2}(p^2+q^2)} e^{\pi ihq(p+iq)+2\pi iqz} e^{-\frac{[z+h(p+iq)]^2\pi}{2h}} \\
&\quad \times \int_{\mathbb{R}^2} e^{-\frac{\pi}{h}(u^2+v^2)} [x+iy+z+h(p+iq)]^k dxdy.
\end{aligned}$$

Note that when $p = q = 0$ we are just projecting $(h/\pi)^k k!\Phi_k(z)$ onto itself. So in this case the value of the integral should be z^k. If we set $z' = z + h(p+iq)$, then the value of the integral should be z'^k. The result of our computation is therefore

$$e^{-\frac{\pi h}{2}(p^2+q^2)} e^{\pi ihq(p+iq)+2\pi iqz} e^{-\frac{[z+h(p+iq)]^2\pi}{2h}} [z+h(p+iq)]^k.$$

We conclude that the Toeplitz operator with symbol $e^{\hbar\Delta/4} e^{2\pi i(-\mathbf{p}^T\mathbf{y}+\mathbf{q}^T\mathbf{x})}$ acts on the elements of the orthonormal basis, and hence on the entire space, by the formula derived in Theorem 2.3. The theorem is proved. □

2.2.3 *Holomorphic quantization in the momentum representation*

There is an annoyance caused by Theorem 2.4, in that the quantization of the function z is the Toeplitz operator with symbol $\bar{z}$, and the quantization of the function $\bar{z}$ is the Toeplitz operator with symbol z. A way around this is to work with anti-holomorphic functions, as Segal has done. Another way, which we will follow, is to choose different identifications for the creation and annihilation operators. This will give a more intuitive picture on how to relate theta functions to knots. The new model will be related via the Segal-Bargmann transform to the quantization in the momentum representation.

We choose complex coordinates such that the momentum is the real part and the position is the imaginary part. As such, $\mathbf{z} = \mathbf{x} + i\mathbf{y}$ with $\mathbf{x} = \boldsymbol{\eta}$ and $\mathbf{y} = \boldsymbol{\xi}$. We use the same convention that

$$a_j = M_{z_j} + 2\hbar\frac{\partial}{\partial z_j}, \quad a_j^\dagger = M_{z_j}.$$

The position and momentum operators are considered in the momentum representation. We will call this new quantization model the **holomorphic quantization in the momentum representation**. The terminology is nonstandard, we use it in this book to emphasize the fact that in this quantization model the **real part of the complex variable is the momentum**. In this case, Theorem 2.2 yields the identifications

$$a_j = M_{z_j} + 2\hbar\frac{\partial}{\partial z_j} = P_j - iQ_j$$

$$a_j^\dagger = M_{z_j} = P_j + iQ_j.$$

Solving for the position and momentum operators we obtain

$$Q_j = i\hbar\frac{\partial}{\partial z_j}, \quad P_j = M_{z_j} + \hbar\frac{\partial}{\partial z_j}.$$

Theorem 2.4 becomes:

Theorem 2.5. *In the case of holomorphic quantization in the momentum representation of n particles in a 1-dimensional space, the quantum observable associated by Weyl quantization to the function f is the Toeplitz operator with symbol $e^{-\hbar\Delta/4}f$.*

In particular we obtain a new representation of the Heisenberg group, similar to the one in Theorem 2.3.

Theorem 2.6. *The Schrödinger representation of the Heisenberg group $\mathbf{H}(\mathbb{R}^n)$ on the Segal-Bargmann space in the momentum representation is given by*

$$\exp(\mathbf{p}^T\mathbf{P} + \mathbf{q}^T\mathbf{Q} + tI)\phi(\mathbf{z}) = e^{-\pi i h\mathbf{p}^T(\mathbf{q}-i\mathbf{p})+2\pi i\mathbf{p}^T\mathbf{z}+2\pi i t}\phi(\mathbf{z} - h(\mathbf{q} - i\mathbf{p}))\,.$$

2.3 Geometric quantization

This section will allow us to place all the quantization models introduced above on equal footing. A detailed account of the methods can be found in [Woodhouse (1980)] or in [Śniatycki (1980)].

2.3.1 *Polarizations*

The quantization models for n particles in a 1-dimensional space described in §2.1.2, §2.2.1, §2.2.3 have in common the fact that their Hilbert spaces are L^2-spaces. But they are not the entire L^2-space of the phase space $\mathbb{R}^n \times \mathbb{R}^n$, instead they are L^2-spaces of functions in the position variables only, in the momentum variables only, or they are holomorphic in complex variables that combine position and momentum. In all these situations, the wave functions depend on a number of variables that is half the dimension of the phase space. Moreover, for the coordinate functions of these variables, the pairwise Poisson brackets are zero. Because we work on a vector space, the choice of global variables with this property is always possible and for this reason the situation is simpler. We discuss this case first. We then address the situation of more general manifolds, which will be needed for the construction of theta functions.

The phase space is the real vector space $\mathbb{R}^n \times \mathbb{R}^n$. The tangent space at a point is also $\mathbb{R}^n \times \mathbb{R}^n$, and all tangent spaces can be identified trivially via translation. Hence we can talk about one tangent space, which we call V for convenience. The space V is endowed with the symplectic form

$$\omega = -\sum_{j=1}^{n} d\xi_j \wedge d\eta_j.$$

For a subspace W of V, we define the orthogonal

$$W^{\perp} = \{v \in V \,|\, \omega(v, w) = 0 \text{ for all } w \in W\}.$$

The subspace W is called isotropic if $W \subset W^{\perp}$ and Lagrangian if it is maximal isotropic. Alternatively, Lagrangian subspaces are isotropic subspaces of dimension n. $\mathbf{L}$ is Lagrangian if and only if $\mathbf{L}^{\perp} = \mathbf{L}$. Note also that the Lie bracket of any two vectors in $\mathbf{L}$ is in $\mathbf{L}$.

We complexify the tangent space to $V \otimes \mathbb{C} = \mathbb{C}^n \times \mathbb{C}^n$, by placing complex coefficients in front of the basis vectors. The form ω extends to a symplectic form on $V \otimes \mathbb{C}$. The notion of Lagrangian subspace extends to the complex situation as well, by considering complex subspaces.

Definition 2.5. A polarization of the symplectic vector space (V, ω) is a Lagrangian subspace $\mathbf{L}$ of $V \otimes \mathbb{C}$.

A polarization of the tangent space is used for selecting which functions on the phase space should play the role of states. As we will later see, these are the functions that are covariantly constant in the direction of the polarization.

Example 2.2. The Lagrangian subspace

$$\mathbf{L} = \operatorname{Span}\left(\frac{\partial}{\partial \eta_1}, \frac{\partial}{\partial \eta_2}, \ldots, \frac{\partial}{\partial \eta_n}\right) \tag{2.14}$$

defines a polarization which will give rise to the quantization in the position representation.

Example 2.3. The polarization defined by the Lagrangian subspace

$$\mathbf{L} = \operatorname{Span}\left(\frac{\partial}{\partial \xi_1}, \frac{\partial}{\partial \xi_2}, \ldots, \frac{\partial}{\partial \xi_n}\right) \tag{2.15}$$

will be associated to the quantization in the momentum representation.

Example 2.4. The space

$$\mathbf{L} = \operatorname{Span}\left(\frac{1}{2}\left(\frac{\partial}{\partial \xi_1} + i\frac{\partial}{\partial \eta_1}\right), \frac{1}{2}\left(\frac{\partial}{\partial \xi_2} + i\frac{\partial}{\partial \eta_2}\right), \ldots, \frac{1}{2}\left(\frac{\partial}{\partial \xi_n} + i\frac{\partial}{\partial \eta_n}\right)\right)$$

is also Lagrangian, because

$$\begin{aligned}
&\omega\left(\frac{1}{2}\left(\frac{\partial}{\partial \xi_j} + i\frac{\partial}{\partial \eta_j}\right), \frac{1}{2}\left(\frac{\partial}{\partial \xi_k} + i\frac{\partial}{\partial \eta_k}\right)\right) \\
&= \frac{1}{4}\left[\omega\left(\frac{\partial}{\partial \xi_j}, \frac{\partial}{\partial \xi_k}\right) + i\omega\left(\frac{\partial}{\partial \xi_j}, \frac{\partial}{\partial \eta_k}\right) + i\omega\left(\frac{\partial}{\partial \eta_j}, \frac{\partial}{\partial \xi_k}\right) - \omega\left(\frac{\partial}{\partial \eta_j}, \frac{\partial}{\partial \eta_k}\right)\right] \\
&= \frac{\delta_{jk} i}{4}\left[\omega\left(\frac{\partial}{\partial \xi_j}, \frac{\partial}{\partial \eta_j}\right) + \omega\left(\frac{\partial}{\partial \eta_j}, \frac{\partial}{\partial \xi_j}\right)\right] = 0.
\end{aligned}$$

Hence it gives rise to a polarization, too. The states associated to it will be functions on the phase space that are holomorphic in the variable $\mathbf{z} = \boldsymbol{\xi} + i\boldsymbol{\eta}$, as in the case of the Segal-Bargmann space.

We now prove a technical result and introduce some notation based on it.

Lemma 2.2. *Let* $\mathbf{L}$ *be a Lagrangian subspace of* V. *Then there exists a basis* $\mathbf{v}_1, \mathbf{v}_2, \ldots, \mathbf{v}_n$ *of* $\mathbf{L}$ *as well as the vectors* $\mathbf{w}_1, \mathbf{w}_2, \ldots, \mathbf{w}_n$ *forming a basis of a Lagrangian subspace* $\mathbf{L}'$ *that is an algebraic complement of* $\mathbf{L}$ *(i.e.* $\mathbf{L} \oplus \mathbf{L}' = V$*) such that*

$$\omega(\mathbf{v}_j, \mathbf{w}_k) = \delta_{jk}.$$

Moreover, starting with any basis $\mathbf{v}_1, \mathbf{v}_2, \ldots, \mathbf{v}_n$ *of* $\mathbf{L}$, *we can find the vectors* $\mathbf{w}_1, \mathbf{w}_2, \ldots, \mathbf{w}_n$ *as above.*

Proof. Start with a basis $\mathbf{u}_1, \mathbf{u}_2, \ldots, \mathbf{u}_n$ of $\mathbf{L}$ and complete it by vectors $\mathbf{u}_{n+1}, \mathbf{u}_{n+2}, \ldots, \mathbf{u}_{2n}$ to a basis of V. Now apply a Gram-Schmidt construction as follows.

At the first step define $\mathbf{v}_1 = \mathbf{u}_1$ and $\mathbf{w}_1 = (\omega(\mathbf{u}_1, \mathbf{u}_{n+1}))^{-1}\mathbf{u}_{n+1}$. Set these vectors aside. Then, for $j = 2, \ldots, n$, let

$$\mathbf{z}_j = \mathbf{u}_j + \omega(\mathbf{w}_1, \mathbf{u}_j)\mathbf{v}_1,$$
$$\mathbf{z}_{n+j} = \mathbf{u}_{n+j} + \omega(\mathbf{w}_1, \mathbf{u}_{n+j})\mathbf{v}_1 - \omega(\mathbf{v}_1, \mathbf{u}_{n+j})\mathbf{w}_1.$$

Repeat the process for the vectors $\mathbf{z}_2, \mathbf{z}_3, \ldots, \mathbf{z}_n, \mathbf{z}_{n+2} \cdots, \mathbf{z}_{2n}$, to produce $\mathbf{v}_2$, $\mathbf{w}_2$. Continue in order to produce the vectors $\mathbf{v}_1, \mathbf{v}_2, \ldots, \mathbf{v}_n, \mathbf{w}_1, \mathbf{w}_2, \ldots, \mathbf{w}_n$. These have the desired property.

For the second part, note that we can start with $\mathbf{u}_j = \mathbf{v}_j$, $j = 1, 2, \ldots, n$. Additionally, $\mathbf{u}_{n+1}$ can be chosen such that $\omega(\mathbf{v}_j, \mathbf{u}_{n+1}) = \delta_{j,1}$. Indeed, we can start with any vectors $\mathbf{u}_{n+1}, \ldots, \mathbf{u}_{2n}$. Then find a vector which is a linear combination of these and has the desired properties, which amounts to solving a linear system of equations in the coefficients. Replace the basis with one that contains this vector. In this new situation we automatically have $\mathbf{z}_j = \mathbf{v}_j$, $j = 2, \ldots, n$. Now we can repeat the procedure and obtain the desired conclusion. □

Definition 2.6. A basis $\mathbf{v}_j, \mathbf{w}_j$, $j = 1, 2, \ldots, n$, with the property that $\omega(\mathbf{v}_j, \mathbf{w}_k) = \delta_{jk}$ is called a **symplectic basis.** The basis provided by Lemma 2.2 is called a **symplectic basis associated to the Lagrangian subspace L.**

Notation: Let x_j, y_j, $j = 1, 2, \ldots, n$,[2] be functions on the phase space $\mathbb{R}^n \times \mathbb{R}^n$ such that

$$X_{x_j} = \mathbf{w}_j, \quad X_{y_j} = -\mathbf{v}_j, \quad j = 1, 2, \ldots, n.$$

These can be used as coordinate functions. We call x_j, y_j, $j = 1, 2, \ldots, n$, the **coordinates associated to the symplectic basis** $\mathbf{v}_j, \mathbf{w}_j$. We set $\mathbf{x} = (x_1, x_2, \ldots, x_n)$ and $\mathbf{y} = (y_1, y_2, \ldots, y_n)$ to be the coordinate vectors.

Note that x_j and y_j are linear in the coordinates ξ_j, η_j, meaning that

$$x_j = \sum_k (a_{jk}\xi_k + b_{jk}\eta_k), \quad y_j = \sum_k (c_{jk}\xi_k + d_{jk}\eta_k).$$

Define the matrix,

$$h = \begin{bmatrix} A & B \\ C & D \end{bmatrix} = \begin{bmatrix} (a_{ij}) & (b_{ij}) \\ (c_{ij}) & (d_{ij}) \end{bmatrix}$$

[2]The choices are made so that in the notation p for the momentum, q for the position, $v = \partial/\partial p$, $w = \partial/\partial q$ and $x = p$, $y = q$.

and for future use, let its inverse be

$$h^{-1} = \begin{bmatrix} \widetilde{A} & \widetilde{B} \\ \widetilde{C} & \widetilde{D} \end{bmatrix} = \begin{bmatrix} (\widetilde{a}_{ij}) & (\widetilde{b}_{ij}) \\ (\widetilde{c}_{ij} & (\widetilde{d}_{ij}) \end{bmatrix}. \tag{2.16}$$

Then

$$\omega = \sum_j dx_j \wedge dy_j.$$

We use the following notation for partial derivatives:

$$\frac{\partial}{\partial x_j} := \mathbf{v}_j, \quad \frac{\partial}{\partial y_j} := \mathbf{w}_j.$$

Example 2.5. For **L** as in (2.14), we can choose the associated symplectic basis to be

$$\mathbf{v}_j = \frac{\partial}{\partial \eta_j}, \quad \mathbf{w}_j = \frac{\partial}{\partial \xi_j}, \quad j = 1, 2, \ldots, n, \tag{2.17}$$

and then $x_j = \eta_j$, $y_j = \xi_j$, $j = 1, 2, \ldots, n$.

Example 2.6. For **L** as in (2.15), we can choose

$$\mathbf{v}_j = -\frac{\partial}{\partial \xi_j}, \quad \mathbf{w}_j = \frac{\partial}{\partial \eta_j}, \quad j = 1, 2, \ldots, n, \tag{2.18}$$

in which case $x_j = -\xi_j, y = \eta_j$. But we can also choose

$$\mathbf{v}_j = -\frac{\partial}{\partial \eta_j}, \quad \mathbf{w}_j = -\frac{\partial}{\partial \xi_j},$$

and then $x = \xi_j$, $y = -\eta_j$.

Polarizations of symplectic manifolds

In the case where the phase space is a more general symplectic manifold, the choice of the n coordinates can no longer be done globally, but at least we can do this locally. The above definition is adapted as follows.

Let (M, ω) be a symplectic $2n$-dimensional manifold. A **complex distribution F** on M is a subbundle of the complexification of the tangent bundle of M. In that sense, at each point $p \in M$, the distribution associates a subspace of $TM \otimes \mathbb{C}$, and these subspaces vary smoothly with p and have all the same dimension. Such a distribution is called **involutive** if the Poisson bracket of any two vector fields in **F** is also in **F** (i.e. $\omega|\mathbf{F} \times \mathbf{F} = 0$).

A complex distribution **F** is called **Lagrangian** if

$$\dim_{\mathbb{C}} \mathbf{F} = \frac{1}{2} \dim_{\mathbb{R}} M \text{ and } \omega|\mathbf{F} \times \mathbf{F} = 0.$$

Definition 2.7. A **polarization of the symplectic manifold** (M, ω) is a complex Lagrangian involutive distribution $\mathbf{F}$ such that the function $p \mapsto \dim(\mathbf{F}_p \cap \overline{\mathbf{F}}_p)$ is constant on M, $\overline{\mathbf{F}}$ being the complex conjugate of $\mathbf{F}$.

The complex distributions $\mathbf{F} \cap \overline{\mathbf{F}}$ and $\mathbf{F} + \overline{\mathbf{F}}$ are complexifications of certain real distributions $\mathbf{D}$ and $\mathbf{E}$ in TM.

Real and Kähler polarizations

The quantization models from §2.1.2 differ from those in §2.2.1 and §2.2.3, in the sense that their state functions depend on real variables. For these we make the following definition.

Definition 2.8. A polarization $\mathbf{F}$ is called **real** if $\mathbf{F} = \overline{\mathbf{F}}$.

Said differently, $\mathbf{F}$ is real if $\mathbf{D} = \mathbf{E}$. In this sense, the polarizations (2.14) and (2.15) are real. A real polarization is spanned locally by vector fields in TM, so we can forget about the complexification and think of the real polarization as being an involutive Lagrangian distribution in TM. It foliates the manifold M by Lagrangian submanifolds.

For the moment we are interested in the case where M is the phase space $\mathbb{R}^n \times \mathbb{R}^n$. As mentioned before, a Lagrangian subspace of the tangent space V is an n-dimensional subspace on which the form ω vanishes. Each such Lagrangian subspace $\mathbf{L}$ of V defines a real distribution. The polarizations defined by (2.14) and (2.15) are particular examples.

We now turn to polarizations that give rise to holomorphic quantization models.

Definition 2.9. A polarization $\mathbf{F}$ is called **Kähler** if $\mathbf{F} \cap \overline{\mathbf{F}} = 0$.

In other words, a polarization is Kähler if $\mathbf{D} = 0$. In this case, at each point $p \in M$,

$$T_pM = \mathbf{F}_p \oplus \overline{\mathbf{F}}_p,$$

and the Hermitian form

$$(\mathbf{v}_p, \mathbf{w}_p) \to i\omega(\mathbf{v}_p, \overline{\mathbf{w}_p})$$

is nondegenerate.

The reason for the name is that on a Kähler manifold M, the subbundles $TM_{(1,0)}$ and $TM_{(0,1)}$ defined in local coordinates by the vector fields $\partial/\partial\mathbf{z}$, respectively $\partial/\partial\bar{\mathbf{z}}$ are Kähler polarizations.

Example 2.7. Consider the phase space $\mathbb{R}^n \times \mathbb{R}^n$. We can identify this space with $\mathbb{C}^n$ by the choice of complex variables $z_j = x_j + iy_j$. Then we consider the vector fields

$$\frac{\partial}{\partial \bar{z}_j} = \frac{1}{2}\left(\frac{\partial}{\partial x_j} + i\frac{\partial}{\partial y_j}\right)$$

in the complexification of $T(\mathbb{R}^n \times \mathbb{R}^n)$. We thus have a polarization

$$\mathbf{L} = \mathrm{Span}\left(\frac{\partial}{\partial \bar{z}_1}, \frac{\partial}{\partial \bar{z}_2}, \ldots, \frac{\partial}{\partial \bar{z}_n}\right)$$

which is Kähler. As we will see below in §2.3.2, the states defined by it are holomorphic functions. The polarization

$$\overline{\mathbf{L}} = \mathrm{Span}\left(\frac{\partial}{\partial z_1}, \frac{\partial}{\partial z_2}, \ldots, \frac{\partial}{\partial z_n}\right)$$

is also Kähler; the states defined by it are anti-holomorphic functions.

Example 2.8. Let us start with a Lagrangian subspace $\mathbf{L}$ of the tangent space V to $\mathbb{R}^n \times \mathbb{R}^n$. Use Lemma 2.2 to obtain the symplectic basis $\mathbf{v}_1, \mathbf{v}_2, \ldots, \mathbf{v}_n, \mathbf{w}_1, \mathbf{w}_2, \ldots \mathbf{w}_n$ associated to $\mathbf{L}$. Each of the following is a Kähler polarization:

$$\mathrm{Span}(\mathbf{w}_j + i\mathbf{v}_j \,|\, j = 1, 2, \ldots, n), \quad \mathrm{Span}(\mathbf{w}_j - i\mathbf{v}_j \,|\, j = 1, 2, \ldots, n),$$
$$\mathrm{Span}(-\mathbf{w}_j + i\mathbf{v}_j \,|\, j = 1, 2, \ldots, n), \quad \mathrm{Span}(-\mathbf{w}_j - i\mathbf{v}_j \,|\, j = 1, 2, \ldots, n).$$

Here is a second list

$$\mathrm{Span}(\mathbf{v}_j + i\mathbf{w}_j \,|\, j = 1, 2, \ldots, n), \quad \mathrm{Span}(\mathbf{v}_j - i\mathbf{w}_j \,|\, j = 1, 2, \ldots, n),$$
$$\mathrm{Span}(-\mathbf{v}_j + i\mathbf{w}_j \,|\, j = 1, 2, \ldots, n), \quad \mathrm{Span}(-\mathbf{v}_j - i\mathbf{w}_j \,|\, j = 1, 2, \ldots, n).$$

The Kähler polarizations from Example 2.7 are particular cases, obtained by starting with the Lagrangian space $\mathbf{L}$ from (2.14) respectively (2.15).

The vectors $\mathbf{v}_1, \mathbf{v}_2, \ldots, \mathbf{v}_n$ define a real polarization. From the above list we distinguish

$$\mathrm{Span}(\mathbf{v}_j + i\mathbf{w}_j \,|\, j = 1, 2, \ldots, n)$$

which we call the **Kähler polarization defined by** $\mathbf{v}_1, \mathbf{v}_2, \ldots, \mathbf{v}_n$.

2.3.2 *The construction of the Hilbert space using geometric quantization*

We recall now the general method of **geometric quantization**, which we use at this moment in order to change the perspective on our prototype. The reader can find a detailed description of this method in [Woodhouse (1980)]. The method will be applied again in §4.2 and §4.3.1 to quantize the Jacobian variety and produce theta functions.

Let M be a $2n$-dimensional manifold with symplectic form ω, which is to be quantized. The Hilbert space of the quantization will consist of sections of a line bundle $\mathcal{L}$ over M. This line bundle is obtained as the tensor product $\mathcal{L}'\otimes\mathcal{L}''$ of two line bundles, defined as follows.

The first line bundle, $\mathcal{L}'$, is a Hermitian line bundle with connection, having the curvature $(1/\hbar)\omega$. This means that if

$$\nabla = d - i\theta$$

is the connection[3], then the connection form θ, called potential, satisfies $d\theta = (1/\hbar)\omega$. The Hermitian structure $[\cdot,\cdot]$ is compatible with the connection, in the sense that for any two sections s and s' of $\mathcal{L}'$,

$$d[s,s'] = [\nabla s, s'] + [s, \nabla s'].$$

Such a line bundle exists if and only if

$$\frac{1}{2\pi\hbar}\omega = \frac{1}{h}\omega \in H^2(M,\mathbb{Z}). \tag{2.19}$$

This is **Weil's integrality condition.**

The line bundle $\mathcal{L}''$ is the metaplectic correction, which contains the information about the inner product of the Hilbert space. To define it, fix a polarization $\mathbf{F}$ of M and take the nth exterior power $\wedge^n\mathbf{F}$ of the bundle of linear frames in $\mathbf{F}$. Then $\mathcal{L}''$ satisfies $\mathcal{L}''\otimes\mathcal{L}'' = \wedge^n\mathbf{F}$, in that sense it is the square root of the volume form.

The Hilbert space will consist of the sections of $\mathcal{L} = \mathcal{L}'\otimes\mathcal{L}''$ that are covariantly constant in the direction of $\mathbf{F}$.

For all examples in this book $\wedge^n\mathbf{F}$, and hence $\mathcal{L}''$, are trivial. Hence we can ignore them, and thus identify $\mathcal{L}$ with $\mathcal{L}'$ (the trivial line bundles $\wedge^n\mathbf{F}$ and $\mathcal{L}''$ do play a small role here: they allow us to scale the inner product but this can be done implicitly without mentioning these bundles).

So, in our particular case, the Hilbert space consists of those sections of $\mathcal{L}'$ that are **covariantly constant** in the direction of $\mathbf{F}$, meaning that

$$\nabla_{\mathbf{v}} s = 0, \text{ for all } \mathbf{v}\in\mathbf{F}.$$

[3]we use the sign convention from [Woodhouse (1980)]

Because the line bundle $\wedge^n \mathbf{F}$ is trivial, from the volume form on M we can define a volume form on $M/\mathbf{D}$, where $\mathbf{D} = \mathbf{F} \cap \bar{\mathbf{F}}$. Using this volume form, we define the inner product on the Hilbert space of the quantization as follows. Let s_1 and s_2 be two sections of $\mathcal{L}$ which are covariantly constant along $\mathbf{F}$. Because the Hermitian inner product of the fibers is compatible with the connection, the function

$$x \mapsto [s(x), s'(x)]$$

is constant along the leaves of the distribution $\mathbf{D}$. We then let

$$\langle s, s' \rangle = \int_{M/\mathbf{D}} [s(x), s'(x)] d\mathrm{vol}_{M/\mathbf{D}}.$$

Example 2.9. Let us examine the quantization of the system of n one-dimensional particles, in a real polarization. Because the line bundle of the quantization lives over $\mathbb{R}^n \times \mathbb{R}^n$, it is trivial, and so we can identify it with $\mathbb{R}^n \times \mathbb{R}^n \times \mathbb{C}$. Hence the states of the Hilbert space of the quantization are functions on the phase space.

The real polarization is defined by a Lagrangian subspace $\mathbf{L}$ of the tangent space V to the phase space.

Let $\mathbf{v}_j, \mathbf{w}_j$, $j = 1, 2, \ldots, n$, be the symplectic basis associated to $\mathbf{L}$ using Lemma 2.2, and let x_j, y_j, $j = 1, 2, \ldots, n$, be the associated coordinates.

Define the potential

$$\theta = -(1/\hbar)\mathbf{y}^T d\mathbf{x} = -(1/\hbar) \sum_{j=1}^{n} y_j dx_j.$$

The functions s that are covariantly constant along $\mathbf{L}$ should satisfy the differential equations

$$\nabla_{\mathbf{v}_j} s = 0, \quad j = 1, 2, \ldots, n,$$

which are explicitly

$$\frac{\partial}{\partial x_j} s + (i/\hbar) y_j s = 0, \quad j = 1, 2, \ldots, n.$$

Solving we obtain

$$s(\mathbf{x}, \mathbf{y}) = \psi(\mathbf{y}) e^{-(i/\hbar)\mathbf{x}^T \mathbf{y}}. \tag{2.20}$$

All information about the state s is contained in ψ, and so we can identify the Hilbert space with the space of functions in the n variables $y_1, y_2, \ldots, y_n$.

A Hermitian form compatible with the connection is

$$[s(x), s'(x)] = s(x)\overline{s'(x)}.$$

We deduce that the measure of integration is the standard Lebesgue measure.

It is important to notice that there are many choices for the Lagrangian subspace $\mathbf{L}$, and each of them yields a different construction. Two particular cases come to mind.

If

$$\mathbf{L} = \mathrm{Span}\left(\frac{\partial}{\partial \eta_1}, \frac{\partial}{\partial \eta_2}, \ldots, \frac{\partial}{\partial \eta_n}\right),$$

then we can choose the basis of V as in (2.17), so $\mathbf{x} = \boldsymbol{\eta}$ and $\mathbf{y} = \boldsymbol{\xi}$, and we have

$$\theta = -(1/\hbar)\boldsymbol{\xi}^T d\boldsymbol{\eta}.$$

The states are functions of the form

$$s(\boldsymbol{\xi}, \boldsymbol{\eta}) = \psi(\boldsymbol{\xi})e^{-(i/\hbar)\boldsymbol{\xi}^T\boldsymbol{\eta}}.$$

This is how one obtains the Hilbert space of the quantization in the position representation, where we identify s with ψ.

Similarly, if

$$\mathbf{L} = \mathrm{Span}\left(\frac{\partial}{\partial \xi_1}, \frac{\partial}{\partial \xi_2}, \ldots, \frac{\partial}{\partial \xi_n}\right),$$

we can choose the basis of V associated to $\mathbf{L}$ as in (2.18), and then

$$\theta = (1/\hbar)\boldsymbol{\eta}^T d\boldsymbol{\xi}.$$

The states are of the form

$$s(\boldsymbol{\xi}, \boldsymbol{\eta}) = \psi(\boldsymbol{\eta})e^{(i/\hbar)\boldsymbol{\xi}^T\boldsymbol{\eta}}.$$

Identifying s with ψ we obtain the Hilbert space $L^2(\mathbb{R}^n, d\boldsymbol{\eta})$ of the quantization in the momentum representation.

Example 2.10. We now examine the case of the quantization of the n one-dimensional particles in a Kähler polarization. We start again with a Lagrangian subspace $\mathbf{L}$ of the tangent space V to the phase space, with associated symplectic basis $\mathbf{v}_j, \mathbf{w}_j$, $j = 1, 2, \ldots, n$, and then consider the polarization

$$\mathbf{F} = \mathrm{Span}\,(\mathbf{v}_j + i\mathbf{w}_j \,|\, j = 1, 2, \ldots, n) = \mathrm{Span}\left(\frac{\partial}{\partial x_j} + i\frac{\partial}{\partial y_j} \,\middle|\, j = 1, 2, \ldots, n\right).$$

Introduce complex coordinates $z_j = x_j + iy_j$, in which the polarization reads

$$\mathbf{F} = \mathrm{Span}\left(\frac{\partial}{\partial \bar{z}_1}, \frac{\partial}{\partial \bar{z}_2}, \ldots, \frac{\partial}{\partial \bar{z}_n}\right).$$

For example, if $\mathbf{L}$ is as in (2.14), $\mathbf{v}_j = -\partial/\partial\eta_j, \mathbf{w}_j = \partial/\partial\xi_j$, and we have the complex coordinates $z_j = \xi_j - i\eta_j$. If $\mathbf{L}$ is as in (2.15) and $\mathbf{v}_j, \mathbf{w}_j$ are chosen as in (2.18), then the complex coordinates are $z_j = \eta_j + i\xi_j$.

Again the line bundle of the quantization is trivial, so the states are functions $s(\mathbf{z}, \bar{\mathbf{z}})$. We need the Hermitian structure and the connection. In this case we demand compatibility of the Hermitian structure, the complex structure, and the connection. The situation is rigid, in view of the following result:

Lemma 2.3. *Let $\mathcal{L}$ be a holomorphic line bundle with a Hermitian structure. Then there is a unique connection compatible with both the Hermitian structure and the complex structure.*

Proof. Locally choose a holomorphic section s with $ds = 0$ (i.e. a holomorphic frame). Let $h = [s, s]$ and $\theta = \sum_j \theta_j dz_j$. Then

$$d[s,s] = dh = [\nabla s, s] + [s, \nabla s] = [ds - i\theta s, s] + [s, ds - i\theta s] = -i\theta h + i\bar{\theta} h.$$

The compatibility with the complex structure means that

$$\frac{\partial h}{\partial z_j} = -i\theta_j h \quad \text{and} \quad \frac{\partial h}{\partial \bar{z}_j} = i\overline{\theta_j} h.$$

Thus

$$\theta_j = i\frac{\partial h}{\partial z_j} h^{-1}, \tag{2.21}$$

which completely determines θ. □

Let us return to our particular situation. The relation (2.21) allows us to determine h. Note first that the right-hand side is the logarithmic derivative. Also, recall that $d\theta = \omega$. We deduce that h satisfies the following equation

$$\bar{\partial}\partial \ln h = (-i/\hbar)\omega = \frac{1}{2\hbar}\sum_j dz_j \wedge d\bar{z}_j,$$

with the standard notation

$$\partial = \sum_j \frac{\partial}{\partial z_j}, \quad \bar{\partial} = \sum_j \frac{\partial}{\partial \bar{z}_j}.$$

The function

$$h(z) = e^{-\|\operatorname{Im} \mathbf{z}\|^2/\hbar}$$

satisfies this equation. In this case

$$\theta = i\sum_j \frac{\partial h}{\partial z_j} h^{-1} dz_j = \frac{i}{2\hbar}\sum_j (z_j - \bar{z}_j) dz_j = \frac{i}{2\hbar}(\mathbf{z} - \bar{\mathbf{z}})^T d\mathbf{z}.$$

The condition for a function to be covariantly constant in the direction of the polarization $\mathbf{F}$ translates to

$$\frac{\partial \phi}{\partial \bar{z}_j} = 0, \quad j = 1, 2, \ldots, n,$$

which means that ϕ is holomorphic. This means that the Hilbert space consists of the holomorphic function on $\mathbb{C}^n = \mathbb{R}^n \times \mathbb{R}^n$, with the inner product

$$\int_{\mathbb{C}^n} \phi_1(\mathbf{z})\overline{\phi_2(\mathbf{z})} h(\mathbf{z}) d\mathbf{x} d\mathbf{y} = \int_{\mathbb{C}^n} \phi_1(\mathbf{z})\overline{\phi_2(\mathbf{z})} e^{-\|\mathbf{z}\|^2/\hbar} d\mathbf{x} d\mathbf{y}.$$

This is precisely the Segal-Bargmann space from §2.2.1.

If we start with

$$\mathbf{L} = \mathrm{Span}\left(\frac{\partial}{\partial \xi_1}, \frac{\partial}{\partial \xi_2}, \ldots, \frac{\partial}{\partial \xi_n}\right),$$

then $\mathbf{v}_j = \partial/\partial\xi_j$, $\mathbf{w}_j = \partial/\partial\eta_j$, and $z_j = \eta_j + i\xi_j$. Then we obtain the Hilbert space of the holomorphic quantization in the momentum representation described in §2.2.3.

On the other hand, if we start with

$$\mathbf{L} = \mathrm{Span}\left(\frac{\partial}{\partial \eta_1}, \frac{\partial}{\partial \eta_1}, \ldots, \frac{\partial}{\partial \eta_1}\right),$$

then we can choose $\mathbf{v}_j = -\partial/\partial\eta_j$, $\mathbf{w}_j = \partial/\partial\xi_j$, and $z_j = \xi_j - i\eta_j$. We thus obtain the Segal-Bargmann space with anti-holomorphic functions, the way it is done in [Segal (1962)].

2.3.3 *The Schrödinger representation from geometric considerations*

Now we explain how to obtain the Schrödinger representation of the Heisenberg Lie algebra from geometric quantization.

Let us recall first the general procedure for quantizing observables using geometric quantization. As before, M is a symplectic manifold with symplectic form ω. Choose Planck's constant h so that Weil's integrality condition is satisfied, and associate to M the Hermitian line bundle with connection $\mathcal{L}$ as in §2.3.2. Next, choose a polarization $\mathbf{F}$ on M and consider the Hilbert space of sections of $\mathcal{L}$ that are covariantly constant in the direction of **F**. Geometric quantization associates to a classical observable $f \in \mathbb{C}^\infty(M, \mathbb{R})$ the quantum observable that acts on sections by

$$\mathrm{op}(f)s = -i\hbar\nabla_{X_f} s + fs = -i\hbar\left[X_f(s) - i(X_f \lrcorner \theta)s\right] + fs. \qquad (2.22)$$

A motivation for this definition is that it satisfies Dirac's third quantization condition, known as the correspondence principle (see §2.1.2).

Return to our prototype of n one-dimensional particles. To describe the representation of the Heisenberg Lie algebra, we apply the formula (2.22) to the position and momentum functions. We will do this for the cases of the real and Kähler polarizations.

The Schrödinger representation in a real polarization

Fix a Lagrangian subspace $\mathbf{L}$ of the tangent space V to the phase space $\mathbb{R}^n \times \mathbb{R}^n$, and consider the real polarization defined by $\mathbf{L}$. Let $\mathbf{v}_j, \mathbf{w}_j$, $j = 1, 2, \ldots, n$, be a symplectic basis associated to $\mathbf{L}$ using Lemma 2.2, then associate to this basis the coordinate functions x_j, y_j, $j = 1, 2, \ldots, n$.

We quantize first the coordinate functions x_j and y_j, $j = 1, 2, \ldots, n$, which form a generating set of the Heisenberg Lie algebra. Recall that

$$X_{x_i} = \frac{\partial}{\partial y_i}, \quad X_{y_i} = -\frac{\partial}{\partial x_i}.$$

Using the formula (2.22) and the fact that $\theta = -(1/\hbar)\mathbf{y}^T d\mathbf{x}$, we compute

$$\begin{aligned} \mathrm{op}(x_j)s &= -i\hbar\frac{\partial s}{\partial y_j} + x_j s \\ \mathrm{op}(y_j)s &= i\hbar\frac{\partial s}{\partial x_j} - y_j x + y_j s = i\hbar\frac{\partial s}{\partial x_j}. \end{aligned}$$

Because the section s is given by (2.20), we have

$$\begin{aligned} \mathrm{op}(x_j)\psi(\mathbf{y})e^{-(i/\hbar)\mathbf{x}^T\mathbf{y}} &= -i\hbar\frac{\partial \psi}{\partial y_j}e^{-(i/\hbar)\mathbf{x}^T\mathbf{y}} \\ \mathrm{op}(y_j)\psi(\mathbf{x})e^{(i/\hbar)\mathbf{x}^T\mathbf{y}} &= y_j e^{(i/\hbar)\mathbf{x}^T\mathbf{y}}. \end{aligned}$$

Treating $\psi(\mathbf{y})$ as the actual state, we see that the operators $\mathrm{op}(x_j)$ and $\mathrm{op}(y_j)$ act as if the coordinates x_j, y_j were the momentum respectively the position, and we were working in the position representation as in (2.4):

$$\mathrm{op}(x_j)\psi = -i\hbar\frac{\partial \psi}{\partial y_j}, \quad \mathrm{op}(y_j)\psi = y_j\psi. \qquad (2.23)$$

Switching to the standard basis, we obtain

$$\begin{aligned}\mathrm{op}(\xi_j)s &= -i\hbar[X_{\xi_j}s - i(X_{\xi_j}\lrcorner\theta)s] + \xi_j s = i\hbar\frac{\partial s}{\partial\eta_j} + \sum_k x_k\frac{\partial y_k}{\partial\eta_j}s + \xi_j s\\ &= i\hbar\frac{\partial s}{\partial\eta_j} + \sum_k d_{kj}x_k s + \xi_j s\\ \mathrm{op}(\eta_j)s &= -i\hbar[X_{\eta_j}s - i(X_{\eta_j}\lrcorner\theta)s] = -i\hbar\frac{\partial s}{\partial\xi_j} - \sum_k x_k\frac{\partial y_k}{\partial\xi_j}s + \eta_j s\\ &= -i\hbar\frac{\partial s}{\partial\xi_j} - \sum_k c_{kj}x_k + \eta_j s.\end{aligned}$$

Example 2.11. For the quantization in the position representation, with **L** given by (2.14), and the basis $\mathbf{v}_j, \mathbf{w}_j$, $j = 1, 2, \ldots, n$, chosen as in (2.17), we have $x_j = \eta_j$ and $y_j = \xi_j$ and hence (2.23) becomes

$$\begin{aligned}\mathrm{op}(\eta_j)\psi &= -i\hbar\frac{\partial\psi}{\partial\xi}\\ \mathrm{op}(\xi_j)\psi &= \xi_j\psi.\end{aligned}$$

We recognize the standard Schrödinger representation of §2.1.2.

Example 2.12. When **L** is given by (2.15) and $\mathbf{v}_j, \mathbf{w}_j$, $j = 1, 2, \ldots, n$ are as in (2.18), then $x_j = \xi_j$, $y_j = -\eta_j$, and we obtain

$$\begin{aligned}\mathrm{op}(\xi_j) &= i\hbar\frac{\partial\psi}{\partial\eta_j}\\ \mathrm{op}(\eta_j) &= \eta_j\psi,\end{aligned}$$

which is the quantization of positions and momenta in the momentum representation.

The Schrödinger representation of the Heisenberg group is obtained by exponentiating the representation of the Heisenberg Lie algebra. This then leads to the Weyl quantization in a given polarization. We obtain the following formula

Proposition 2.5.

$$op\left(e^{2\pi i(\mathbf{p}^T\mathbf{x}+\mathbf{q}^T\mathbf{y})}\right)\psi(\mathbf{y}) = e^{2\pi i\mathbf{q}^T\mathbf{y}-\pi ih\mathbf{p}^T\mathbf{q}}\psi(\mathbf{y}+h\mathbf{p}). \tag{2.24}$$

To switch to the notation for the Heisenberg group, note that the quantizations of x_j and y_j yield operators in the Heisenberg Lie algebra $\mathfrak{H}(\mathbb{R}^n)$.

Denote these operators by X_j and Y_j, $j = 1, 2, \ldots, n$. They are defined by the equations

$$\begin{bmatrix} \mathbf{X} \\ \mathbf{Y} \end{bmatrix} = h \begin{bmatrix} \mathbf{Q} \\ \mathbf{P} \end{bmatrix} = \begin{bmatrix} A & B \\ C & D \end{bmatrix} \begin{bmatrix} \mathbf{Q} \\ \mathbf{P} \end{bmatrix}, \tag{2.25}$$

where $\mathbf{X} = (X_1, X_2, \ldots, X_n)$, $\mathbf{Y} = (Y_1, Y_2, \ldots, Y_n)$. Then any element of the Heisenberg group can be written in the form $\exp(\mathbf{p}^T\mathbf{X} + \mathbf{q}^T\mathbf{Y} + tI)$, and we have

$$\exp\left(\mathbf{p}^T\mathbf{X} + \mathbf{q}^T\mathbf{Y} + tI\right)\psi(\mathbf{y}) = e^{2\pi i\mathbf{q}^T\mathbf{y} - \pi i h\mathbf{p}^T\mathbf{q} + 2\pi i t}\psi(\mathbf{y} + h\mathbf{p}). \tag{2.26}$$

In particular

$$\begin{aligned} \exp(p_jX_j)\psi(y_1, \ldots, y_n) &= \psi(y_1, \ldots, y_j + p_jh, \ldots, y_n) \\ \exp(q_jY_j)\psi(y_1, \ldots, y_n) &= e^{2\pi i q_j y_j}\psi(y_1, \ldots, y_n). \end{aligned}$$

The Schrödinger representation in a Kähler polarization

Starting with the same Lagrangian subspace $\mathbf{L}$, with associated basis $\mathbf{v}_j, \mathbf{w}_j$, $j = 1, 2, \ldots, n$, we consider the Kähler polarization

$$\mathbf{F} = \operatorname{Span}\left(\mathbf{w}_j + i\mathbf{v}_j \,|\, j = 1, 2, \ldots, n\right) = \operatorname{Span}\left(\frac{\partial}{\partial x_j} + i\frac{\partial}{\partial y_j} \,\middle|\, j = 1, 2, \ldots, n\right),$$

as in Example 2.10.

Again, we work first in the coordinates x_j, y_j, $j = 1, 2, \ldots, n$, with $z_j = x_j + iy_j$. The states are holomorphic functions $\phi(\mathbf{z})$. We have

$$X_{x_j} = \frac{\partial}{\partial y_j} = i\left(\frac{\partial}{\partial z_j} - \frac{\partial}{\partial \bar{z}_j}\right), \quad X_{y_j} = -\frac{\partial}{\partial x_j} = -\left(\frac{\partial}{\partial z_j} + \frac{\partial}{\partial \bar{z}_j}\right).$$

Using the formula for the geometric quantization of observables we obtain

$$\begin{aligned} \operatorname{op}(x_j)\phi = &- i\hbar\left[i\left(\frac{\partial}{\partial z_j} - \frac{\partial}{\partial \bar{z}_j}\right) + \frac{i}{2\hbar}(z_j - \bar{z}_j)\left(\frac{\partial}{\partial z_j} - \frac{\partial}{\partial \bar{z}_j}\right)\lrcorner dz_j\right]\phi \\ &+ \frac{z_j + \bar{z}_j}{2}\phi = \hbar\frac{\partial\phi}{\partial z_j} + z_j\phi \end{aligned}$$

and

$$\begin{aligned} \operatorname{op}(y_j)\phi = &- i\hbar\left[-\left(\frac{\partial}{\partial z_j} + \frac{\partial}{\partial \bar{z}_j}\right) - \frac{1}{2\hbar}(z_j - \bar{z}_j)\left(\frac{\partial}{\partial z_j} + \frac{\partial}{\partial \bar{z}_j}\right)\lrcorner dz_j\right]\phi \\ &+ \frac{z_j - \bar{z}_j}{2i}\phi = i\hbar\frac{\partial\phi}{\partial z_j}. \end{aligned}$$

If we treat the coordinates x_j, y_j, $j = 1, 2, \ldots, n$, as momenta respectively positions (e.g η_j and ξ_j), we recognize the action of the Heisenberg Lie algebra from §2.2.3:

$$\mathrm{op}(x_j) = \hbar\frac{\partial}{\partial z_j} + M_{z_j}, \quad \mathrm{op}(y_j) = i\hbar\frac{\partial}{\partial z_j}. \tag{2.27}$$

For the quantizations of the standard coordinate functions we use the change of coordinates from (2.16) to obtain

$$\mathrm{op}(\xi_j)\phi = \sum_k \left[(\widetilde{d}_{jk} + i\widetilde{c}_{jk})\hbar\frac{\partial\phi}{\partial z_k} + \widetilde{b}_{jk}z_k\phi\right]$$

$$\mathrm{op}(\eta_j)\phi = \sum_k \left[(\widetilde{a}_{jk} + i\widetilde{b}_{jk})\hbar\frac{\partial\phi}{\partial z_k} + \widetilde{d}_{jk}z_k\phi\right].$$

Example 2.13. If we choose

$$\mathbf{L} = \mathrm{Span}\left(\frac{\partial}{\partial\xi_1}, \frac{\partial}{\partial\xi_2}, \ldots, \frac{\partial}{\partial\xi_n}\right),$$

with the polarization

$$\mathbf{F} = \mathrm{Span}\left(\frac{\partial}{\partial\eta_j} + i\frac{\partial}{\partial\xi_j}\,\middle|\, j = 1, 2, \ldots, n\right),$$

and let $z_j = \eta_j + i\xi_j$, then we obtain the quantization model from §2.2.3.

Example 2.14. On the other hand, if we start with

$$\mathbf{L} = \mathrm{Span}\left(\frac{\partial}{\partial\eta_1}, \frac{\partial}{\partial\eta_2}, \ldots, \frac{\partial}{\partial\eta_n}\right),$$

with the polarization

$$\mathbf{F} = \mathrm{Span}\left(\frac{\partial}{\partial\xi_j} - i\frac{\partial}{\partial\eta_j}\,\middle|\, j = 1, 2, \ldots, n\right),$$

and let $z_j = \xi_j - i\eta_j$, then we obtain the standard Segal-Bargmann model but with anti-holomorphic functions.

By exponentiating the representation of the Heisenberg Lie algebra given by (2.27) we obtain the representation of the Heisenberg group in holomorphic setting, which in the notation from (2.25) is

$$\exp\left(\mathbf{p}^T\mathbf{X} + \mathbf{q}^T\mathbf{Y} + tI\right)\phi(\mathbf{z}) = e^{2\pi i\mathbf{p}^T\mathbf{z} - \pi i h\mathbf{p}^T(\mathbf{q}-i\mathbf{p}) + 2\pi i t}\phi(\mathbf{z} - h(\mathbf{q} - i\mathbf{p})).$$

2.3.4 *Passing from real to Kähler polarizations*

In general, for a symplectic manifold M, there are many choices of a polarization, and hence many quantization models. One wishes to identify these in a canonical way. This is indeed possible, and is done using a Blattner-Kostant-Sternberg kernel (see [Śniatycki (1980)] for a definition). The Segal-Bargmann transform is an instance of this identification.

Start with a Lagrangian subspace $\mathbf{L}$ of the tangent space to $\mathbb{R}^n \times \mathbb{R}^n$, and associate to it the basis $\mathbf{v}_j$, $\mathbf{w}_j$, and coordinates x_j, y_j, $j = 1, 2, \ldots, n$, as in Example 2.9.

Consider the real polarization defined by $\mathbf{L}$ it we associate as above the Hilbert spaces $L^2(\mathbb{R}^n, d\mathbf{x})$ of square integrable functions in the variables x_j, $j = 1, 2, \ldots, n$.

Then consider the Kähler polarization defined by the vectors $\mathbf{w}_j + i\mathbf{v}_j$, $j = 1, 2, \ldots, n$, respectively $\mathcal{H}L(\mathbb{C}^n, d\mu_\hbar)$ of functions that are holomorphic in the variables $z_j = x_j + iy_j$ and square integrable with respect to the measure

$$d\mu_\hbar = e^{-\|\operatorname{Im}\mathbf{z}\|^2/\hbar} d\mathbf{x}d\mathbf{y}$$

The proof of Theorem 2.2 translates to yield the following result.

Proposition 2.6. *The Segal-Bargmann transform*

$$\mathcal{T}_{SB}^{\mathbf{L}} : L^2(\mathbb{R}, d\mathbf{x}) \to \mathcal{H}L(\mathbb{C}^n, d\mu_\hbar)$$

defined as

$$\mathcal{T}_{SB}^{\mathbf{L}}\psi(\mathbf{z}) = (2\pi\hbar)^{-n/2} \int_{\mathbb{R}^n} e^{(\mathbf{z}-\widetilde{\mathbf{x}})^T(\mathbf{z}-\widetilde{\mathbf{x}})/\hbar}\psi(\widetilde{\mathbf{x}})d\widetilde{\mathbf{x}},$$

is a unitary isomorphism that identifies the representation of the Heisenberg Lie algebra defined by (2.23) with that defined by (2.27).

2.4 The Schrödinger representation as an induced representation

The Stone-von Neumann Theorem 2.1 implies that all information about the Schrödinger representation is contained in the Heisenberg group, so one should be able to recover this representation directly from the group. In this section we explain how this is done using a standard construction in group representation theory, the **induced representation**. We follow the exposition from [Lion and Vergne (1980)].

The general idea is as follows. Given a subgroup H of a group G, and a representation ρ of H on a vector space W, the goal is to produce a representation of G that restricts to ρ on H. This representation is realized as the regular action of G on the space of W-valued functions on G which transform under H in an equivariant fashion defined by ρ.

To define the subgroup H suited for our needs, we start with a real polarization of the phase space, in the guise of a Lagrangian subspace $\mathbf{L}$ of the tangent space to the phase space. To it we associate the Lie algebra $\mathfrak{L}(\mathbf{L})$ of the functions $f \in C^\infty(\mathbb{R}^n \times \mathbb{R}^n, \mathbb{R})$ with the property that $X_f \in \mathbf{L}$. Then $\mathfrak{L}(\mathbf{L}) + \mathbb{R}I$ is a maximal abelian Lie subalgebra of the Heisenberg Lie algebra $\mathfrak{H}(\mathbb{R}^n)$. By exponentiating its elements we obtain a maximal abelian subgroup of the Heisenberg group $\mathbf{H}(\mathbb{R}^n)$, which we denote by $\exp(\mathfrak{L}(\mathbf{L}) + \mathbb{R}I)$. This is the subgroup of interest to us.

Example 2.15. If

$$\mathbf{L} = \mathrm{Span}\left(\frac{\partial}{\partial \xi_1}, \frac{\partial}{\partial \xi_2}, \ldots, \frac{\partial}{\partial \xi_n}\right),$$

then $\mathfrak{L}(\mathbf{L})$ is generated by the functions $\eta_1, \eta_2, \ldots, \eta_n$. It follows that

$$\exp(\mathfrak{L}(\mathbf{L}) + \mathbb{R}I) = \left\{\exp(\mathbf{p}^T\mathbf{P} + tI) \,|\, \mathbf{p} \in \mathbb{R}^n\right\}.$$

Example 2.16. Similarly, if

$$\mathbf{L} = \mathrm{Span}\left(\frac{\partial}{\partial \eta_1}, \frac{\partial}{\partial \eta_2}, \ldots, \frac{\partial}{\partial \eta_n}\right),$$

then $\mathfrak{L}(\mathbf{L})$ is generated by the functions $\xi_1, \xi_2, \ldots, \xi_n$, and hence

$$\exp(\mathfrak{L}(\mathbf{L}) + \mathbb{R}I) = \left\{\exp(\mathbf{q}^T\mathbf{Q} + tI) \,|\, \mathbf{q} \in \mathbb{R}^n\right\}.$$

The group $\exp(\mathfrak{L}(\mathbf{L}) + \mathbb{R}I)$, being abelian, has only one-dimensional irreducible representations. Thus the irreducible representations of this group are its characters. Of these, we consider the character

$$\chi_{\mathbf{L}} : \exp(\mathfrak{L}(\mathbf{L}) + \mathbb{R}I) \to U(1)$$

obtained by restricting to the maximal abelian subgroup the function $\chi : \mathbf{H}(\mathbb{R}^n) \to U(1)$,

$$\chi(\exp(\mathbf{p}^T\mathbf{P} + \mathbf{q}^T\mathbf{Q} + tI)) = e^{2\pi i t}, \quad \mathbf{p}, \mathbf{q} \in \mathbb{R}^n.$$

To the maximal abelian subgroup $\exp(\mathfrak{L}(\mathbf{L}) + \mathbb{R}I)$ and character $\chi_{\mathbf{L}}$ we associate the induced representation of the Heisenberg group as follows.

Definition 2.10. The Hilbert space of the induced representation, $\mathcal{H}(\mathbf{L})$, is the space of functions

$$\psi : \mathbf{H}(\mathbb{R}^n) \to \mathbb{C}$$

satisfying the equivariance condition

$$\psi(uu') = \chi_{\mathbf{L}}(u')^{-1}\psi(u), \text{ for all } u' \in \exp(\mathfrak{L}(\mathbf{L}) + \mathbb{R}I),$$

and for which the map $u \bmod \exp(\mathfrak{L}(\mathbf{L})+\mathbb{R}I) \mapsto |\psi(u)|$ is square integrable with respect to the invariant measure on the set of the left equivalence classes modulo $\exp(\mathfrak{L}(\mathbf{L}) + \mathbb{R}I)$ induced by the Haar measure of $\mathbf{H}(\mathbb{R}^n)$.

Let us explain how the measure of integration is defined. The Haar measure on $\mathbf{H}(\mathbb{R}^n)$ is translation invariant, and is unique up to a multiplication by a constant. It is defined by a volume form that lies in the highest exterior power of the cotangent space to the identity element. Identifying the tangent space at the identity with $\mathbb{R}^{2n+1}$, the volume form is up to multiplication by a constant equal to $dx_1 \wedge \cdots \wedge dx_n \wedge dy_1 \cdots \wedge dy_n \wedge dt$, and the measure it induces is up to multiplication by a constant the Lebesgue measure $dx_1 \cdots dx_n dy_1 \cdots dy_n dt$.

The space of equivalence classes $\mathbf{H}(\mathbb{R}^n)/\exp(\mathfrak{L}(\mathbf{L}) + \mathbb{R}I)$ is a quotient manifold of dimension n. The tangent space at the equivalence class of the identity element is isomorphic to $\mathbb{R}^n$, which can be thought of as an algebraic complement of $\mathbf{L}$. Consider a measure that is translation invariant under the action of the Heisenberg group. It is induced, up to multiplication by a constant, by the Lebesgue measure dm on $\mathbb{R}^n$.

The inner product

$$\langle \psi_1, \psi_2 \rangle = \int_{\mathbf{H}(\mathbb{R}^n)/\exp(\mathfrak{L}(\mathbf{L})+\mathbb{R}I)} \psi_1 \overline{\psi_2} dm$$

makes $\mathcal{H}(\mathbf{L})$ into a Hilbert space.

Definition 2.11. The **abstract Schrödinger representation** of the Heisenberg group $\mathbf{H}(\mathbb{R}^n)$ is the action on $\mathcal{H}(\mathbf{L})$ given by

$$u_0\psi(u) = \psi(u_0^{-1}u). \tag{2.28}$$

It is not hard to check that this representation satisfies the conditions of the Stone-von Neumann Theorem, hence it should be unitary equivalent to the standard representation. To explicate this equivalence, associate to $\mathbf{L}$ the symplectic basis $\mathbf{v}_j, \mathbf{w}_j$, $j = 1, 2, \ldots, n$, as in Lemma 2.2, and let $\mathbf{L}^\natural$ be the space with basis $\mathbf{w}_j$, $j = 1, 2, \ldots, n$. Introduce also coordinates $\mathbf{x}, \mathbf{y}$ on the phase space as before.

For each $u \in \mathbf{H}(\mathbb{R}^n)$ there are unique $\mathbf{l} \in \mathfrak{L}(\mathbf{L})$, $\mathbf{l}^\natural \in \mathfrak{L}(\mathbf{L}^\natural)$, and real number t such that $u = \exp(\mathbf{l}^\natural)\exp(\mathbf{l}+tI)$. For a function ψ, if we know its value on $\mathbf{l}^\natural$, then we can recover its value on $u = \exp(\mathbf{l}^\natural)\exp(\mathbf{l}+tI)$ using the equivariance condition (2.28). It follows that the map

$$\mathcal{H}(\mathbf{L}) \to L^2(\mathbf{L}^\natural, d\mathbf{x})$$

defined by

$$(u \mapsto \psi(u)) \mapsto (\mathbf{l}^\natural \mapsto \psi(\exp(\mathbf{l}^\natural))$$

is an isomorphism. It is not hard to see that it is also unitary. So we can identify the Hilbert space $\mathcal{H}(\mathbf{L})$ with $L^2(\mathbf{L}^\natural, d\mathbf{x})$.

We compute explicitly the action defined by (2.28) as follows:

$$\begin{aligned}\exp(q_jY_j)\psi(\exp(\sum_k x_kX_k)) &= \psi(\exp(-q_jY_j)\exp(\sum_k x_kX_k))\\ = \psi(\exp(\sum_k x_kX_k)\exp(Y_j + hq_jx_jI)) &= e^{-2\pi i h q_j x_j}\psi(\exp(\sum_k x_kX_k))\end{aligned}$$

and

$$\begin{aligned}\exp(p_jX_j)\psi(\exp(\sum_k x_kX_k)) &= \psi(\exp(-p_jX_j)\exp(\sum_k x_kX_k))\\ = \psi(\exp(x_1X_1 + \ldots + (x_j - p_j)X_j, \ldots, X_n)).\end{aligned}$$

Using the identification of $\mathcal{H}(\mathbf{L})$ with $L^2(\mathbf{L}^\natural, d\mathbf{x})$ we obtain the action of the Heisenberg group

$$\exp(\mathbf{p}^T\mathbf{X} + \mathbf{q}^T\mathbf{Y} + tI)\psi(\mathbf{x}) = e^{-2\pi i h\mathbf{q}^T\mathbf{x} - \pi i h\mathbf{p}^T\mathbf{q} + 2\pi i t}\psi(\mathbf{x} - \mathbf{p}).$$

This is not quite the action from (2.26), because negative signs appear in this formula and because the Planck's constant is in the wrong place. The two representations are nevertheless unitary equivalent. A possible unitary operator that establishes this equivalence is the Fourier transform

$$\mathcal{F}\psi(\mathbf{y}) = \int_{\mathbb{R}^n} \psi(\mathbf{x})e^{-2\pi i\mathbf{x}^T\mathbf{y}}d\mathbf{x}$$

followed by the change of variables $\mathbf{y} \to -\mathbf{y}$, which turns this into the action in the momentum representation from Proposition 2.2, and then apply $\mathcal{F}_h^{-1}$ to turn this into the action in the position representation.

The description of the Schrödinger representation as an induced representation has the advantage that it does not depend on any smooth or complex structure on the phase space, it is a purely algebraic object. And it facilitates tremendously the definition of the Fourier transform associated to a symplectomorphism.

To summarize, for each Lagrangian subspace of the phase space we have three representations of the Heisenberg group:

- A representation obtained by exponentiating the geometric quantization of the Heisenberg Lie algebra in the real polarization defined by the Lagrangian space.
- A representation obtained by exponentiating the geometric quantization of the Heisenberg Lie algebra in a Kähler polarization associated to the Lagrangian subspace.
- An abstract representation of the Heisenberg group on a space of equivariant functions on the Heisenberg group.

The first and the second are related by the Segal-Bargmann transform, and the first and the third are related by the standard Fourier transform.

2.5 The Fourier transform and the representation of the symplectic group $Sp(2n, \mathbb{R})$

2.5.1 *The Fourier transform defined by a pair of Lagrangian subspaces*

The Fourier transform is another instance of the identification of quantization models given by a Blattner-Kostant-Sternberg kernel; in this case the Hilbert spaces of the quantizations in two different real polarizations are identified. Inspired by [Lion and Vergne (1980)], we will work in the framework of §2.4. This construction is **particular** to the system of n one-dimensional particles that is our prototype.

Each Lagrangian subspace $\mathbf{L}$ of the tangent space to the phase space defines a different representation of the Heisenberg group (in a real polarization) as shown in §2.3.3. By the Stone-von Neumann theorem, any two such representations are unitarily equivalent. By Schur's lemma, the unitary equivalence is unique up to multiplication by a constant.

There is a tautological way for identifying the two representations. Let $\mathbf{L}$ and $\mathbf{L}'$ be the Lagrangian subspaces. Associate to them the symplectic bases $\mathbf{v}_j, \mathbf{w}_j$ respectively $\mathbf{v}'_j, \mathbf{w}'_j$, with coordinate functions $(\mathbf{x}, \mathbf{y})$, respectively $(\mathbf{x}', \mathbf{y}')$. The identification $\mathbf{x}' = \mathbf{x}$, $\mathbf{y}' = \mathbf{y}$ renders the two quantization models identical.

However, in the representations defined by $\mathbf{L}$ and $\mathbf{L}'$, the operators associated to the standard coordinate functions act differently on the space $L^2(\mathbb{R}^2, d\mathbf{x})$. By exponentiation the coordinate functions define two different representations of the Heisenberg group. It is these two representations that we want to identify, and we can identify them because of the Stone-von Neumann Theorem.

Definition 2.12. The **Fourier transform** associated to the pair of Lagrangian subspaces $\mathbf{L}$, $\mathbf{L}'$ is the unitary map $\mathcal{F}_{\mathbf{L},\mathbf{L}'}$ on $L^2(\mathbb{R}^2, d\mathbf{x})$ that identifies the Schrödinger representation defined by $\mathbf{L}$ with the one defined by $\mathbf{L}'$.

We insist, the Fourier transform is well defined up to multiplication by a constant.

The simplest way to derive a formula for the Fourier transform is by making use of the abstract construction of the Schrödinger representation given in §2.4. To this end, let $\psi \in \mathcal{H}(\mathbf{L})$, meaning that ψ is a function on the Heisenberg group that is equivariant with respect to the right action of $\exp(\mathfrak{L}(\mathbf{L}) + \mathbb{R}I)$. If ψ were also equivariant with respect to the action of $\exp(\mathfrak{L}(\mathbf{L}') + \mathbb{R}I)$, then we could simply map ψ to itself, and the Heisenberg group would act the same way in both situations. Unfortunately this is not the case, but we can use ψ to produce an equivariant map with respect to the action of $\exp(\mathfrak{L}(\mathbf{L}') + \mathbb{R}I)$ by a process of averaging.

As such, we let the Fourier transform of ψ be

$$(\mathcal{F}_{\mathbf{L},\mathbf{L}'}\psi)(u) = \int_{\exp(\mathfrak{L}(\mathbf{L}')+\mathbb{R}I)} \psi(uu')\chi_{\mathbf{L}'}(u')du'$$

where du' is a Haar measure on $\mathbf{L}'$. Because for $u' \in \exp(\mathfrak{L}(\mathbf{L}) + \mathbb{R}I) \cap \exp(\mathfrak{L}(\mathbf{L}') + \mathbb{R}I) = \exp(\mathfrak{L}(\mathbf{L}) \cap \mathfrak{L}(\mathbf{L}') + \mathbb{R}I)$ one has

$$\psi(uu')\chi(u') = \psi(u)\chi(u')^{-1}\chi(u') = \psi(u),$$

it follows that $\psi(uu')\chi(u')$ is constant on each equivalence class modulo $\exp(\mathfrak{L}(\mathbf{L}) \cap \mathfrak{L}(\mathbf{L}') + \mathbb{R}I)$. So we can improve our definition to the following

$$(\mathcal{F}_{\mathbf{L},\mathbf{L}'}\psi)(u) = \int_{\exp(\mathfrak{L}(\mathbf{L}'))/\exp(\mathfrak{L}(\mathbf{L})\cap\mathfrak{L}(\mathbf{L}'))} \psi(uu')du'. \tag{2.29}$$

Here the factor $\chi_{\mathbf{L}'}(u')$ is equal to 1 when $u' \in \exp(\mathfrak{L}(\mathbf{L}'))$.

The map defined by (2.29) interpolates the actions of $\mathbf{H}(\mathbb{R}^n)$ on $\mathcal{H}(\mathbf{L})$ and $\mathcal{H}(\mathbf{L}')$ because

$$\begin{aligned} u_0(\mathcal{F}_{\mathbf{L},\mathbf{L}'}\psi)(u) &= (\mathcal{F}_{\mathbf{L},\mathbf{L}'}\psi)(u_0^{-1}u) \\ &= \int_{\exp(\mathfrak{L}(\mathbf{L}'))/\exp(\mathfrak{L}(\mathbf{L})\cap\mathfrak{L}(\mathbf{L}'))} \psi(u_0^{-1}uu')du' \\ &= (\mathcal{F}_{\mathbf{L},\mathbf{L}'}(u_0\psi))(u). \end{aligned}$$

We would like to find a more explicit formula for the Fourier transform, which will motivate the name, and also explain how to choose the integration measure on $\exp(\mathfrak{L}(\mathbf{L}') + \mathbb{R}I)$ and $\exp(\mathfrak{L}(\mathbf{L}'))/\exp(\mathfrak{L}(\mathbf{L}) \cap \mathfrak{L}(\mathbf{L}'))$ so as to make the Fourier transform unitary. For that we need the following result, which is a stronger version of Lemma 2.2.

Lemma 2.4. *Given the Lagrangian subspaces* $\mathbf{L}$ *and* $\mathbf{L}'$ *of the tangent space* V *to the phase space* $\mathbb{R}^n \times \mathbb{R}^n$*, and a basis* $\mathbf{v}_1, \mathbf{v}_2, \dots, \mathbf{v}_n$ *of* $\mathbf{L}$ *whose last* $n-k$ *vectors form a basis of* $\mathbf{L} \cap \mathbf{L}'$ *for some* $0 \leq k \leq n$*. Then there exist vectors* $\mathbf{w}_1, \mathbf{w}_2, \dots, \mathbf{w}_n$ *that complete* $\mathbf{v}_1, \mathbf{v}_2, \dots, \mathbf{v}_n$ *to a basis of* V*, such that*
(a) $\omega(\mathbf{v}_j, \mathbf{w}_j) = \delta_{ij}$*;*
(b) $\mathbf{w}_1, \mathbf{w}_2, \dots, \mathbf{w}_k, \mathbf{v}_{k+1}, \dots, \mathbf{v}_n$ *is a basis of* $\mathbf{L}'$*.*

Proof. First, let us consider the case where $\mathbf{L} \cap \mathbf{L}' = \{0\}$. We follow the same steps as in the proof of Lemma 2.2, and we start with a basis $\mathbf{v}_1, \mathbf{v}_2, \dots, \mathbf{v}_n, \mathbf{u}_{n+1}, \dots \mathbf{u}_{2n}$, such that the last n vectors are a basis for $\mathbf{L}'$. At the first step, $\mathbf{v}_1 \in \mathbf{L}$ and $\mathbf{w}_1 \in \mathbf{L}'$. Also, because $\mathbf{L}'$ is Lagrangian, $\omega(\mathbf{w}_1, \mathbf{u}_{n+j}) = 0$, $j = 2, 3, \dots, n$. So the vectors $\mathbf{z}_{n+j}$ lie in $\mathbf{L}'$. Therefore, the algorithm produces a basis with the desired properties.

If $\mathbf{L} \cap \mathbf{L}'$ is nontrivial, consider the vector space $(\mathbf{L} + \mathbf{L}')/\mathbf{L} \cap \mathbf{L}'$, and let $2k$ be its dimension. Because both $\mathbf{L}$ and $\mathbf{L}'$ are Lagrangian, ω factors to a symplectic form on this space. Now we are in the previous situation, and we produce a basis for this space, whose first k elements are a basis for $\mathbf{L}/\mathbf{L} \cap \mathbf{L}'$ and the last k elements are a basis for $\mathbf{L}'/\mathbf{L} \cap \mathbf{L}'$. Lift these elements to the vectors $\mathbf{v}_1, \mathbf{v}_2, \dots, \mathbf{v}_k$ in $\mathbf{L}$ respectively $\mathbf{w}_1, \mathbf{w}_2, \dots, \mathbf{w}_k$ in $\mathbf{L}'$. Complete these with a basis of $\mathbf{L} \cap \mathbf{L}'$, and then to a basis of the whole space. Applying the Gram-Schmidt procedure to the vectors listed in this order produces the desired bases. □

Example 2.17. Using the construction provided by Lemma 2.4 and following the prescription from § 2.4, we can associate to $\mathbf{L}$ and $\mathbf{L}'$ the spaces $\mathbf{L}^\natural$ and $\mathbf{L}'^\natural$, with bases

$$\mathbf{w}_1, \mathbf{w}_2, \dots, \mathbf{w}_n \text{ respectively } -\mathbf{v}_1, \dots, -\mathbf{v}_k, \mathbf{w}_{k+1}, \dots, \mathbf{w}_n. \quad (2.30)$$

If we introduce the coordinates $\mathbf{x}, \mathbf{y}$ on the phase space corresponding the basis $\mathbf{v}_j, \mathbf{w}_j$, $j = 1, 2, \dots, n$, then we can identify $\mathcal{H}(\mathbf{L})$ and $\mathcal{H}(\mathbf{L}')$ with the space of square integrable functions in the variables $x_1, x_2, \dots, x_n$ respectively with the space of square integrable functions in the variables

$y_1, \ldots, y_k, x_{k+1}, \ldots, x_n$. For the second set of variables the identification is

$$\psi\left(\exp\left(\sum_{j=1}^{k} -y_j Y_j + \sum_{j=k+1}^{n} x_j X_j\right)\right) \mapsto \psi(x_1, \ldots, x_k, y_{k+1}, \ldots, y_n)$$

because of the negative signs in (2.30).

The integration measure on $\exp(\mathfrak{L}(\mathbf{L}') + \mathbb{R}I)$ is obtained by pushing forward to the Heisenberg group the measure $dy_1 \cdots dy_k dx_{k+1} \cdots dx_n dt$ on $\mathbf{L}'^{\natural} \times \mathbb{R}$, and the measure on $\exp(\mathfrak{L}(\mathbf{L}'))/\exp(\mathfrak{L}(\mathbf{L}) \cap \mathfrak{L}(\mathbf{L}'))$ is obtained by pushing forward to the Heisenberg group the measure $dx_1 dx_2 \cdots dx_n$ on $\mathbf{L}'/\mathbf{L} \cap \mathbf{L}'$.

Translating the definition (2.29) into the new notation we deduce that the Fourier transform of the function $\psi(x_1, x_2, \ldots, x_n)$ is a function in the variables $y_1, \ldots, y_k, x_{k+1}, \ldots, x_n$, given by the formula

$$\int \psi\left(\exp\left(-\sum_{j=1}^{k} y_j Y_j + \sum_{j=k+1}^{n} x_j X_j\right)\exp\left(\sum_{j=1}^{k} x_j X_j\right)\right) dx_1 \cdots dx_k$$

$$= \int \psi\left(\exp\left(\sum_{j=1}^{n} x_j X_j\right)\exp\left(-\sum_{j=1}^{k} y_j Y_j\right)\exp\left(h\sum_{j=1}^{k} x_j y_j I\right)\right) dx_1 \ldots dx_k$$

$$= \int \psi\left(\exp\left(\sum_{j=1}^{n} x_j X_j\right)\right) e^{-2\pi i h \sum_{j=1}^{k} x_j y_j} dx_1 \cdots dx_k$$

$$= \int \psi(x_1, x_2, \ldots, x_n) e^{-2\pi i h \sum_{j=1}^{k} x_j y_j} dx_1 \cdots dx_k,$$

and we recognize the standard Fourier transform in the first k variables.

2.5.2 *The Maslov index*

The Stone-von Neumann Theorem 2.1 implies that if we quantize our system of n one-dimensional particles in two different coordinate systems, we obtain equivalent models. This means that changes of coordinates are quantizable. As such, to each linear symplectic map there is an associated unitary map on the Hilbert space of the quantization. By Schur's Lemma, each such map is uniquely defined up to multiplication by a constant, hence we have a projective representation of the symplectic group $Sp(2n, \mathbb{R})$ on the Hilbert space of the quantization. In this section we introduce a technical device, the Maslov index, which will be used for resolving the projective ambiguity of this representation.

There are many ways in which the Maslov index can be defined, the approach from [Kashiwara and Schapira (1990)] being one of the most popular. We follow [Wall (1969)], but adopt the standard sign convention, which is opposite to Wall's.

Let $\mathbf{L}_1, \mathbf{L}_2, \mathbf{L}_3$ be three Lagrangian subspaces of the symplectic vector space V, which in our situation is the tangent space to the phase space $\mathbb{R}^n \times \mathbb{R}^n$. The set of vectors $\mathbf{l}_1 \in \mathbf{L}_1$ with the property that there exist $\mathbf{l}_2 \in \mathbf{L}_2$ and $\mathbf{l}_3 \in \mathbf{L}_3$ such that

$$\mathbf{l}_1 + \mathbf{l}_2 + \mathbf{l}_3 = 0$$

is the subspace $\mathbf{L}_1 \cap (\mathbf{L}_2 + \mathbf{L}_3)$. The correspondence

$$\mathbf{l}_1 \mapsto \mathbf{l}_2$$

yields a well defined homomorphism

$$\mathbf{L}_1 \cap (\mathbf{L}_2 + \mathbf{L}_3) \to \frac{\mathbf{L}_2 \cap (\mathbf{L}_1 + \mathbf{L}_3)}{\mathbf{L}_2 \cap \mathbf{L}_3},$$

whose kernel is $\mathbf{L}_1 \cap \mathbf{L}_3$. Hence

$$\frac{\mathbf{L}_1 \cap (\mathbf{L}_2 + \mathbf{L}_3)}{\mathbf{L}_1 \cap \mathbf{L}_3} \text{ and } \frac{\mathbf{L}_2 \cap (\mathbf{L}_1 + \mathbf{L}_3)}{\mathbf{L}_2 \cap \mathbf{L}_3},$$

are isomorphic. The two can be further factored by $\mathbf{L}_1 \cap \mathbf{L}_2$ to obtain again isomorphic spaces. From here we conclude that the spaces

$$\frac{\mathbf{L}_1 \cap (\mathbf{L}_2 + \mathbf{L}_3)}{\mathbf{L}_1 \cap \mathbf{L}_2 + \mathbf{L}_1 \cap \mathbf{L}_3}, \quad \frac{\mathbf{L}_2 \cap (\mathbf{L}_1 + \mathbf{L}_3)}{\mathbf{L}_2 \cap \mathbf{L}_1 + \mathbf{L}_2 \cap \mathbf{L}_3}, \quad \frac{\mathbf{L}_3 \cap (\mathbf{L}_1 + \mathbf{L}_2)}{\mathbf{L}_3 \cap \mathbf{L}_1 + \mathbf{L}_3 \cap \mathbf{L}_2}$$

are isomorphic, where the isomorphism with the third space follows by exchanging $\mathbf{L}_2$ with $\mathbf{L}_3$ in the above argument. Next, recall the symplectic form ω and define $B_0 : \mathbf{L}_1 \cap (\mathbf{L}_2 + \mathbf{L}_3) \times \mathbf{L}_1 \cap (\mathbf{L}_2 + \mathbf{L}_3) \to \mathbb{R}$,

$$B_0(\mathbf{l}_1, \mathbf{l}_1') = -\omega(\mathbf{l}_1, \mathbf{l}_2'),$$

where $\mathbf{l}_2'$ is chosen such that it belongs to $\mathbf{L}_2$ and there is $\mathbf{l}_3' \in \mathbf{L}_3$ with $\mathbf{l}_1' + \mathbf{l}_2' + \mathbf{l}_3' = 0$.[4] Note that

$$B_0(\mathbf{l}_1, \mathbf{l}_1') = -\omega(\mathbf{l}_1, \mathbf{l}_2') = \omega(\mathbf{l}_1, \mathbf{l}_1 + \mathbf{l}_3') = \omega(\mathbf{l}_1, \mathbf{l}_3'). \tag{2.31}$$

There could exist several possible choices for $\mathbf{l}_2'$ and $\mathbf{l}_3'$, but if $\mathbf{l}_2''$ and $\mathbf{l}_3''$ is another choice, then

$$\omega(\mathbf{l}_1, \mathbf{l}_2') = -\omega(\mathbf{l}_2 + \mathbf{l}_3, \mathbf{l}_2') = -\omega(\mathbf{l}_3, \mathbf{l}_2') = \omega(\mathbf{l}_3, \mathbf{l}_1' + \mathbf{l}_3') = \omega(\mathbf{l}_3, \mathbf{l}_1'),$$

[4]The negative sign in front of ω is chosen so as to agree with the standard sign convention for the Maslov index!

and for the same reason $\omega(\mathbf{l}_1, \mathbf{l}_2'') = \omega(\mathbf{l}_3, \mathbf{l}_1')$. This proves that B_0 is well defined. Also if $\mathbf{l}_1 + \mathbf{l}_2 + \mathbf{l}_3 = 0$ and $\mathbf{l}_1' + \mathbf{l}_2' + \mathbf{l}_3' = 0$, then

$$\begin{aligned} B_0(\mathbf{l}_1, \mathbf{l}_1') - B_0(\mathbf{l}_1', \mathbf{l}_1) &= -\omega(\mathbf{l}_1, \mathbf{l}_2') + \omega(\mathbf{l}_1', \mathbf{l}_2) = -\omega(\mathbf{l}_1, \mathbf{l}_2') - \omega(\mathbf{l}_2, \mathbf{l}_1') \\ &= -\omega(\mathbf{l}_1 + \mathbf{l}_2, \mathbf{l}_1' + \mathbf{l}_2') + \omega(\mathbf{l}_1, \mathbf{l}_1') + \omega(\mathbf{l}_2, \mathbf{l}_2') \\ &= -\omega(\mathbf{l}_3, \mathbf{l}_3') = 0. \end{aligned}$$

It follows that B_0 is a symmetric bilinear form.

Furthermore, if $\mathbf{l}_1 \in \mathbf{L}_1 \cap \mathbf{L}_2$, then $B_0(\mathbf{l}_1, \mathbf{l}_1') = -\omega(\mathbf{l}_1, \mathbf{l}_2') = 0$, and the same is true for $\mathbf{l}_1 \in \mathbf{L}_1 \cap \mathbf{L}_3$ by (2.31). This shows that B_0 induces a symmetric bilinear form

$$B_{\mathbf{L}_1,\mathbf{L}_2,\mathbf{L}_3} : \frac{\mathbf{L}_1 \cap (\mathbf{L}_2 + \mathbf{L}_3)}{\mathbf{L}_1 \cap \mathbf{L}_2 + \mathbf{L}_1 + \mathbf{L}_3} \times \frac{\mathbf{L}_1 \cap (\mathbf{L}_2 + \mathbf{L}_3)}{\mathbf{L}_1 \cap \mathbf{L}_2 + \mathbf{L}_1 + \mathbf{L}_3} \to \mathbb{R}.$$

Definition 2.13. The **Maslov index** $\tau(\mathbf{L}_1, \mathbf{L}_2, \mathbf{L}_3)$ of the triple of Lagrangian subspaces $(\mathbf{L}_1, \mathbf{L}_2, \mathbf{L}_3)$ is equal to the signature of the bilinear form $B_{\mathbf{L}_1,\mathbf{L}_2,\mathbf{L}_3}$.

Recall that the **signature** of a symmetric bilinear form is obtained from the diagonalization of the form, as the difference between the number of positive diagonal entries and the number of negative diagonal entries. The factorization of $\mathbf{L}_1 \cap (\mathbf{L}_2 \cap \mathbf{L}_3)$ by $\mathbf{L}_1 \cap \mathbf{L}_2 + \mathbf{L}_1 \cap \mathbf{L}_3$ is done in order to remove the degeneracy of the bilinear form B_0, so the Maslov index is equal also to the signature of B_0.

Remark 2.1. Kashiwara [Kashiwara and Schapira (1990)] defines the Maslov index as the signature of the bilinear form $Q : \mathbf{L}_1 \oplus \mathbf{L}_2 \oplus \mathbf{L}_3 \to \mathbb{R}$, $Q(\mathbf{l}_1 + \mathbf{l}_2 + \mathbf{l}_3) = \omega(\mathbf{l}_1, \mathbf{l}_2) + \omega(\mathbf{l}_2, \mathbf{l}_3) + \omega(\mathbf{l}_3, \mathbf{l}_1)$.

Example 2.18. Consider the case of just one particle, when the phase space is $\mathbb{R} \times \mathbb{R}$, and let

$$\mathbf{L}_1 = \mathbb{R}\frac{\partial}{\partial \xi}, \quad \mathbf{L}_2 = \mathbb{R}\frac{\partial}{\partial \eta}, \quad \mathbf{L}_3 = \mathbb{R}\left(\frac{\partial}{\partial \xi} + \frac{\partial}{\partial \eta}\right).$$

Then $\mathbf{L}_1 \cap (\mathbf{L}_2 + \mathbf{L}_3) = \mathbf{L}_1$, and $\mathbf{L}_1 \cap \mathbf{L}_2 = \mathbf{L}_1 \cap \mathbf{L}_3 = \{0\}$. Hence $B_{\mathbf{L}_1,\mathbf{L}_2,\mathbf{L}_3} = B_0$ on the space $\mathbb{R}(1,0)$. If $\mathbf{l}_1 = a(1,0)$, $\mathbf{l}_1' = b(1,0)$, then we are forced to choose $\mathbf{l}_2 = a(0,1)$, $\mathbf{l}_3 = -a(1,1)$, and $\mathbf{l}_2' = b(0,1)$, $\mathbf{l}_3' = -b(1,1)$. We compute

$$B_{\mathbf{L}_1,\mathbf{L}_2,\mathbf{L}_3}(\mathbf{l}_1, \mathbf{l}_1') = -\omega(a(1,0), b(0,1)) = -ab\omega((1,0),(0,1)) = ab.$$

The matrix of this bilinear form is the scalar 1, and so the index of the form itself is 1. We conclude that

$$\tau(\mathbf{L}_1, \mathbf{L}_2, \mathbf{L}_3) = 1.$$

Example 2.19. Again in the case of a single particle, let

$$\mathbf{L}_1 = \mathbb{R}\frac{\partial}{\partial \xi}, \quad \mathbf{L}_2 = \mathbb{R}\frac{\partial}{\partial \eta}, \quad \mathbf{L}_3 = \mathbb{R}\left(\frac{\partial}{\partial \xi} - \frac{\partial}{\partial \eta}\right).$$

Then

$$\tau(\mathbf{L}_1, \mathbf{L}_2, \mathbf{L}_3) = -1.$$

The following result shows that these two situations are in fact generic. We say that two Lagrangian subspaces are transverse if their intersection is trivial.

Proposition 2.7. *Let $\mathbf{L}_1, \mathbf{L}_2, \mathbf{L}_3$ be three pairwise transverse Lagrangian subspaces. Then there is a symplectic basis $\mathbf{v}_j, \mathbf{w}_j$, $j = 1, 2, \ldots, n$ and an integer m, $0 \leq m \leq n$, such that*

$$\mathbf{L}_1 = \mathbb{R}\mathbf{v}_1 \oplus \mathbb{R}\mathbf{v}_2 \oplus \cdots \oplus \mathbb{R}\mathbf{v}_n, \quad \mathbf{L}_2 = \mathbb{R}\mathbf{w}_1 \oplus \mathbb{R}\mathbf{w}_2 \oplus \cdots \oplus \mathbb{R}\mathbf{w}_n,$$
$$\mathbf{L}_3 = \mathbb{R}(\mathbf{v}_1 + \mathbf{w}_1) \oplus \cdots \oplus \mathbb{R}(\mathbf{v}_m + \mathbf{w}_m) \oplus \mathbb{R}(\mathbf{v}_{m+1} - \mathbf{w}_{m+1}) \oplus \cdots \oplus \mathbb{R}(\mathbf{v}_n - \mathbf{w}_n).$$

In this case

$$\tau(\mathbf{L}_1, \mathbf{L}_2, \mathbf{L}_3) = n - 2m.$$

Proof. Because $V = \mathbf{L}_2 \oplus \mathbf{L}_3$, for every $\mathbf{l}_1 \in \mathbf{L}_1$ there are unique $\mathbf{l}_2 \in \mathbf{L}_2$ and $\mathbf{l}_3 \in \mathbf{L}_3$ such that $\mathbf{l}_1 + \mathbf{l}_2 + \mathbf{l}_3 = 0$. Define the bilinear form $S : \mathbf{L}_1 \times \mathbf{L}_1 \to \mathbb{R}$, $S(\mathbf{l}_1, \mathbf{l}_1') = \omega(\mathbf{l}_2, \mathbf{l}_3)$. This form is nondegenerate, so there is a basis $\mathbf{v}_1, \mathbf{v}_2, \ldots, \mathbf{v}_n$ of $\mathbf{L}_1$ such that $S(\mathbf{v}_j, \mathbf{v}_k) = \delta_{jk}\epsilon_j$, for some $\epsilon_j = \pm 1$. Using Lemma 2.4 we construct a basis $\mathbf{w}_1, \mathbf{w}_2, \ldots, \mathbf{w}_n$ of $\mathbf{L}_2$, such that $\mathbf{v}_j, \mathbf{w}_j$ is a symplectic basis of $V = \mathbf{L}_1 \oplus \mathbf{L}_2$.

For $\mathbf{v}_j$, let $\sum_l c_{jl}\mathbf{w}_l$ be the unique element in $\mathbf{L}_2$ such that $\mathbf{v}_j + \sum_l c_{jl}\mathbf{w}_l \in \mathbf{L}_3$. Then

$$S(\mathbf{v}_j, \mathbf{v}_k) = \omega\left(\sum_l c_{jl}\mathbf{w}_l, -\mathbf{v}_k - \sum_l c_{kl}\mathbf{w}_l\right) = \sum_l c_{jl}\omega(\mathbf{w}_l, -\mathbf{v}_k) = c_{jk}.$$

Hence $c_{jk} = \delta_{jk}\epsilon_j$, which implies that $\mathbf{v}_j, \mathbf{w}_j$, $j = 1, 2, \ldots, n$, have the desired property.

We compute

$$B_{\mathbf{L}_1,\mathbf{L}_2,\mathbf{L}_3}(\mathbf{v}_j, \mathbf{v}_k) = -\omega(\mathbf{v}_j, \epsilon_k\mathbf{w}_k) = -\delta_{jk}\epsilon_k.$$

The signature of this form is $(n - m) - m = n - 2m$, as claimed. □

Proposition 2.8. *The Maslov index has the following properties:*

(i) $\tau(\mathbf{L}_1,\mathbf{L}_2,\mathbf{L}_3) = -\tau(\mathbf{L}_2,\mathbf{L}_1,\mathbf{L}_3) = -\tau(\mathbf{L}_1,\mathbf{L}_3,\mathbf{L}_2)$.

(ii) *If h is a linear symplectomorphism of V, then*

$$\tau(h(\mathbf{L}_1),h(\mathbf{L}_2),h(\mathbf{L}_3)) = \tau(\mathbf{L}_1,\mathbf{L}_2,\mathbf{L}_3).$$

(iii) *If $\mathbf{L}_1,\mathbf{L}_2,\mathbf{L}_3,\mathbf{L}_4$ are four Lagrangian subspaces, then the Maslov index satisfies the cocycle condition*

$$\tau(\mathbf{L}_1,\mathbf{L}_2,\mathbf{L}_3) - \tau(\mathbf{L}_1,\mathbf{L}_2,\mathbf{L}_4) + \tau(\mathbf{L}_1,\mathbf{L}_3,\mathbf{L}_4) - \tau(\mathbf{L}_2,\mathbf{L}_3,\mathbf{L}_4) = 0.$$

Proof. The first property follows from

$$B_{\mathbf{L}_1,\mathbf{L}_2,\mathbf{L}_3} = -B_{\mathbf{L}_2,\mathbf{L}_1,\mathbf{L}_3} = -B_{\mathbf{L}_1,\mathbf{L}_3,\mathbf{L}_2}$$

under the canonical identification of the spaces where these forms are defined, which can be checked easily using the definition. The second property is a consequence of the fact that symplectomorphisms preserve the symplectic form ω. We prove (iii) following the idea from [Turaev (1994)], but with a few modifications.

We first prove the identity with the additional assumption that

$$\mathbf{L}_1 \cap \mathbf{L}_3 = \mathbf{L}_2 \cap \mathbf{L}_4 = \{0\}.$$

Let

$$E = \{(\mathbf{l}_1,\mathbf{l}_2,\mathbf{l}_3,\mathbf{l}_4) \in \mathbf{L}_1 \oplus \mathbf{L}_2 \oplus \mathbf{L}_3 \oplus \mathbf{L}_4 \mid \mathbf{l}_1+\mathbf{l}_2+\mathbf{l}_3+\mathbf{l}_4 = 0\}.$$

Let also $E_j \subset E$ consist of those elements with the property that $\mathbf{l}_j = 0$, $j = 1,2,3,4$. Define the bilinear symmetric form $B_E : E \times E \to \mathbf{R}$,

$$\begin{aligned} B_E((\mathbf{l}_1,\mathbf{l}_2,\mathbf{l}_3,\mathbf{l}_4),(\mathbf{l}_1',\mathbf{l}_2',\mathbf{l}_3',\mathbf{l}_4')) &= -\omega(\mathbf{l}_1,\mathbf{l}_2') - \omega(\mathbf{l}_2,\mathbf{l}_3') - \omega(\mathbf{l}_3,\mathbf{l}_4') - \omega(\mathbf{l}_4,\mathbf{l}_1') \\ &\quad -\omega(\mathbf{l}_1',\mathbf{l}_2) - \omega(\mathbf{l}_2',\mathbf{l}_3) - \omega(\mathbf{l}_3',\mathbf{l}_4) - \omega(\mathbf{l}_4',\mathbf{l}_1). \end{aligned}$$

We claim that $\tau(\mathbf{L}_1,\mathbf{L}_2,\mathbf{L}_3)$ is equal to the signature of B_E restricted to E_4. Because for an element $(\mathbf{l}_1,\mathbf{l}_2,\mathbf{l}_3,0)$ in E_4 one has $\mathbf{l}_1 = -\mathbf{l}_2 - \mathbf{l}_3$, we have an isomorphism $E_4 \to \mathbf{L}_2 \cap (\mathbf{L}_1 + \mathbf{L}_3)$ given by

$$(\mathbf{l}_1,\mathbf{l}_2,\mathbf{l}_3,0) \mapsto \mathbf{l}_2.$$

Note that because $\mathbf{L}_1$ and $\mathbf{L}_3$ are transverse, $\mathbf{l}_2$ uniquely determines $\mathbf{l}_1$ and $\mathbf{l}_3$. For $(\mathbf{l}_1,\mathbf{l}_2,\mathbf{l}_3,0)$ and $(\mathbf{l}_1',\mathbf{l}_2',\mathbf{l}_3',0)$ in E_4, we compute

$$\omega(\mathbf{l}_1,\mathbf{l}_2') = \omega(\mathbf{l}_1,-\mathbf{l}_1'-\mathbf{l}_3') = -\omega(\mathbf{l}_1,\mathbf{l}_3') = \omega(\mathbf{l}_2+\mathbf{l}_3,\mathbf{l}_3') = \omega(\mathbf{l}_2,\mathbf{l}_3').$$

A similar computation shows that $\omega(\mathbf{l}_2',\mathbf{l}_3) = \omega(\mathbf{l}_2,\mathbf{l}_3')$. We obtain

$$B_E((\mathbf{l}_1,\mathbf{l}_2,\mathbf{l}_3,0),(\mathbf{l}_1',\mathbf{l}_2',\mathbf{l}_3',0)) = -4\omega(\mathbf{l}_2,\mathbf{l}_3') = 4B_0((\mathbf{l}_1,\mathbf{l}_2,\mathbf{l}_3),(\mathbf{l}_1',\mathbf{l}_2',\mathbf{l}_3')).$$

Hence the equality of signatures. The same argument can be applied to show that the signatures of $B_E|E_1$, $B_E|E_2$, and $B_E|E_3$ are respectively $\tau(\mathbf{L}_2, \mathbf{L}_3, \mathbf{L}_4)$, $\tau(\mathbf{L}_1, \mathbf{L}_3, \mathbf{L}_4)$, and $\tau(\mathbf{L}_1, \mathbf{L}_2, \mathbf{L}_4)$, because in each triple there is a pair of transverse Lagrangian subspaces.

Next, notice that

$$E_1 \cap E_3 = \{(0, a, 0, -a) \,|\, a \in \mathbf{L}_2 \cap \mathbf{L}_4\} = \{0\}$$
$$E_2 \cap E_4 = \{(a, 0, -a, 0) \,|\, a \in \mathbf{L}_1 \cap \mathbf{L}_3\} = \{0\}.$$

Also, if $\mathbf{l} \in E_j$ and $\mathbf{l}' \in E_{j+2}$, then $B_E(\mathbf{l}, \mathbf{l}') = 0$, for $j = 1, 2, 3, 4$, with indices taken modulo 4. It follows that

$$B_E = B_E|E_1 \oplus B_E|E_3 = B_E|E_2 \oplus B_E|E_4.$$

Hence

$$\begin{aligned}\sigma(B_E) &= \sigma(B_E|E_1) + \sigma(B_E|E_3) = \sigma(B_E|E_2) + \sigma(B_E|E_4)\\ &= \tau(\mathbf{L}_2, \mathbf{L}_3, \mathbf{L}_4) + \tau(\mathbf{L}_1, \mathbf{L}_2, \mathbf{L}_3) = \tau(\mathbf{L}_1, \mathbf{L}_3, \mathbf{L}_4) + \tau(\mathbf{L}_1, \mathbf{L}_2, \mathbf{L}_3).\end{aligned}$$

We obtain the cocycle relation in this particular situation.

Let us now remove the additional conditions on the Lagrangian subspaces. First, let us show that we can remove the condition $\mathbf{L}_1 \cap \mathbf{L}_3 = \{0\}$. Choose a Lagrangian subspace $\mathbf{L}_5$ that is transversal to $\mathbf{L}_1, \mathbf{L}_2, \mathbf{L}_3$, and $\mathbf{L}_4$. Using what we already proved, we can write

$$\begin{aligned}\tau(\mathbf{L}_1, \mathbf{L}_2, \mathbf{L}_3) &= \tau(\mathbf{L}_1, \mathbf{L}_2, \mathbf{L}_5) - \tau(\mathbf{L}_1, \mathbf{L}_3, \mathbf{L}_5) + \tau(\mathbf{L}_2, \mathbf{L}_3, \mathbf{L}_5)\\ \tau(\mathbf{L}_1, \mathbf{L}_2, \mathbf{L}_4) &= \tau(\mathbf{L}_1, \mathbf{L}_2, \mathbf{L}_5) - \tau(\mathbf{L}_1, \mathbf{L}_4, \mathbf{L}_5) + \tau(\mathbf{L}_2, \mathbf{L}_4, \mathbf{L}_5)\\ \tau(\mathbf{L}_1, \mathbf{L}_3, \mathbf{L}_4) &= \tau(\mathbf{L}_1, \mathbf{L}_3, \mathbf{L}_5) - \tau(\mathbf{L}_1, \mathbf{L}_4, \mathbf{L}_5) + \tau(\mathbf{L}_3, \mathbf{L}_4, \mathbf{L}_5)\\ \tau(\mathbf{L}_2, \mathbf{L}_3, \mathbf{L}_4) &= \tau(\mathbf{L}_2, \mathbf{L}_3, \mathbf{L}_5) - \tau(\mathbf{L}_2, \mathbf{L}_4, \mathbf{L}_5) + \tau(\mathbf{L}_3, \mathbf{L}_4, \mathbf{L}_5),\end{aligned}$$

because in each quadruple there are two pairs of mutually transversal Lagrangian subspaces. By adding the first and second equation, subtracting the second and the third, and using (i), we obtain the cocycle condition, because the right-hand side becomes 0. Repeating the trick we remove the condition $\mathbf{L}_2 \cap \mathbf{L}_4 = 0$, and the result is proved. □

Given an isotropic subspace W and a Lagrangian subspace $\mathbf{L}$ of V, define

$$\mathbf{L}^W = (\mathbf{L} \cap W^\perp) + W = (\mathbf{L} + W) \cap W^\perp.$$

Proposition 2.9. *Let $\mathbf{L}_1, \mathbf{L}_2, \mathbf{L}_3$ be Lagrangian subspaces and $W \subset \mathbf{L}_1 \cap \mathbf{L}_2 + \mathbf{L}_2 \cap \mathbf{L}_3 + \mathbf{L}_3 \cap \mathbf{L}_1$ be an isotropic subspace. Then*

$$\tau(\mathbf{L}_1, \mathbf{L}_2, \mathbf{L}_3) = \tau(\mathbf{L}_1^W, \mathbf{L}_2^W, \mathbf{L}_3^W).$$

Proof. Note first that $\mathbf{L}_1\cap\mathbf{L}_2+\mathbf{L}_2\cap\mathbf{L}_3+\mathbf{L}_3\cap\mathbf{L}_1$ is isotropic, so $\mathbf{L}_1\cap\mathbf{L}_2$, $\mathbf{L}_2\cap\mathbf{L}_3$, and $\mathbf{L}_3\cap\mathbf{L}_1$ are in $W^\perp$.

If $\mathbf{l}\in\mathbf{L}_1\cap\mathbf{L}_2+\mathbf{L}_2\cap\mathbf{L}_3+\mathbf{L}_3\cap\mathbf{L}_1$ then $\mathbf{l}=\mathbf{l}_{12}+\mathbf{l}_{23}+\mathbf{l}_{31}$, with $\mathbf{l}_{12}\in\mathbf{L}_1\cap\mathbf{L}_2$, $\mathbf{l}_{23}\in\mathbf{L}_2\cap\mathbf{L}_3$ and $\mathbf{l}_{31}\in\mathbf{L}_3\cap\mathbf{L}_1$. We have

$$\mathbf{l}_{23}=\mathbf{l}-\mathbf{l}_{12}-\mathbf{l}_{31}\in W+(\mathbf{L}_1\cap W^\perp)=\mathbf{L}_1^W.$$

But $\mathbf{l}_{23}\in\mathbf{L}_2\cap\mathbf{L}_3\subset W^\perp$. Hence

$$\mathbf{l}_{23}\in\mathbf{L}_1^W\cap(\mathbf{L}_2\cap W^\perp). \tag{2.32}$$

Using the fact that $\mathbf{L}_1^W\cap(\mathbf{L}_2\cap W^\perp)=\mathbf{L}_1^W\cap\mathbf{L}_2$ and (2.32) we can write

$$\mathbf{L}_1^W=(\mathbf{L}_1\cap\mathbf{L}_1^W)+\mathbf{L}_1\cap(\mathbf{L}_2\cap W^\perp)=(\mathbf{L}_1\cap\mathbf{L}_1^W)+(\mathbf{L}_2\cap\mathbf{L}_1^W).$$

It follows that $\tau(\mathbf{L}_1,\mathbf{L}_1^W,\mathbf{L}_2)=0$, because the signature is computed on a trivial vector space. The same argument works to show that this is true if we replace the indices 1 and 2 by any $i,j=1,2,3$. Using the cocycle formula proved in Proposition 2.8, we obtain

$$\begin{aligned}\tau(\mathbf{L}_1,\mathbf{L}_2,\mathbf{L}_3)&=\tau(\mathbf{L}_1,\mathbf{L}_2,\mathbf{L}_1^W)+\tau(\mathbf{L}_2,\mathbf{L}_3,\mathbf{L}_1^W)+\tau(\mathbf{L}_3,\mathbf{L}_1,\mathbf{L}_1^W)\\&=\tau(\mathbf{L}_2,\mathbf{L}_3,\mathbf{L}_1^W)=\tau(\mathbf{L}_1^W,\mathbf{L}_2,\mathbf{L}_3).\end{aligned}$$

Now $W,\mathbf{L}_1^W,\mathbf{L}_2,\mathbf{L}_3$ satisfy the conditions from the statement, so we can repeat to show that this is further equal to $\tau(\mathbf{L}_1^W,\mathbf{L}_2^W,\mathbf{L}_3)$. Repeating once more we conclude that this is equal to $\tau(\mathbf{L}_1^W,\mathbf{L}_2^W,\mathbf{L}_3^W)$, as desired. □

2.5.3 *The resolution of the projective ambiguity of the representation of $Sp(2n,\mathbb{R})$*

In this section we resolve the projectivity of the representation of the symplectic group $Sp(2n,\mathbb{R})$ defined by Fourier transforms, by first choosing an integration measure to make the Fourier transform precise, and then passing to a $\mathbb{Z}$-extension of the symplectic group.

The choice of the integration measure

We need an explicit, unambiguous formula for the Fourier transform. The formula provided by Lemma 2.4 has the downside that it depends on the basis $\mathbf{v}_1,\mathbf{v}_2,\ldots,\mathbf{v}_n$ and on the way it is completed to a basis of V by $\mathbf{w}_1,\mathbf{w}_2,\ldots,\mathbf{w}_n$. The fact is that we cannot eliminate this dependency completely, but at least we can reduce it to a minimal amount of necessary data.

It would be ideal if each Lagrangian subspace came with its own basis, is such a way that when taking a pair of Lagrangian subspaces, the corresponding bases would fit the requirements of Lemma 2.4. A brief investigation shows that this is too much to ask. But we are not really concerned with the vectors $\mathbf{v}_1, \mathbf{v}_2, \ldots, \mathbf{v}_n, \mathbf{w}_1, \mathbf{w}_2, \ldots, \mathbf{w}_n$, just with the translation invariant measure they define on $\mathbf{L}'/\mathbf{L}\cap\mathbf{L}'$.

We endow a priori each Lagrangian subspace $\mathbf{L}$ with a translation invariant measure $dm_{\mathbf{L}}$, which is unique up to multiplication by a constant. Given a pair of Lagrangian subspaces $\mathbf{L}$ and $\mathbf{L}'$, we need to endow $\mathbf{L}'/\mathbf{L}\cap\mathbf{L}'$ with a translation invariant measure defined in a canonical way in terms of $dm_{\mathbf{L}}$ and $dm_{\mathbf{L}'}$.

The key observation is that if W' is a subspace of a vector space W, then at the level of the maximal exterior product we have a canonical identification

$$\wedge^{\max} W = \wedge^{\max} W' \oplus \wedge^{\max}(W/W').$$

Hence a translation invariant measure on W is the product of translation invariant measures on W' and W/W'.

The symplectic form ω defines a volume form $\omega \wedge \omega \wedge \cdots \wedge \omega$ on the tangent space V to the phase space, where there are n factors in the wedge product. As such, ω defines a translation invariant measure on V. Also, ω induces a symplectic form ω_0 on $\mathbf{L}+\mathbf{L}'/\mathbf{L}\cap\mathbf{L}'$ which in turns defines an integration measure dm on this space. In this space $\mathbf{L}/\mathbf{L}\cap\mathbf{L}'$ and $\mathbf{L}'/\mathbf{L}\cap\mathbf{L}'$ are transverse Lagrangian subspaces. Then

$$dm = d\mu d\mu', \tag{2.33}$$

where $d\mu$ and $d\mu'$ are translation invariant measures on $\mathbf{L}/\mathbf{L}\cap\mathbf{L}'$ and $\mathbf{L}'/\mathbf{L}\cap\mathbf{L}'$ respectively. Also

$$dm_{\mathbf{L}} = d\mu d\nu \tag{2.34}$$

$$dm_{\mathbf{L}'} = d\mu' d\nu, \tag{2.35}$$

where $d\nu$ is some translation invariant measure on $\mathbf{L}\cap\mathbf{L}'$. Given that $d\mu, d\mu', d\nu$ are the Lebesgue measures multiplied by some positive constants, the equations (2.33), (2.34), (2.35) yield a system of equations in these constants that determine them uniquely. Consequently, the measure $d\mu'$ on $\mathbf{L}'/\mathbf{L}\cap\mathbf{L}'$ is uniquely determined by $dm_{\mathbf{L}}$, $dm_{\mathbf{L}'}$ and ω.

Now, as said above, we associate to each Lagrangian subspace $\mathbf{L}$ a measure $dm_{\mathbf{L}}$. We define the Fourier transform $\mathcal{F}_{\mathbf{L},\mathbf{L}'}$ by the formula

$$(\mathcal{F}_{\mathbf{L},\mathbf{L}'}\psi)(u) = h^{k/2} \int_{\exp(\mathfrak{L}(\mathbf{L}'))/\exp(\mathfrak{L}(\mathbf{L})\cap\mathfrak{L}(\mathbf{L}'))} \psi(uu')d\mu', \tag{2.36}$$

where $k = \dim \mathbf{L}'/\mathbf{L}\cap\mathbf{L}'$ and the measure on $\exp(\mathfrak{L}(\mathbf{L}'))/\exp(\mathfrak{L}(\mathbf{L})\cap\mathfrak{L}(\mathbf{L}'))$ is induced by the measure $d\mu'$ on $\mathbf{L}'/\mathbf{L}\cap\mathbf{L}'$ defined above in terms of $dm_{\mathbf{L}}$

and $dm_{\mathbf{L}'}$. This choice of the measure $d\mu'$ makes the Fourier transform unitary and also

$$\mathcal{F}_{\mathbf{L},\mathbf{L}'} = (\mathcal{F}_{\mathbf{L}',\mathbf{L}})^{-1}. \tag{2.37}$$

Example 2.20. If $\mathbf{L} = \mathbf{L}'$, then $\mathbf{L}'/\mathbf{L} \cap \mathbf{L}' = \{0\}$, and the measure induced on it is 1. The Fourier transform is the identity operator.

Example 2.21. Recall the situation of Example 2.17. The measures on $\mathbf{L}$ and $\mathbf{L}'$ are just the Lebesgue measures $dx_1dx_2\dots dx_n$ respectively $dy_1dy_2\cdots dy_n$ respectively $dx_1\cdots dx_kdy_{k+1}\cdots dy_n$. The measure of integration of the Fourier transform is $h^{k/2}dx_1dx_2\dots dx_k$.

The Fourier transform and the Maslov index

Given three Lagrangian subspaces $\mathbf{L}_1, \mathbf{L}_2, \mathbf{L}_3$, there is a constant $a(\mathbf{L}_1, \mathbf{L}_2, \mathbf{L}_3)$ such that

$$\mathcal{F}_{\mathbf{L}_3,\mathbf{L}_1}\mathcal{F}_{\mathbf{L}_2,\mathbf{L}_3}\mathcal{F}_{\mathbf{L}_1,\mathbf{L}_2} = a(\mathbf{L}_1, \mathbf{L}_2, \mathbf{L}_3)I.$$

This constant does not depend on the choices of $dm_{\mathbf{L}_1}$, $dm_{\mathbf{L}_2}$, $dm_{\mathbf{L}_3}$, since if we change one of these measures by a constant, then in the two Fourier transforms defined using it, one time shows up the constant and one time shows up its reciprocal. The following theorem and its proof are from [Lion and Vergne (1980)].

Theorem 2.7. *Let* $\mathbf{L}_1, \mathbf{L}_2, \mathbf{L}_3$ *be three Lagrangian subspaces. Then*

$$\mathcal{F}_{\mathbf{L}_3,\mathbf{L}_1}\mathcal{F}_{\mathbf{L}_2,\mathbf{L}_3}\mathcal{F}_{\mathbf{L}_1,\mathbf{L}_2} = e^{-\frac{\pi i}{4}\tau(\mathbf{L}_1,\mathbf{L}_2,\mathbf{L}_3)}I.$$

Proof. **Case 1: $\mathbf{L}_1, \mathbf{L}_2, \mathbf{L}_3$** pairwise transverse. We employ Proposition 2.7. The bases of these three Lagrangian spaces allow us to identify them with $\mathbb{R}^n$, and we let the measures associated to them be the ones pulled back from $\mathbb{R}^n$ under this identification.

As algebraic complements of $\mathbf{L}_1, \mathbf{L}_2, \mathbf{L}_3$ we take $\mathbf{L}_2, \mathbf{L}_1, \mathbf{L}_2$, respectively. The symplectic bases of the pairs are respectively $\mathbf{v}_j, \mathbf{w}_j$, $j = 1, 2, \dots, n$, $\mathbf{w}_j, -\mathbf{v}_j$, $j = 1, 2, \dots, n$, and $\mathbf{v}_j + \epsilon_k\mathbf{w}_j, \mathbf{w}_j$, $j = 1, 2, \dots, n$. Then in the coordinate system $(\mathbf{x}, \mathbf{y})$ induced by the symplectic basis $\mathbf{v}_j, \mathbf{w}_j$, $j = 1, 2, \dots, n$, the Hilbert spaces $\mathcal{H}(\mathbf{L}_1)$, $\mathcal{H}(\mathbf{L}_2)$, and $\mathcal{H}(\mathbf{L}_3)$ are identified respectively with $L^2(\mathbf{R}, d\mathbf{x})$, $L^2(\mathbf{R}, d\mathbf{y})$, and $L^2(\mathbf{R}, d\mathbf{x})$ via the identifications

$$\psi(x_1, x_2, \dots, x_n) \mapsto \psi\left(\exp\left(\sum_k x_k X_k\right)\right)$$

$$\psi(y_1, y_2, \dots, y_n) \mapsto \psi\left(\exp\left(-\sum_k y_k Y_k\right)\right)$$

$$\psi(x_1, x_2, \dots, x_n) \mapsto \psi\left(\exp\left(\sum_k x_k X_k\right)\right).$$

Examples 2.17 and 2.21 show that

$$(\mathcal{F}_{\mathbf{L}_1,\mathbf{L}_2}\psi)(y_1, \dots, y_n) = h^{n/2} \int \psi(\mathbf{x}) e^{-2\pi i h \mathbf{x}^T \mathbf{y}} d\mathbf{x},$$

the Fourier transform in n variables.

We compute

$$\begin{aligned}
&(\mathcal{F}_{\mathbf{L}_2,\mathbf{L}_3}\psi)(x_1, x_2, \dots, x_n) \\
&= h^{n/2} \int \psi\left(\exp\left(\sum_{k=1}^n x_k X_k\right) \exp\left(\sum_{k=1}^n y_k (X_k + \epsilon_k Y_k)\right)\right) d\mathbf{y} \\
&= h^{n/2} \int \psi\left(\exp\left(\textstyle\sum_{k=1}^n \epsilon_k y_k Y_k\right) \exp\left(\textstyle\sum_{k=1}^n (x_k + y_k) X_k + h\epsilon_k (y_k^2/2 + x_k y_k) I\right)\right) d\mathbf{y} \\
&= h^{n/2} \int \psi(-\epsilon_1 y_1, \dots, -\epsilon_n y_n) e^{-2\pi i h \sum_k \epsilon_k (x_k y_k + y_k^2/2)} d\mathbf{z} \\
&= h^{n/2} (-1)^m \int \psi(\mathbf{y}) e^{-\pi i h \sum_k \epsilon_k y_k^2} e^{2\pi i h \mathbf{y}^T \mathbf{x}} d\mathbf{y}.
\end{aligned}$$

Finally,

$$\begin{aligned}
&(\mathcal{F}_{\mathbf{L}_3,\mathbf{L}_1}^{-1}\psi)(x_1, x_2, \dots, x_n) = (\mathcal{F}_{\mathbf{L}_1,\mathbf{L}_3}\psi)(x_1, x_2, \dots, x_n) \\
&= h^{n/2} \int \psi\left(\exp\left(\sum_k x_k X_K\right) \exp\left(\sum_k z_k (X_k + \epsilon_k Y_k)\right)\right) d\mathbf{z} \\
&= h^{n/2} \int \psi\left(\exp\left(\textstyle\sum_k (x_k + z_k) X_k\right) \exp\left(\textstyle\sum_k \epsilon_k z_k Y_k - h \sum_k (\epsilon_k z_k^2/2) I\right)\right) d\mathbf{z} \\
&= h^{n/2} \int \psi(\mathbf{x} + \mathbf{z}) e^{\pi i h \sum_k \epsilon_k z_k^2} d\mathbf{z} = h^{n/2} \int \psi(\mathbf{z}) e^{\pi i h \sum_k \epsilon_k (z_k - x_k)^2} d\mathbf{z}.
\end{aligned}$$

Because of (2.37), it suffices to show that

$$\mathcal{F}_{\mathbf{L}_2,\mathbf{L}_3} \mathcal{F}_{\mathbf{L}_1,\mathbf{L}_2} = e^{\frac{\pi i}{4}(2m-n)} \mathcal{F}_{\mathbf{L}_1,\mathbf{L}_3}.$$

For this we use the following result.

Lemma 2.5. *Let w be a complex number with positive imaginary part, and let t be a complex number. Then*

$$h \int e^{-2\pi i h t y} e^{\pi i h w y^2} dy = \left(\frac{w}{i}\right)^{-1/2} e^{-\pi i h w^{-1} t^2},$$

where the branch of the function $w \mapsto (w/i)^{-1/2}$ on $\operatorname{Im} w > 0$ is chosen so that it is equal to 1 when $w = i$.

Proof. Write the identity as

$$h \int e^{i\pi h w (y - t/w)^2} dy = \left(\frac{w}{i}\right)^{-1/2}.$$

Because the integrand on the left-hand side is an analytical function in both z and t, using the principle of analytical continuation it suffices to prove the equality for w and t imaginary. Then it becomes a standard Gaussian integral, and is evaluated as such. □

Returning to the proof of the theorem, we have that $\mathcal{F}_{\mathbf{L}_2,\mathbf{L}_3}\mathcal{F}_{\mathbf{L}_1,\mathbf{L}_2}\psi(\mathbf{z})$ is equal to

$$h^{n/2}(-1)^m \int h^{n/2} \int \psi(\mathbf{z}) e^{-2\pi i h \mathbf{z}^T \mathbf{y}} d\mathbf{z} e^{-\pi i h \sum_k \epsilon_k y_k^2} e^{2\pi i h \mathbf{y}^T \mathbf{x}} d\mathbf{y}$$

$$= \lim_{\boldsymbol{\delta} \to \mathbf{0}} h^n (-1)^m \iint \psi(\mathbf{z}) e^{-2\pi i h (\mathbf{z}^T - \mathbf{x}^T)\mathbf{y}} e^{\pi i h \sum_K (-\epsilon_k + \delta_k i) y_k^2} d\mathbf{z} d\mathbf{y}$$

where the limit is taken over the vectors $\boldsymbol{\delta} = (\delta_1, \delta_2, \ldots, \delta_k)$ with positive entries. We change the order of integration and use Lemma 2.5 to obtain

$$\lim_{\boldsymbol{\delta} \to \mathbf{0}} h^{n/2}(-1)^m \int \psi(\mathbf{z}) h^{n/2} \int e^{-2\pi i h (\mathbf{z}^T - \mathbf{x}^T)\mathbf{y}} e^{\pi i h \sum_K (-\epsilon_k + \delta_k i) y_k^2} d\mathbf{y} d\mathbf{z}$$

$$= \lim_{\boldsymbol{\delta} \to \mathbf{0}} h^{n/2} \left(\prod_{k=1}^{n} (\delta_k + i\epsilon_k)^{-1/2} \right) \int \psi(\mathbf{z}) e^{\pi i h (\epsilon_k - i\delta_k)^{-1} \sum_k (z_k - x_k)^2} d\mathbf{z}.$$

Passing to the limit and using the fact that $(\pm i)^{-1/2} = \mp e^{\pm \frac{\pi i}{4}}$, we obtain

$$h^{n/2} e^{\frac{\pi i}{4}(2m-n)} \int \psi(\mathbf{z}) e^{\pi i h \epsilon_k \sum_k (z_k - x_k)^2} d\mathbf{z},$$

as desired.

Case 2: $\mathbf{L}_3 = (\mathbf{L}_1 \cap \mathbf{L}_3) + (\mathbf{L}_2 \cap \mathbf{L}_3)$. In this case we trivially have $\tau(\mathbf{L}_1, \mathbf{L}_2, \mathbf{L}_3) = 0$. Hence we have to show that

$$\mathcal{F}_{\mathbf{L}_1,\mathbf{L}_2} = \mathcal{F}_{\mathbf{L}_3,\mathbf{L}_2}\mathcal{F}_{\mathbf{L}_1,\mathbf{L}_3}.$$

Note first that $\mathbf{L}_1 \cap \mathbf{L}_2 \cap \mathbf{L}_3 = \mathbf{L}_1 \cap \mathbf{L}_2$. Indeed, if $\mathbf{l} \in \mathbf{L}_1 \cap \mathbf{L}_2$, then $\mathbf{l} \in \mathbf{L}_1^{\perp}$ and $\mathbf{l} \in \mathbf{L}_2^{\perp}$, and hence $\mathbf{l} \in \mathbf{L}_3^{\perp} = \mathbf{L}_3$.

Write

$$\mathbf{L}_3 = (\mathbf{L}_1 \cap \mathbf{L}_2) \oplus W_1 \oplus W_2$$
$$\mathbf{L}_1 = (\mathbf{L}_1 \cap \mathbf{L}_2) \oplus W_1 \oplus W_1'$$
$$\mathbf{L}_2 = (\mathbf{L}_1 \cap \mathbf{L}_2) \oplus W_2 \oplus W_2'.$$

Then $\mathcal{F}_{\mathbf{L}_1,\mathbf{L}_3}$ is a standard Fourier transform between W_1' and W_2, while $\mathcal{F}_{\mathbf{L}_3,\mathbf{L}_2}$ is a standard Fourier transform between W_1 and W_2'. Their composition is the standard Fourier transform between $W_1' \oplus W_1$ and $W_2 \oplus W_2'$, which is equal to $\mathcal{F}_{\mathbf{L}_1,\mathbf{L}_2}$.

Case 3: $\mathbf{L}_1, \mathbf{L}_2, \mathbf{L}_3$ in general position. We induct on the dimension of V. For V of dimension 2, which is the base case of our induction, only Cases 1 or 2 can occur, and these have been checked.

The property has been checked when $\mathbf{L}_1, \mathbf{L}_2, \mathbf{L}_3$ are pairwise transverse, so let us assume this is not the case. By relabeling the spaces, we can assume that $\mathbf{L}_1 \cap \mathbf{L}_2$ is nontrivial, and let W be the intersection. Then W is an isotropic subspace. We have

$$\mathbf{L}_1 = \mathbf{L}_1^W \text{ and } \mathbf{L}_2 = \mathbf{L}_2^W. \tag{2.38}$$

Also

$$\mathbf{L}_3^W = (\mathbf{L}_3 \cap \mathbf{L}_3^W) + (\mathbf{L}_1 \cap \mathbf{L}_3^W) = (\mathbf{L}_3 \cap \mathbf{L}_3^W) + (\mathbf{L}_2 \cap \mathbf{L}_3^W).$$

Hence, by using Case 2, we can write

$$\mathcal{F}_{\mathbf{L}_1,\mathbf{L}_3} = \mathcal{F}_{\mathbf{L}_3^W,\mathbf{L}_3}\mathcal{F}_{\mathbf{L}_1,\mathbf{L}_3^W}, \quad \text{and} \quad \mathcal{F}_{\mathbf{L}_3,\mathbf{L}_2} = \mathcal{F}_{\mathbf{L}_3^W,\mathbf{L}_2}\mathcal{F}_{\mathbf{L}_3,\mathbf{L}_3^W}.$$

We compute

$$\mathcal{F}_{\mathbf{L}_3,\mathbf{L}_1}\mathcal{F}_{\mathbf{L}_2,\mathbf{L}_3}\mathcal{F}_{\mathbf{L}_1,\mathbf{L}_2} = \mathcal{F}_{\mathbf{L}_3^W,\mathbf{L}_1}\mathcal{F}_{\mathbf{L}_3,\mathbf{L}_3^W}\mathcal{F}_{\mathbf{L}_3^W,\mathbf{L}_3}\mathcal{F}_{\mathbf{L}_2,\mathbf{L}_3^W}\mathcal{F}_{\mathbf{L}_1,\mathbf{L}_2} \tag{2.39}$$

$$= \mathcal{F}_{\mathbf{L}_3^W,\mathbf{L}_1}\mathcal{F}_{\mathbf{L}_2,\mathbf{L}_3^W}\mathcal{F}_{\mathbf{L}_1,\mathbf{L}_2}. \tag{2.40}$$

The three Lagrangian subspaces $\mathbf{L}_1, \mathbf{L}_2, \mathbf{L}_3^W$ are contained in $W^\perp$, which is strictly included in V. The computation of the expression in (2.40) can therefore be carried out in $V' = W^\perp/W$, with $\mathbf{L}_1, \mathbf{L}_2, \mathbf{L}_3^W$ substituted by the Lagrangian subspaces of V': $\mathbf{L}_1/W$, $\mathbf{L}_2/W$, $\mathbf{L}_3^W/W$. Using the fact that, by Proposition 2.9 and (2.38), we obtain

$$\tau(\mathbf{L}_1, \mathbf{L}_2, \mathbf{L}_3) = \tau(\mathbf{L}_1^W, \mathbf{L}_2^W, \mathbf{L}_3^W), = \tau(\mathbf{L}_1, \mathbf{L}_2, \mathbf{L}_3^W)$$

and we obtain the identity from the statement. □

The Fourier transform defined by a linear symplectomorphism

Let us now turn to the case of the Fourier transform defined by a linear symplectomorphism. Let $h \in Sp(2n, \mathbb{R})$ be such a symplectomorphism. Then h acts on the Heisenberg group $\mathbf{H}(\mathbb{R}^n)$ by

$$h \cdot \exp(\mathbf{p}^T\mathbf{P} + \mathbf{q}^T\mathbf{Q} + tI) = \exp\left(\left(h\begin{bmatrix}\mathbf{p}\\ \mathbf{q}\end{bmatrix}\right)^T (\mathbf{P}, \mathbf{Q}) + tI\right).$$

More explicitly, if

$$h = \begin{bmatrix} A & B \\ C & D \end{bmatrix},$$

then

$$h \cdot \exp(\mathbf{p}^T \mathbf{P} + \mathbf{q}^T \mathbf{Q} + tI) = \exp((A\mathbf{p} + B\mathbf{q})^T \mathbf{P} + (C\mathbf{p} + D\mathbf{q})^T \mathbf{Q} + tI).$$

Now consider a "standard" Schrödinger representation of the Heisenberg group. To this end, we fix a Lagrangian subspace $\mathbf{L}$ of V, a symplectic basis $\mathbf{v}_j, \mathbf{w}_j$, $j = 1, 2, \ldots, n$, associated to it and coordinates $\mathbf{x}, \mathbf{y}$, and consider the algebraic complement $\mathbf{L}^\natural$ of $\mathbf{L}$ defined as in § 2.4. Associate to $\mathbf{L}$ the translation invariant measure $dm_{\mathbf{L}}$ as explained in the beginning of this section. With all these choices, the Schrödinger representation defined by $\mathbf{L}$ is identified with a representation of $\mathbf{H}(\mathbb{R}^n)$ on $L^2(\mathbb{R}^n, d\mathbf{x})$.

If $h \in Sp(2n, \mathbb{R})$, then h transforms this entire construction by

$$\mathbf{L} \mapsto h(\mathbf{L}), \quad \mathbf{L}^\natural \mapsto h(\mathbf{L}^\natural) = h(\mathbf{L})^\natural,$$
$$(\mathbf{x}, \mathbf{y}) \to h(\mathbf{x}, \mathbf{y}), \quad dm_{\mathbf{L}} \mapsto dh(m_{\mathbf{L}}).$$

where the measure $dh(m_{\mathbf{L}})$ is defined so that the measure of a Borel set E equal the measure of $h^{-1}(E)$ with respect to $dm_{\mathbf{L}}$. The map $A(h)$ acting on functions on $\mathbf{H}(\mathbb{R}^n)$ by

$$(A(h)\psi)(u) = \psi(h^{-1} \cdot u)$$

defines a unitary operator between $\mathcal{H}(\mathbf{L})$ with measure $dm_{\mathbf{L}}$ and $\mathcal{H}(h(\mathbf{L}))$ with measure $dh(m_{\mathbf{L}})$, which interpolates the Schrödinger representation on $\mathcal{H}(\mathbf{L})$ with that on $\mathcal{H}(h(\mathbf{L}))$. We now identify $\mathcal{H}(h(\mathbf{L}))$ with $\mathcal{H}(\mathbf{L}) \cong L^2(\mathbb{R}^n, d\mathbf{x})$ using the Fourier transform defined by a pair of Lagrangian subspaces, to obtain the following definition.

Definition 2.14. The **Fourier transform defined by the symplectomorphism** h is

$$\mathcal{F}_{\mathbf{L}}(h) = \mathcal{F}_{h(\mathbf{L}),\mathbf{L}} \circ A(h).$$

$\mathcal{F}_{\mathbf{L}}(h)$ is a unitary map of $\mathcal{H}(\mathbf{L}) \cong L^2(\mathbb{R}^n, d\mathbf{x})$ onto itself. It satisfies

$$\mathcal{F}_{\mathbf{L}}(h) u (\mathcal{F}_{\mathbf{L}}(h))^{-1} = h \cdot u.$$

In other words, if $u = \mathrm{op}(f)$, where f is an exponential function, then

$$\mathcal{F}_{\mathbf{L}}(h) \mathrm{op}(f) (\mathcal{F}_{\mathbf{L}}(h))^{-1} = \mathrm{op}(f \circ h^{-1}).$$

Replacing h by its inverse, we rewrite this in the form it is usually encountered in the literature as

$$\text{op}(f \circ h) = (\mathcal{F}_{\mathbf{L}}(h))^{-1}\text{op}(f)\mathcal{F}_{\mathbf{L}}(h). \tag{2.41}$$

Since any function can be expressed as an integral of exponentials, this property of the Fourier transform holds for general observables f.

Other quantization models satisfy this equality up to an error in the size of Planck's constant, meaning that

$$\text{op}(f \circ h) = (\mathcal{F}_{\mathbf{L}}(h))^{-1}\text{op}(f)\mathcal{F}_{\mathbf{L}}(h) + O(\hbar).$$

This is a consequence of Egorov's theorem on pseudodifferential operators [Egorov (1969)]. Weyl quantization is the only quantization model that satisfies this condition exactly. In what follows, we will call (2.41) the **exact Egorov identity**. We will see in Chapter 7 that its discrete analogue plays an important role in low dimensional topology.

A different way to phrase the definition of the Fourier transform $\mathcal{F}_{\mathbf{L}}(h)$ is as follows. There are two representations of $\mathbf{H}(\mathbb{R}^n)$, one is the standard representation, and one is a representation in which $\exp(\mathbf{p}^T\mathbf{P} + \mathbf{q}^T\mathbf{Q} + tI)$ acts the way

$$\exp((A\mathbf{p} + B\mathbf{q})^T\mathbf{P} + (C\mathbf{p} + D\mathbf{q})^T\mathbf{Q} + tI)$$

acts in the standard representation. By the Stone-von Neumann Theorem, the two representations are unitary equivalent. The Fourier transform is the unitary operator that establishes this unitary equivalence.

Theorem 2.8. *Let h, h' be elements of $Sp(2n, \mathbb{R})$. Then*

$$\mathcal{F}_{\mathbf{L}}(h'h) = e^{\frac{i\pi}{4}\tau(h'h(\mathbf{L}),h'(\mathbf{L}),\mathbf{L})}\mathcal{F}_{\mathbf{L}}(h')\mathcal{F}_{\mathbf{L}}(h).$$

Proof. First, note that $A(h'h) = A(h')A(h)$, and also that for every $h \in Sp(2n, \mathbb{R})$ and Lagrangian subspaces $\mathbf{L}_1, \mathbf{L}_2$, one has

$$\mathcal{F}_{h(\mathbf{L}_1),h(\mathbf{L}_2)} = A(h)\mathcal{F}_{\mathbf{L}_1,\mathbf{L}_2}A(h)^{-1}.$$

We compute

$$\begin{aligned}\mathcal{F}_{\mathbf{L}}(h')\mathcal{F}_{\mathbf{L}}(h) &= \mathcal{F}_{h'(\mathbf{L}),\mathbf{L}}A(h')\mathcal{F}_{h(\mathbf{L}),\mathbf{L}}A(h)\\ &= \mathcal{F}_{h'(\mathbf{L}),\mathbf{L}}A(h')A(h')^{-1}\mathcal{F}_{h'h(\mathbf{L}),h'(\mathbf{L})}A(h')A(h)\\ &= \mathcal{F}_{h'(\mathbf{L}),\mathbf{L}}\mathcal{F}_{h'h(\mathbf{L}),h'(\mathbf{L})}A(h'h).\end{aligned}$$

Using Theorem 2.7, we obtain that this equals

$$e^{\frac{i\pi}{4}\tau(\mathbf{L},h'(\mathbf{L}),h'h(\mathbf{L}))}\mathcal{F}_{\mathbf{L}}(h'h) = e^{-\frac{i\pi}{4}\tau(h'h(\mathbf{L}),h'(\mathbf{L}),(\mathbf{L}))}\mathcal{F}_{\mathbf{L}}(h'h),$$

as desired. □

In view of this result, the map

$$h \mapsto \mathcal{F}_{\mathbf{L}}(h)$$

is only a **projective** representation of the symplectic group $Sp(2n,\mathbb{R})$ on $L^2(\mathbb{R}, d\mathbf{x})$.

To produce a true representation, we define the group

$$\widetilde{Sp(2n,\mathbb{R})}_{\mathbf{L}} = Sp(2n,\mathbb{R}) \times \mathbb{Z},$$

with the multiplication rule[5]

$$(h',n')(h,n) = (h'h, n' + n - \tau(h'h(\mathbf{L}), h'(\mathbf{L}), \mathbf{L}).$$

Said differently, the group $\widetilde{Sp(2n,\mathbb{R})}_{\mathbf{L}}$ is a $\mathbb{Z}$-extension of the symplectic group by the cocycle $c(h',h) = \tau(h'h(\mathbf{L}), h'(\mathbf{L}), \mathbf{L})$. Note that c does indeed satisfy the cocycle condition

$$c(h''h',h) + c(h'',h') = c(h'',h'h) + c(h',h)$$

because of Proposition 2.8 (iii) showing that $\widetilde{Sp(2n,\mathbb{R})}_{\mathbf{L}}$ defined this way is a group.

Define the representation of $\widetilde{Sp(2n,\mathbb{R})}_{\mathbf{L}}$ by

$$(h,n) \mapsto e^{\frac{i\pi}{4}n}\mathcal{F}_{\mathbf{L}}(h).$$

Theorem 2.8 implies that this is a true representation of $\widetilde{Sp(2n,\mathbb{R})}_{\mathbf{L}}$.

One should point out that if $\mathbf{L}'$ is another Lagrangian subspace, then the group isomorphism $\widetilde{Sp(2n,\mathbb{R})}_{\mathbf{L}} \to \widetilde{Sp(2n,\mathbb{R})}_{\mathbf{L}'}$ given by

$$(h,n) \mapsto (h, n - \tau(\mathbf{L}, h(\mathbf{L}'), \mathbf{L}') - \tau(\mathbf{L}, h(\mathbf{L}), h(\mathbf{L}'))$$

identifies the representation defined by $\mathbf{L}$ with that defined by $\mathbf{L}'$.

Another observation is that a $\mathbf{Z}_2$-extension is sufficient to resolve the projective ambiguity, in which case one obtains the metaplectic representation, also known as the Segal-Shale-Weil representation. We will not discuss it in this book because we are really focused on techniques that might extend to other cases of Chern-Simons theory beyond the abelian case which is the main topic.

[5]The sign convention is so that later this will look the same as Wall's formula for the nonadditivity of the signature.

Conclusion to the second chapter

To a system of n free one-dimensional particles, we associate a quantum system consisting of:

1. **A Hilbert space of quantum states.** This Hilbert space depends on a polarization given in the guise of a Lagrangian subspace of the tangent space. There are three versions of this construction: one in a real polarization, where the Hilbert space is $L^2(\mathbb{R}^n)$, one in a Kähler polarization, where it is a Segal-Bargmann space of holomorphic functions over $\mathbb{C}^n$, and an abstract version, where it is defined as the space of equivariant functions on the Heisenberg group with real entries $H(\mathbb{R}^n)$.
2. **The operators which are Weyl quantizations of (periodic) exponential functions on the phase space of the system.** These operators form a group, the Heisenberg group $\mathbf{H}(\mathbb{R}^n)$, which acts on the Hilbert space of the quantization giving rise to the Schrödinger representation. The Schrödinger representation is identified with the representation induced by a character of a maximal abelian subgroup of the Heisenberg group.
3. **A projective representation of the symplectic group on the Hilbert space of the quantization.** This representation interpolates the action of the symplectic group on the symbols of operators (the exact Egorov identity). It becomes a true representation if we pass to a $\mathbb{Z}$-extension of the symplectic group defined in terms of the Maslov index.

Chapter 3

Surfaces and curves

In this chapter we review a few facts about surfaces, and about curves embedded in surfaces. These are needed for both the construction of the Jacobian variety, and hence for the definition of Riemann's theta functions, and for establishing the relationship between theta functions and knots, which is the topic of this book. Besides the case of surfaces and curves, we address briefly the case of 3-dimensional manifolds that are cylinders over surfaces. We will return to general 3-dimensional manifolds in Chapter 6. The results that are not part of the buildup of a mathematician without a particular interest in topology are proved.

We are therefore concerned with compact 1, 2, and 3-dimensional manifolds with or without boundary. A property that distinguishes these manifolds from those of higher dimensions is that they can all be endowed with the structure of smooth manifolds. Moreover, any two smooth structures on the same manifold are diffeomorphic. So two manifolds of dimension at most 3 are homeomorphic if and only if they are diffeomorphic.

Additionally, every compact manifold of dimension at most 3 can be endowed with a piece-wise linear structure. This means that the manifold is homeomorphic with one that is obtained by gluing together finitely many polyhedra along faces by affine maps. So, every 1-dimensional manifold is homeomorphic to one obtained by joining together several segments at endpoints, every surface is homeomorphic to one obtained by gluing together polygons along their sides, and every 3-dimensional manifold is homeomorphic to one obtained by gluing together several solid polyhedra along their faces.

In particular all these manifolds are triangulable. This means that 1-dimensional manifolds can be decomposed into segments, 2-dimensional manifolds can be decomposed into triangles, and 3-dimensional manifolds can be decomposed into tetrahedra. Given two piece-wise linear structures on the same manifold, they can be identified by a piece-wise linear homeomorphism, after an eventual subdivision of the polyhedra.

It follows that for our purposes, we can consider either smooth 1, 2, or 3-dimensional manifolds or piece-wise linear ones. For the construction of the Jacobian variety we need Riemann surfaces and smooth curves on them. Therefore the case of smooth structures should prevail.

We assume the reader is familiar with the basic facts about algebraic and differential topology that can be found in [Munkres (2000)], [Spanier (1966)], [Hirsch (1976)], and [Brendon (1993)]. There is some input of more advanced facts in differential topology from [Cerf (1970)].

3.1 The topology of surfaces

3.1.1 *The classification of surfaces*

All surfaces considered in this book are **orientable and compact**. If the surface has no boundary then it is called **closed**.

Theorem 3.1. *(Classification of Surfaces) Every compact, orientable surface is homeomorphic to a sphere with $g \geq 0$ handles attached and with finitely many disjoint disks removed. The number g, called genus, together with the number of boundary components classify these surfaces.*

A few examples of closed surfaces are shown in Figure 3.1. Surfaces with

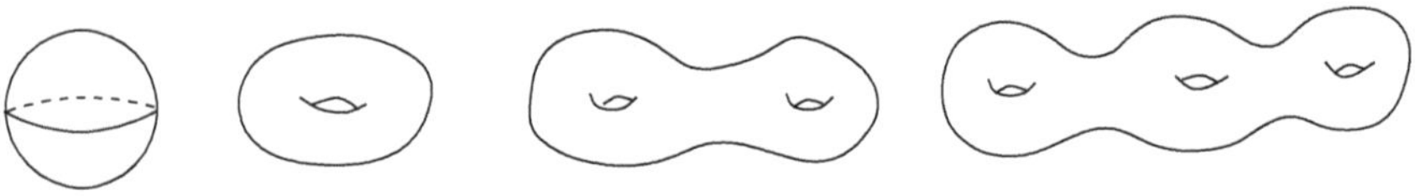

Fig. 3.1 Surfaces of genera $0, 1, 2$, and 3

boundary are obtained from these by removing a finite number of disjoint open disks. A closed genus g surface will be denoted by Σ_g, while a genus g surface with n boundary components will be denoted by $\Sigma_{g,n}$ (in this sense Σ_g stands for $\Sigma_{g,0}$). Σ_0 is the **sphere** which we also denote by S^2, $\Sigma_{0,1}$ is the **disk** which we denote by B^2, $\Sigma_{0,2}$ is the **annulus**, and Σ_1 is the **torus** $S^1 \times S^1$.

Given a surface, we can endow it with a smooth structure. An orientation of the surface can be defined by choosing at each point a 2-dimensional frame of its tangent space in such a way that these frames vary smoothly with the point.

Each closed genus g surface Σ_g bounds a genus g **handlebody** H_g. This handlebody is the 3-dimensional manifold that lies inside the surface when it is embedded in the 3-dimensional space as depicted in Figure 3.1. Rigorously H_g is obtained by gluing to the 3-dimensional ball

$$B^3 = \{(x, y, z) \,|\, x^2 + y^2 + z^2 \leq 1\}$$

g 1-handles. A 1-handle is homeomorphic to $B^2 \times [0, 1]$ where B^2 is a closed disk; it is glued to B^3 so that $B^2 \times \{0\}$ and $B^2 \times \{1\}$ are identified via homeomorphisms with two disjoint disks in the boundary ∂B^3. Moreover, the spheres bounding B^3 and $B^2 \times [0, 1]$ should be oriented, and the gluing is via orientation-reversing homeomorphisms.

The handlebody H_g is homeomorphic to the cylinder over a genus zero surface with $g + 1$ boundary components, $\Sigma_{0,g+1} \times [0, 1]$.

3.1.2 *The fundamental group*

In the general situation, the **fundamental group** is defined as follows. Let X be a path connected topological space. Fix a base point x_0 in X. Consider the set of all loops

$$\{\gamma : [0, 1] \to X \mid \gamma(0) = \gamma(1) = x_0\},$$

equipped with a binary operation defined by

$$(\gamma_1 * \gamma_2)(t) = \begin{cases} \gamma_1(2t) & \text{if } 0 \leq t \leq 1/2 \\ \gamma_2(2t - 1) & \text{if } 1/2 \leq t \leq 1. \end{cases} \tag{3.1}$$

Introduce an equivalence relation on loops, called homotopy, such that two loops γ_1 and γ_2 are homotopic if there exists a continuous map $H : [0, 1] \times X \to X$ such that $H(s, 0) = H(s, 1) = x_0$, $H(0, t) = \gamma_1(t)$ and $H(1, t) = \gamma_2(t)$. Denote the set of equivalence classes by $\pi_1(X, x_0)$. The binary operation (3.1) factors to a multiplication in $\pi_1(X, x_0)$, which turns it into a group. This is the fundamental group of X.

If x_1 is another base point, then $\pi_1(X, x_0)$ is isomorphic to $\pi_1(X, x_1)$, though not canonically. The isomorphism is defined by a path from x_0 to x_1 which is used for transforming paths based at x_1 into paths based at x_0. Because of this isomorphism most of the time we ignore the base point, and denote the fundamental group simply by $\pi_1(X)$.

A loop can also be viewed as a map

$$\gamma : S^1 \to X. \tag{3.2}$$

The loop is called smooth if this map γ is smooth. By abuse of language, we call the image of γ itself the loop. The notation γ will stand for both the map and its image.

Example 3.1. The genus g closed surface Σ_g can be obtained from a regular $2g$-gon by gluing pairs of sides as shown in Figure 3.2. An application of the Seifert-van Kampen theorem implies that the fundamental group $\pi_1(\Sigma_g)$ has the presentation

$$\langle a_1, b_1, a_2, b_2, \ldots, a_g, b_g | a_1 b_1 a_1^{-1} b_1^{-1} a_2 b_2 a_2^{-1} b_2^{-1} \cdots a_g b_g a_g^{-1} b_g^{-1} = 1 \rangle. \quad (3.3)$$

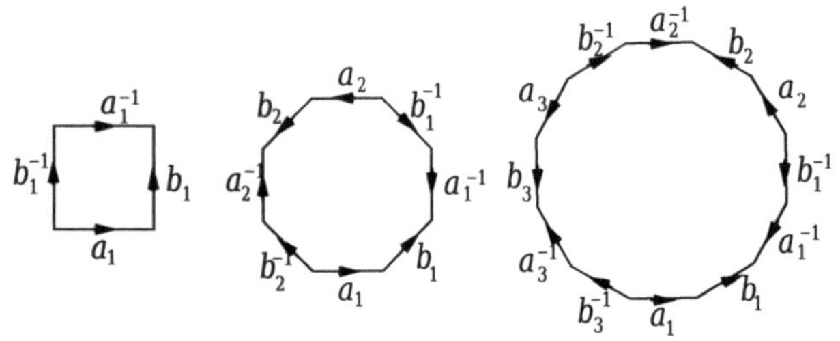

Fig. 3.2 Obtaining the genus 1, 2, and 3 surfaces from a polygon

Definition 3.1. A **simple closed curve** in a surface Σ is a loop $\gamma : S^1 \to \Sigma$ that is an embedding.

We say that a simple closed curve γ is parallel to a boundary component δ if there is an embedding of an annulus into the surface such that one boundary component of the annulus is mapped to γ and the other to δ.

Definition 3.2. A simple closed curve that is not null-homotopic and is not parallel to a boundary component is called **essential**. A simple closed curve γ is called **separating** if $\Sigma \backslash \gamma$ is not connected. Otherwise the curve is called **non-separating**.

As a corollary of the classification of surfaces, a simple closed curve on a surface is null-homotopic if and only if it bounds a disk. So a simple closed curve is not essential either if it bounds a disk, or if together with a boundary component bounds an annulus.

If the surface has a smooth structure, then every loop is homotopic to a smooth loop, so we can define the fundamental group using only smooth loops. Moreover, in the definition, the homotopies can be taken to be smooth. For this reason we will almost always consider smooth curves on surfaces.

The same considerations apply in the case of a piece-wise linear structure - the fundamental group can be described using piece-wise linear loops and piece-wise linear homotopies.

3.1.3 *The homology and cohomology groups*

The definition of homology groups

To aid reader's intuition, we introduce the **homology groups** using simplicial homology. For that we need the notion of an n-dimensional simplex, which is a space homeomorphic to

$$\{(x_0, x_1, \ldots, x_n) \in \mathbb{R}^{n+1} \,|\, x_j \geq 0, \sum_{j=0}^{n} x_j = 1\}.$$

The faces of this simplex are the subsets with the property that some of the coordinates are zero. When all but one of the coordinates are zero, the face is called a vertex.

A topological space X can be triangulated if it is obtained by gluing together finitely many simplices such that any two are glued by a homeomorphism along a face. As explained above, compact 1, 2, and 3-dimensional manifolds can be triangulated. Figure 3.3 shows how a triangulated torus could look like.

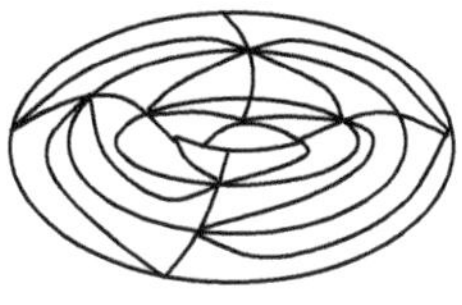

Fig. 3.3 Triangulated torus

Label the vertices of the triangulation of X as $v_1, v_2, \ldots, v_M$. If vertices v_i, v_j determine an edge, denote the edge by $< v_i, v_j >$. We view edges as oriented pairs with the convention that $< v_i, v_j > = - < v_j, v_i >$. If vertices $v_{j_0}, v_{j_1}, \ldots v_{j_n}$ determine a n-dimensional simplex of the triangulation, we denote that simplex by $< v_{j_0}, v_{j_1}, \ldots, v_{j_n} >$. We orient simplices so that

$$< v_{j_0}, v_{j_1}, \ldots, v_{j_n} > = \pm < v_{\sigma(j_0)}, v_{\sigma(j_1)}, \ldots, v_{\sigma(j_n)} >,$$

where σ is a permutation, and the sign is positive if σ is even, and negative if σ is odd.

Fix a triangulation, and let $C_n(X,\mathbb{Z})$ be the free abelian group with basis the n-dimensional simplices of the triangulation. In other words, $C_n(X,\mathbb{Z})$ is the free $\mathbb{Z}$-module with basis the n-dimensional simplices. The elements of $C_n(X,\mathbb{Z})$ are called n-dimensional chains. Define the complex of groups

$$\cdots \stackrel{\partial}{\to} C_n(X,\mathbb{Z}) \stackrel{\partial}{\to} \cdots \stackrel{\partial}{\to} C_2(X,\mathbb{Z}) \stackrel{\partial}{\to} C_1(X,\mathbb{Z}) \stackrel{\partial}{\to} C_0(X,\mathbb{Z}) \to 0, \quad (3.4)$$

in which the group homomorphisms ∂, called **boundary operators**, are defined by

$$\partial < v_{j_1}, v_{j_2}, \ldots, v_{j_n} >= \sum_j (-1)^k < v_{j_1}, \ldots, \widehat{v}_{j_k}, \ldots, v_{j_n} >$$

where the hat means that v_{j_k} is missing. Let $B_n(X,\mathbb{Z})$ be the image of $\partial : C_{n+1}(X,\mathbb{Z}) \to C_n(X,\mathbb{Z})$, and let $Z_n(X,\mathbb{Z})$ be the kernel of $\partial : C_n(X,\mathbb{Z}) \to C_{n-1}(X,\mathbb{Z})$. The elements of $B_n(X,\mathbb{Z})$ are called n-boundaries, while the elements of $Z_n(X,\mathbb{Z})$ are called n-cycles. Every boundary is a cycle, thus one can make the following definition.

Definition 3.3. The nth **homology group** of X with integer coefficients is

$$H_n(X,\mathbb{Z}) = Z_n(X,\mathbb{Z})/B_n(X,\mathbb{Z}), \quad n = 0, 1, 2, \ldots.$$

A result that lies at the heart of algebraic topology shows that the groups $H_n(X,\mathbb{Z})$, $n \geq 0$, do not depend on the triangulation; they are topological invariants of X. If X is connected, then the group $H_1(X,\mathbb{Z})$ is the abelianization of $\pi_1(X)$.

A continuous map $f : X \to X'$ induces a family of homomorphisms at the level of homology groups

$$f_* : H_n(X,\mathbb{Z}) \to H_n(X',\mathbb{Z}), \quad n = 0, 1, 2.$$

The correspondence $f \mapsto f_*$ is functorial in the sense that

$$(f \circ g)_* = f_* \circ g_* \text{ and } (1_X)_* = 1_{H_n(X,\mathbb{Z})}.$$

Example 3.2. If M is a manifold and ∂M is its boundary, then the inclusion map $i : \partial M \to M$ induces a group homomorphism $i_* : H_1(\partial M,\mathbb{Z}) \to H_1(M,\mathbb{Z})$.

One can consider the real or complex vector spaces with bases the vertices, edges, and triangles and then repeat the entire construction to obtain the **homology groups with real and complex coefficients**

$$H_n(X,\mathbb{R}) \text{ and } H_n(X,\mathbb{C}), \quad n = 0, 1, 2, \ldots.$$

We can also let the coefficients lie in the ring $\mathbb{Z}_N$ for some integer $N \geq 2$. To define the **homology groups with $\mathbb{Z}_N$ coefficients**, $H_n(X, \mathbb{Z}_N)$, take the coefficients of chains in $C_n(X, \mathbb{Z})$ modulo N.

There is also the notion of **relative homology**. To introduce it, let X be a topological space and A a subset of X. With the definition of simplicial homology given above, X is triangulated and A a subset of X obtained by selecting some of the simplices (not necessarily of the highest dimension) in the triangulation of X. Then the complex of groups (3.4) factors to a complex

$$\cdots \stackrel{\partial}{\to} C_n(X,\mathbb{Z})/C_n(A,\mathbb{Z}) \stackrel{\partial}{\to} \cdots \stackrel{\partial}{\to} C_2(X,\mathbb{Z})/C_2(A,\mathbb{Z}) \stackrel{\partial}{\to} \\ C_1(X,\mathbb{Z})/C_1(A,\mathbb{Z}) \stackrel{\partial}{\to} C_0(X,\mathbb{Z})/C_0(A,\mathbb{Z}) \to 0.$$

Using this complex we define the relative homology groups, $H_n(X, A, \mathbb{Z})$. Geometrically, the elements of $H_n(X, A, \mathbb{Z})$ are represented by chains in X which have the boundary in A.

There is a long exact sequence in homology

$$\cdots \to H_n(A,\mathbb{Z}) \to H_n(X,\mathbb{Z}) \to H_n(X,A,\mathbb{Z}) \to H_{n-1}(A,\mathbb{Z}) \to \cdots .$$

In this sequence, the maps $H_n(A,\mathbb{Z}) \to H_n(X,\mathbb{Z})$ and $H_n(X,\mathbb{Z}) \to H_n(X,A,\mathbb{Z})$ are defined by the inclusions $A \hookrightarrow X$ and $X = (X,\emptyset) \hookrightarrow (X,A)$, while the map $\delta : H_n(X,A,\mathbb{Z}) \to H_{n-1}(A,\mathbb{Z})$ is defined by mapping a chain in X, whose boundary is in A, to its boundary.

Again the situation translates to the coefficient rings $\mathbb{R}$, $\mathbb{C}$, and $\mathbb{Z}_N$. We will use these groups when X is a manifold M and A is its boundary ∂M.

Another exact sequence in homology is the **Mayer-Vietoris sequence**, which is a very efficient tool in computing homology. It is associated to a triple (X, A, B) where $A, B \subset X$ such that the interiors of A and B cover X. This sequence is

$$\cdots \to H_{n+1}(X,\mathbb{Z}) \xrightarrow{\partial_*} H_n(A\cap B,\mathbb{Z}) \xrightarrow{(i_*,i'_*)} H_n(A,\mathbb{Z}) \oplus H_n(B,\mathbb{Z}) \\ \xrightarrow{j_*-j'_*} H_n(X,\mathbb{Z}) \xrightarrow{\partial_*} H_{n-1}(A\cap B,\mathbb{Z}) \to \cdots ,$$

where $i : A \cap B \hookrightarrow A$, $i' : A \cap B \hookrightarrow B$, $j : A \hookrightarrow X$, $j' : B \hookrightarrow X$ are inclusions, and the boundary map ∂_* is defined as follows: every element x in $H_n(X)$ can be represented by a cycle which is the sum of two chains one lying in A and the other in B. These two chains have the same boundary, which is a cycle in $A \cap B$. The homology class of this cycle is $\partial_* x$. There are Meyer-Vietoris sequences corresponding to the coefficient rings $\mathbb{R}, \mathbb{C}, \mathbb{Z}_N$.

The intersection form

Let M be a compact, smooth, oriented manifold of even dimension, say $\dim M = 2k$, where k is a positive integer. Assume also that every homology class of M can be represented by an embedded smooth submanifold. Given the orientation there is a bilinear form, called **intersection form** (or **pairing**),

$$\cdot : H_k(M, \mathbb{Z}) \times H_k(M, \mathbb{Z}) \to \mathbb{Z},$$

defined as follows:

Let N and N' be k-dimensional compact, smooth, oriented submanifolds of M representing homology classes $[N]$ and $[N']$ in $H_k(M, \mathbb{Z})$. Bring N_1 and N_2 in general position so that they intersect transversally at finitely many points. At each intersection point x of N and N', consider a frame $\mathbf{v}_1, \mathbf{v}_2, \ldots, \mathbf{v}_k$ giving the orientation of N, and a frame $\mathbf{v}'_1, \mathbf{v}'_2, \ldots, \mathbf{v}'_k$ giving the orientation of N'. Then the orientation of the frame $\mathbf{v}_1, \mathbf{v}_2, \ldots, \mathbf{v}_k, \mathbf{v}'_1, \mathbf{v}'_2, \ldots, \mathbf{v}'_k$ could be that of M or not. In the first case let the sign $\sigma(x)$ of the crossing be $+1$, in the second case let $\sigma(x) = -1$. Define the intersection of $[N]$ and $[N']$ to be

$$[N] \cdot [N'] = \sum_{x \in N \cap N'} \sigma(x).$$

It turns out that this definition does not depend on the choices of representatives N and N', just on the homology classes they define.

The intersection form can be similarly defined when replacing $\mathbb{Z}$ by $\mathbb{R}$ or $\mathbb{C}$, and also for the relative homology groups $H_n(X, A)$.

The definition of cohomology groups

If in the chain complex

$$\cdots \xrightarrow{\partial} C_n(X, \mathbb{Z}) \xrightarrow{\partial} \cdots \xrightarrow{\partial} C_2(X, \mathbb{Z}) \xrightarrow{\partial} C_1(X, \mathbb{Z}) \xrightarrow{\partial} C_0(X, \mathbb{Z}) \to 0,$$

we replace each $C_n(X, \mathbb{Z})$ with its dual

$$C^n(X, \mathbb{Z}) = \mathrm{Hom}(C_n(X, \mathbb{Z}), \mathbb{Z}),$$

and the maps ∂ by their duals ∂^*, we obtain the simplicial cochain complex

$$\cdots \xleftarrow{\partial^*} C^n(X, \mathbb{Z}) \xleftarrow{\partial^*} \cdots \xleftarrow{\partial^*} C^2(X, \mathbb{Z}) \xleftarrow{\partial^*} C^1(X, \mathbb{Z}) \xleftarrow{\partial^*} C^0(X, \mathbb{R}) \leftarrow 0.$$

Let $B^n(X, \mathbb{Z})$ be the image of $\partial^* : C^{n-1}(X, \mathbb{Z}) \to C^n(X, \mathbb{Z})$, and let $Z^n(X, \mathbb{Z})$ be the kernel of $\partial^* : C^n(X, \mathbb{Z}) \to C_{n+1}(X, \mathbb{Z})$. The elements of $B^n(X, \mathbb{Z})$ are called n-coboundaries, while the elements of $Z^n(X, \mathbb{Z})$ are called n-cycles. Every coboundary is a cocycle and we have the following definition.

Definition 3.4. The nth **cohomology group** of X with integer coefficients is

$$H^n(X,\mathbb{Z}) = Z^n(X,\mathbb{Z})/B^n(X,\mathbb{Z}), \quad n = 0,1,2,\ldots.$$

Again, the cohomology groups do not depend on the triangulation of X. As with homology, we can also define the relative cohomology groups $H^n(X,A,\mathbb{Z})$. The same construction can be repeated by replacing the coefficient ring by $\mathbb{R}$, $\mathbb{C}$, and $\mathbb{Z}_N$. Note that

$$H^n(X,\mathbb{C}) = H^n(X,\mathbb{R}) \otimes_{\mathbb{R}} \mathbb{C}.$$

Cohomology with complex coefficients will play an important role in Chapter 4, the one with real coefficients in Chapter 6, while the one with coefficients in $\mathbb{Z}_N$ in Chapter 7.

Let us point out that

$$C^n(X,\mathbb{R}) \times C_n(X,\mathbb{R}) \to \mathbb{R},$$

which consists of evaluating a cochain on a chain, descends to a nondegenerate bilinear pairing

$$H^n(X,\mathbb{R}) \times H_n(X,\mathbb{R}) \to \mathbb{R}.$$

As such the vector space $H^n(X,\mathbb{R})$ is the dual of the vector space $H_n(X,\mathbb{R})$. The same holds true if we replace $\mathbb{R}$ by $\mathbb{C}$.

Poincaré duality

Let M be an n-dimensional oriented compact manifold without boundary (which we assume to be triangulable so that our definitions work). Poincaré duality states that there is a canonical isomorphism

$$H^k(M,\mathbb{Z}) \cong H_{n-k}(M,\mathbb{Z}).$$

If the n-dimensional manifold M has boundary, then we are in the instance of the Poincaré-Lefschetz duality, which in its simplest form reads

$$H^k(M,\partial M,\mathbb{Z}) \cong H_{n-k}(M,\mathbb{Z}) \quad H^k(M,\mathbb{Z}) \cong H_{n-k}(M,\partial M,\mathbb{Z}).$$

In particular the long exact sequence in homology

$$\cdots \to H_k(\partial M,\mathbb{Z}) \to H_k(M,\mathbb{Z}) \to H_k(M,\partial M,\mathbb{Z}) \to H_{k-1}(\partial M,\mathbb{Z}) \to \cdots.$$

becomes via Poincaré duality the long exact sequence in cohomology

$$\cdots \to H^{n-k-1}(\partial M, \mathbb{Z}) \to H^{n-k}(M, \partial M, \mathbb{Z}) \to H^{n-k}(M, \mathbb{Z}) \\ \to H^{n-k}(\partial M, \mathbb{Z}) \to \cdots .$$

Poincaré duality turns the Meyer-Vietoris sequence in homology into the Meyer-Vietoris sequence in cohomology

$$\cdots \to H^{n-1}(A \cap B, \mathbb{Z}) \stackrel{\partial^*}{\to} H^n(X, \mathbb{Z}) \xrightarrow{j^*-j'^*} H^n(A, \mathbb{Z}) \oplus H^n(B, \mathbb{Z}) \\ \xrightarrow{(i^*, i'^*)} H^n(A \cap B, \mathbb{Z}) \stackrel{\partial^*}{\to} H^{n+1}(X, \mathbb{Z}) \to \cdots .$$

Everything said above stays true if we replace the coefficient ring by $\mathbb{R}, \mathbb{C}$, or $\mathbb{Z}_N$.

3.1.4 *The homology groups of a surface and the intersection form*

The homology groups of a compact orientable surface

Let Σ be a compact orientable surface. Then

$$H_0(\Sigma, \mathbb{Z}) = \mathbb{Z}$$

and if Σ is closed then

$$H_2(\Sigma, \mathbb{Z}) = \mathbb{Z}.$$

We are mostly concerned with the first homology group $H_1(\Sigma, \mathbb{Z})$, for which we have

$$H_1(\Sigma_g, \mathbb{Z}) = \mathbb{Z}^{2g},$$

and if $n \geq 1$, $H_1(\Sigma_{g,n}, \mathbb{Z}) = \mathbb{Z}^{2g+n-1}$.

An element of $H_1(\Sigma, \mathbb{Z})$ can be represented as a finite set of disjoint oriented simply connected curves on Σ.

Definition 3.5. A finite set (possibly empty) of disjoint oriented essential simple closed curves is called a **multicurve**.

We denote curves and multicurves by Greek letters. Given a multicurve γ, we denote its homology class by $[\gamma]$.

Using the triangulation, we can choose the multicurve defining a homology class to consist of piece-wise linear curves. Endowing the surface with a smooth structure, we can choose these curves to be smooth. The homology class of a multicurve does not change under continuous deformations.

Example 3.3. A particular situation is that of the genus zero surface with $n+1$ holes, $\Sigma_{0,n+1}$. This surface can be represented as a disk with n holes. A basis of $H_1(\Sigma_{0,n+1}, \mathbb{Z})$ is given by the oriented curves

$$a_1, a_2, \ldots, a_n$$

depicted in Figure 3.4. The element $(k_1, k_2, \ldots, k_n) \in \mathbb{Z}^n = H_1(\Sigma_{0,n+1}, \mathbb{Z})$

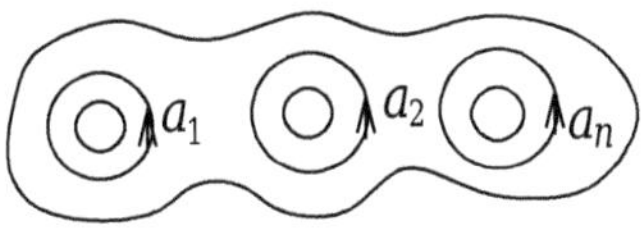

Fig. 3.4 A basis for $H_1(\Sigma_{0,n+1}, \mathbb{Z})$.

is represented by a multicurve $a_1^{k_1} a_2^{k_2} \ldots a_n^{k_n}$ which consists of $|k_1|$ parallel copies of a_1, $|k_2|$ parallel copies of a_2, ..., $|k_n|$ parallel copies of a_n, where the parallel copies are oriented the same way as a_j if $k_j \geq 0$ and have orientation opposite to that of a_j if $k_j < 0$. Figure 3.5 shows the multicurve that represents $(3, 0, -1) \in H_1(\Sigma_{0,4}, \mathbb{Z})$.

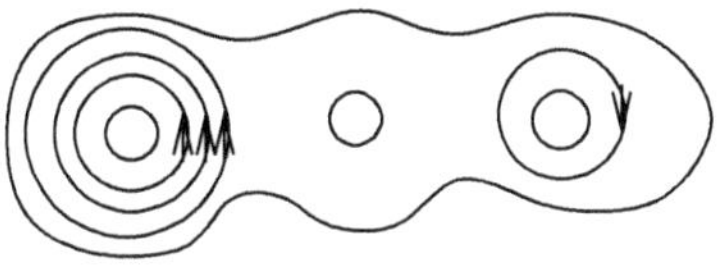

Fig. 3.5 An example of a multicurve on $\Sigma_{0,4}$.

For the surface Σ_g, $g \geq 1$, if $f : \Sigma_g \to \Sigma_g$, is a homeomorphism, then f_* is an automorphism of $H_1(\Sigma_g, \mathbb{Z}) = \mathbb{Z}^{2g}$. So f_* is represented by a matrix in $SL(2g, \mathbb{Z})$.

Unlike the case of more general topological spaces, for orientable surfaces there is no difference between working with homology with coefficients in $\mathbb{Z}$ and working with coefficients in $\mathbb{R}$ or $\mathbb{C}$. This means that for the coefficient rings $\mathbb{R}$ and $\mathbb{C}$, the above statements remain true if we replace everywhere $\mathbb{Z}$ by $\mathbb{R}$ and $\mathbb{C}$ respectively. In fact

$$H_1(\Sigma, \mathbb{R}) = H_1(\Sigma, \mathbb{Z}) \otimes_{\mathbb{Z}} \mathbb{R} \text{ and } H_1(\Sigma, \mathbb{C}) = H_1(\Sigma, \mathbb{Z}) \otimes_{\mathbb{Z}} \mathbb{C}. \quad (3.5)$$

The same holds true if we let the coefficients lie in the ring $\mathbb{Z}_N$ for some integer $N \geq 2$. In this case it suffices to take the coefficients of equivalence classes in $H_1(\Sigma, \mathbb{Z})$ modulo N, and hence

$$H_1(\Sigma_g, \mathbb{Z}_N) = \mathbb{Z}_N^{2g}.$$

The intersection form

Let Σ be a smooth **oriented** surface Σ. Given the orientation we can define the **intersection form**,

$$\cdot : H_1(\Sigma, \mathbb{Z}) \times H_1(\Sigma, \mathbb{Z}) \to \mathbb{Z}.$$

In this particular case it is defined as follows: Consider two homology classes in $H_1(\Sigma, \mathbb{Z})$. Each can be represented by oriented smooth multicurves. Perturb them so that they intersect only at finitely many points and the intersections are transverse. To every intersection point associate either $+1$ or -1 depending on whether the tangent vectors c_1', c_2' form a frame that agrees with the orientation of Σ or not. The **algebraic intersection number** of the multicurves is the sum of these numbers over all intersection points. It depends only on the homology classes of the two multicurves, a fact to be proved in Chapter 4. The value of the intersection form on the homology classes defined by the two multicurves is this algebraic intersection number.

Example 3.4. For the oriented simple closed curves c_1 and c_2 on the genus 2 surface shown in Figure 3.6, the intersection number is ± 2, where the sign depends on the orientation. If we orient the surface such that the frame at one point on the visible part of the surface is parallel to the $\partial/\partial x$, $\partial/\partial y$ frame of the plane of the paper, then the sign is negative.

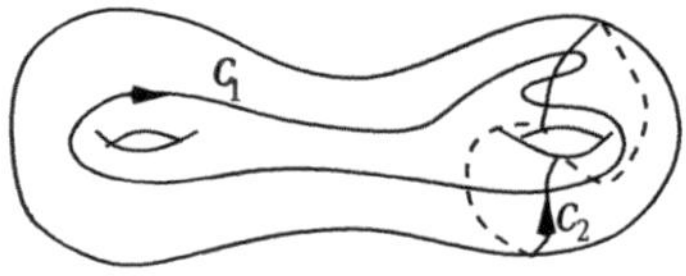

Fig. 3.6 Two intersecting curves

This method of computing the intersection number parallels that for computing the linking number in §1.2.1. In the next chapter (Theorem 4.4) we will exhibit an analytical formula for computing the algebraic intersection number, and, like for the linking number, its topological invariance is a consequence Stokes' Theorem.

Canonical bases of a closed surface

Definition 3.6. Given a closed oriented genus g surface Σ_g, a **canonical basis** $a_1, a_2, \ldots, a_g, b_1, b_2, \ldots, b_g$ of $H_1(\Sigma_g, \mathbb{Z})$ is a basis of oriented smooth essential simple closed curves such that

- a_j and a_k are disjoint for every j and k,
- b_j and b_k are disjoint for every j and k,
- a_j and b_k do not intersect if $j \neq k$,
- for each j, a_j and b_j intersect transversely at one point and the frame determined by the tangent vectors to a_j respectively b_j at the intersection point is oriented the same way as the frame that determines the orientation of Σ_g.

From the point of view of the intersection form, the elements of the canonical basis satisfy

$$a_j \cdot a_k = b_j \cdot b_k = 0, \quad a_j \cdot b_k = \delta_{jk}, \quad j, k = 1, 2, \ldots, g.$$

Remark 3.1. If $a_1, a_2, \ldots, a_g, b_1, b_2, \ldots, b_g$ is a canonical basis, then $b_1, b_2, \ldots, b_g, -a_1, -a_2, \ldots, -a_g$ is a canonical basis, where $-a_j$ is a_j with orientation reversed.

An example of a canonical basis is shown in Figure 3.7. The curves a_j, b_j, $j = 1, 2, \ldots, g$, are the generators of the fundamental group from (3.3) (to be rigorous we should transform them into loops based at the same point by connecting them by arcs to this base point).

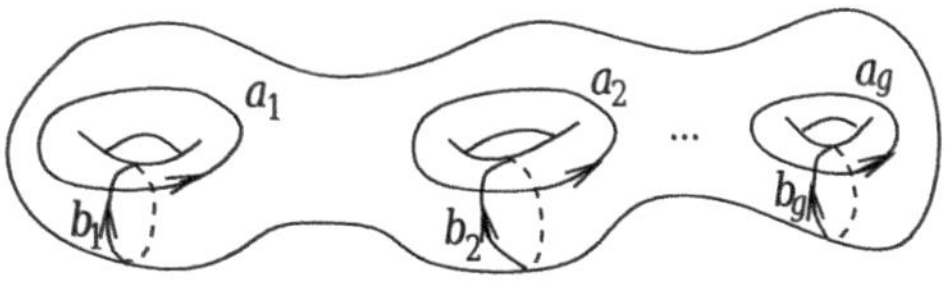

Fig. 3.7 A canonical basis of $H_1(\Sigma_g, \mathbb{Z})$

The next result shows that a surface has infinitely many canonical bases.

Proposition 3.1. *Each element of a canonical basis is non-separating. Given a non-separating simple closed curve γ, there is a canonical basis $a_1, a_2, \ldots, a_g, b_1, b_2, \ldots, b_g$ with $b_1 = \gamma$.*

Proof. If γ_1 is a separating curve, and γ_2 intersects γ_1 transversally at finitely many points, then γ_1 and γ_2 necessarily intersect at an even number of points. This is because at each intersection point γ_2 crosses from one connected component of Σ_g to the other. This proves the first half of the statement.

For the second, $\Sigma_g \backslash \gamma$ is a connected surface with two boundary components. The classification of orientable surfaces shows that $\Sigma_g \backslash \gamma$ is homeomorphic to the surface $\Sigma_{g-1,2}$. Put $\Sigma_{g-1,2}$ in standard position, as shown in Figure 3.8 in the case $g = 3$. Glue back along γ and use the canonical basis from Figure 3.7.

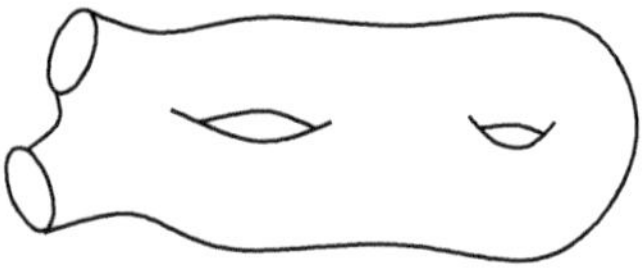

Fig. 3.8 Standard form of a surface with two boundary components

□

The intersection form is extended to the homology groups with real and complex coefficients using (3.5), to obtain bilinear forms

$$\cdot : H_1(\Sigma, \mathbb{R}) \times H_1(\Sigma, \mathbb{R}) \to \mathbb{R} \text{ and } \cdot : H_1(\Sigma, \mathbb{C}) \times H_1(\Sigma, \mathbb{C}) \to \mathbb{C}.$$

Proposition 3.2. *Let Σ_g be a genus g surface, $g \geq 1$. Then the intersection form*

$$\cdot : H_1(\Sigma_g, \mathbb{R}) \times H_1(\Sigma_g, \mathbb{R}) \to \mathbb{R}$$

is a symplectic form on $H_1(\Sigma_g, \mathbb{R})$.

Proof. The intersection form is skew symmetric. Since canonical bases exist, this form is nondegenerate. Hence it is a symplectic form. □

A canonical basis is a symplectic basis for the intersection form.

3.2 Curves on surfaces

3.2.1 *Isotopy versus homotopy*

In this section we recall a notion that is more intuitive than homotopy, that of isotopy, in which a simple closed curve slides without self-intersections into another.

Definition 3.7. Given a topological space X, an isotopy is a continuous map $H : [0, 1] \times X \to X$ such that for each $t \in [0, 1]$, $H(t, \cdot)$ is a homeomorphism.

In other words, an isotopy is a continuous path of homeomorphisms.

Definition 3.8. Two simple closed curves γ_1 and γ_2 in a path connected space X are called **ambient isotopic** if there is an isotopy $H : [0,1] \times X \to X$ such that $H|\{0\} \times X = 1_X$ and $(H|\{1\} \times X) \circ \gamma_1 = \gamma_2$ The map H is called an **ambient isotopy** between the curves.

The map

$$H_0 = H \circ (1_{[0,1]}, \gamma_1) : [0,1] \times S^1 \to X$$

defines a homotopy between γ_1 and γ_2 such that $H_0|\{s\} \times S^1 \to X$ is a homeomorphism onto the image for all s. This leads to a weaker notion.

Definition 3.9. Two simple closed curves γ_1 and γ_2 in X are called **isotopic** if there is a homotopy $H : [0,1] \times S^1 \to X$ between them such that for each $s \in [0,1]$ the map $H|\{s\} \times S^1$ is a homeomorphism onto the image. The map H is called an **isotopy** between the two curves.

The following implications are obvious

$$\text{ambient isotopic} \Rightarrow \text{isotopic} \Rightarrow \text{homotopic}.$$

It will be shown below that two homotopic simple closed curves on a surface are necessarily ambient isotopic. Hence for simple closed curves on surfaces the three notions are equivalent. We assume all maps are smooth.

Definition 3.10. The **minimal intersection number** of two simple closed curves γ_1 and γ_2 on a surface is the minimal number of intersections of two curves γ_1' and γ_2' such that γ_1' is homotopic to γ_1 and γ_2' is homotopic to γ_2. Two curves are said to have **minimal intersection** if their intersection number equals their minimal intersection number.

Definition 3.11. A **bigon** determined by two curves on a surface is a region homeomorphic to a disk, whose boundary is the union of two arcs each of which belonging to one of the curves. If the interior of the bigon does not intersect the curves, the bigon is called **minimal**.

Lemma 3.1. *Two smooth transverse simple closed curves on a surface have minimal intersection if and only if there is no embedded minimal bigon between them.*

Proof. Denote the surface by Σ. Suppose there is an embedded minimal bigon. Then by pushing the curves away from each other in this minimal

Fig. 3.9 Elimination of a bigon

bigon, as shown in Figure 3.9, we can reduce the number of intersections by 2. So the curves do not have minimal intersection.

Conversely, assume that γ_1 and γ_2 do not have minimal intersection. Then the curves necessarily intersect. If the surface is a sphere, then there is a bigon. Otherwise, the universal covering space of the surface is homeomorphic to a disk D. Consider the universal cover,

$$\pi : D \to \Sigma.$$

Let $\tilde{\gamma_1}$ and $\tilde{\gamma_2}$ be lifts of the two curves. Because the curves γ_1 and γ_2 do not have minimal intersection, $\tilde{\gamma_1}$ and $\tilde{\gamma_2}$ intersect in more than one point. Both $\tilde{\gamma_1}$ and $\tilde{\gamma_2}$ separate D, so there is a minimal bigon B between them. Project this bigon back to the surface Σ_g. In the projection the boundary of the bigon might cross itself, dividing the image of the bigon into finitely many regions. At least one of these regions is itself a minimal bigon.

Here is the explanation, which can be followed on Figure 3.10. Color all lifts of γ_1 red and all lifts of γ_2 blue (red arcs are represented by continuous lines in Figure 3.10 and blue arcs by dotted lines). Inside B there are several red and blue arcs, which divide B into finitely many polygons of two or more sides. The colors of the sides of each polygon alternate because arcs of the same color never cross. We will show that at least one of the polygons is a bigon.

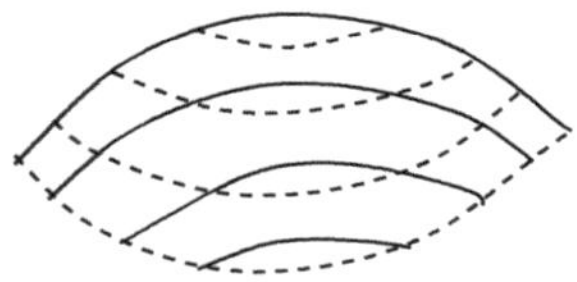

Fig. 3.10 The existence of the bigon in the lift

We argue by induction on the number of polygons. If there is just one polygon, then it is a bigon. For the induction step, pick an interior red arc, which joins two points on the blue boundary. This arc and the blue boundary determine a smaller bigon, whose interior is divided by the remaining arcs. We can pick for $\tilde{\gamma_1}$ the lift that contains this arc, and use the induction hypothesis to deduce the existence of a bigon.

The projection of this bigon is the desired minimal bigon on Σ, and we are done. □

Lemma 3.2. *For every two smooth simple closed curves in the interior of a surface there is an ambient isotopy that maps the first curve to a curve that has minimal intersection number with the second.*

Proof. Let γ_1 and γ_2 be the two curves. Standard results in differential topology (see [Brendon (1993)] and [Hirsch (1976)]) show that there is an ambient isotopy that makes γ_1 and γ_2 transverse. Because the images of γ_1 and γ_2 are compact and they intersect transversely at each intersection point, they intersect only finitely many times. If γ_1 and γ_2 do not have minimal intersection, then by Lemma 3.1 there is an embedded minimal bigon between them. By eliminating the minimal bigon we can reduce their intersection number. This process can be repeated only finitely many times, so after a few steps we must reach the minimal intersection number of the two curves. □

Lemma 3.3. *Two disjoint homotopic essential simple closed curves on a compact orientable surface bound an annulus.*

Proof. Because any hole on a surface can be pushed off an annulus, we may assume that the surface is closed. Let therefore γ_1 and γ_2 be two homotopic curves on the surface Σ_g, $g \geq 0$. The curves are homologous, which means that they bound a subsurface $\Sigma \subset \Sigma_g$. We claim that either Σ or its complement is homeomorphic to an annulus. Assume that this is not true. We distinguish two situations: where the curves are null-homologous, and where they are not.

If the curves are null-homologous then each of them bounds a 2-cycle, so each cuts the surface into two connected components as can be seen in Figure 3.11. Glue a disk to γ_2 so as to make it null-homotopic, and collapse the part of the surface bounded by γ_2 that does not contain γ_1 to a point. We obtain a new surface $\Sigma_{g'}$ with $g' < g$. In this surface γ_1 is not null-homotopic, because for example it can be expressed as

$$\gamma_1 = a_1 b_1 a_1^{-1} b_1^{-1} \cdots a_k b_k a_k^{-1} b_k^{-1}$$

for some $k < g'$ in the fundamental group of $\Sigma_{g'}$. This is impossible because γ_1 and γ_2 were homotopic in the original surface, hence they must be homotopic in $\Sigma_{g'}$.

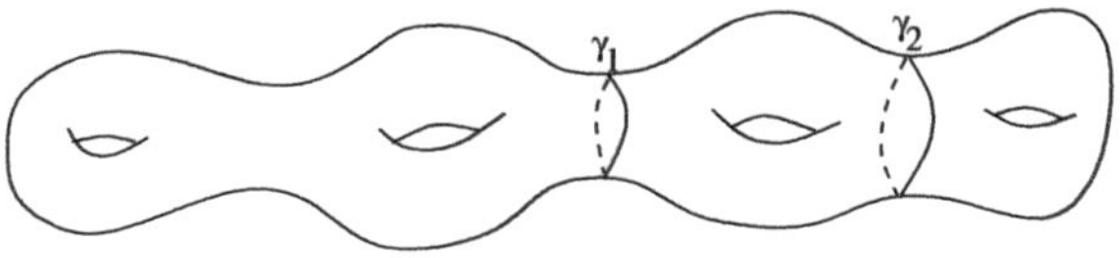

Fig. 3.11 Two disjoint null homologous simple closed curves

If the curves are not null-homologous, then neither separates the surface. Being homologous, they can be made to look as in Figure 3.12 using the classification of surfaces. Shrink γ_2 and the dotted curve to a point to make the handle containing γ_2 disappear. Now we are again in the previous situation, with γ_1 a separating curve that is not null-homotopic. This is again impossible. Hence our assumption was false and the conclusion follows.

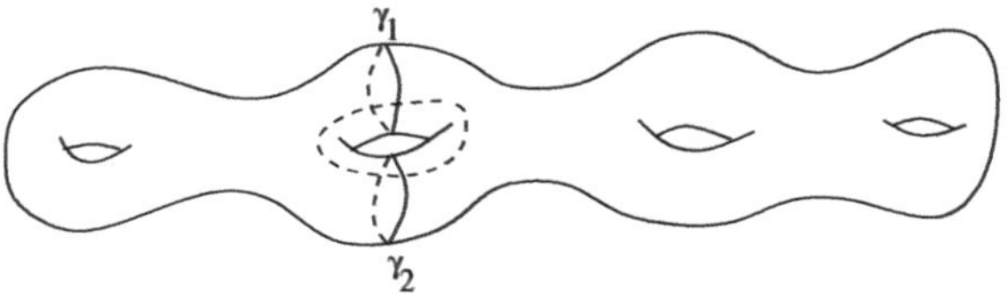

Fig. 3.12 Two disjoint homologous simple closed curves

□

Theorem 3.2. *Two smooth essential simple closed curves in the interior of a compact orientable surface are homotopic if and only if they are ambient isotopic.*

Proof. Of course, we only need to show the direct implication, namely that if the two curves are homotopic then they are ambient isotopic. The minimal intersection number of the two curves is 0 because there is a homotopy from the first curve to a parallel copy of the second. By Lemma 3.2 there is an ambient isotopy that makes the two curves disjoint. By Lemma 3.3 they bound an annulus. Add to this annulus tubular neighborhoods of the curves, so that the two curves lie inside an annulus as shown in Figure 3.13. Note that because the curves are homologous, they are oriented the same way. Without a loss of generality we may assume that the annulus is

$$A = \{(re^{2\pi it} \,|\, 1 \leq r \leq 4,\, 0 \leq t \leq 1\},$$

and the curves are

$$\gamma_1(t) = 2e^{2\pi it}, \quad \gamma_2(t) = 3e^{2\pi it}, \quad 0 \leq t \leq 1.$$

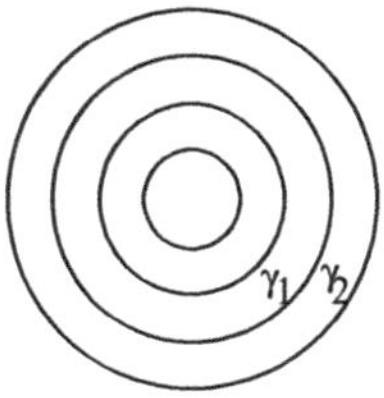

Fig. 3.13 Two disjoint homotopic simple closed curves in an annulus

Define the ambient isotopy H to be the identity outside of A, while on A,

$$H(re^{2\pi it}) = f(r)e^{2\pi it}, \quad 1 \leq r \leq 4, 0 \leq t \leq 1,$$

where f is a smooth increasing function such that $f(1) = 1$, $f(2) = 3$, $f(4) = 4$. The theorem is proved. □

3.2.2 *Multicurves on a torus*

The torus is the closed genus 1 surface Σ_1. We will use it repeatedly in this book to illustrate our ideas, and for that reason we give it a particular treatment. We identify Σ_1 with the quotient of $\mathbb{R}^2$ under the map

$$\pi : \mathbb{R}^2 \to \mathbb{R}^2/\mathbb{Z}^2.$$

The map π defines the universal covering of the torus. If we make the identification $\Sigma_1 = S^1 \times S^1$, where $S^1 = \{z \,|\, |z| = 1\}$, then the covering map is

$$\pi : \mathbb{R}^2 \to S^1 \times S^1, \quad \pi(s,t) = (e^{2\pi is}, e^{2\pi it}).$$

Formula (3.3) applied to the case where $g = 1$ shows that the fundamental group is abelian, and so it is isomorphic to the first homology group with integer coefficients. We have

$$\pi_1(\Sigma_g) = H_1(\Sigma_g, \mathbb{Z}) = \mathbb{Z}^2.$$

The standard orientation of $\mathbb{R}^2$ induces an orientation of Σ_1. Any pair of oriented simply connected curves that have algebraic intersection number 1 defines a basis for $H_1(\Sigma_g, \mathbb{Z})$ and gives a pair of generators of the fundamental group. Fix a choice of two such curves, and define the projection map π so that the curves are the images of the x- respectively y-axis.

Definition 3.12. For $p, q \in \mathbb{Z}$ such that $\gcd(p, q) = 1$, we denote by (p, q) the oriented simple closed curve on the torus that is the image through the projection π of the line of slope q/p passing through the origin in the universal covering space $\mathbb{R}^2$, oriented from the $(0,0)$ towards the point (p,q). For general $p, q \in \mathbb{Z}$, we denote by (p,q) the oriented multicurve consisting of n parallel copies of the curve $(p/n, q/n)$, where $n = \gcd(p,q)$. We call this the **(multi)curve** (p,q).

Figure 3.14 shows schematically the curve $(4,3)$, as seen in the universal covering space of the torus.

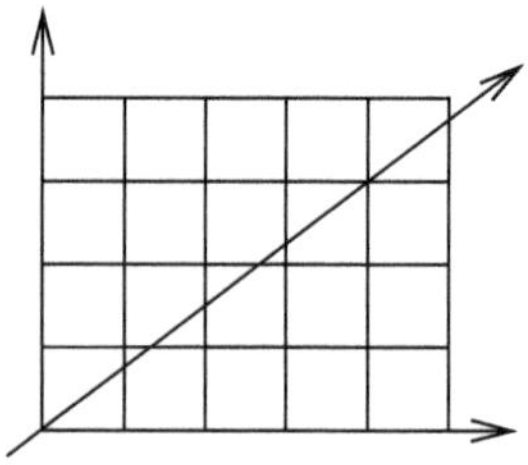

Fig. 3.14 The simple closed curve $(4,3)$

Theorem 3.3. *Every element of $H_1(\Sigma_1, \mathbb{Z})$ can be represented uniquely by an oriented multicurve of the form (p,q). Moreover, the map that associates to each $(p,q) \in \mathbb{Z}^2$ the multicurve (p,q) is an isomorphism between $\mathbb{Z}^2$ and $H_1(\Sigma_1, \mathbb{Z})$.*

Proof. Choose $\pi((0,0))$ to be the basepoint of the fundamental group. It is standard (see [Munkres (2000)]) that the fundamental group can be put in one-to-one correspondence with $\mathbb{Z}^2 = \pi^{-1}(\pi((0,0)))$. Thus $\pi_1(\Sigma_1) = H_1(\Sigma_1, \mathbb{Z}) = \mathbb{Z}^2$. We deduce that every homology class can be represented as the image of the line segment that joins $(0,0)$ with some $(p,q) \in \mathbb{Z}^2$. If $gcd(p,q) = n$, then in $H_1(\Sigma_1, \mathbb{Z})$, $(p,q) = n(p/n, q/n)$, and we can turn the latter into a multicurve by pushing the n curves away from each other. □

A canonical basis consists of a pair of curves $a_1 = (p,q)$, $b_1 = (p', q')$ such that $a_1 \cdot b_1 = pq' - qp' = 1$. Two examples are shown in Figure 3.15 The standard choice is $a_1 = (1,0)$, $b_1 = (0,1)$, and we always represent this basis as in the example on the left of Figure 3.15.

Understanding an arbitrary surface using the torus

If we remove a disk from the torus Σ_1, we obtain the torus with one boundary component $\Sigma_{1,1}$. Because the hole can slide freely on the surface of the

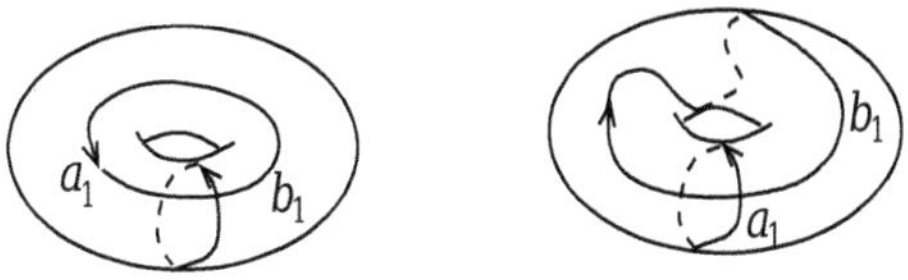

Fig. 3.15 Two canonical bases on the torus

torus, most of what was said above about the torus remains true for the punctured torus. In particular,

$$H_1(\Sigma_{1,1}, \mathbb{Z}) = \mathbb{Z}^2.$$

Cap the hole by the disk to produce the torus Σ_1, and fix a canonical basis of Σ_1. The inclusion map $i : \Sigma_{1,1} \hookrightarrow \Sigma_1$ induces an isomorphism $i_* : H_1(\Sigma_{1,1}, \mathbb{Z}) \to H_1(\Sigma_1, \mathbb{Z})$. This isomorphism is canonical, because any two locations of the hole are equivalent, for the hole can be moved continuously from one location to another without crossing the curve, as suggested in Figure 3.16.

The inclusion i maps the canonical basis of $H_1(\Sigma_1, \mathbb{Z})$ to a basis of $H_1(\Sigma_{1,1}, \mathbb{Z})$. We call it a canonical basis of the punctured torus. Using this basis, we can represent every homology class in $H_1(\Sigma_{1,1}, \mathbb{Z})$ by a multicurve (p, q), $p, q \in \mathbb{Z}$. This is done by representing the multicurve on the torus without boundary, then removing a disk that does not intersect it.

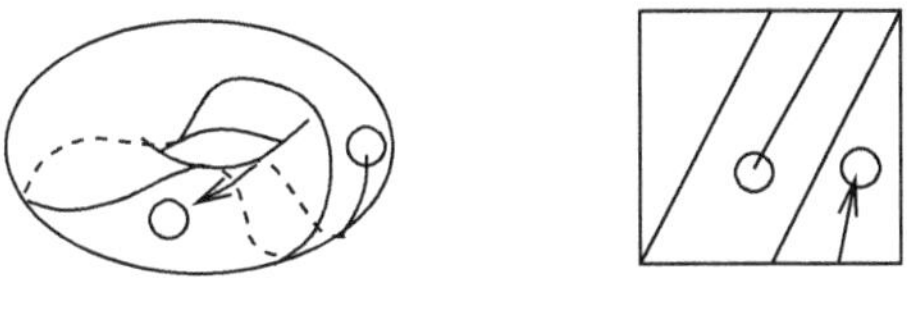

Fig. 3.16 Sliding the hole without crossing the curve

For $g > 1$, the closed genus g surface Σ_g can be obtained by gluing g punctured tori to a sphere with g open disks removed, as suggested in Figure 3.17. In fact, given a canonical basis a_j, b_j, $j = 1, 2, \ldots, g$, of Σ_g, for each j a regular neighborhood of the union of curves a_j, b_j is a punctured torus. If we choose the g regular neighborhoods to be disjoint, then the complement of the union of these tori is a sphere with g punctures.

Note that a_j, b_j is a canonical basis of the jth punctured torus. Then the multicurves that lie entirely in the jth punctured torus can be parametrized by pairs of the form (p_j, q_j), $p_j, q_j \in \mathbb{Z}$. In this setting, a pair $(\mathbf{p}, \mathbf{q})$,

$\mathbf{p}, \mathbf{q} \in \mathbb{Z}^g$ defines an oriented multicurve on Σ_g. We call it the **multicurve** $(\mathbf{p}, \mathbf{q})$ on the surface Σ_g.

Remark 3.2. The representation of a multicurve as $(\mathbf{p}, \mathbf{q})$ depends on the choice of the canonical basis. Not every multicurve that represents a nontrivial homology class is of the form $(\mathbf{p}, \mathbf{q})$. This holds true only for the torus.

Proposition 3.3. *The map that associates to each pair* $(\mathbf{p}, \mathbf{q}) \in \mathbb{Z}^{2g}$ *the multicurve* $(\mathbf{p}, \mathbf{q})$ *defines an isomorphism between* $\mathbb{Z}^{2g}$ *and* $H_1(\Sigma_g, \mathbb{Z})$.

Proof. The Mayer-Vietoris sequence in homology implies that

$$H_1(\Sigma_g, \mathbb{Z}) = \oplus_{j=1}^{g} H_1(\Sigma_{1,1}, \mathbb{Z}).$$

Every homology class in $H_1(\Sigma_g, \mathbb{Z})$ can then be represented by an oriented multicurve whose curves lie entirely in some of the punctured tori in the desired way. □

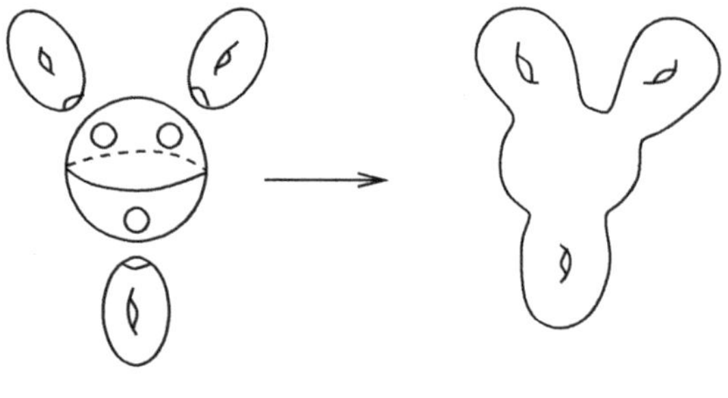

Fig. 3.17 Gluing punctured tori to a puncture sphere

3.2.3 *The first homology group of a surface as a group of curves*

The homology group with coefficients in $\mathbb{Z}$

In this section all curves, surfaces, and maps, including isotopies, are smooth. Cycles in $H_1(\Sigma_g, \mathbb{Z})$ can be represented by multicurves on Σ_g, hence $H_1(\Sigma_g, \mathbb{Z})$ is a group of equivalence classes of multicurves. To facilitate the introduction of group algebras later on, we write the operation in $H_1(\Sigma_g, \mathbb{Z})$ **multiplicatively**.

For a heuristical analysis, let us look at the torus Σ_1, and recall the considerations from §3.2.2. Examining closely the multiplication in homology:

$$(1,0)\cdot(0,1) = (1,1)$$

shown pictorially in Figure 3.18, we notice that the curve $(1,1)$ can be obtained by drawing the curve $(1,0)$ on top of the curve $(0,1)$, erasing the crossing, and joining the curves so that the orientations agree.

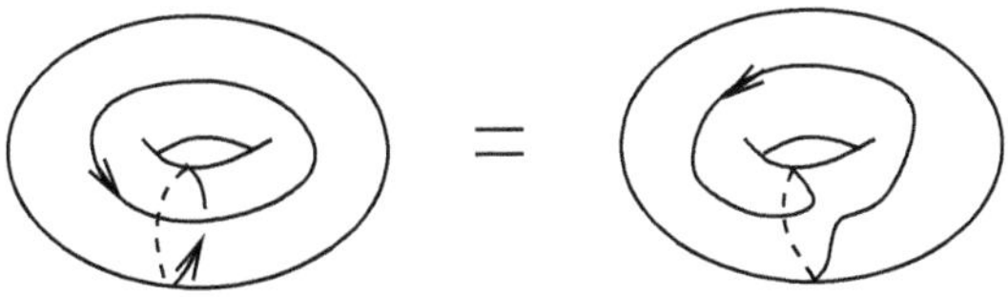

Fig. 3.18 The multiplication of $(1,0)$ and $(0,1)$

Definition 3.13. Given two oriented curves on a surface that cross transversally at some point, the procedure of erasing the crossing and then joining the curves so that the orientations agree and the curves no longer cross at that point is called **smoothing of the crossing**.

This operation is described in Figure 3.19.

Fig. 3.19 The smoothing of a crossing

Looking at other examples on the torus we deduce that the multiplication law in $H_1(\Sigma_1, \mathbb{Z})$,

$$(p,q)\cdot(p',q') = (p+p', q+q')$$

can be described by drawing the multicurve (p,q) on top of the multicurve (p',q'), then smoothing all crossings. Not surprisingly, the same works for all surfaces.

There is a *caveat* – the multicurves must have minimal intersection number. Otherwise, by Lemma 3.1 bigons are present. Sometimes these bigons disappear in the process of smoothing, sometimes they introduce circles that bound disks, as shown in Figure 3.20. We call such circles trivial and delete them.

Therefore, to compute the product of two homology classes in $H_1(\Sigma_g, \mathbb{Z})$, we represent them as multicurves, draw the first on top of the second and then smoothen all crossings and delete all trivial circles that arise. The homology class of the resulting multicurve is the product of the two homology classes.

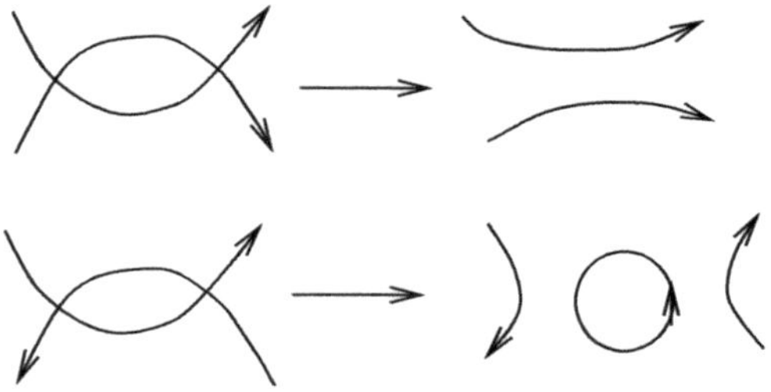

Fig. 3.20 The elimination of a bigon by smoothing crossings

The representation of a homology class by a multicurve is not unique. Figure 3.21 shows two multicurves that represent the same homology class, one consisting of a single curve, and one consisting of two curves.

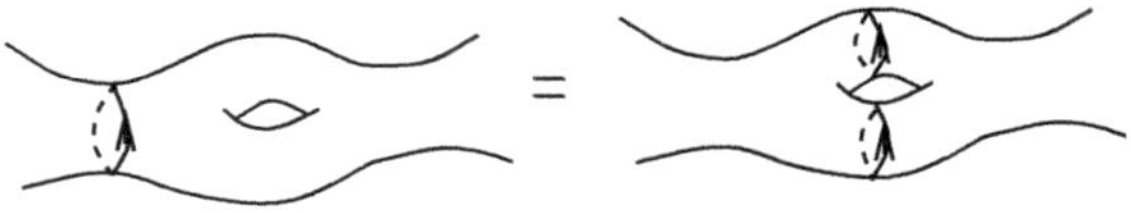

Fig. 3.21 Two multicurves representing the same homology class

If we allow the curves to move on the surface by isotopies[1], then we can transform the multicurve on the left into the one on the right by applying the operation of smoothing crossings and deleting trivial circles. This is done in Figure 3.22.

We are now in position to state rigorous results. First, some notation. Let $\mathrm{Mul}(\Sigma)$ be the set of oriented multicurves on the orientable compact surface Σ. Recall that $\Sigma = \Sigma_{g,n}$ for some $g \geq 0, n \geq 0$. Endow this set with multiplication defined by drawing two multicurves one on top of the other, smoothing all crossings and deleting all trivial circles that arise. The multiplication on $\mathrm{Mul}(\Sigma)$ has all properties of an abelian group structure, except for the existence of the inverse. The identity element is the **empty** multicurve $\emptyset$ (the one with no components).

[1]Isotopies not ambient isotopies!

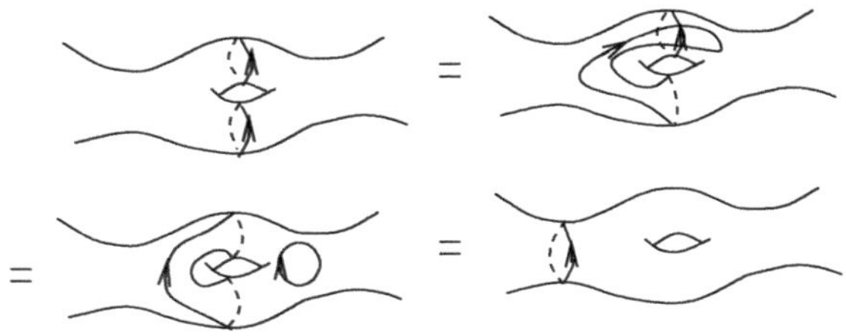

Fig. 3.22 Transforming two multicurves one into the other

On Mul(Σ) introduce the relation $\gamma \sim \gamma'$ if γ can be transformed into γ' by a finite sequence of the following moves:

- transform one of the curves of γ under an isotopy (which can allow it to cross other curves of γ);
- smoothen a crossing;
- add or delete a trivial circle.

Lemma 3.4. *The relation $\sim$ is an equivalence relation.*

Proof. The fact that $\gamma \sim \gamma'$ and $\gamma' \sim \gamma''$ implies $\gamma \sim \gamma''$ follows from the definition.

Let us show that $\gamma \sim \gamma'$ implies $\gamma' \sim \gamma$. When γ is transformed under isotopy, each time it crosses γ' it introduces a bigon. The smoothings of crossings amount to the elimination of bigons, as in Figure 3.20. The moves from Figure 3.20 are reversible, since we can cross back the curves and then smoothen the crossings to reverse the isotopy. Hence the conclusion. □

We set

$$\mathfrak{M}(\Sigma) = \mathrm{Mul}(\Sigma)/\sim .$$

Theorem 3.4. *The multiplication rule on Mul(Σ) descends to $\mathfrak{M}(\Sigma)$ inducing an abelian group structure. As a group, $\mathfrak{M}(\Sigma)$ is isomorphic to $H_1(\Sigma, \mathbb{Z})$, with isomorphism defined by the map that associates to a multicuve its homology class.*

Proof. The multiplication rule is compatible with the quotient map because both are defined by smoothings of crossings and deletions of trivial circles.

Consider the map

$$F : \mathrm{Mul}(\Sigma) \to H_1(\Sigma, \mathbb{Z}),$$

which associates to each multicurve its homology class. The map F is onto. It also satisfies

$$F(\gamma\gamma') = F(\gamma)F(\gamma').$$

Because the homology class is not changed under isotopy or under smoothings of crossings, F descends to a map

$$\hat{F} : \mathfrak{M}(\Sigma) \to H_1(\Sigma, \mathbb{Z}).$$

Let us show first that if a multicurve γ is null-homologous, then in $\mathrm{Mul}(\Sigma)$ it is equivalent to the empty multicurve (the one with no components).

If γ consists of just one component, then it is a separating curve on the surface. Applying the computation from Figure 3.22 in reverse, then applying the computation directly, but on the other side, we can move the null homologous curve past a handle on the surface. Applying the computation several times, we can move the curve past all handles, transforming it into a trivial circle. This circle can then be deleted.

Next assume that the null-homologous multicurve γ consists of several curves, not all of them null-homologous themselves. Then two curves of γ are part of the boundary of the same connected 2-chain in $C_2(\Sigma, \mathbb{Z})$. Trace a path in the 2-chain that connects them, as shown in Figure 3.23. Consider the isotopy of one of the curves obtained by shrinking this path to a point, and then crossing the other curve into a bigon. This bigon is necessarily of the form shown at the bottom of Figure 3.20, because the curves are boundary components of the same connected 2-chain. Consequently, with the elimination of the bigons we transform the curves into a single curve, thus reducing the number of curves. Repeating we can transform γ into a multicurve, all of whose components are null-homologous. But then γ can be further transformed into a multicurve with no components.

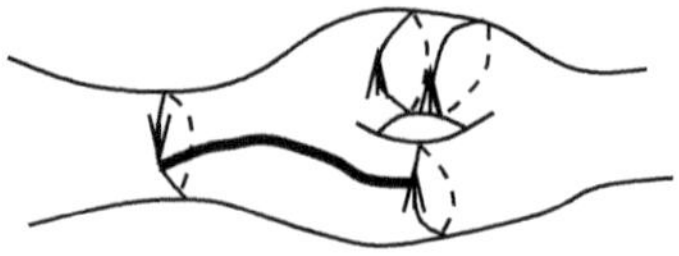

Fig. 3.23 A path connecting two simple closed curves

As a byproduct of this discussion, we obtain that if γ^{-1} is the multicurve γ with all orientations reversed, then $\gamma\gamma^{-1}$ is equivalent to the empty multicurve in $\mathfrak{M}(\Sigma)$. Hence $\mathfrak{M}(\Sigma)$ is a group. Consequently, $\hat{F}$ is a group homomorphism. And we have just shown that the kernel of $\hat{F}$ is $\{0\}$. Then $\hat{F}$ is an isomorphism and the theorem is proved. □

Remark 3.3. As a corollary of the proof, if two multicurves are equivalent, then one can be transformed into the other by isotopies, the smoothing of crossings in pairs that arise from the introduction of bigons, and addition or deletion of trivial circles.

If for a closed surface Σ_g, $g \geq 1$, we choose a canonical basis a_j, b_j, $j = 1, 2, \ldots, g$, of its first homology group, then we can identify $H_1(\Sigma_g, \mathbb{Z})$ with $\mathbb{Z}^{2g}$. As explained in the previous section, each homology class $(\mathbf{p}, \mathbf{q})$ can be represented by the oriented multicurve denoted also by $(\mathbf{p}, \mathbf{q})$. So the multicurves $(\mathbf{p}, \mathbf{q})$, $\mathbf{p}, \mathbf{q} \in \mathbb{Z}^g$ parametrize the equivalence classes of the group $\mathfrak{M}(\Sigma)$, and the multiplication rule in $\mathfrak{M}(\Sigma)$ reads

$$(\mathbf{p}, \mathbf{q})(\mathbf{p}', \mathbf{q}') = (\mathbf{p} + \mathbf{p}', \mathbf{q} + \mathbf{q}').$$

It is important to keep in mind that this identification of $\mathfrak{M}(\Sigma)$ with $\mathbb{Z}^{2g}$, which allows us to single out the $(\mathbf{p}, \mathbf{q})$-multicurves, depends on the choice of the canonical basis!

The homology group with coefficients in $\mathbb{Z}_N$

We also need to understand the homology group $H_1(\Sigma, \mathbb{Z}_N)$, because as we will see in Chapter 5, it is related to a finite Heisenberg group. In this case we introduce the relation $\sim_N$ on Mul(Σ) by $\gamma \sim_N \gamma'$ if and only if $\gamma' = \gamma(\gamma'')^N$ for some multicurve γ''. It is not hard to realize that $\sim_N$ descends to an equivalence relation on $\mathfrak{M}(\Sigma)$, because $\gamma' = \gamma(\gamma'')^N$ implies $\gamma = \gamma'(\gamma''^{-1})^N$, where γ''^{-1} is the multicurve γ'' with all orientations reversed. Set

$$\mathfrak{M}_N(\Sigma) = \mathfrak{M}(\Sigma)/\sim_N.$$

Theorem 3.5. *The multiplication on $\mathfrak{M}(\Sigma)$ descends to a multiplication on $\mathfrak{M}_N(\Sigma)$, and $\mathfrak{M}_N(\Sigma)$ endowed with this multiplication is a group. Moreover, the map Mul$(\Sigma) \to H_1(\Sigma, \mathbb{Z}_N)$ that associates to a multicurve its homology class descends to a group isomorphism $\mathfrak{M}_N(\Sigma) \to H_1(\Sigma, \mathbb{Z}_N)$.*

Proof. The conclusion follows from

$$H_1(\Sigma, \mathbb{Z}_N) = H_1(\Sigma, \mathbb{Z}_N)/NH_1(\Sigma, \mathbb{Z}_N).$$

□

3.2.4 *Links in the cylinder over a surface*

Definition 3.14. A **link** in a 3-dimensional manifold is an embedding of a disjoint union of finitely many circles. If the link is the embedding of just one circle, it is called a **knot**.

To avoid the case of "wild" knots and links, which leave the realm of combinatorial topology, we endow the 3-dimensional manifold with a smooth structure and allow only smooth knots or links.

The next result relates isotopy to ambient isotopy in the case of multicurves. We embed a surface Σ in the cylinder $\Sigma \times [0,1]$ as the surface $\Sigma \times \{1/2\}$. Under this embedding, a multicurve on Σ becomes a link in $\Sigma \times [0,1]$.

Proposition 3.4. *Let γ and γ' be two smooth multicurves in a compact orientable smooth surface Σ, which are also viewed as oriented links in $\Sigma \times [0,1]$. Then γ can be transformed into γ' by a sequence of smooth isotopies in Σ of its components if and only if γ can be transformed into γ' by a smooth ambient isotopy of $\Sigma \times [0,1]$.*

It might not be true that γ and γ' are ambient isotopic in Σ, because one of the isotopies might take one of the components of γ over another. An example is when γ consists of n parallel components and γ' is obtained by permuting these components.

Proof. Let c be a component of γ and let N_c be a tubular neighborhood of c in Σ that does not intersect the other components of γ. We claim that there is an ambient isotopy that equals the identity map outside $N_c \times [0,1]$ and that moves c to $\Sigma \times \{\epsilon_c\}$ where ϵ_c is an arbitrary number in $(0,1)$. Indeed, modulo a homeomorphism

$$N_c = \{z \in \mathbb{C} \,|\, 1 \leq |z| \leq 3\}, \quad c = \{z \in \mathbb{C}, |z| = 2\}.$$

Define $H : N_c \times [0,1] \times [0,1] \to N_c \times [0,1]$, such that for a fixed θ and fixed s we have

$$H(re^{i\theta}, t, s) = (e^{i\theta}, t, s), \quad r \in \{1,3\}, t \in \{0,1\}, s \in [0,1]$$
$$H(2e^{i\theta}, 1/2, s) = (2e^{i\theta}, s/2 + (1-s)\epsilon_c),$$

then extend H smoothly. H is the identity on $\partial N_c \times [0,1]$ so it can be extended to $\Sigma \times [0,1]$. By perturbing it slightly along $\partial N_c \times [0,1]$, we can make sure that this extension is also smooth.

We conclude that there is an ambient isotopy that moves the components of γ at different levels ϵ_c in the cylinder. Transform γ' by an ambient isotopy so that its components are at the same level with the components of γ that are transformed into them. Let H_c, where c ranges among the components of γ, be the set of isotopies that change γ to γ'. By Theorem 3.2, we may assume that H_c is an ambient isotopy in $\Sigma \times \{\epsilon_c\}$. We want to construct an ambient isotopy whose restriction to each slice $\Sigma \times \{\epsilon_c\}$ is H_c. For this it suffices to construct for each small $\delta > 0$ an ambient isotopy of $\Sigma \times [\epsilon_c - \delta, \epsilon_c + \delta]$ that restricts to H_c and is identity on the boundary. This can be done by setting

$$H : \Sigma \times [\epsilon_c - \delta, \epsilon_c + \delta] \times [0,1] \to \Sigma \times [\epsilon_c - \delta, \epsilon_c + \delta],$$
$$H(x, \epsilon_c \pm t\delta, s) = H_c(x, (1-t)s) \quad s, t \in [0,1].$$

Again, there is a problem with H being smooth along $\partial\Sigma \times [\epsilon_c - \delta, \epsilon_c + \delta]$, but again this can be achieved by perturbing it. The lemma is proved. □

Recall the basis $a_1, a_2, \ldots, a_g$ of $H_1(\Sigma_{0,g+1}, \mathbb{R})$ introduced in Example 3.3 and depicted in Figure 3.4 (in the present situation $n = g$). The curves a_j have the property that each is isotopic to a boundary component of $\Sigma_{0,g+1}$. Because $\Sigma_{0,g+1}$ is a strong deformation retract of H_g, the same curves are a basis of H_g.

Definition 3.15. A basis of $H_1(H_g, \mathbb{Z})$ consisting of oriented simple closed curves $a_1, a_2, \ldots, a_g$ is called **canonical** if there is a homeomorphism of H_g to $\Sigma_{0,g+1} \times [0,1]$ such that the images of $a_1, a_2, \ldots, a_g$ lie in $\Sigma_{0,g+1} \times \{1/2\}$, and in this surface they are isotopic to g of the boundary components.

Example 3.5. An example consists of the curves $a_1, a_2, \ldots, a_g$ in Figure 3.7, after we fill in the handlesbody in the standard way.

3.3 The mapping class group of a surface

In this section we review the facts about the mapping class group of a surface that are needed in this book. For a detailed exposition of the subject we recommend [Farb and Margalit (2011)].

3.3.1 *The definition of the mapping class group*

Let Σ be an **oriented smooth compact surface**. Denote by $\mathrm{Diff}^+(\Sigma, \partial\Sigma)$ the group of **orientation preserving diffeomorphisms** that map Σ to itself and are identity on the boundary. Endow $\mathrm{Diff}^+(\Sigma, \partial\Sigma)$ with the compact-open topology, which is the smallest topology in which the sets $\{h \,|\, h(K) \subset U\}$ are open, where K ranges among all compact and U among all open subsets of Σ. This turns $\mathrm{Diff}^+(\Sigma, \partial\Sigma)$ into a topological group, the operations of composition and taking the inverse being continuous.

Two diffeomorphisms h_1 and h_2 are called **isotopic** if there is a path that connects them in $\mathrm{Diff}^+(\Sigma, \partial\Sigma)$. This means that there is a smooth **isotopy** $H : [0,1] \times \Sigma \to \Sigma$ such that $H(0,\cdot) = h_1$ and $H(1,\cdot) = h_2$. Let $\mathrm{Diff}_0(\Sigma, \partial\Sigma)$ be the path connected component of the identity map, it consists of the diffeomorphisms that are isotopic to the identity. Then $\mathrm{Diff}_0(\Sigma, \partial\Sigma)$ is a closed subgroup of $\mathrm{Diff}^+(\Sigma, \partial\Sigma)$.

Definition 3.16. The mapping class group of of the oriented smooth compact surface Σ is

$$\mathrm{MCG}(\Sigma) = \mathrm{Diff}^+(\Sigma, \partial\Sigma)/\mathrm{Diff}_0(\Sigma, \partial\Sigma).$$

The mapping class group of a surface is sometimes referred to as the **modular group**.

So the mapping class group is the group of isotopy classes of diffeomorphisms of the surface. We are only interested in the mapping class group $\mathrm{MCG}(\Sigma_g)$ of a closed genus g surface Σ_g, $g \geq 1$, but in order to study this group we will need to look at surfaces with boundary as well, in which case the mapping class group is defined using homeomorphisms (and isotopies of homeomorphisms) that are fixed on the boundary. We need surfaces with marked points too, with the marked points interpreted also as punctures. If we mark n points on Σ_g, we denote by $\mathrm{PMCG}(\Sigma_g, n)$ the subgroup of $\mathrm{MCG}(\Sigma_g)$ consisting of diffeomorphisms that keep the n points fixed. For two different choices of the n marked points, the corresponding groups are isomorphic.

A theorem proved by Munkres [Munkres (1960)], Smale, and Whitehead [Whitehead (1961)] shows that every orientation preserving homeomorphism of Σ is isotopic to a diffeomorphism of Σ. As a corollary, if $\mathrm{Homeo}^+(\Sigma, \partial\Sigma)$ denotes the group of orientation preserving homeomorphisms of Σ that restrict to the identity on the boundary, and $\mathrm{Homeo}_0(\Sigma, \partial\Sigma)$ is the path connected component of the identity, then

$$\mathrm{MCG}(\Sigma) = \mathrm{Homeo}^+(\Sigma, \partial\Sigma)/\mathrm{Homeo}_0(\Sigma, \partial\Sigma).$$

So when analyzing the mapping class group, we can work with either diffeomorphisms or homeomorphisms.

Example 3.6. An example of a nontrivial element of the mapping class group is the **Dehn twist**. It is defined by an oriented smooth simple closed curve γ in the surface that does not bound a disk (and as such is homotopically nontrivial). Consider a closed annulus A that is a tubular neighborhood of γ, and give it coordinates $(e^{2\pi is}, t)$, $0 \leq s < 1$, $0 \leq t \leq 1$, such that the oriented curve γ is parametrized by $(e^{2\pi is}, 1/2)$. The Dehn twist about γ is defined to be the isotopy class of the homeomorphism of the surface that is identity outside of A, and on A it is

$$(e^{2\pi is}, t) \mapsto (e^{2\pi i(s+t)}, t).$$

The local behavior of the Dehn twist in the neighborhood of c is pictured in Figure 3.24.

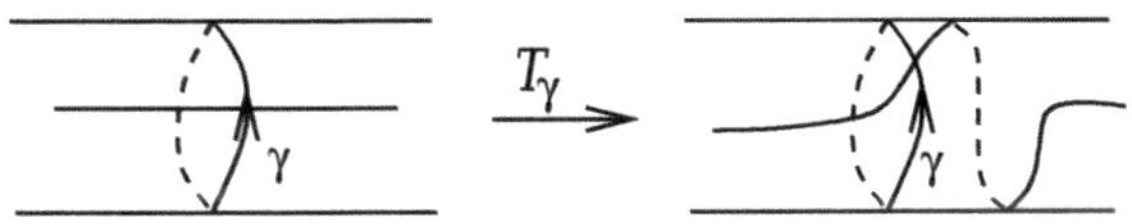

Fig. 3.24 The Dehn twist

Max Dehn [Dehn (1987)] discovered that the mapping class group of an oriented smooth closed genus g surface Σ_g, $g \geq 1$, is generated by finitely many Dehn twists about nonseparating curves. Later W.B.R. Lickorish [Lickorish (1962)] refined this result by showing that the Dehn twists about the $3g - 1$ curves $a_1, a_2, \ldots, a_g, b_1, b_2, \ldots, b_g, c_1, c_2, \ldots, c_{g-1}$ shown in Figure 3.25 are a family of generators for $\text{MCG}(\Sigma_g)$. The result was further improved by S. Humphries [Humphries (1977)], who showed that the minimal number of generators of $\text{MCG}(\Sigma_g)$ is $2g + 1$ [Humphries (1977)], and that the Dehn twists about $a_1, a_2, \ldots, a_g, b_1, b_2, c_1, c_2, \ldots, c_{g-1}$ are such a family of generators. For our purpose we only need a simplified version of these results, that the mapping class group of a closed surface is generated by Dehn twists about nonseparating curves. Before proceeding with the proof, let us examine a few simple situations.

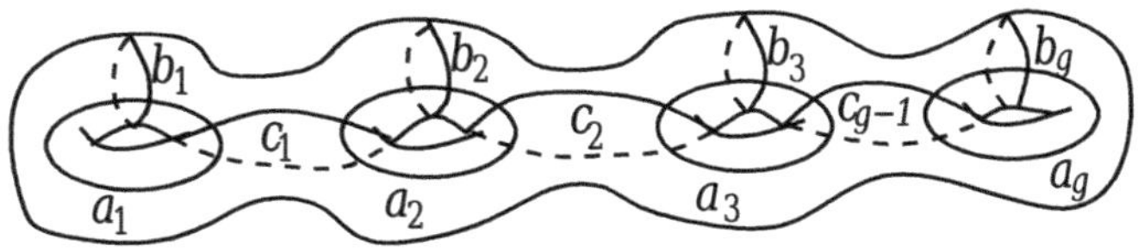

Fig. 3.25 The Dehn twists generating the mapping class group

3.3.2 Particular cases of mapping class groups

The mapping class groups of the disk and the sphere

Proposition 3.5. *(Alexander Lemma) The mapping class group of the disk is trivial.*

Proof. For $h \in \mathrm{MCG}(\Sigma_{0,1})$, define

$$H : [0,1] \times \Sigma_{0,1} \to \Sigma_{0,1}, \quad H(s,z) = \begin{cases} sh(z/s) & \text{if } 0 \le |z| \le s \\ z & \text{if } s \le |z| \le 1. \end{cases}$$

Then H is an isotopy between the identity homeomorphism and h fixing the boundary. □

Proposition 3.6. *The mapping class group of the 2-dimensional sphere is trivial.*

Proof. Let h be a homeomorphism of the sphere Σ_0, and let γ be a smooth simple closed curve. Then γ and $h(\gamma)$ are homotopic, so they are ambient isotopic by Theorem 3.2. Modifying h by this isotopy, we may assume that $h(\gamma) = \gamma$. Note that $h|\gamma$ is orientation preserving, so there is an isotopy of γ, $H : [0,1] \times \gamma \to \gamma$ that maps $h|\gamma$ to the identity. Choose a regular neighborhood of γ and identify it with the annulus $\{z \mid 1 \le |z| \le 3\}$ so that γ is the circle $\{z \mid |z| = 2\}$. Extend H to an isotopy H' of the sphere such that H' is the identity outside the annulus, and inside the annulus it is given by

$$H'(s, re^{2\pi it}) = rH((1 - |r-2|)s, e^{2\pi it}).$$

This isotopy changes h to a homeomorphism that is the identity map on γ. By the Jordan Curve Theorem, γ divides the sphere into two disks. Using the Alexander Lemma (Proposition 3.5), we can isotope h to the identity in each disk, and we are done. □

The mapping class group of the annulus

Proposition 3.7. *The mapping class group of the annulus*

$$A = \{z \mid 1 \le |z| \le 3\}$$

is isomorphic to $\mathbb{Z}$ and is generated by the Dehn twist T about the curve $t \mapsto 2e^{2\pi it}$.

Proof. We want to show that every homeomorphism of the annulus is a power of the Dehn twist T. We focus on the curve $\gamma : [1,3] \to A, \gamma(s) = h(s)$, which is the image through h of the line segment l joining 1 and 3. It has the property that $\gamma(1) = 1$ and $\gamma(3) = 3$. Consider the universal cover

$$\pi : [1,3] \times \mathbb{R} \to A, \quad \pi(r,t) = re^{2\pi it}.$$

Let $\tilde{\gamma}$ be the lift of γ such that $\tilde{\gamma}(1) = (0,1)$. Then $\tilde{\gamma}(3) = n$ for some integer n. We want to show that h is isotopic to T^n.

Let γ' be the curve $\gamma : [1,3] \to A$, $\gamma'(s) = T^n(s)$. Then γ and γ' are homotopic relative to the boundary. By removing bigons, we may assume that they do not intersect except at the boundary, so they bound a disk (which itself is a bigon). We can therefore map γ to γ' by an ambient isotopy, and hence we may assume that h coincides with T^n on the line segment l. If we cut the annulus along this line segment we obtain a disk, and because of the Alexander Lemma (Proposition 3.5) we can isotope h to T^n on this disk.

Also, because of the homotopy lifting lemma, homeomorphisms for which $\tilde{\gamma}(3)$ are different are not isotopic. It follows that the group generated by T is free. □

The mapping class group of the torus

We introduce the homeomorphisms S and T of the torus described in Figure 3.26. S is induced by the counterclockwise 90°-rotation of the plane about the origin, via the universal covering map $\pi : \mathbb{R}^2 \to \Sigma_1$, $\pi(s,t) = (e^{2\pi is}, e^{2\pi it})$, while T is the Dehn twist about the $(0,1)$ curve.

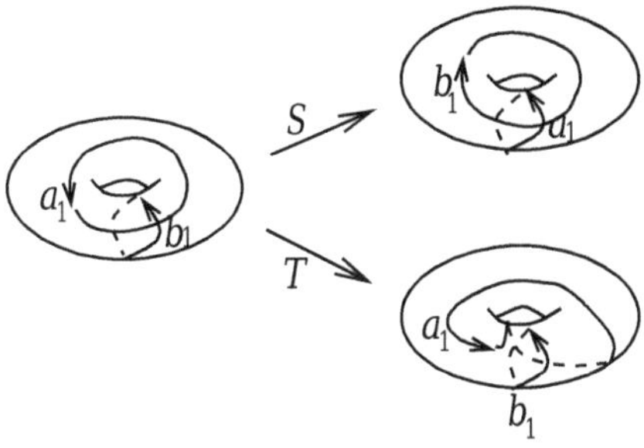

Fig. 3.26 The S and T maps

Proposition 3.8. *The mapping class group of the torus is generated by the (equivalence classes of the) maps S and T, and is isomorphic to the* *modular group $SL(2,\mathbb{Z})$ with isomorphism defined by*

$$S \mapsto \begin{pmatrix} 0 & -1 \\ 1 & 0 \end{pmatrix} \text{ and } T \mapsto \begin{pmatrix} 1 & 0 \\ 1 & 1 \end{pmatrix}.$$

Proof. Fix a canonical basis $a_1 = (1,0), b_1 = (0,1)$ of $H_1(\Sigma_1,\mathbb{Z})$. Let $h_* : H_1(\Sigma_1,\mathbb{Z}) \to H_1(\Sigma_1,\mathbb{Z})$,

$$h_*(a_1) = aa_1 + cb_1, \quad h_*(b_1) = ba_1 + db_1, \tag{3.6}$$

be the map induced by h in homology. Because h is invertible,

$$\begin{pmatrix} a & b \\ c & d \end{pmatrix} \in SL(2,\mathbb{Z}).$$

Note that the matrices of S and T are as given in the statement of the proposition, so in particular the matrices of S and T generate $SL(2,\mathbb{Z})$. We thus produced a homomorphism $\mathrm{MCG}(\Sigma_1) \to SL(2,\mathbb{Z})$ which is onto. To show that it is one-to-one, assume that the matrix associated to h_* is the identity. Then $h(a_1)$ is homological to a_1 and $h(b_1)$ is homological to b_1. Because $\pi_1(\Sigma_1)$ is commutative, equality in $H_1(\Sigma_1,\mathbb{Z})$ implies equality in $\pi_1(\Sigma_1)$. Hence $h(a_1)$ and a_1 are homotopic, and $h(b_1)$ and b_1 are homotopic. By Theorem 3.2, there is an ambient isotopy that maps $h(b_1)$ to b_1, so we may assume that $h(b_1) = b_1$. Like in the proof of Proposition 3.6, we may actually assume that $h|b_1$ is the identity.

Cut open along b_1 to obtain an annulus. In this annulus h is isotopic relative to the boundary to a power of the Dehn twist T. But then it is isotopic to that power of T in the original torus. However that can only be the 0th power, or else the matrix of h in homology is nontrivial. This completes the proof. □

Remark 3.4. If T' is the Dehn twist about the curve $a_1 = (1,0)$, then its matrix is

$$\begin{pmatrix} 1 & 1 \\ 0 & 1 \end{pmatrix}.$$

This matrix and that of the Dehn twist T generate $SL(2,\mathbb{Z})$. We conclude that $\mathrm{MCG}(\Sigma_1)$ is generated by Dehn twists.

3.3.3 *Elements of Morse and Cerf theory*

The general case

In this unit we recall a few facts from Morse and Cerf theory. They will be used for proving that the mapping class group is generated by Dehn twists about nonseparating curves, and later, in Chapter 6, for showing that two surgery diagrams of the same 3-dimensional manifold can be changed into one another by Kirby moves.

Morse theory, initiated by Marston Morse, is mainstream and can be found in many textbooks on differential topology. Cerf theory, developed by Jean Cerf in [Cerf (1970)], concerns the natural stratification (in the sense of Thom) of the space of differentiable functions from a smooth compact n-dimensional manifold without boundary M to $\mathbb{R}$.

A critical point p of a differentiable function

$$f : M \to \mathbb{R}$$

is **non-degenerate** if there is a local chart $\mathbf{x}$ at p such that $\mathbf{x}(p) = \mathbf{0}$, that turns f into

$$f(\mathbf{x}) = f(p) - x_1^2 - \cdots - x_k^2 + x_{k+1}^2 + \cdots + x_n^2.$$

The number k is called the **index** of p. A function that has only non-degenerate critical points is called Morse.

If f is a Morse function on M, then its gradient ∇f is a smooth vector field on M. For each critical point p, the **ascending manifold** of p consists of p and the union of all integral lines of ∇f that start at p. Similarly, the **descending manifold** of p consists of p and the union of all integral lines of ∇f that end at p.

Let us recall the fundamental theorem of Morse theory, which is phrased here in terms of handle addition, rather than cell addition (for those familiar only with the cell addition picture, the handle is a neighborhood of the cell). In the n-dimensional case, a k-**handle** is $B^k \times B^{n-k}$; such a handle is attached to an n-dimensional manifold M with boundary by gluing it along $S^{k-1} \times B^{n-k} = \partial B^k \times B^{n-k}$ to ∂M. In this terminology, the handles attached to B^3 to obtain a 3-dimensional handlebody are 1-handles.

Theorem 3.6. *(Morse) Let p be a critical point of index k of the Morse function $f : M \to \mathbb{R}$, and set $f(p) = q$. Choose $\epsilon > 0$ such that $f^{-1}([q-\epsilon, q+\epsilon])$ is compact and contains no other critical points. Then the manifold $f^{-1}((-\infty, q+\epsilon])$ is diffeomorphic to $f^{-1}((-\infty, q-\epsilon])$ with a k-handle attached.*

Combining this with the fact that Morse functions exist, we can build any smooth manifold by adding handles. A consequence of this fact is that for every smooth manifold, the homology classes can be represented by smoothly embedded submanifolds. Hence we can use the definition of the intersection number from §3.1.3 for all smooth manifolds.

Let us now turn to Cerf theory. The strata $\mathcal{F}^j$, $j \geq 0$, consist of the functions of "codimension" j, where the codimension of a function is the sum of the "codimensions" of its critical points and critical values. For the purpose of this book, we are mostly interested in the strata $\mathcal{F}^0$ and $\mathcal{F}^1$, and in the relationship between them.

A critical point p of f is called a **birth-death point** if there is a local chart $\mathbf{x}$ at p such that $\mathbf{x}(p) = \mathbf{0}$, and in this chart

$$f(\mathbf{x}) = f(p) - x_1^2 - \cdots - x_k^2 + x_{k+1}^2 + \cdots + x_{n-1}^2 + x_n^3.$$

Note that it is irrelevant whether the sign in front of x_n^3 is $+$ or $-$, since we can change the coordinates by $x_n \mapsto -x_n$.

- The stratum $\mathcal{F}^0$ consists of the functions whose critical points are non-degenerate and critical values are distinct.
- The stratum $\mathcal{F}^1$ consists of the functions f that have finitely many critical points such that one of the following two conditions hold:
  - all critical points are non-degenerate except for one, which is a birth-death point, and all critical values are distinct;
  - all critical points are non-degenerate and all critical values are distinct, except for two which are equal.

Endow the space of differentiable functions on M with the topology induced by the sup norm. The following results hold.

Theorem 3.7. *(Morse) $\mathcal{F}^0$ is open and dense in the space of all differentiable functions on M.*

Theorem 3.8. *(Cerf) Given two functions $f_0, f_1 \in \mathcal{F}^0$, there is a continuous path $F : [0,1] \to \mathcal{F}^0 \cup \mathcal{F}^1$ with $F(0) = f_0$, $F(1) = f_1$, and $F(t) \in \mathcal{F}^0$ except at finitely many points where it traverses $\mathcal{F}^1$.*

To understand such a path, it is customary to draw the graphic of the critical values of the functions $F(t)$, $t \in [0,1]$. We interpret these values as the heights of the critical points.

Example 3.7. An example is shown in Figure 3.27. From that graphic we can infer that both f_0 and f_1 have 5 critical points. The first event is when two critical points switch heights, passing through a point where two critical values are equal. Next, two critical points cancel each other at a death point. The last event is a birth point, where two critical points are born.

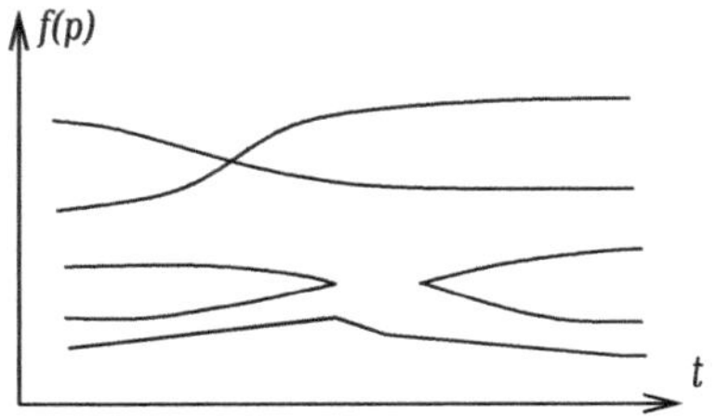

Fig. 3.27 The graphic of the heights of critical points

Here is the description of what happens at a death point. The path $F(t)$ can be made to pass through the function

$$-x_1^2 - \cdots - x_k^2 + x_{k+1}^2 + \cdots + x_{n-1}^2 + x_n^3$$

at $t = t_0$ as

$$-x_1^2 - \cdots - x_k^2 + x_{k+1}^2 + \cdots + x_{n-1}^2 + x_n^3 + (t - t_0)x_n.$$

Slicing in the nth coordinate we obtain the cubic $x_n^3 + (t - t_0)x_n$. When $t < t_0$, this cubic has two critical points, those for which $3x_n^2 + (t - t_0) = 0$. One of them is a maximum and one is a minimum. The critical points of the cubic become non-degenerate critical points of $F(t)$; the maximum has index $k+1$ and the minimum has index k.

When $t > t_0$, namely beyond the death point, the cubic no longer has critical points. So for $t > t_0$, $F(t)$ has no critical points; the two critical points canceled each other at $t = t_0$.

A similar description can be given for birth points. In particular we deduce that the critical points that are born or canceled at a birth-death point have indices that differ by ± 1.

Later in the book we will need a few results about deforming one path $F : [0,1] \to \mathcal{F}^0 \cup \mathcal{F}^1$ into another. In this case we have to cross $\mathcal{F}^2$, and the results explain how this may happen. In all figures, the number next to the graphic of a critical point specifies the index of the critical point. We will say that the graphic of one path in $\mathcal{F}^0 \cup \mathcal{F}^1$ can be changed to another if the path can be deformed so that it crosses transversally $\mathcal{F}^2$ at one point.

Lemma 3.5. *(Triangle Lemma)*[2] *Consider the configurations from Figure 3.28.*

1. *If (a) $i_1 + i_3 \leq n-1$, or (b) $\min(i_1, i_3) \leq i_2 - 1$, or (c) $i_1 = i_2 = i_3 \leq n-2$, then one can pass from the configuration on the left to the one on the right.*

[2] *Lemme du triangle* on page 78 in [Cerf (1970)]

2. *If (a)* $i_1+i_3 \geq n+1$, *or (b)* $\max(i_1, i_3) \geq i_2+1$, *or (c)* $i_1 = i_2 = i_3 \geq 2$, *then one can pass from the configuration on the right to the one on the left.*

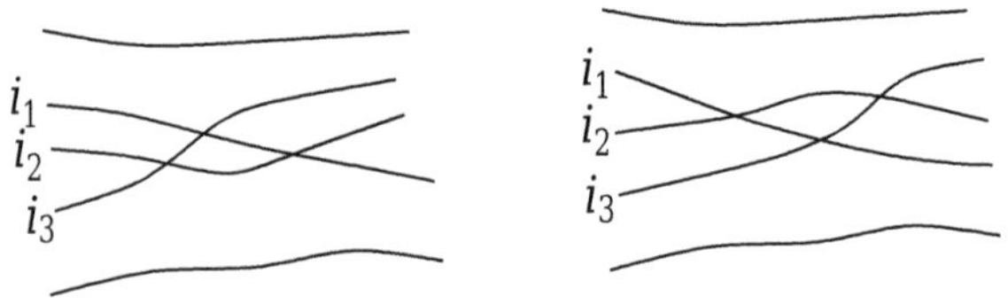

Fig. 3.28 Triangle Lemma

Lemma 3.6. *(Beak Lemma)*[3] *In Figure 3.29, one can always change the graphic on the left to the graphic on the right top if* $i_1 > 0$ *and one can always change the graphic on the left to the to the graphic on the right bottom if* $i_2 < n$.

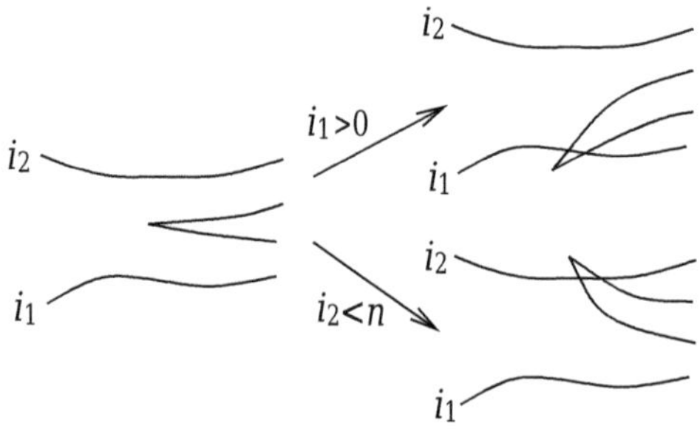

Fig. 3.29 Beak Lemma

The next result is the Swallow Tail Lemma (*Lemme de la queue d'aronde* on page 93 in [Cerf (1970)]). In [Cerf (1970)] it is stated for manifolds of dimension at least 6, while later we will apply it to manifolds of dimension 4. Actually we only need a very particular case of this result, which we state and prove below. The stratum $\mathcal{F}^2$ will be crossed at a swallow tail singularity, which is of the form

$$-x_1^2 - x_2^2 - \cdots - x_k^2 + x_{k+1}^2 + \cdots + x_{n-1}^2 + x_n^4.$$

Lemma 3.7. *(Swallow Tail Lemma) In both cases from Figure 3.30, one can change the graphic on the left to the one on the right.*

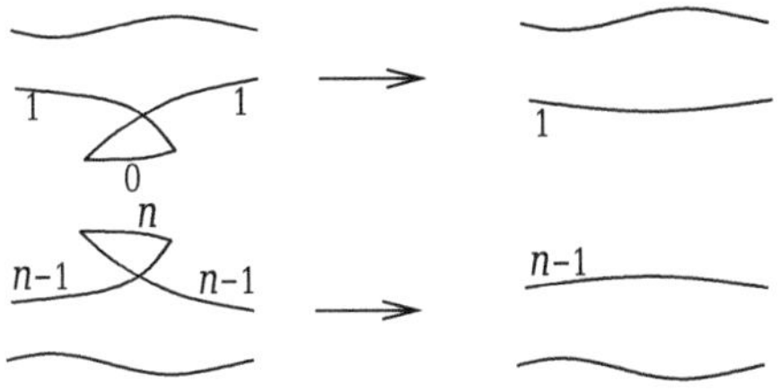

Fig. 3.30 Swallow Tail Lemma

The name is motivated by the shape illustrated on the left-hand side of Figure 3.30. We will always refer to such a configuration as a **swallow tail** (sometimes it is called a dove tail).

Proof. The first case is obtained by turning the second upside down, namely by changing the path F to $-F$. So let us prove the second case.

It is easy to understand what happens when $n = 1$. Then M is a segment or a circle, and since we are only interested in the local picture, we can assume that M is an open interval. Then the path whose graphic is shown on the left looks (locally near the critical points in question) as depicted in Figure 3.31. Because we are in the 1-dimensional situation, we have a lot of flexibility in choosing coordinates, and we can make this path be the swallow tail path defined in Lemma 1 on page 87 in [Cerf (1970)]. For this we set

$$F_\epsilon(t) = x^4 + (\epsilon \cos 2\pi t)x^2 + (\epsilon \sin 2\pi t)x,\ 0 \leq t \leq 1,$$

were ϵ is any positive number. That this path satisfies the desired condition follows from analyzing the roots of the equation

$$F_\epsilon'(t) = 4x^3 + 2\alpha x + \beta.$$

The equation has double roots when (α, β) belongs to the cubic $8\xi^3+27\eta^2 = 0$. This curve partitions the circle $|z| = \epsilon$ into two arcs symmetric with respect to the x-axis. On one arc the equation has three distinct real roots, on the other it has one real root and two complex roots. Precisely when $t = 1/2$, $F_\epsilon(t)$ has two minima at the same level.

When $\epsilon \to 0$, the loop in Figure 3.30 shrinks to a point, which is the swallow tail singularity $F_\epsilon(0) = x^4$. We extend the definition to F_ϵ to negative values of ϵ by $x^4 - \epsilon x^3 + \epsilon^2 x^2$, which has only one critical point (of index 0). So when ϵ passes from positive values to negative values, the graphic of $F_\epsilon(t)$ changes as in Figure 3.30.

[3] *Lemme d'apparition d'un bec* on page 83 in [Cerf (1970)])

Fig. 3.31 Swallow tail path

For general n, since we only consider the case of a birth/death with indexes $n-1$ and n, we can change coordinates (by pushing the birth-death to happen in the last coordinate) to have

$$F(t) = -x_1^2 - \cdots - x_{n-1}^2 + x_n^4 + (\epsilon \cos 2\pi t)x_n^2 + (\epsilon \sin 2\pi t)x_n,\ 0 \leq t \leq 1. \square$$

The last result we will need is the following.

Lemma 3.8. *(Independent Singularities Lemma)*[4] *Given two parts of the graphic, suppose that the ascending and descending manifolds of the critical points of one part do not intersect those from the other part. Then we may deform one part relative to the other.*

Cerf theory on a surface

In the case of a closed surface Σ, the non-degenerate critical points of a differentiable function

$$f : \Sigma \to \mathbb{R}$$

can have index 0, 1, or 2. We can embed neighborhoods of the critical points in $\mathbb{R}^3$ with coordinates x, y, z in such a way that f is the z-coordinate. Then the points of index 0, 1, and 2 look respectively as depicted in Figure 3.32.

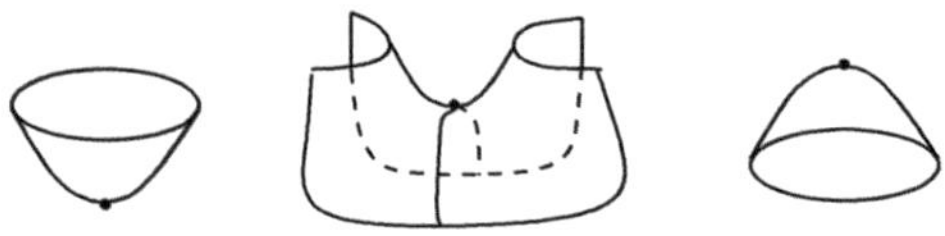

Fig. 3.32 Critical points of index 0, 1, and 2 on a surface

The occurrence of a birth/death point along a path is shown in Figure 3.33, traveled from the left to the right for birth, and from right to left for death.

We want to understand how two critical values cross, when both critical points have index 1. For this, we look at the semi-local picture of a critical

[4] *Lemme des Singularités Indépendentes* on page 75 in [Cerf (1970)]

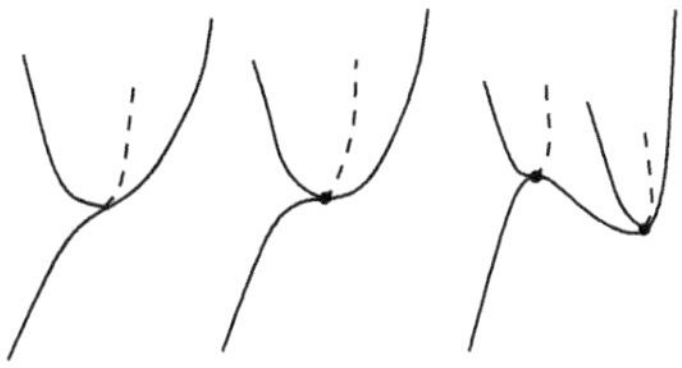

Fig. 3.33 A birth/death event on a path

point. Let f be a Morse function, with all critical values distinct. The semi-local neighborhood of a critical point p is the connected component containing p of the set

$$f^{-1}([f(p)-\epsilon, f(p)+\epsilon]),$$

where ϵ is so chosen, that $[f(p)-\epsilon, f(p)+\epsilon]$ contains no other critical points. Note that if y is a regular value of f, then $f^{-1}(y)$ is the disjoint union of finitely many circles. It follows that the semilocal picture of a point of index 1 resembles a "pair of pants".

The ascending (respectively, descending) manifold of a critical point is the manifold along which f increases (respectively, decreases) the fastest. In the case of a index 1 critical point, these manifolds are curves. We will use the descending manifolds as reference. In Figure 3.34 we show the two possible semi-local neighborhoods of an index 1 critical point, together with the combinatorial picture of the corresponding descending manifolds.

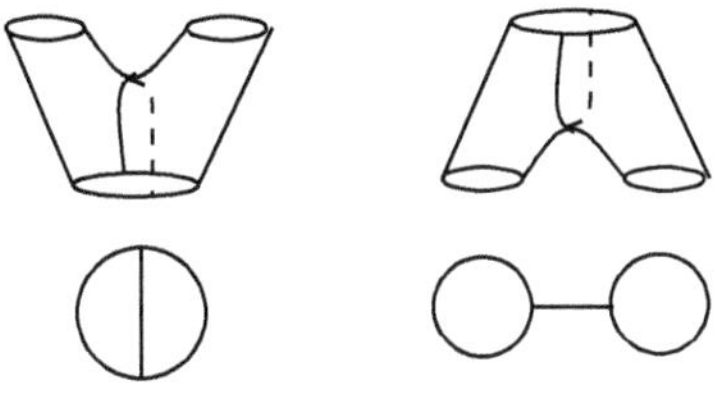

Fig. 3.34 Semi-local picture of an index 1 critical point

The situation of interest to us is when the critical points whose critical values cross have overlapping semi-local neighborhoods. The semi-local picture of the crossing will consist of the union of the two semi-local neighborhoods of the critical points. In this case we depict the descending manifolds up to where they intersect the lowest circle (or circles). There are five possible connected diagrams made out of two arcs, and several circles, and hence there are five types of crossings. They are shown in Figure 3.35.

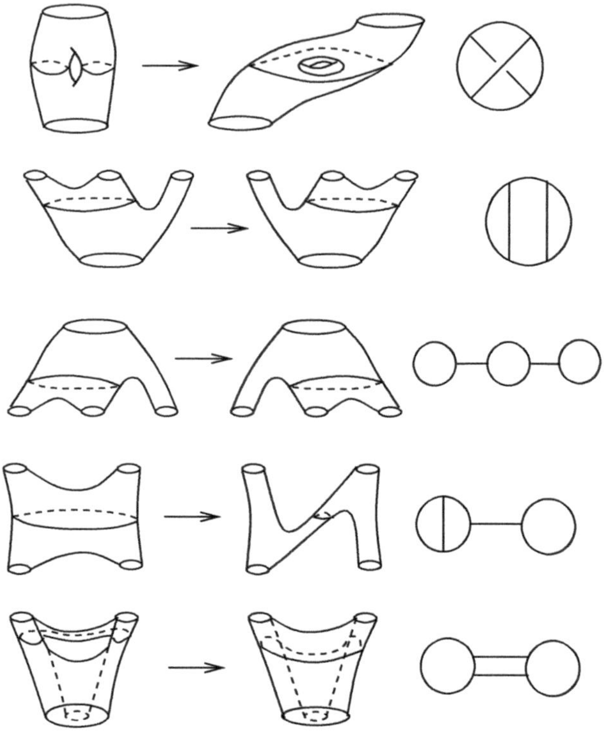

Fig. 3.35 Semi-local picture of the crossing of two critical values

3.3.4 *The mapping class group of a closed surface is generated by Dehn twists*

It is now time to prove the main result.

Theorem 3.9. *(Dehn) The mapping class group of an oriented smooth closed genus g surface Σ_g, $g \geq 1$, is generated by Dehn twists along nonseparating simple closed curves.*

Cut systems

For the first step of the proof we follow closely the method of [Hatcher and Thurston (1980)], which is based on Cerf theory.

Definition 3.17. A collection of g disjoint nonseparating simple closed curves on the closed orientable genus g surface, Σ_g, is called a **cut system** if $\Sigma_g \backslash (\gamma_1 \cup \gamma_2 \cup \cdots \cup \gamma_g)$ is a sphere with $2g$ holes.

Theorem 3.10. *[Hatcher and Thurston (1980)] Any two cut systems can be transformed into one another by a finite sequence of simple moves, in which just one C_j changes at a time to a nonseparating simple closed curve that intersects it transversely in one point and is disjoint from the other C_k's.*

Proof. Let $f : \Sigma_g \to \mathbb{R}$ be a differentiable function whose critical points are non-degenerate and have distinct critical values, namely $f \in \mathcal{F}^0$. Define the equivalence relation on Σ_g by $x \sim_f y$ if $f(x) = f(y)$ and x and y lie in the same connected component of $f^{-1}(f(x))$. Then $\Sigma_g/ \sim_f$ is a graph whose vertices are the critical points of f. We denote this graph by $\Gamma(f)$. An example is shown in Figure 3.36.

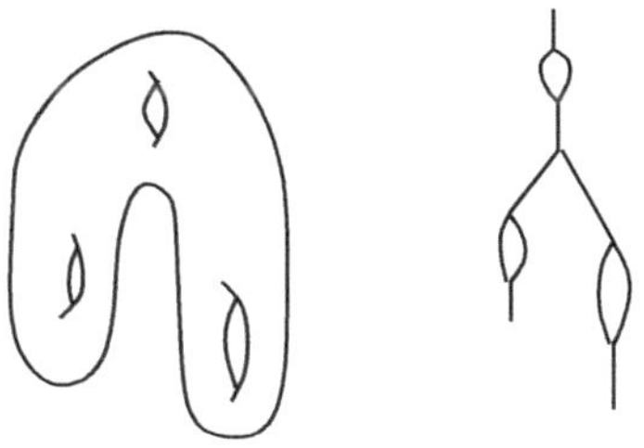

Fig. 3.36 Graph associated to a height function

Consider a maximal tree T of $\Gamma(f)$. Let $e_1, e_2, \ldots, e_g$ be the edges that do not belong to T. For each edge e_j, let $m(e_j) = \min f|e_j$, in other words, $m(e_j)$ is the value of f at the lower of the two critical points that are the endpoints of e_j. Order the edges by the numbers $m(e_j)$. Let γ_j be an equivalence class that corresponds to an interior point of e_j, $j = 1, 2, \ldots, g$, and order γ_j the way the e_j's are ordered. Then $\gamma_1, \gamma_2, \ldots, \gamma_g$ is a cut system (see Figure 3.37, in which the edges e_1, e_2, e_3 are represented by dotted lines).

Moreover, for every cut system there is a function $f \in \mathcal{F}^0$ and a maximal tree T such that the cut system arises in this way. This can be done by embedding the surface into $\mathbb{R}^3$ such that the circles of the cut system are horizontal and at different heights, taking f to be the height (i.e. the z-coordinate), and choosing T to be the complement of the edges that correspond to the cut system. The function f can be made to belong to $\mathcal{F}^0$, because $\mathcal{F}^0$ is dense.

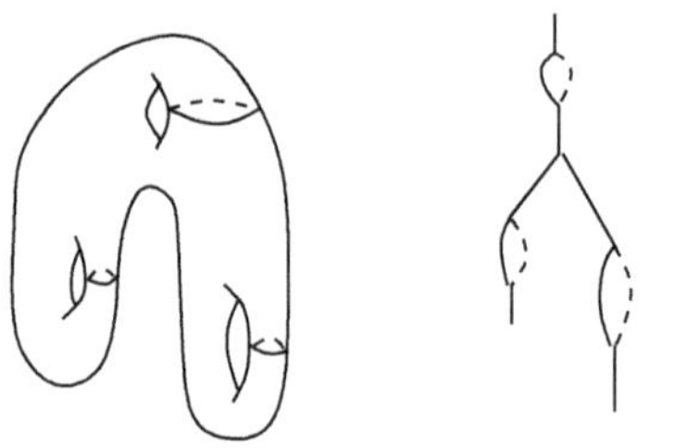

Fig. 3.37 Cut system associated to a tree

Now let $\gamma_1^0, \gamma_2^0, \ldots, \gamma_g^0$ and $\gamma_1^1, \gamma_2^1, \ldots, \gamma_g^1$ be two cut systems, which arise from $(f_0, T_0 \subset \Gamma(f_0))$ and $(f_1, T_1 \subset \Gamma(f_1))$. By Theorem 3.8, there is a path

$$F : [0,1] \to \mathcal{F}^0 \cup \mathcal{F}^1$$

joining f_0 and f_1 that traverses $\mathcal{F}^1$ at finitely many points.

The cut system is kept fixed as long as $F(t)$ evolves inside $\mathcal{F}^0$, but it might change once we cross $\mathcal{F}^1$. In fact, when traversing $\mathcal{F}^1$ we might be forced to remove one curve of the cut system, then replace it with a new level curve. And there are more choices of this new level curve. Thus the evolution of the tree T is ambiguous. The next two results show that we can proceed each time with whatever choice we make.

Lemma 3.9. *Let T and T' be two maximal trees of the same graph. Then one can pass from T to T' by finite sequence of steps, each of which producing a maximal tree by replacing one edge of T by an edge of T'.*

Proof. Let a be an edge of $T' \backslash T$. Then $T \cup a$ has a loop, which therefore must have an edge b of T'. Substitute a by b to obtain a maximal tree that is closer to T'. We can repeat this procedure until we reach the tree T'. □

Lemma 3.10. *Let $\gamma_1, \gamma_2, \ldots, \gamma_g$ and $\gamma_1', \gamma_2', \ldots, \gamma_g'$ be two cut systems associated to the maximal trees T and T' of the graph Γ_f. Then one can pass from $\gamma_1, \gamma_2, \ldots, \gamma_g$ to $\gamma_1', \gamma_2', \ldots, \gamma_g'$ by a finite sequence of moves, each of which consists of replacing one curve by a curve that intersects it at exactly one point.*

Proof. By Lemma 3.9 we only have to check the case where we replace one edge of the tree with another, such that the two edges belong to a loop. Let γ_j and γ_k' be the elements of the cut systems corresponding to these edges.

There is a curve γ on the surface such that the map $\Sigma_g \to \Gamma(f)$ is a homeomorphism from γ onto the loop. Then γ_j and γ'_k intersect γ at exactly one point, and $\gamma_j \to \gamma \to \gamma'_k$ is the desired sequence of moves. $\square$

In view of these results, as $F(t)$ evolves in $\mathcal{F}^0 \cup \mathcal{F}^1$, we may allow the maximal tree T to change finitely many times. Let us analyze what happens as we proceed along the path.

It is important to observe that the elements of a cut system come from edges that run between vertices corresponding to critical points of index 1. At a birth/death point, $\Gamma(F(t))$ changes by adding one vertex in the middle of an edge and then adding one edge at that vertex. Then new edge ends at an index 0 or 2 critical point. So nothing interesting happens at a birth/death point. The same is true when critical points cross and one has index 0 or 2.

Therefore the only interesting situation is where two critical points of index 1 cross, and each of them belongs to one of the edges e_j that are removed in order to produce the tree. If the semi-local pictures of the two critical points are disjoint the cut system does not change. If the semi-local pictures overlap, then we are in one of the situations from Figure 3.35.

In the first situation the graph does not change, but one curve of the cut system is replaced by a curve that intersects it at one point (see Figure 3.38). This move satisfies the requirement of the theorem.

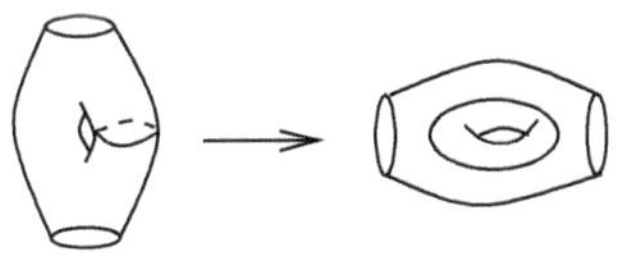

Fig. 3.38 Replacing one curve by another that intersects it at one point

In each of the remaining four situations, one edge is collapsed at the moment of the crossing, as illustrated in Figure 3.39. Because we are allowed to modify the maximal tree, we include this collapsed edge into the tree both before and after the crossing. Then the crossing does not alter the cut system.

To conclude, along the path we change the cross system either when the first crossing depicted in Figure 3.35 happens or when the maximal tree is changed. Both situations can be resolved by a sequence of transformations of the cut system that satisfies the requirement of the theorem. The concatenation of all these sequences yields the sequence of transformations

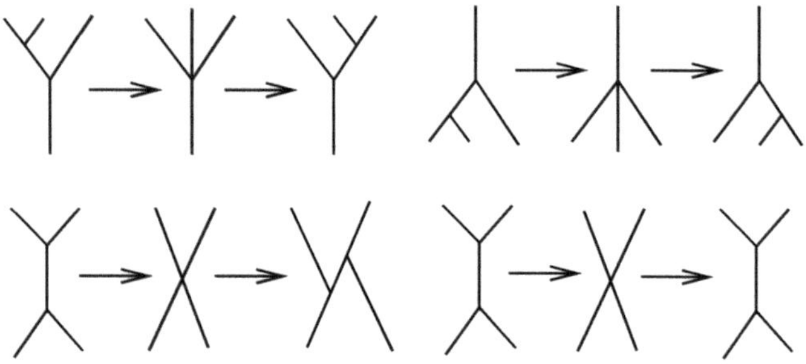

Fig. 3.39 Behavior of the graph at a crossing

that changes $\gamma_1^0, \gamma_2^0, \ldots, \gamma_g^0$ to $\gamma_1^1, \gamma_2^1, \ldots, \gamma_g^1$ with the desired property. This completes the proof of the theorem. □

Reduction to the case of the sphere with 2g boundary components

The next result shows that the process of changing one cut system into another can be realized by using Dehn twists.

Lemma 3.11. *If γ_1 and γ_2 are oriented nonseparating simple closed curves on a surface that intersect at exactly one point, then $T_{\gamma_1}^{-1}T_{\gamma_2}(\gamma_1) = \gamma_2$.*

Proof. A pictorial proof is shown in Figure 3.40. □

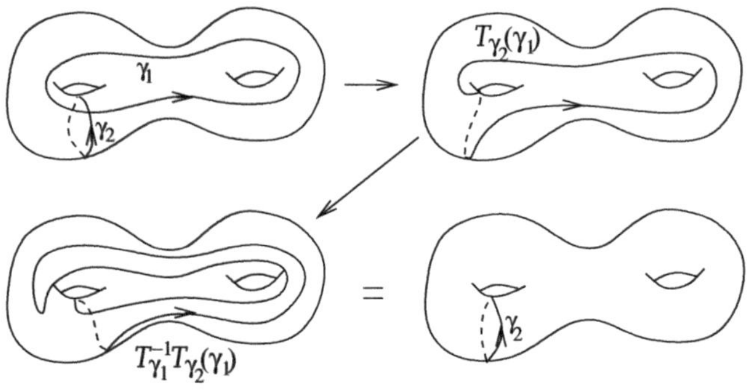

Fig. 3.40 Mapping one curve to another by Dehn twists

Assume that we are given two cut systems, $\gamma_1^0, \gamma_2^0, \ldots, \gamma_g^0$ and $\gamma_1^1, \gamma_2^1, \ldots, \gamma_g^1$, and assume also that the curves are oriented. The above lemma shows that we can change one cut system into the other by Dehn twists. We wish to make sure that additionally, for each j, γ_j^0 is mapped to γ_j^1, and that the orientations agree. This can be arranged by using the following two results.

Lemma 3.12. *Let γ_j and γ_k be two curves in a cut system $\gamma_1, \gamma_2, \ldots, \gamma_g$. Then there is a finite sequence of Dehn twists that permute γ_j and γ_k while keeping all the other elements of the cut system fixed.*

Proof. Using the classification of surfaces, we can map Σ_g such that γ_j and γ_k lie in a genus 2 subsurface $\Sigma_{2,1}$ with one boundary component in the position described in Figure 3.41, while all other curves are outside this surface. There is a curve γ in $\Sigma_{2,1}$ that intersects γ_j as well as γ_k in exactly one point.

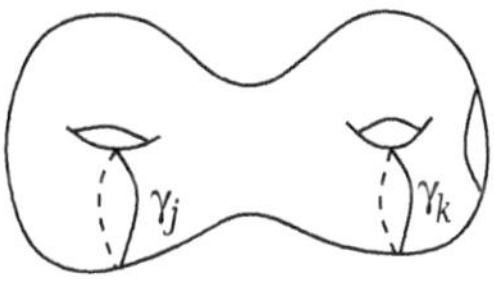

Fig. 3.41 Standard position of two curves in a cut system

To specify the Dehn twists, orient the curves such that $\gamma \cdot \gamma_j = \gamma \cdot \gamma_k = 1$. The computation in Figure 3.42 shows that the composition of Dehn twists

$$T_\gamma^{-1} T_{\gamma_j}^{-1} T_{\gamma_k}^{-1} T_\gamma$$

switches the two curves. Because this map is supported inside $\Sigma_{2,1}$, it leaves the other elements of the cut system invariant. □

Lemma 3.13. *Let $\gamma_1, \gamma_2, \ldots, \gamma_g$ be a cut system and assume that for some j, γ_j is oriented. Then there is a composition of Dehn twists that maps γ_j into itself but with orientation reversed and that leaves all other elements of the cut system invariant.*

Proof. By the classification of surfaces there is a torus with one boundary component embedded in the surface that contains γ_j as the $(1, 0)$ curve and contains no other curve of the cut system. Let γ be the $(0, 1)$ curve of this torus. Then

$$T_\gamma^{-1} T_{\gamma_j}^{-2} T_\gamma$$

has the desired property. □

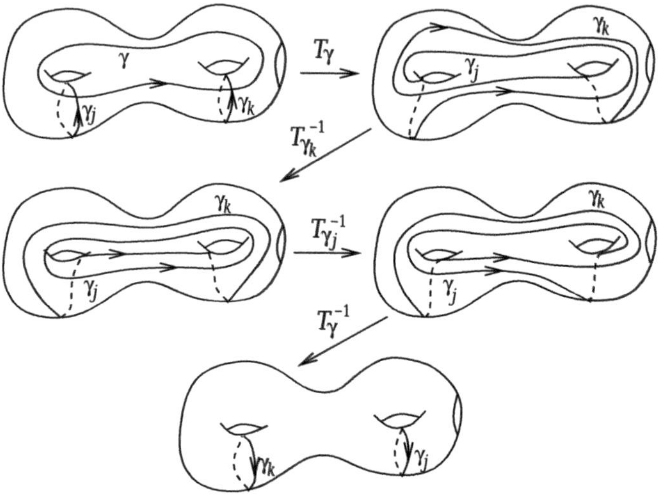

Fig. 3.42 Switching two curves in a cut system

So if h is a diffeomorphism of Σ_g, and $\gamma_1, \gamma_2, \ldots, \gamma_g$ is a cut system, then there is a sequence of Dehn twists $T_1, T_2, \ldots, T_g$ such that the diffeomorphism

$$h_0 = T_1 \circ T_2 \circ \cdots \circ T_n \circ h$$

maps each γ_j into itself. Like in the proof of Proposition 3.6, we can change h_0 by an isotopy such that it is the identity map on γ_j for each $j = 1, 2, \ldots, g$. We are left to show that every such diffeomorphism is generated by Dehn twists.

Cut open Σ_g into a sphere with $2g$ boundary components. We will show that the mapping class group of this sphere with holes is generated by Dehn twists, which will then prove that the mapping class group of Σ_g is generated by Dehn twists.

The case of the sphere with n holes

We will now prove that the mapping class group of a sphere with finitely many boundary components is generated by Dehn twists. For this we follow closely the methods of [Farb and Margalit (2011)].

Let $\Sigma_{0,n}$ be the sphere with n boundary components. To show that $\mathrm{MCG}(\Sigma_{0,n})$ is generated by Dehn twists, we reduce the problem to $\mathrm{PMCG}(\Sigma_0, n)$.

Let $T_1, T_2, \ldots, T_n$ be Dehn twists along curves that are respectively parallel (i.e. isotopic) to the n boundary components of $\Sigma_{0,n}$. Embed $\Sigma_{0,n}$ into Σ_0 and consider n marked points (or punctures) $P_1, P_2, \ldots, P_n$ of Σ_0 one in each connected component of $\Sigma_0 \backslash \Sigma_{0,n}$.

Each $h \in \mathrm{Homeo}^+(\Sigma_{0,n})$, being the identity on the boundary, can be extended to a homeomorphism of Σ_0 which keeps the n marked points fixed. Consequently we obtain a group homomorphism

$$\eta : \mathrm{MCG}(\Sigma_{0,n}) \to \mathrm{PMCG}(\Sigma_0, n).$$

Proposition 3.9. *The kernel of η is generated by $T_1, T_2, \ldots, T_n$.*

Proof. Let $h \in \mathrm{Homeo}^+(\Sigma_{0,n}, \partial\Sigma_{0,n})$ be a homeomorphism that represents an element of $\ker(\eta)$. Let $\hat{h}$ be its extension by identity to Σ_0. Because h is in the kernel, $\hat{h}$ is isotopic to the identity homeomorphism via an isotopy that keeps P_j fixed, $j = 1, 2, \ldots, n$.

Next, choose a collection of curves $\delta_1, \delta_2, \ldots, \delta_m$ such that they have pairwise minimal intersection number and intersect transversally, are pairwise nonisotopic, and for distinct j, k, l, at least one of $\gamma_j \cup \gamma_k$, $\gamma_k \cup \gamma_l$, or $\gamma_l \cup \gamma_j$ is empty, and such that the surface obtained by cutting $\Sigma_{0,n}$ along these curves is a collection of disks and annular neighborhoods of the boundary components.

Because $\hat{h}$ is isotopic to the identity, it follows that, for each j, $h(\delta_j) = \hat{h}(\delta_j)$ is ambient isotopic to δ_j in $\Sigma_0 \backslash \{P_1, P_2, \ldots, P_n\}$. It is not hard to see that the ambient isotopy can be performed entirely inside $\Sigma_{0,n}$ by keeping the boundary fixed. Indeed, if $h(\delta_j)$ and δ_j do not intersect, then by Lemma 3.3 they bound an annulus in $\Sigma_0 \backslash \{P_1, P_2, \ldots, P_n\}$, and since this annulus has its boundaries in $\Sigma_{0,n}$ it must lie entirely in $\Sigma_{0,n}$. If $h(\delta_j)$ and $\delta(j)$ intersect, then they form bigons, which do not contain the punctures, and again must lie in $\Sigma_{0,n}$.

So $h(\delta_j)$ and δ_j are ambient isotopic in $\Sigma_{0,n}$ relative to the boundary for all $j = 1, 2, \ldots, m$. What is usually referred to as the "Alexander Method" implies that $h(\cup_j \delta_j)$ is ambient isotopic relative to the boundary to $\cup_j \delta_j$. Here is a short proof of this fact due to Allen Hatcher.

The proof is by induction on n. The base case is obvious. For the inductive step, assume that the property is true for the curves $\delta_1, \delta_2, \ldots, \delta_{k-1}$ and let us show that it remains true when we add δ_k. Set

$$\Delta_{k-1} = \delta_1 \cup \delta_2 \cup \cdots \cup \delta_{k-1} = h(\delta_1) \cup h(\delta_2) \cup \cdots \cup h(\delta_{k-1}).$$

First, perform an ambient isotopy that fixes Δ_{k-1} and perturbs $h(\delta_k)$ to intersect δ_k transversally at finitely many points. This can be done because

$h(\delta_k)$ does not pass through the vertices of Δ_{k-1} and because it crosses Δ_{k-1} transversally, and so we can make $h(\delta_k)$ disjoint from δ_k along the edges of Δ_{k-1} and then perturb them in the complement to be in general position.

Then perform an ambient isotopy relative to the boundary that fixes Δ_{k-1} and make $h(\delta_k)$ disjoint from δ_k. If the curves are not already disjoint, then this is done as follows. Choose a bigon formed by the two curves. By the hypothesis on triples, the intersection of Δ_{k-1} with this bigon is a collection of disjoint arcs. By the assumption on minimal intersection, each such arc connects the δ_k part of the boundary of the bigon with the $h(\delta_k)$ part of the boundary of the bigon. Thus there is an ambient isotopy that removes the bigon, and which maps Δ_{k-1} into itself (make the bigon be the unit disk with the upper semicircle δ_k, the lower semicircle $h(\delta_k)$, and the arcs in Δ_{k-1} being vertical, and take the ambient isotopy to move vertically).

So now $h(\delta_k)$ and δ_k are disjoint, which means by Lemma 3.3 that they bound an annulus. The intersection of Δ_{k-1} with the annulus is either empty, or a collection of disjoint arcs each connecting $h(\delta_k)$ and δ_k. By the same argument as above, there is an ambient isotopy relative to the boundary that fixes Δ_{k-1} and takes $h(\delta_k)$ to δ_k.

Consequently, h is isotopic to a homeomorphism of $\Sigma_{0,n}$ that fixes the δ_j's and the boundary of $\Sigma_{0,n}$. The complement of all these curves consists of disks and annuli. Restricted to each disk h is isotopic to the identity. Restricted to each annulus, h is isotopic to the power of a Dehn twist. The conclusion follows. $\square$

We should remark that the kernel is the free group generated by $T_1, T_2, \ldots, T_n$. In view of this result, if we show that $\text{PMCG}(\Sigma_0, n)$ is generated by Dehn twists, then so is $\text{MCG}(\Sigma_{0,n})$.

Proposition 3.10. *For $n \geq 0$, $PMCG(\Sigma_0, n)$ is generated by Dehn twists.*

Proof. The proof is by induction on n. The base case $n = 0$ is just Proposition 3.6. The induction step starts with the construction of a fiber bundle. Let $P_1, P_2, \ldots, P_n$ be the marked points and let $\text{Homeo}^+(\Sigma_0, k)$, $k \geq 0$, be the subgroup of $\text{Homeo}^+(\Sigma_0)$ consisting of the homeomorphisms that keep $P_1, P_2, \ldots, P_k$ fixed. Define

$$\mathcal{E} : \text{Homeo}^+(\Sigma_0, n-1) \to \Sigma_0 \backslash \{P_1, P_2, \ldots, P_{n-1}\}, \quad \mathcal{E}(f) = f(P_n).$$

We claim that $\mathcal{E} : \text{Homeo}^+(\Sigma_0, n-1) \to \Sigma_0$ is a fiber bundle with fiber $\text{Homeo}^+(\Sigma_0, n)$. To see this, let U be an open neighborhood of P_n in Σ_0

that does not contain any of the other points and is homeomorphic to a disk. For $P \in U$, choose $h_P \in \mathrm{Homeo}^+(U)$ such that $h_P(P) = P_n$ and so that h_P varies continuously with P. Then there is a homeomorphism

$$F : U \times \mathrm{Homeo}^+(\Sigma_0, n) \to \mathcal{E}^{-1}(U), \quad F(P, h) = h_P \circ h.$$

Note that $F^{-1}(h) = (h(P_n), h^{-1}_{h(P_n)} \circ h)$. If P'_n is another point, choose a homeomorphism h_0 of the sphere that keeps $P_1, P_2, \ldots, P_n$ invariants and that maps P_n to P'_n. Then $h \to h_0 \circ h$ is a homeomorphism between $\mathcal{E}^{-1}(U)$ and $\mathcal{E}^{-1}(h_0(U))$. This proves that we have indeed a fiber bundle.

The long exact sequence in homotopy of this fiber bundle ends in

$$\begin{aligned} \cdots \to \pi_1(\mathrm{Homeo}^+(\Sigma_0, n-1)) &\to \pi_1(\Sigma_0 \backslash \{P_1, P_2, \ldots, P_{n-1}\}) \\ \to \pi_0(\mathrm{Homeo}^+(\Sigma_0, n)) &\to \pi_0(\mathrm{Homeo}^+(\Sigma_0, n-1)) \to 1. \end{aligned}$$

Of course,

$$\pi_0(\mathrm{Homeo}^+(\Sigma_0, n)) = \mathrm{PMCG}(\Sigma_0, n)$$

and

$$\pi_0(\mathrm{Homeo}^+(\Sigma_0, n-1)) = \mathrm{PMCG}(\Sigma_0, n-1).$$

One should note that, for $n \geq 3$,

$$\pi_1(\mathrm{Homeo}^+(\Sigma_0, n-1)) = 1,$$

a result due to Mary-Elizabeth Hamstrom [Hamstrom (1962)], so for $n \geq 3$ we have the following particular case of the Birman exact sequence

$$1 \to \pi_1(\Sigma_{0,n-1}) \to \mathrm{PMCG}(\Sigma_0, n) \to \mathrm{PMCG}(\Sigma_0, n-1) \to 1.$$

But we do not need Hamstrom's result in order to complete the proof of the proposition. All we need is to understand the two morphisms

$$\pi_1(\Sigma_0 \backslash \{P_1, P_2, \ldots, P_n\}) \to \mathrm{PMCG}(\Sigma_0, n) \to \mathrm{PMCG}(\Sigma_0, n-1).$$

The second is easy, it is the map that forgets the point P_n (note that it is surjective since every homeomorphism is isotopic to one that fixes P_n). Following [Farb and Margalit (2011)], we denote this homomorphism by *Forget*, and we denote the first homomorphism by *Push*.

If we show that the image of $\pi_1(\Sigma_0 \backslash \{P_1, P_2, \ldots, P_{n-1}\})$ in $\mathrm{PMCG}(\Sigma_0, n)$ is generated by Dehn twists then by using the induction hypothesis we are done. To describe explicitly the map *Push*, we choose the base point of the fundamental group at P_n. If $\gamma : [0,1] \to \Sigma_0 \backslash \{P_1, P_2, \ldots, P_n\}$ is a loop representing an element of the mapping class group, let H be an ambient isotopy of the surface such that $H(s, P_n) = \gamma(s)$. Then

$$\mathit{Push}(\gamma) = H(1, \cdot).$$

A way to see this is by using the equality im(*Push*) = ker(*Forget*). If h represents an element in the kernel of *Forget*, then h is isotopic to the identity via an isotopy that leaves $P_1, P_2, \ldots, P_{n-1}$ fixed. The loop γ is traced by P_n during this isotopy.

We can give an explicit construction of the isotopy H. Identify a neighborhood of γ with the annulus $\{z \,|\, 1 \leq |z| \leq 3$ such that γ is $\{z \,|\, |z| = 2\}$ and $P_n = 2$. Inside the annulus, choose

$$H(s, re^{2\pi it}) = re^{2\pi i[t+(1-|r-2|)s]}.$$

When $r = 1$ or 3, $H(s, re^{2\pi it}) = re^{2\pi it}$, while when $r = 2$, then $H(s, 2) = H(s, P_n) = re^{2\pi is}$. So we can extend H outside the annulus by the identity map, to obtain an isotopy with the desired property. Let γ_1 and γ_3 be the boundary curves $e^{2\pi it}$ respectively $3e^{2\pi it}$ of the annulus. Then *Push*(γ) = $H(1, \cdot)$ is equal to $T_{\gamma_1} T_{\gamma_3}^{-1}$. This shows that the image of *Push* is generated by Dehn twists and the proposition is proved. □

The completion of the proof

We have shown that in order to prove that the mapping class group of a closed genus g surface Σ_g is generated by Dehn twists, it suffices to prove that the mapping class group of the sphere with $2g$ boundary components is generated by Dehn twists. And then we proved this latter fact. This means that up to this point we have proved that the mapping class group of Σ_g is generated by Dehn twists along curves which might be separating or not. The following result completes the proof of Theorem 3.9.

Lemma 3.14. *A Dehn twist about a separating curve is a product of Dehn twists about nonseparating curves.*

Proof. Let γ be the separating curve. Using the classification of surfaces, we can place the separating curve on a genus 2 subsurface in standard position as shown in Figure 3.43.

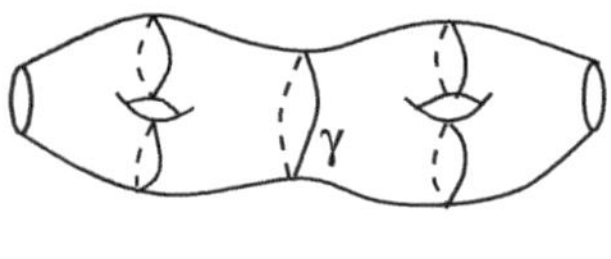

Fig. 3.43 Separating curve in standard position

We choose a nonseparating curve γ' as reference, such that γ' and γ intersect at 2 points, and also use the curves $\gamma_1, \gamma_2, \gamma_3$ as depicted on the left of Figure 3.44. The map

$$T_{\gamma_3}^{-1}T_{\gamma_2}T_{\gamma_1}^{-1}T_{\gamma_3}$$

maps the configuration on the left to the one on the right. It is not hard to notice that this acts on γ' like half of the Dehn twist about γ.

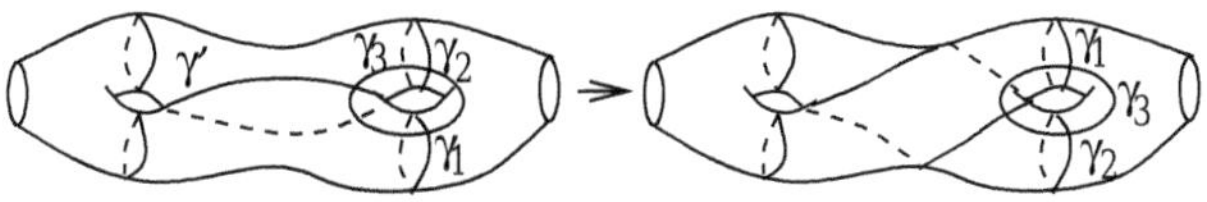

Fig. 3.44 Half of a Dehn twist

Apply this twice (see Figure 3.45) to produce a configuration in which the curve γ' is transformed by a homeomorphism h that behaves on γ' as if it were the Dehn twist. The other four nonseparating curves depicted are invariant under h, and h is the identity outside of the genus 2 surface. We have shown that the mapping class group of the sphere with 3 boundary components is generated by Dehn twists, and because any simple closed curve is parallel to the boundary, these Dehn twists are about curves parallel to the boundary components. We deduce that up to multiplication by Dehn twists about γ' and the other four nonseparating curves, h is the Dehn twist about γ. Hence the conclusion.

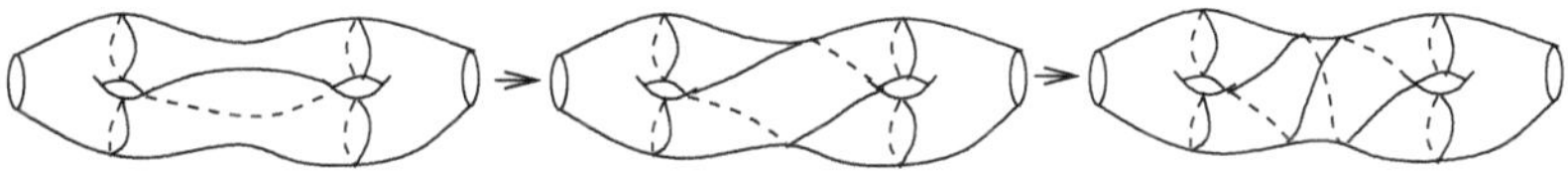

Fig. 3.45 Dehn twist about separating curve written as Dehn twists about nonseparating curves

□

Conclusions to the third chapter

In this chapter we recalled the classification of compact orientable surfaces, the fundamental group, the first homology group, and the mapping class group of a surface. We discussed the relationship between homotopy and isotopy of curves on a surface and described the structure of the mapping class group. We also introduced a few elements of Cerf theory.

The essential facts to be used in the following chapters are:

1. The **combinatorial description of the first homology group** in terms of multicurves with multiplication defined by the smoothing of crossings.

2. The fact that the **mapping class group** of a closed oriented surface of genus at least 1 **is generated by Dehn twists** about nonseparating curves.

Chapter 4

The theta functions associated to a Riemann surface

In short, this chapter is about the quantization of a system of finitely many one-dimensional particles with **periodic positions and momenta**. But there is more to the story. Such a system is less likely to occur in physics, but is a natural occurrence in algebraic geometry. The torus that constitutes the phase space is the Jacobian variety of a Riemann surface. Hence, by contrast with Chapter 2, there are three players in this game:

- the Riemann surface,
- the Jacobian variety of the Riemann surface, and
- the quantum system.

And, as we will see in Chapter 5, the quantum system has a topological description in terms of the original Riemann surface.

The quantum physical point of view was introduced to the theory of theta functions by Yuri Manin [Manin (1990)] (see also [Manin (2001)] and [Manin (2004)]). Below we use his framework and present the theory of Riemann's theta functions following [Gelca and Uribe (2014)], with inputs from [Bloch *et al.* (2003)], [Andersen (2005)], and [Manoliu (1997)].

4.1 The Jacobian variety

We start by reviewing standard facts about Riemann surfaces. We are particularly intersted in the procedure of associating to a Riemann surface its Jacobian variety. Our treatment follows [Farkas and Kra (1991)] closely. Another text that the reader can consult is [Griffiths (1994)]. We assume that the reader is familiar with the basic facts from complex analysis, nevertheless we review all necessary definitions and constructions.

4.1.1 *De Rham cohomology*

Let M be a smooth n-dimensional real manifold. On M we introduce differential forms as follows. The space of 0-forms on M is

$$\Omega^0(M,\mathbb{R}) = \{f : M \to R \,|\, f \text{ smooth}\}\,.$$

For $1 \leq k \leq n$, let T^*M be the cotangent bundle of M and let $\Omega^k(M,\mathbb{R})$ be the smooth sections of the kth exterior power of T^*M. In local coordinates, a k-form looks like

$$\sum_{i_1,i_2,\ldots,i_k} f_{i_1 i_2 \ldots i_k} dx_{i_1} \wedge dx_{i_2} \wedge \cdots \wedge dx_{i_k}\,.$$

The **exterior derivative** is the differential operator

$$d : \Omega^k(M,\mathbb{R}) \to \Omega^{k+1}(M,\mathbb{R})$$

that in local coordinates looks like

$$df_{i_1 i_2 \ldots i_k} dx_{i_1} \wedge \cdots \wedge dx_{i_k} = \sum_{j=1}^{n} \frac{\partial f_{i_1 i_2 \ldots i_k}}{\partial x_j} dx_j \wedge dx_{i_1} \wedge \cdots \wedge dx_{i_k}\,.$$

There is a cochain complex of exterior differential forms, called the **de Rham complex**

$$0 \to \Omega^0(M,\mathbb{R}) \stackrel{d}{\to} \Omega^1(M,\mathbb{R}) \stackrel{d}{\to} \Omega^2(M,\mathbb{R}) \stackrel{d}{\to} \cdots \stackrel{d}{\to} \Omega^n(M,\mathbb{R}) \to 0.$$

Definition 4.1. The elements of the image of $d : \Omega^{k-1}(M,\mathbb{R}) \to \Omega^k(M,\mathbb{R})$ are called **exact differential k-forms**. The elements of the kernel of the map $d : \Omega^{k-1}(M,\mathbb{R}) \to \Omega^k(M,\mathbb{R})$ are called **closed differential k-forms**.

It is not hard to check that $d^2 = 0$, so exact forms are closed.

Definition 4.2. The kth **de Rham cohomology group** is the quotient of the space of closed forms by the space of exact forms.

Theorem 4.1. *(de Rham) The kth de Rham cohomology group of M is isomorphic, in a canonical way, to $H^k(M,\mathbb{R})$.*

Of course, with the definition of cohomology groups given in the previous chapter one might want to assume that M is triangulable as well. This will always be the case with the manifolds of interest in this book.

The pairing

$$H^k(M,\mathbb{R}) \times H_k(M,\mathbb{R}) \to \mathbb{R} \tag{4.1}$$

can be described as follows. Start with the pairing

$$\Omega^n(M,\mathbb{R}) \times C_n(M,\mathbb{R}) \to \mathbb{R}, \tag{4.2}$$

defined by

$$(\alpha,\sigma) = \int_\sigma \alpha.$$

Because of Stokes' Theorem, this descends to a well defined pairing

$$H^n(M,\mathbb{R}) \times H_n(M,\mathbb{R}) \to \mathbb{R}, \tag{4.3}$$

and this is the same as (4.1). This pairing is non-degenerate and identifies $H^n(M,\mathbb{R})$ with the dual of $H_n(M,\mathbb{R})$.

If M has no boundary and has even dimension, say $n = 2m$, then Poincaré duality turns the intersection pairing in $H_m(M,\mathbb{R})$ into the pairing

$$H^m(M,\mathbb{R}) \times H^m(M,\mathbb{R}) \to \mathbb{R},$$

given by

$$(\alpha,\beta) \mapsto \int_M \alpha \wedge \beta.$$

In this chapter we are interested in surfaces endowed with a complex structure. For this reason we introduce also the de Rham cohomology using differential forms with complex coefficients. Thus we work with the de Rham complex of the spaces

$$\Omega^k(M,\mathbb{C}) = \Omega^k(M,\mathbb{R}) \otimes_{\mathbb{R}} \mathbb{C}.$$

Everything said above translates to this situation, and the de Rham cohomology groups with complex coefficients are canonically isomorphic with the cohomology groups $H^n(M,\mathbb{C})$.

Notation: For convenience, throughout this chapter we will denote $H^n(M,\mathbb{C})$ by $H^n(M)$.

4.1.2 *Hodge theory on a Riemann surface*

Let Σ_g be a closed genus g Riemann surface, $g \geq 1$. This means that Σ_g has a complex structure, which is defined by a system of complex charts on it such that the coordinate changes are holomorphic functions. The complex structure endows the surface automatically with an orientation, by considering in a local chart the complex coordinate $z = x + iy$, and letting the frame defining the orientation be $\frac{\partial}{\partial x}, \frac{\partial}{\partial y}$.

We want to associate to this surface a $2g$-dimensional torus – its Jacobian variety. For that we need a few facts from Hodge theory.

Differential forms

Our goal is to understand $H^1(\Sigma_g)$. To this end we will examine 1-forms in detail, and then we will compute this cohomology group and give a detailed description of Poincaré duality for this situation.

We consider differential forms on Σ_g. As said before, a 0-form is just a function on Σ_g. A 1-form in local coordinates looks like

$$\alpha = f(x,y)dx + g(x,y)dy,$$

or

$$\alpha = u(z)dz + v(z)d\bar{z},$$

depending whether we work in real or complex coordinates, with f, g, u, v functions on the local chart (and of course $f = u + v$ and $g = i(u - v)$). Passing from one coordinate chart to another changes the f and g coefficients according to the rule

$$\begin{bmatrix} f(x,y) \\ g(x,y) \end{bmatrix} \mapsto \begin{bmatrix} \dfrac{\partial x}{\partial x'} & \dfrac{\partial y}{\partial x'} \\ \dfrac{\partial x}{\partial y'} & \dfrac{\partial y}{\partial y'} \end{bmatrix} \begin{bmatrix} f(x(x',y'), y(x',y')) \\ g(x(x',y'), y(x',y')) \end{bmatrix}.$$

A differential 2-form in local coordinates looks like

$$f dx \wedge dy = \frac{i}{2} f dz \wedge d\bar{z},$$

and changes from one coordinates system to another by the rule

$$f(x,y) \mapsto \det \begin{bmatrix} \dfrac{\partial x}{\partial x'} & \dfrac{\partial y}{\partial x'} \\ \dfrac{\partial x}{\partial y'} & \dfrac{\partial y}{\partial y'} \end{bmatrix} f(x(x',y'), y(x',y')),$$

or in complex coordinates

$$f(z) \mapsto \left| \frac{dz}{dz'} \right|^2 f(z(z')).$$

In our discussion, besides differential forms, which have smooth functions as coefficients, we might also consider C^1 forms, whose coefficients are just C^1 functions.

Recall that the exterior differentiation operator d acts on 0-forms by

$$df = \frac{\partial f}{\partial x} dx + \frac{\partial f}{\partial y} dy,$$

on 1-forms by

$$d(f dx + g dy) = df \wedge dx + dg \wedge dy = \left(\frac{\partial g}{\partial x} - \frac{\partial f}{\partial y} \right) dx \wedge dy,$$

and on 2-forms is equal to zero.

We will be mainly concerned with 1-forms. A differential 1-form α is **closed** if $d\alpha = 0$ and **exact** if there is a smooth function f on Σ_g such that $\alpha = df$.

We introduce a few more operators. First, the operators ∂ and $\bar{\partial}$, which on 0-forms are defined locally by

$$\partial f = \frac{\partial f}{\partial z} dz, \quad \bar{\partial} f = \frac{\partial f}{\partial \bar{z}} d\bar{z},$$

on 1-forms are defined by

$$\partial(udz + vd\bar{z}) = \frac{\partial v}{\partial z} dz \wedge d\bar{z}, \quad \bar{\partial}(udz + vd\bar{z}) = -\frac{\partial u}{\partial \bar{z}} dz \wedge d\bar{z},$$

and on 2-forms are both zero. Note that the condition that a function f is **holomorphic** is that $\bar{\partial} f = 0$.

The **conjugation operator** $*$ is defined on 1-forms[1] by

$$*(fdx + gdy) = -gdx + fdy.$$

In complex notation this reads

$$*(udz + vd\bar{z}) = -iudz + ivd\bar{z}.$$

It is not hard to check that ∂, $\bar{\partial}$, and $*$ are well defined, in the sense that their definition does not depend on the coordinate chart. The operator $*$ can be extended to all forms, since in its definition we did not use the fact that f and g are differentiable. In this more general setting, this operator can be used to define an inner product on 1-forms by

$$\langle \alpha, \beta \rangle = \iint_{\Sigma_g} \alpha \wedge *\bar{\beta}. \tag{4.4}$$

Since Σ_g has a fixed orientation, the sign of the double integral is unambiguous, so this inner product is well defined. We define the norm

$$\|\alpha\| = \sqrt{\langle \alpha, \alpha \rangle},$$

and let $L^2(\Sigma_g)$ be the space of 1-forms for which this norm is finite, endowed with the inner product (4.4). Differential 1-forms and C^1-form form dense subspaces of $L^2(\Sigma_g)$.

If we define the **Laplacian operator** Δ in local coordinates by

$$\Delta = \frac{\partial^2}{\partial x^2} + \frac{\partial^2}{\partial y^2},$$

[1]The $*$ operator is defined on 0- and 2-forms, as well, but we do not need that here.

as it was done in §2.2.2, then

$$\Delta = d * d = -2i\bar{\partial}\partial,$$

so the Laplacian is well defined for functions on the Riemann surface. This version of the Laplace operator differs slightly from that of the general de Rham theory, but in this particular case it is simpler and serves the same purpose. As we will later see, for general smooth manifolds the Laplace operator takes a more complicated form.

Definition 4.3. A differential 1-form α is called **co-closed** if $*\alpha$ is closed and **co-exact** if $*\alpha$ is exact. A function f is called harmonic if $\Delta f = 0$. A differential 1-form α is called **harmonic** if locally there is a harmonic function f such that $\alpha = df$.

Proposition 4.1. *A differential* 1*-form* α *is harmonic if and only if it is closed and co-closed.*

Proof. If $\alpha = df$, with f harmonic, then on the one hand $d\alpha = d^2 f = 0$, and on the other hand $d * \alpha = d * df = 0$. For the converse, if α is closed then locally there exists f such that $\alpha = df$. And if α is also co-closed, $d * \alpha = d * df = 0$, which proves that f is harmonic. □

Definition 4.4. A differential 1-form α is called **holomorphic** if locally there is a holomorphic function with $\alpha = df$.

Proposition 4.2. *The* 1*-form* α *is holomorphic if and only if* $\alpha = udz$ *with* u *holomorphic. If* f *is a harmonic function on* Σ_g*, the* ∂f *is a holomorphic* 1*-form.*

Proof. Let $\alpha = udz + vd\bar{z}$. If α is holomorphic, say $\alpha = df$ with f holomorphic, then $u = \partial f$ is holomorphic and $v = \bar{\partial} f = 0$. Conversely, if $\alpha = udz$, then choose coordinate charts such that u has a power series expansion in its coordinate chart. The function f is obtained by integrating the power series term by term.

For the second part, note that if f is harmonic, then $\bar{\partial}\partial f = 0$. Because $\partial f = udz$, this implies $\partial u = 0$, so u is holomorphic. Apply the first part of the proposition to obtain the conclusion. □

Theorem 4.2. *A differential form* α *is holomorphic if and only if* $\alpha = \beta + i * \beta$*, where* β *is a harmonic differential.*

Proof. Let $\beta = udz + vd\bar{z}$ be a 1-form. Then

$$d\beta = -\left(\frac{\partial u}{\partial \bar{z}} - \frac{\partial v}{\partial z}\right) dz \wedge d\bar{z}$$

and

$$d * \beta = i\left(\frac{\partial u}{\partial \bar{z}} + \frac{\partial v}{\partial z}\right) dz \wedge d\bar{z}.$$

Then β is harmonic if and only if both u and $\bar{v}$ are holomorphic.

So if β is harmonic, then

$$\beta + i * \beta = udz + vd\bar{z} + udz - vd\bar{z} = 2udz.$$

Hence $\beta + i * \beta$ is holomorphic.

Conversely, if α is holomorphic, meaning that $\alpha = udz$ with u holomorphic, then α and $\bar{\alpha} = \bar{u}d\bar{z}$ are harmonic, and so is

$$\beta = \frac{1}{2}(\alpha - \bar{\alpha}) = \frac{1}{2}(udz + \bar{u}d\bar{z}).$$

It is not hard to check that $\beta + i * \beta = \alpha$, and the theorem is proved. □

The 1*-form associated to a curve*

We explain now how to associate to a simple closed curve on a surface a differential 1-form. As it will become clear, this procedure makes explicit the Poincaré duality between $H_1(M, \mathbb{C})$ and $H^1(M)$. The construction uses only of the smooth structure on Σ_g.

Let c be an oriented simple closed curve on Σ_g and let $j : S^1 \times (-1, 1) \to \Sigma_g$ be an embedding of an open annulus such that $j(S^1 \times \{0\})$ coincides with c and the orientation of c is obtained by traveling on S^1 in the counterclockwise direction. Consider a smooth function

$$f : \Sigma_g \backslash c \to \mathbb{R}$$

such that

$$f(P) = \begin{cases} 1 & \text{if } P \in j(S^1 \times (-1/2, 0)) \\ 0 & \text{if } P \in \Sigma_g \backslash j(S^1 \times (-1, 0)) \end{cases}$$

and define

$$\eta_c = \begin{cases} df & \text{on } \Omega \backslash c \\ 0 & \text{on } c. \end{cases}$$

Then η_c is a closed, smooth, real differential 1-form. But it is not exact.

Recall the non-degenerate pairing (4.3), and note that the intersection pairing on $H_1(\Sigma_g, \mathbb{C})$ is also non-degenerate. Consequently, both $H^1(\Sigma_g)$

and $H_1(\Sigma_g, \mathbb{C})$ itself can be viewed as duals of $H_1(\Sigma_g, \mathbb{C})$. So there should be an isomorphism that maps curves to 1-forms. The next result shows that η_c is a good candidate for the form associated to c.

Theorem 4.3. *Let $\alpha \in L^2(\Sigma_g)$ be a closed form of class C^1, and let c be an oriented simple closed curve. Then*

$$\int_c \alpha = \langle \alpha, *\eta_c \rangle .$$

Proof. Consider $0 < \epsilon < 1/2$ and let $c_{\pm\epsilon} = j|S^1 \times \{\pm\epsilon\}$ with orientations given by traveling counterclockwise on S^1. Note that $*^2$, the square of the $*$ operator, is the negative of the identity operator, and also note that η_c is a real 1-form. Thus we have

$$\langle \alpha, *\eta_c \rangle = \iint_{\Sigma_g} \alpha \wedge ** \eta_c = -\iint_{\Sigma_g} \alpha \wedge \eta_c = -\lim_{\epsilon \to 0} \iint_{\Sigma_g \setminus j(S^1 \times [-\epsilon,\epsilon])} \alpha \wedge df.$$

Because α is closed, this is further equal to

$$\lim_{\epsilon \to 0} \iint_{\Sigma_g \setminus j(S^1 \times [-\epsilon,\epsilon])} df \wedge \alpha + f d\alpha = \lim_{\epsilon \to 0} \iint_{\Sigma_g \setminus j(S^1 \times [-\epsilon,\epsilon])} d(f\alpha).$$

By Stokes' Theorem, this is equal to

$$\lim_{\epsilon \to 0} \int_{c_{-\epsilon}} f\alpha - \int_{c_{+\epsilon}} f d\alpha.$$

Using the definition of f, we see that this is

$$\lim_{\epsilon \to 0} \int_{c_{-\epsilon}} \alpha = \int_c \alpha,$$

as desired. □

The result shows that the map $c \mapsto *\eta_c$ induces a homomorphism between $H_1(\Sigma_g, \mathbb{C})$ and $H^1(\Sigma_g)$, which transforms the pairing (4.3) into the inner product on the space of 1-forms. The intersection in $H_1(\Sigma_g, \mathbb{Z})$ can also be modeled using 1-forms, as the next result shows.

Theorem 4.4. *Given oriented simple closed curves c_1 and c_2 on Σ_g,*

$$c_1 \cdot c_2 = \iint_{\Sigma_g} \eta_{c_1} \wedge \eta_{c_2}.$$

Proof. Let $\eta_{c_1} = df$. By Theorem 4.3,

$$\iint_{\Sigma_g} \eta_{c_1} \wedge \eta_{c_2} = -\int_{c_2} \eta_{c_1} = -\int_{c_2} df.$$

If f had no discontinuities, the integral would be equal to 0. However, f has discontinuities exactly where the two curves intersect, and the jump is -1 at positive crossings, and $+1$ one at negative crossings. It follows that the integral is equal to the algebraic intersection number. □

Let us return to the pairing (4.3). Note that if the integral of a 1-form on every closed curve is zero then the form itself is zero. In view of Theorem 4.4, if the integrals of all forms on a curve are zero then the curve is null-homologous. It follows that the pairing (4.3) is nondegenerate, and the map $c \to \eta_c$ defines an isomorphism $H_1(\Sigma_g, \mathbb{C}) \to H^1(\Sigma_g)$. We conclude that $H^1(\Sigma_g)$ is $2g$-dimensional and has the basis

$$\eta_{a_j}, \eta_{b_j}, \quad j = 1, 2, \dots g,$$

where a_j, b_j, $j = 1, 2, \dots, g$, is a canonical basis of $H_1(\Sigma_g, \mathbb{Z})$.

The space of square integrable forms

Recall the Hilbert space $L^2(\Sigma_g)$ of 1-forms endowed with the inner product (4.4). Differential 1-forms and C^1-forms form dense subspaces of $L^2(\Sigma_g)$. Denote by E the closure of the subspace of exact forms on Σ_g, and by E^* the closure of the subspace of co-exact forms.

Lemma 4.1. *Let α be a C^1-form. Then $\alpha \in E^{*\perp}$ (respectively $E^\perp$) if and only if α is closed (respectively co-closed).*

Proof. Assume that α is closed. In E^*, the forms $*df$ with f differentiable are dense. Because $** = -1$, we have

$$\langle \alpha, *df \rangle = \iint_{\Sigma_g} \alpha \wedge \overline{**df} - \iint_{\Sigma_g} \alpha \wedge d\bar{f}.$$

This is further equal to

$$-\iint [d(\alpha \bar{f}) - d\alpha \wedge \bar{f}] = -\iint_{\Sigma_g} d(\alpha \bar{f}),$$

where we used the fact that α is closed. Because Σ_g has no boundary, by Stokes' Theorem this is equal to zero. Hence $\alpha \in E^{*\perp}$.

Conversely, if $\alpha \in E^{*\perp}$, then $\langle \alpha, *df \rangle = 0$ for all smooth functions f. Redoing the above computation we deduce that

$$\iint_{\Sigma_g} d\alpha \wedge \bar{f} = 0.$$

Since this holds for all smooth functions f, it follows that $d\alpha = 0$, as desired.

The case of $E^\perp$ is analogous. □

Theorem 4.5. *One has the orthogonal decomposition*

$$L^2(\Sigma_g) = E \oplus E^* \oplus H(\Sigma_g),$$

where $H(\Sigma_g)$ is the space of harmonic differentials.

Proof. By Lemma 4.1, E and E^* are orthogonal, so if we denote $H = E^{\perp} \cap E^{*\perp}$, then

$$L^2(\Sigma_g) = E \oplus E^* \oplus H.$$

It remains to show that $H(\Sigma_g) = H$. That $H(\Sigma_g) \subset H$ is a corollary of Lemma 4.1 and Proposition 4.1. Let us prove the converse inclusion, namely let us show that if α is orthogonal to both E and E^*, then α is harmonic.

If α were a C^1-form, then α would be both closed and co-closed, by Lemma 4.1, and hence by Proposition 4.1 it would be harmonic. Let us show that α is C^1. This property can be checked locally, so let us fix a coordinate disk D, and let $z = x + iy$ be the coordinates in this disk. Write $\alpha = pdx + qdy$, and choose an arbitrary smooth real-valued function f that is supported in D. We compute

$$0 = \left\langle \alpha, d\left(\frac{\partial f}{\partial x}\right)\right\rangle = \iint_D \left(p\frac{\partial^2 f}{\partial x^2} + q\frac{\partial^2 f}{\partial y \partial x}\right) dx \wedge dy,$$

$$0 = \left\langle \alpha, *d\left(\frac{\partial f}{\partial x}\right)\right\rangle = \iint_D \left(-p\frac{\partial^2 f}{\partial y^2} + q\frac{\partial^2 f}{\partial x \partial y}\right) dx \wedge dy.$$

Subtracting we obtain

$$0 = \iint_D p\left(\frac{\partial^2 f}{\partial x}^2 + \frac{\partial^2 f}{\partial y^2}\right) dx \wedge dy = \iint_D p\Delta f dxdy.$$

In the terminology of differential equations, this means that p is weakly harmonic. Working with $*\alpha$ (which is also in H), we deduce that q is weakly harmonic. The fact that p and q are C^1-functions follows from a fundamental result, Weyl's Lemma, which we state and prove in the general case below. □

Weyl's Lemma

Before proceeding with the discussion of Weyl's Lemma, let us recall a fundamental result about harmonic functions.

Theorem 4.6. *(The Mean-Value Property) Let f be a harmonic function in the n-dimensional unit ball centered at the origin, and let $0 < r < 1$. Then*

$$f(0) = A(S_r^{n-1})^{-1} \int_{S_r^{n-1}} f d\sigma_r,$$

where S_r^{n-1} is the sphere of radius r centered on the origin, $A(S_r^{n-1})$ is the area of the sphere, and σ_r is the surface area element.

Proof. Denote the integral by $M(r)$. Changing the variable we obtain

$$M(r) = A(S_r^{n-1})^{-1}\int_{S_r^{n-1}} f(r\mathbf{n})r^n d\sigma_1 = A(S_1^{n-1})^{-1}\int_{S_1^{n-1}} f(r\mathbf{n})d\sigma_1.$$

We compute

$$\frac{\partial M}{\partial r} = A(S_1^{n-1})\int \nabla f(r\mathbf{n})\cdot n d\sigma_1 = A(S_r^{n-1})^{-1}\int_{S_r^{n-1}}\frac{\partial f}{\partial n}d\sigma_r,$$

where $\partial f/\partial n$ is the derivative of f in the normal direction to the sphere. By Stokes' Theorem this is equal to

$$A(S_r^{n-1})^{-1}\int_{B_r^n}\Delta f dVol = 0,$$

where B_r^n is the ball of radius r centered at the origin.

This shows that $r \mapsto M(r)$ is a constant function. The conclusion follows from the fact that $\lim_{r\to 0} M(r) = f(0)$. □

Theorem 4.7. *(Weyl's Lemma) Let U be an open subset of $\mathbb{R}^n$ and let $f \in L^2(U)$. Then*

$$\int_U f\,\Delta\phi = 0$$

for all smooth functions ϕ with compact support in U if and only if f is harmonic, meaning that $\Delta f = 0$.

Proof. Without loss of generality we may assume that $U = B_1^n$ is the unit ball centered at the origin. We can extend all functions on B_1^n to the entire $\mathbf{R}^n$ by setting them equal to zero outside B_1^n.

First note that if f is at least twice differentiable, then by Green's theorem

$$\int_{B_1^n} f\,\Delta\phi - \int_{B_1^n}\phi\,\Delta f = \int_{\partial B_1^n} f\frac{\partial\phi}{\partial n} - \phi\frac{\partial f}{\partial n} = 0.$$

Hence for all smooth functions ϕ with compact support in B_1^n,

$$\int_{B_1^n}\Delta f\,\phi = 0,$$

showing that $\Delta f = 0$, namely that f is harmonic.

But f is not necessarily a C^2-function. We resolve this issue by convoluting f with a mollifier that turns it into a smooth function. To this end, consider the bump function

$$\rho(\mathbf{x}) = \begin{Bmatrix} Ce^{-1/(1-\|\mathbf{x}\|^2)} & \text{if } \|\mathbf{x}\| \leq 1 \\ 0 & \text{if } \|\mathbf{x}\| \geq 1, \end{Bmatrix}$$

where $C = (\int_U e^{-1/(1-\|\mathbf{x}\|^2)}d\mathbf{x})^{-1}$. Define the family of mollifiers

$$\rho_\epsilon(\mathbf{x}) = \epsilon^{-n}\rho\left(\frac{\mathbf{x}}{\epsilon}\right), \quad \epsilon > 0.$$

These functions have integrals equal to 1, are supported in the ball B_ϵ^n of radius ϵ centered at the origin and converge, in distributional sense, to Dirac's delta function as $\epsilon \to 0$. Let us convolute f with these functions to obtain

$$f_\epsilon(\mathbf{x}) = (f * \rho_\epsilon)(\mathbf{x}) = \int_{\mathbf{R}^n} f(\mathbf{y})\rho_\epsilon(\mathbf{x}-\mathbf{y})d\mathbf{y}.$$

Since f is an L^2 function, f_ϵ is smooth because we can differentiate under the integral sign (using the Dominated Convergence Theorem). By Young's inequality for convolutions,

$$\|f_\epsilon\|_2 \leq \|\rho_\epsilon\|_1\|f\|_2 = \|f\|_2. \tag{4.5}$$

Next, we will show that

$$\|f - f_\epsilon\|_2 \to 0, \text{ as } \epsilon \to 0. \tag{4.6}$$

If g is another function, then

$$\begin{aligned}\|f - f_\epsilon\|_2 &\leq \|f-g\|_2 + \|g-g_\epsilon\|_2 + \|g_\epsilon - f_\epsilon\|_2 \\ &= \|f - f_\epsilon\|_2 \leq \|f-g\|_2 + \|g-g_\epsilon\|_2 + \|(g-f)_\epsilon\|_2.\end{aligned}$$

Combining this fact with (4.5), we see that if the convergence property (4.6) is true for a sequence of functions that approximates f in L^2, then it is true for f itself. We can choose the function g that approximates f to be continuous and compactly supported in B_1^n. We have

$$\begin{aligned}(g-g_\epsilon)(\mathbf{x}) &= \int_{\mathbb{R}^n}(g(\mathbf{x}) - g(\mathbf{x}-\mathbf{y}))\rho_\epsilon(\mathbf{y})d\mathbf{y} \\ &= \int_{\mathbb{R}^n}(g(\mathbf{x}) - g(\mathbf{x}-\epsilon\mathbf{z}))\rho(\mathbf{z})d\mathbf{z}.\end{aligned}$$

The integrand is compactly supported, tends pointwise to zero almost everywhere and is bounded from above by a constant that depends on ρ only (and not on ϵ). Hence by the Dominated Convergence Theorem $g - g_\epsilon$ converges to zero in L^2. This proves (4.6).

For $\epsilon < 1$, and $\mathbf{y}$ in the ball $B_{1-\epsilon}^n$, the function $\mathbf{x} \mapsto \rho_\epsilon(\mathbf{x}-\mathbf{y})$ is smooth and compactly supported in B_1^n. Using the hypothesis, we deduce that

$$\Delta f_\epsilon = \int_{\mathbb{R}^n} f(\mathbf{y})\Delta\rho_\epsilon(\mathbf{x}-\mathbf{y})d\mathbf{y} = 0.$$

This shows that f_ϵ is harmonic in $B_{1-\epsilon}^n$.

Now let $0 < \epsilon < \delta$. Then f_ϵ and f_δ are both harmonic in $B^n_{1-\delta}$. Let us examine $(f_\epsilon * \rho_\delta)(\mathbf{x})$. We have

$$(f_\epsilon * \rho_\delta)(\mathbf{x}) = \int_{B^n_{\mathbf{x},1-\delta}} f_\epsilon(\mathbf{y})\rho_\delta(\mathbf{x}-\mathbf{y})d\mathbf{y}$$

where $B^n_{\mathbf{x},1-\delta}$ is the ball of radius $1-\delta$ centered at $\mathbf{x}$. Switching to spherical coordinates this integral becomes

$$\int_0^{1-\delta} \rho_\delta(r) \int_{S^{n-1}_r} f_\epsilon(\mathbf{z})d\mathbf{z}dr,$$

because ρ_δ is constant on spheres centered at the origin. The inner integral is $A(S^{n-1}_r)f_\epsilon(\mathbf{x})$, by Theorem 4.6. Thus the integral is equal to

$$f(\mathbf{x})\int_0^{1-\delta} \rho_\delta(r)A(S^{n-1}_r)dr = \int_{\mathbb{R}^n} \rho_\delta(\mathbf{x})d\mathbf{x} = 1.$$

It follows that $f_\epsilon = f_\epsilon * \rho_\delta$ on $B^n_{1-\delta}$. For the same reason $f_\delta = f_\delta * \rho_\epsilon$ on $B^n_{1-\delta}$. We conclude that for $\epsilon < \delta$,

$$f_\delta = f_\delta * \rho_\epsilon = f * \rho_\delta * \rho_\epsilon = f * \rho_\epsilon * \rho_\delta = f_\epsilon * \rho_\delta = f_\epsilon \text{ on } B^n_{1-\delta}.$$

It follows that we can define a function f_h such that $f_h = f_\epsilon$ on $B^n_{1-\epsilon}$ for all $0 < \epsilon < 1$. This function is harmonic in the unit ball. Moreover

$$\|f_h - f_\epsilon\|_2 \to 0 \text{ when } \epsilon \to 0.$$

Combining this with (4.6), we deduce that $f = f_h$, so f is harmonic. The theorem is proved. □

4.1.3 *The construction of the Jacobian variety*

A harmonic basis of the first cohomology group

From Theorem 4.5 we deduce that

$$H = (E \oplus H)/E = (E^*)^\perp/E = H^1(\Sigma_g).$$

It follows that every homology class can be represented by a harmonic form. Consider the harmonic forms α_j, β_j, $j = 1, 2, \ldots, g$ such that, in $H^1(\Sigma_g)$,

$$\alpha_j = \eta_{b_j}, \quad \beta_j = -\eta_{a_j}, \quad j = 1, 2, \ldots, g.$$

By slightly modifying the entire framework, and working over $\mathbb{R}$ instead of $\mathbb{C}$, we can make sure that α_j, β_j are real forms. By Theorem 4.4,

$$\int_{a_j} \alpha_k = \int_{b_j} \beta_k = \delta_{jk}, \quad \int_{b_j} \alpha_k = \int_{a_j} \beta_k = 0. \tag{4.7}$$

The basis α_j, β_j, $j = 1, 2, \ldots, g$, satisfying (4.7) is called **dual** to the canonical basis a_j, b_j, $j = 1, 2, \ldots, g$.

We start with some computations. First, using (4.7) and (4.3) we obtain

$$\iint_{\Sigma_g} \alpha_j \wedge \alpha_k = \iint_{\Sigma_g} \beta_j \wedge \beta_k = 0, \quad \iint_{\Sigma_g} \alpha_j \wedge \beta_k = \delta_{jk}, \tag{4.8}$$

for $j, k = 1, 2, \ldots, g$. In other words,

$$\langle \alpha_j, *\alpha_k \rangle = \langle \beta_j, *\beta_k \rangle = 0, \quad \langle \alpha_j, *\beta_k \rangle = -\delta_{jk}.$$

and $\langle \alpha_j, \beta_k \rangle$.

Proposition 4.3. *If γ_1 and γ_2 be two closed 1-forms on Σ_g. Then*

$$\iint_{\Sigma_g} \gamma_1 \wedge \gamma_2 = \sum_{j=1}^{g} \left(\int_{a_j} \gamma_1 \int_{b_j} \gamma_2 - \int_{b_j} \gamma_1 \int_{a_j} \gamma_2 \right).$$

Proof. Stokes' Theorem implies that the integral on the left does not change if we replace γ_1 and γ_2 by forms homologous to them, hence we may assume that γ_1 and γ_2 are linear combinations of the α_j, β_j. Let

$$\gamma_k = \sum_{j=1}^{g} \mu_j^k \alpha_j + \sum_{j=1}^{g} \nu_j^k \beta_j, \quad k = 1, 2.$$

Then by (4.8)

$$\iint_{\Sigma_g} \gamma_1 \wedge \gamma_2 = \sum_{j=1}^{g} (\mu_j^1 \nu_j^2 - \mu_j^2 \nu_j^1),$$

and by (4.7),

$$\mu_j^k = \int_{a_j} \gamma_k, \quad \nu_j^k = \int_{b_j} \gamma_k, \quad k = 1, 2.$$

This proves the proposition. □

Corollary 4.1. *If γ is a harmonic 1-form, then*

$$\|\gamma\|^2 = \langle \gamma, *\bar{\gamma} \rangle = \sum_{j=1}^{g} \left(\int_{a_j} \gamma \int_{b_j} *\bar{\gamma} - \int_{b_j} \gamma \int_{a_j} *\bar{\gamma} \right).$$

Proof. Since γ is harmonic, both γ and $*\bar{\gamma}$ are closed. Hence we can apply the proposition. □

Using Proposition 4.3 we can write

$$\iint_{\Sigma_g} \alpha_j \wedge *\alpha_k = \int_{b_j} *\alpha_k, \quad \iint_{\Sigma_g} \alpha_j \wedge *\beta_k = \int_{b_j} *\beta_k,$$

$$\iint_{\Sigma_g} \beta_j \wedge *\alpha_k = -\int_{a_j} *\alpha_k, \quad \iint_{\Sigma_g} \beta_j \wedge *\beta_k = -\int_{a_j} *\beta_k, \; j,k = 1,2,\ldots,g.$$

Let the values of these integrals be a_{jk}, b_{jk}, c_{jk}, and d_{jk} respectively. Consider the matrices $A = (a_{jk})$, $B = (b_{jk})$, $C = (c_{jk})$, $D = (d_{jk})$, and arrange them into a matrix

$$\Gamma = \begin{pmatrix} A & B \\ C & D \end{pmatrix}.$$

Theorem 4.8. *The matrix Γ is real, symmetric, and positive definite.*

Proof. Γ is real because the α_j and β_j are real valued forms. Also $A^T = A$ and $D^T = D$ by their definition, and $C^T = B$ because

$$\iint_{\Sigma_g} \beta_j \wedge *\alpha_k = \iint_{\Sigma_g} *\beta_j \wedge ** \alpha_k = -\iint_{\Sigma_g} *\beta_j \wedge \alpha_k = \iint_{\Sigma_g} \alpha_k \wedge \beta_j.$$

So Γ is symmetric. To check that it is positive define, we show that for every nozero vector $(\boldsymbol{\mu}, \boldsymbol{\nu}) = (\mu_1, \mu_2, \ldots, \mu_g, \nu_1, \nu_2, \ldots, \nu_g)$ we have

$$(\boldsymbol{\mu}^T, \boldsymbol{\nu}^T)\Gamma \begin{pmatrix} \boldsymbol{\mu} \\ \boldsymbol{\nu} \end{pmatrix} > 0.$$

To this end we introduce the (non-zero) harmonic 1-form

$$\gamma = \sum_{j=1}^{g} \mu_j \alpha_j + \sum_{j=1}^{g} \nu_j \beta_j.$$

Using Corollary 4.1 we obtain

$$\begin{aligned} 0 < \|\gamma\|^2 &= \sum_{j=1}^{g} \left(\int_{a_j} \gamma \int_{b_j} *\gamma - \int_{b_j} \gamma \int_{a_j} *\gamma \right) \\ &= \sum_{k,l=1}^{g} \mu_k \mu_l \sum_{j=1}^{g} \left(\int_{a_j} \alpha_k \int_{b_j} *\alpha_l - \int_{b_j} \alpha_k \int_{a_j} *\alpha_l \right) \\ &\quad + \sum_{k,l=1}^{g} \mu_k \nu_l \sum_{j=1}^{g} \left(\int_{a_j} \alpha_k \int_{b_j} *\beta_l - \int_{b_j} \alpha_k \int_{a_j} *\beta_l \right) \\ &\quad + \sum_{k,l=1}^{g} \nu_k \mu_l \sum_{j=1}^{g} \left(\int_{a_j} \beta_k \int_{b_j} *\alpha_l - \int_{b_j} \alpha_k \int_{a_j} *\beta_l \right) \\ &\quad + \sum_{k,l=1}^{g} \nu_k \nu_l \sum_{j=1}^{g} \left(\int_{a_j} \beta_k \int_{b_j} *\beta_l - \int_{b_j} \beta_k \int_{a_j} *\beta_l \right). \end{aligned}$$

Using (4.7) we conclude that this is equal to

$$\boldsymbol{\mu}^T A\boldsymbol{\mu} + \boldsymbol{\mu}^T B\boldsymbol{\nu} + \boldsymbol{\nu}^T C\boldsymbol{\mu} + \boldsymbol{\nu}^T D\boldsymbol{\nu} = (\boldsymbol{\mu}^T, \boldsymbol{\nu}^T)\Gamma(\boldsymbol{\mu}, \boldsymbol{\nu}).$$

Hence Γ is positive definite. □

Holomorphic 1-forms and the period matrix

Next, we introduce the 1-forms

$$\widetilde{\alpha}_j = \alpha_j + i * \alpha_j, \quad \widetilde{\beta}_j = \beta_j + i * \beta_j, \quad j = 1, 2 \ldots, g,$$

which, by Theorem 4.2, are holomorphic.

Proposition 4.4. *The following equalities hold for* $j, k = 1, 2, \ldots, g$*:*

$$\int_{a_j} \widetilde{\beta}_k = -id_{jk}, \quad \int_{b_j} \widetilde{\alpha}_k = \delta_{jk} + ib_{jk}.$$

Proof. We compute

$$\int_{a_j} \widetilde{\beta}_k = \int_{a_j} (\beta_k + i * \beta_k) = \int_{a_j} \beta_k + i \int_{a_j} *\beta_k = 0 - id_{jk} = -id_{jk}$$

and

$$\int_{b_j} \widetilde{\beta}_k = \int_{b_j} (\beta_k + i * \beta_k) = \int_{b_j} \beta_k + i \int_{b_j} *\beta_k = \delta_{jk} + ib_{jk}.$$

□

Theorem 4.9. *The vector space of holomorphic* 1*-forms on* Σ_g *has dimension* g*, and a basis for this space is given by the forms* $\widetilde{\beta}_j$*,* $j = 1, 2, \ldots, g$*.*

Proof. Let $\mathscr{H}(\Sigma_g)$ be the space of holomorphic 1-forms, and $\bar{\mathscr{H}}(\Sigma_g)$ the space of anti-holomorphic forms (i.e. those whose conjugates are holomorphic). Then $\mathscr{H}(\Sigma_g) \cap \bar{\mathscr{H}}(\Sigma_g) = \{0\}$. The equality

$$\alpha = \frac{1}{2}(\alpha + i * \alpha) + \frac{1}{2}(\alpha - i * \alpha)$$

shows that every harmonic form is the sum of a holomorphic and an anti-holomorphic form. Consequently

$$H^1(\Sigma_g) = \mathscr{H}(\Sigma_g) \oplus \bar{\mathscr{H}}(\Sigma_g).$$

Because the conjugation operator is an isomorphism, it follows that the dimension of $\mathscr{H}(\Sigma_g)$ is half the dimension of $H^1(\Sigma_g)$, thus it is equal to g.

By Proposition 4.4,

$$\int_{a_j} \widetilde{\beta}_k = -id_{jk}, \quad j, k = 1, 2, \ldots, g.$$

The matrix $D = (d_{jk})$ is positive definite, hence invertible, so its columns are linearly independent. This implies that $\widetilde{\beta}_k$, $k = 1, 2, \ldots, g$ are linearly independent, hence they form a basis. □

Theorem 4.10. *There is a unique basis ζ_j, $j = 1, 2, \ldots, g$, of the space of holomorphic 1-forms $\mathscr{H}(\Sigma_g)$ with the property that*

$$\int_{a_j} \zeta_k = \delta_{jk}. \tag{4.9}$$

Moreover, in this basis, the matrix $\Pi = (\pi_{jk})$ defined by

$$\pi_{jk} = \int_{b_j} \zeta_k \tag{4.10}$$

is symmetric with positive definite imaginary part.

Proof. Define

$$(\zeta_1, \zeta_2, \ldots, \zeta_g)^T = (iD)^{-1}(\widetilde{\beta}_1, \widetilde{\beta}_2, \ldots, \widetilde{\beta}_g)^T.$$

Using Proposition 4.4, we see that (4.9) is satisfied, and that

$$\Pi = -D^{-1}B^T + iD^{-1} = -D^{-1}C + iD^{-1}.$$

We have $\operatorname{Im}\Pi = D^{-1}$, which is positive definite, since D is positive definite. Note that D^{-1} is also symmetric, so we need to show that $D^{-1}C$ is symmetric as well. Because

$$(D^{-1}C)^T = B(D^{-1})^T = BD^{-1},$$

we have to show that $D^{-1}C = BD^{-1}$, which is equivalent to $DB = CD$. Because α_j, β_j, $j = 1, 2, \ldots, g$, is a basis of $H^1(\Sigma_g)$, we can write

$$*\alpha_j = \sum_{j=1}^{g} \mu_{jk}\alpha_k + \sum_{j=1}^{g} \nu_{jk}\beta_k \tag{4.11}$$

$$*\beta_j = \sum_{j=1}^{g} \mu'_{jk}\alpha_k + \sum_{j=1}^{g} \nu'_{jk}\beta_k. \tag{4.12}$$

Then using (4.8) we obtain

$$a_{jk} = \iint_{\Sigma_g} \alpha_j \wedge *\alpha_k = \iint_{\Sigma_g} \alpha_j \wedge \left(\sum_{k=l}^{g} \mu_{kl}\alpha_l + \sum_{l=1}^{g} \nu_{kl}\beta_l \right) = \nu_{kj}.$$

This combined with the fact that $a_{jk} = a_{kj}$ gives $a_{jk} = \nu_{jk}$. Computing the same for the others, we deduce that the matrix of $*$ in the basis α_j, β_j, $j = 1, 2, \ldots, g$, is

$$\begin{pmatrix} -B & A \\ -D & C \end{pmatrix}.$$

And we have

$$*^2 = \begin{pmatrix} -I & 0 \\ 0 & -I \end{pmatrix} = \begin{pmatrix} B^2 - AD & -BA + AC \\ DB - CD & -DA + C^2 \end{pmatrix}.$$

Hence $DB - CD = 0$, as desired.

Finally, note that the functionals

$$\zeta_j \mapsto \int_{a_j} \zeta_j, \quad j = 1, 2, \ldots, g,$$

are linearly independent, and hence they form a basis for the dual space of $\mathscr{H}(\Sigma_g)$. This implies that every holomorphic form is completely determined by the values that these functionals take on it. This proves the uniqueness of the basis. □

The definition of the Jacobian variety

Let us summarize what we have done so far. Given a closed genus g surface, Σ_g, and a canonical basis a_j, b_j, $j = 1, 2, \ldots, g$, of $H_1(\Sigma_g, \mathbb{Z})$ we proved that there is a unique basis ζ_j, $j = 1, 2, \ldots g$, of the space of holomorphic 1-forms $\mathscr{H}(\Sigma_g)$ such that

$$\int_{a_j} \zeta_k = \delta_{jk}. \tag{4.13}$$

To this basis we associated the matrix $\Pi = (\pi_{jk})$ defined by

$$\pi_{jk} = \int_{b_j} \zeta_k. \tag{4.14}$$

We put all the integrals from (4.13) and (4.14) into one matrix (I, Π), called **period matrix**. The column vectors of this matrix are called periods.[2] We denote the column vectors by $\boldsymbol{\lambda}_1, \boldsymbol{\lambda}_2, \ldots, \boldsymbol{\lambda}_{2g}$. Note that for $j = 1, 2, \ldots, g$,

$$\boldsymbol{\lambda}_j = \mathbf{e}_j,$$

where $\mathbf{e}_j$ denotes the **standard basis vector**, whose kth coordinate is δ_{jk}, $j, k = 1, 2, \ldots, g$.

The matrix Π is uniquely determined by the complex structure on Σ_g and the canonical basis. Theorem 4.10 shows that Π is symmetric with positive definite imaginary part. The space of matrices with this property is called the **Siegel upper half-space**. Write

$$\Pi = X + iY,$$

[2] Recall the discussion in §1.1.

with X and Y real matrices. Then X and Y are symmetric, and Y is positive definite.

If $g = 1$, then $Y = \tau$, a complex number in the upper half-plane $\operatorname{Im}\tau > 0$. Moreover, every such number arises from a complex structure. The torus endowed with the complex structure and the canonical basis of the first homology defines a point in the Teichmüller space, and we obtain the standard identification between the Teichmüller space of the torus and the upper half-plane. If $g > 1$, characterizing which matrices Π arise through the above construction makes the object of Schottky's problem.

Using the fact that Y is positive definite, and hence invertible, we deduce that the vectors $\boldsymbol{\lambda}_1, \boldsymbol{\lambda}_2, \ldots, \boldsymbol{\lambda}_{2g}$ are linearly independent over $\mathbb{R}$, hence the form a basis of $\mathbb{C}^g$ viewed as a real vector space. The subgroup of $\mathbb{C}^g$ that they generate is a **lattice**. We denote this lattice by $\boldsymbol{\Lambda}(I, \Pi)$.

Definition 4.5. The **Jacobian variety** associated to the Riemann surface Σ_g endowed with canonical basis a_j, b_j, $j = 1, 2, \ldots, g$, is

$$\mathcal{J}(\Sigma_g) = \mathbb{C}^g / \boldsymbol{\Lambda}(I, \Pi).$$

The complex structure on $\mathbb{C}^g$ induces a complex structure on $\mathcal{J}(\Sigma_g)$. Hence $\mathcal{J}(\Sigma_g)$ is a complex torus. We let $\mathbf{z} = (z_1, z_2, \ldots, z_g)$ be the complex coordinates inherited from $\mathbb{C}^g$. To be rigorous, we should keep in mind that these coordinates are defined modulo the period lattice.

Remark 4.1. The group structure of $\mathbb{C}^g$ induces a group structure on $\mathcal{J}(\Sigma_g)$ for which both operations

$$(\mathbf{z}, \mathbf{w}) \mapsto \mathbf{z} + \mathbf{w} \text{ and } \mathbf{z} \mapsto -\mathbf{z}$$

are holomorphic. It follows that the Jacobian variety is what is called an **abelian variety** (for more details see [Mumford (1970)]).

Topologically, the Jacobian variety is homeomorphic to the $2g$-dimensional torus $(S^1)^{2g}$.

4.2 The quantization of the Jacobian variety of a Riemann surface in a real polarization

4.2.1 *Classical mechanics on the Jacobian variety*

We now introduce a new set of real coordinates on the Jacobian variety of the Riemann surface Σ_g. For this, consider the dual to the space of holomorphic 1-forms $\mathscr{H}(\Sigma_g)$, which we denote by $\mathscr{H}(\Sigma_g)^\#$. As a vector space over $\mathbb{C}$ it has a basis consisting of the functionals

$$\zeta_j^\#, j = 1, 2, \ldots, g, \text{ defined by } \zeta_j^\#(\zeta_k) = \delta_{jk}. \tag{4.15}$$

As a real vector space it has dimension $2g$, and basis $\zeta_j^\#, i\zeta_j^\#$, $j = 1, 2, \ldots, g$. Another basis is given by the functionals

$$\zeta \mapsto \int_{a_j} \zeta, \quad \zeta \mapsto \int_{b_j} \zeta, \quad j = 1, 2, \ldots, g. \tag{4.16}$$

The coordinates of these functionals in the basis $\zeta_j^\#, i\zeta_j^\#$ are the column vectors of the period matrix.

If we identify $\mathscr{H}(\Sigma_g)^\#$ with $\mathbb{C}^g$ using the basis (4.15), and factor it by the lattice defined by the functionals from (4.16), then we obtain the definition of the Jacobian variety from the previous section.

But we can also identify $\mathscr{H}(\Sigma_g)^\#$, as a real vector space, with $\mathbb{R}^g \times \mathbb{R}^g$ using the basis (4.16). Then the factorization by the same lattice identifies the Jacobian variety with a standard real torus as

$$\mathcal{J}(\Sigma_g) = \mathbb{R}^{2g}/\mathbb{Z}^{2g} = (S^1)^{2g}. \tag{4.17}$$

The map

$$a_j \mapsto (\zeta \mapsto \int_{a_j} \zeta), \quad b_j \mapsto (\zeta \mapsto \int_{b_j} \zeta), \quad j = 1, 2, \ldots, g,$$

defines an isomorphism between $H_1(\Sigma_g, \mathbb{R})$ and $\mathscr{H}(\Sigma_g)^\#$. Hence we can make the identification

$$\mathcal{J}(\Sigma_g) = H_1(\Sigma_g, \mathbb{R})/H_1(\Sigma_g, \mathbb{Z}). \tag{4.18}$$

Let $(\mathbf{x}, \mathbf{y})$ be the coordinates on $\mathcal{J}(\Sigma_g)$ induced by the coordinates on $\mathbb{R}^g \times \mathbb{R}^g$ under the identification (4.17). Each coordinate is defined modulo $\mathbb{Z}$. Recall the complex coordinates on the Jacobian variety from §4.1.3. Then

$$\mathbf{z} = \mathbf{x} + \Pi\mathbf{y}.$$

If we write $\mathbf{z} = \mathbf{u} + i\mathbf{v}$, then

$$\mathbf{u} = \mathbf{x} + X\mathbf{y}, \quad \mathbf{v} = Y\mathbf{y},$$

and

$$\mathbf{x} = \mathbf{u} - XY^{-1}\mathbf{v}, \quad \mathbf{y} = Y^{-1}\mathbf{v}.$$

We transform the Jacobian variety into the **phase space** of a classical mechanical system by introducing the symplectic form

$$\omega = \sum_{j=1}^{g} dx_j \wedge dy_j. \tag{4.19}$$

In the $(\mathbf{u}, \mathbf{v})$ coordinates, this symplectic form is

$$\begin{aligned}(d\mathbf{x})^T \wedge d\mathbf{y} &= [(d\mathbf{u})^T - (d\mathbf{v})^T Y^{-1} X) \wedge Y^{-1} d\mathbf{v} \\ &= (d\mathbf{u})^T \wedge Y^{-1} d\mathbf{v} - (d\mathbf{v})^T Y^{-1} X Y^{-1} \wedge d\mathbf{v}.\end{aligned}$$

Because X and Y are symmetric, so is $Y^{-1}XY^{-1}$. This implies that the second term is zero, since the $dv_j \wedge dv_k$ cancel in pairs. Hence

$$\omega = (d\mathbf{u})^T \wedge Y^{-1} d\mathbf{v},$$

or, in complex coordinates,

$$\begin{aligned}\omega &= -\frac{i}{4}(d\mathbf{z} + d\bar{\mathbf{z}})(d\mathbf{z} - d\bar{\mathbf{z}}) \\ &= \frac{i}{2}(d\mathbf{z})^T \wedge Y^{-1} d\bar{\mathbf{z}} + \frac{i}{4}(d\mathbf{z})^T \wedge Y^{-1} d\mathbf{z} - \frac{i}{4}(d\bar{\mathbf{z}})^T \wedge Y^{-1} d\bar{\mathbf{z}}.\end{aligned}$$

Again because Y^{-1} is symmetric, the second and third terms are zero, so

$$\omega = \frac{i}{2}(d\mathbf{z})^T \wedge Y^{-1} d\bar{\mathbf{z}}.$$

The **classical observables** of this mechanical system are functions on the torus. Each such observable can be written as a Fourier series of exponentials. For that reason it suffices to concentrate only on the exponential functions

$$\exp(2\pi i(\mathbf{p}^T\mathbf{x} + \mathbf{q}^T\mathbf{y})), \quad \mathbf{p} \text{ and } \mathbf{q} \in \mathbb{Z}^g.$$

Using the identification (4.18), we can view $(\mathbf{p}, \mathbf{q})$ as an element of $H_1(\Sigma_g, \mathbb{Z})$. As such,

$$(\mathbf{p}, \mathbf{q}) \mapsto \exp(2\pi i(\mathbf{p}^T\mathbf{x} + \mathbf{q}^T\mathbf{y})), \quad \mathbf{p} \text{ and } \mathbf{q} \in \mathbb{Z}^g,$$

defines an isomorphism between $H_1(\Sigma_g, \mathbb{Z})$ and the multiplicative group of periodic exponential functions on $\mathcal{J}(\Sigma_g)$.

By analogy with §1.3, we can interpret $\mathcal{J}(\Sigma_g)$ as a space of g one-dimensional particles with periodic positions and momenta. In this perspective, the coordinates $\mathbf{x} = (x_1, x_2, \ldots, x_g)$ are the **momenta** and the coordinates $\mathbf{y} = (y_1, y_2, \ldots, y_g)$ are the **positions**, so that the form $\omega = \sum_j dx_j \wedge dy_j$ is the standard symplectic form in classical mechanics.

4.2.2 *The Hilbert space of the quantization of the Jacobian variety in a real polarization*

The curvature of a line bundle

Because the prequantization line bundle lives over a torus, let us explain a few facts about line bundles over general manifolds.

Definition 4.6. A (complex) **line bundle** $\mathcal{L}$ on a manifold M is defined by a cover $(U_j)_{j\in J}$ of M by contractible open sets such that $U_j \cap U_k$ is either empty or contractible for all j and k, and for every pair (j,k) with $U_j \cap U_k \neq \emptyset$ a smooth map

$$c_{jk} : U_j \cap U_k \to \mathbb{C}\backslash\{0\}.$$

The maps c_{jk} should satisfy the conditions

$$c_{kj} = (c_{jk})^{-1} \text{ and } c_{jk}c_{kl}c_{lj} = 1, \text{ for all } j,k,l \in J. \tag{4.20}$$

The line bundle itself is the quotient of the disjoint union of the sets $U_j \times \mathbb{C}$ by the equivalence relation which identifies $(x,z) \in U_j \times \mathbb{C}$ with $(x, c_{jk}z)$ for all $x \in U_j \cap U_k$, $z \in \mathbb{C}$.

A line bundle is called **holomorphic** if the transition functions c_{jk} are holomorphic.

The conditions (4.20) mean that c_{jk}, $j,k \in J$, is a **Čech cocycle**, more precisely a Čech 2-cocycle. If we consider a family of functions

$$d_j : U_j \to \mathbb{C}\backslash\{0\},$$

then the cocycle $d_j c_{jk}$ defines an equivalent line bundle.

Cocycles form an abelian group under the multiplication,

$$((c_{jk}),(c'_{jk})) \mapsto (c_{jk}c'_{jk})$$

which we denote by $\check{Z}^2((U_j),\mathbb{C})$. Let $\check{B}^2((U_j),\mathbb{C})$ be the subgroup consisting of cocycles of the form $d_j d_k^{-1}$, $j,k \in J$, where the functions d_j are as above. If the quotient of two cocycles is in $\check{B}^2((U_j),\mathbb{C})$ then they define the same line bundle. In fact, it can be proved (see [Griffiths and Harris (1994)] or [Mumford (1970)]) that this is a necessary and sufficient condition.

The tensor product of two line bundles over the same manifold is obtained by considering a collection of charts common to both and the associated cocycles. Then the cocycle of the tensor product line bundle is the product of the cocycles of the two line bundles.

The quotient group

$$\check{H}^2((U_j),\mathbb{C}) = \check{Z}^2((U_j),\mathbb{C})/\check{B}^2((U_j),\mathbb{C})$$

is called the second **Čech cohomology group**. Each element of $\check{H}^2((U_j), \mathbb{C})$ defines a line bundle up to equivalence.

Now let us consider the case where M is endowed with a symplectic 2-form ω. We want to explain how a line bundle with curvature ω is constructed, and when this is possible.

The form ω being symplectic is closed, and so on each open set U_j it is exact, by Poincaré's Lemma. Hence there are real 1-forms θ_j on U_j called potentials[3], such that

$$d\theta_j = \omega.$$

We have $d(\theta_j - \theta_k) = 0$ on $U_j \cap U_k$, so by using again Poincaré's Lemma on the contractible domains $U_j \cap U_k$, we deduce that there are smooth functions

$$f_{jk} : U_j \cap U_k \to \mathbb{R}$$

such that $f_{jk} = \theta_j - \theta_k$.

Assume that

$$c_{jk} = e^{if_{jk}} \tag{4.21}$$

satisfies the cocycle condition (4.20). Then it defines a line bundle $\mathcal{L}$. The line bundle $\mathcal{L}$ is said to have curvature ω.

The 1-forms $-i\theta_j$ are the local expressions of a connection form on M. Indeed, by differentiating the relation (4.21) we obtain

$$dc_{jk} = ie^{if_{jk}} df_{jk} = ic_{jk}(\theta_j - \theta_k).$$

This can be rewritten as

$$-i\theta_k = c_{jk}^{-1} dc_{jk} + c_{jk}^{-1}(-i\theta_j)c_{jk},$$

and we recognize the formula expressing the change of the connection form under changes of coordinates.

A deep theorem [Weil (1958)] states that such a line bundle exists if and only if the **Weil integrality condition**

$$\frac{1}{2\pi}\omega \in H^2(M, \mathbb{Z})$$

is satisfied. This condition means that the integral of ω over any closed oriented surface is an integral multiple of 2π. The form $1/2\pi\omega$ is the first Chern class of the line bundle.

We do not prove Weil's theorem. Instead we impose the integrality condition and then construct the line bundle explicitly.

[3] Not to be confused with the theta functions to be defined later.

The prequantization line bundle

Let us first quantize the Jacobian variety in a real polarization using the methods of §2.3.2. We work in the real coordinates $(\mathbf{x}, \mathbf{y})$, and so we make the identification

$$\mathcal{J}(\Sigma_g) = (S^1)^{2g} = \mathbb{R}^{2g}/\mathbb{Z}^{2g},$$

where the lattice $\mathbf{\Lambda}(I, \Pi)$ is identified with $\mathbb{Z}^{2g}$ and is now generated by The lattice is now generated by the standard vectors $\mathbf{e}_j$, $j = 1, 2, \ldots, 2g$ (where the vector $\mathbf{e}_j$ has all coordinates equal to zero except for the jth which is equal to 1).

The symplectic form ω was chosen so that in order for

$$\frac{1}{h}\omega = \frac{1}{2\pi\hbar}\omega$$

to define an integral class in cohomology, **Planck's constant h must be the reciprocal of an integer**. We additionally impose the condition that this integer is **even**, for reasons that will become clear later. Hence we choose Planck's constant

$$h = \frac{1}{N}, \quad N \text{ even.}$$

The prequantization line bundle is a line bundle $\mathcal{L}$ on $\mathcal{J}(\Sigma_g)$ with curvature $2\pi N\omega$.

Consider the quotient map

$$\pi : \mathbb{R}^{2g} \to \mathcal{J}(\Sigma_g).$$

We can define the line bundle $\mathcal{L}$ on $\mathcal{J}(\Sigma_g)$ by taking as the open contractible sets U_j, $j \in J$ to be the 4^{2g} images through π of the translates of $(0, 1/2)^{2g}$ by vectors whose coordinates range in the set $\{0, 1/4, 1/2, 3/4\}$, and then constructing the corresponding cocycle.

We can simplify the description of the line bundle $\mathcal{L}$ by using the covering map π. Let $\pi^*(\mathcal{L})$ be the pull back of $\mathcal{L}$ to $\mathbb{R}^{2g}$. Because every line bundle on $\mathbb{R}^{2g}$ is trivial,

$$\pi^*(\mathcal{L}) = \mathbb{R}^{2g} \times \mathbb{C}.$$

The quotient map

$$\pi^*(\mathcal{L}) = \mathbb{R}^{2g} \times \mathbb{C} \to \mathcal{L}$$

is defined by a map

$$\Lambda : \mathbb{C}^g \times \mathbb{Z}^{2g} \to \mathbb{C}$$

that tells how fibers that lie over the same point are identified. This means that $((\mathbf{x},\mathbf{y}),\zeta)$ is identified with $((\mathbf{x}',\mathbf{y}'),\zeta')$ if and only if

$$(\mathbf{x}',\mathbf{y}') = (\mathbf{x},\mathbf{y}) + \boldsymbol{\lambda}, \quad \zeta' = \Lambda((\mathbf{x},\mathbf{y}),\boldsymbol{\lambda})\zeta.$$

Consistency (namely the fact that the above identification of fibers is an equivalence relation) forces Λ to satisfy the cocycle condition

$$\Lambda((\mathbf{x},\mathbf{y}),\boldsymbol{\lambda})\Lambda((\mathbf{x},\mathbf{y})+\boldsymbol{\lambda},\boldsymbol{\lambda}') = \Lambda((\mathbf{x},\mathbf{y})\boldsymbol{\lambda}+\boldsymbol{\lambda}'),$$

for all $(\mathbf{x},\mathbf{y}) \in \mathbb{R}^{2g}, \boldsymbol{\lambda}, \boldsymbol{\lambda}' \in \mathbb{Z}^{2g}$.

The pull back $\pi^*(2\pi N\omega)$ of the form $2\pi N\omega$ is still closed, because closeness is a local condition and the lattice $\mathbb{Z}^{2g}$ acts discretely. Because of the Poincaré Lemma it is exact, so we can find a global potential for it on $\mathbb{R}^{2g}$. In fact

$$\theta = -2\pi N \frac{1}{\hbar}\mathbf{y}^T d\mathbf{x} = -2\pi N \sum_{l=1}^{g} y_l dx_l$$

is such a global potential. Let θ_j be the restriction of this potential to one of the sets U_j. Then on $U_j \cap U_k$

$$\theta_j - \theta_k = 2\pi N \sum_{l\in I(j,k)} \epsilon(l) dx_l$$

where $U_k = \mathbf{v} + U_j$ and $I(j,k)$ is the set of l, $1 \leq l \leq g$ such that the coordinate v_{g+l} is equal to $\frac{3}{4}\epsilon(l)$ with $\epsilon(l) = \pm 1$. This formula is motivated by the fact that for such an l, a point common to $U_j \cap U_k$ has the lth y-coordinate equal y_l in the chart U_j, and to $y_l + \epsilon(l)$ in the chart U_k (Figure 4.1 should help this intuition). In particular, if no coordinate of $\mathbf{v}$ is $\pm\frac{3}{4}$, then on $U_j \cap U_k$ one has $\theta_j = \theta_k$.

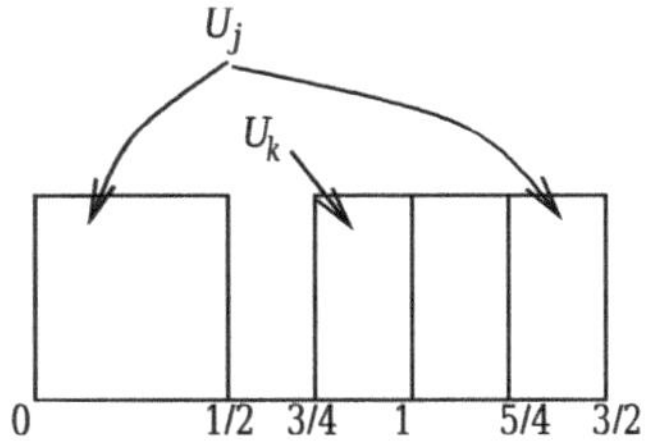

Fig. 4.1 The overlap of two charts on the torus

We can choose $f_{jk} = 2\pi N \sum_{l\in I(j,k)} x_l$, and hence the cocycle of the line bundle is given by

$$\Lambda((\mathbf{x},\mathbf{y}),\mathbf{e}_j) = 1, \quad \Lambda((\mathbf{x},\mathbf{y}),\mathbf{e}_{g+j}) = e^{-2\pi i N x_j}, \quad j = 1,2,\ldots,g.$$

We view the sections of the line bundle as functions s on $\mathbb{R}^g \times \mathbb{R}^g$ satisfying the following **periodicity conditions**

$$\begin{aligned} s(x_1,\ldots,x_j+1,\ldots,y_g) &= s(x_1,\ldots,x_j\ldots,y_g), \\ s(x_1,\ldots,y_j+1,\ldots,y_g) &= e^{-2\pi i N x_j} s(x_1,\ldots,y_j,\ldots,y_g), \end{aligned} \tag{4.22}$$

for $j = 1,2,\ldots,g$.

Note that the cocycle preserves the standard length in $\mathbb{C}$. So we can use the standard Euclidean metric as the metric in the fibers of the line bundle. Then the measure of integration, used for defining the inner product of sections, is the standard Lebesgue measure on the fundamental domain $[0,1]^{2g}$.

The Hilbert space of the quantization

We treat the y_j's as the positions and the x_j's as the momenta (ξ_j and η_j in the notation from Chapter 2). We adapt Example 2.9 to the case of the torus. Consider the real polarization $\mathbf{F}$ on $\mathcal{J}(\Sigma_g)$ defined by the vector fields

$$\frac{\partial}{\partial x_1}, \frac{\partial}{\partial x_2}, \ldots, \frac{\partial}{\partial x_g}.$$

This means that we work in the position representation.

Lemma 4.2. *The covariantly constant sections of the prequantization line bundle are of the form*

$$s(\mathbf{x},\mathbf{y}) = \psi(\mathbf{y})e^{-2\pi i N \mathbf{x}^T \mathbf{y}},$$

where ψ has the support in the submanifold

$$S = \left\{(\mathbf{x},\mathbf{y}) \,|\, \mathbf{y} + \frac{\boldsymbol{\mu}}{N} = 0, \boldsymbol{\mu} \in \mathbb{Z}_N^g\right\}.$$

Proof. Again we view sections as functions on $\mathbb{R}^g \times \mathbb{R}^g$ satisfying the periodicity conditions (4.22). The condition that a section s is covariantly constant along the polarization is expressed by the system of differential equations

$$\nabla_{\partial/\partial x_j} s = 0, \quad j = 1,2,\ldots,g.$$

This means that

$$\frac{\partial}{\partial x_j} s + 2\pi i N y_j s = 0.$$

Solving the system of differential equations we obtain

$$s(\mathbf{x},\mathbf{y}) = \psi(\mathbf{y})e^{-2\pi i N \mathbf{x}^T \mathbf{y}}.$$

The conditions (4.22) yield

$$\psi(\mathbf{y})e^{-2\pi i N\mathbf{x}^T\mathbf{y}-2\pi i N y_j} = \psi(\mathbf{y})e^{-2\pi i N\mathbf{x}^T\mathbf{y}}$$
$$\psi(\mathbf{y})e^{-2\pi i N\mathbf{x}^T\mathbf{y}-2\pi i N x_j} = e^{-2\pi i N x_j}\psi(\mathbf{y})e^{-2\pi i N\mathbf{x}^T\mathbf{y}},$$

for $j = 1, 2, \ldots, g$. The second condition is automatically satisfied. The first condition implies that that $\psi(\mathbf{y}) = 0$ unless $N\mathbf{y} \in \mathbb{Z}^g$. We conclude that the sections must be supported in the submanifold

$$S = \left\{(\mathbf{x}, \mathbf{y}) \,|\, \mathbf{y} + \frac{\boldsymbol{\mu}}{N} = 0, \boldsymbol{\mu} \in \mathbb{Z}_N^g\right\}.$$

□

The situation differs significantly from that in §2.3.2 in the sense that the sections have support in the submanifold S, which has Lebesgue measure zero. So if we want to identify the sections with the functions $\psi(\mathbf{y})$, as we did before, then these functions are supported on a finite set. At first glance, the Hilbert space is trivial!

Such situations appear often when applying the method of geometric quantization. They have been resolved by Śniatycki in full generality, and the procedure has been outlined in [Śniatycki (1980)]. We thus apply Śniatycki's method.

The Hilbert space consists therefore of the covariantly constant sections, which are supported in the submanifold S. The submanifold S is called the **Bohr-Sommerfeld variety**. Its connected components

$$S_{\boldsymbol{\mu}} = \left\{(\mathbf{x}, \mathbf{y}) \,|\, \mathbf{y} + \frac{\boldsymbol{\mu}}{N} = 0\right\}$$

are called Bohr-Sommerfeld fibers. They are Lagrangian submanifolds of the Jacobian variety.

Let us look for sections supported entirely in one Bohr-Sommerfeld fiber. Modulo multiplication by a constant, there is exactly one such section for each fiber $S_{\boldsymbol{\mu}}$, namely

$$s_{\boldsymbol{\mu}}(\mathbf{x}, \mathbf{y}) = \begin{cases} e^{-2\pi i N\mathbf{x}^T\mathbf{y}} & \text{if } \mathbf{y} + \frac{\boldsymbol{\mu}}{N} = \mathbf{0}, \\ 0 & \text{otherwise.} \end{cases} \tag{4.23}$$

Any other covariantly constant section is a linear combination of these. We conclude that $s_{\boldsymbol{\mu}}$, $\boldsymbol{\mu} \in \mathbb{Z}_N^g$, is a basis of the Hilbert space of the quantization. The only difficulty lies in defining the inner product.

We point out that exactly like in §2.3.2, the bundle $\Lambda^n\mathbf{F}$ of linear frames in $\mathbf{F}$ is trivial, and the trivialization is

$$\mathcal{J}(\Sigma_g) \times C \to \Lambda^n\mathbf{F}, \quad (p, z) \mapsto z\mathrm{Vol}_p,$$

where

$$\mathrm{Vol} = dx_1 \wedge dx_2 \wedge \cdots \wedge dx_g.$$

For that reason the bundle $\mathcal{L}''$ such that $\mathcal{L}'' \otimes \mathcal{L}'' = \wedge^n \mathbf{F}$ can be chosen to be trivial. Hence the choice of a half-density necessary for the definition of the inner product is just the choice of a scaling constant for the integral in the fiber. According to [Śniatycki (1980)], the inner product is defined as

$$\langle s, s' \rangle = \sum_{\boldsymbol{\mu} \in \mathbb{Z}_N^g} \int_{S_{\boldsymbol{\mu}}} s(\mathbf{x}, \mathbf{y}) \overline{s'(\mathbf{x}, \mathbf{y})} c_{\boldsymbol{\mu}} d\mathbf{x}.$$

where $c_{\boldsymbol{\mu}}$ are the scaling constants. We set $c_{\boldsymbol{\mu}} = 1$, $\boldsymbol{\mu} \in \mathbb{Z}_N^g$. With this choice of the inner product, the sections

$$s_{\boldsymbol{\mu}}, \quad \boldsymbol{\mu} \in \mathbb{Z}_N^g,$$

form an orthonormal basis.

We convene that whenever we write $s_{\mathbf{n}}$ where $\mathbf{n}$ has integer entries, these entries are automatically considered modulo N.

4.2.3 *The Schrödinger representation of the finite Heisenberg group*

The Jacobian variety $\mathcal{J}(\Sigma_g)$ is a quotient of $\mathbb{R}^g \times \mathbb{R}^g$, and so for quantizing the observables we can apply the method of Weyl quantization in an equivariant fashion. We use the results of §2.3.3, pointing out that we are only interested in the quantization of the exponential functions

$$\exp(2\pi i(\mathbf{p}^T \mathbf{x} + \mathbf{q}^T \mathbf{y})), \quad \mathbf{p} \text{ and } \mathbf{q} \in \mathbb{Z}^g.$$

In the quantization of the phase space $\mathbb{R}^g \times \mathbb{R}^g$, the functions x_j and y_j, $j = 1, 2, \ldots, g$, are quantized as

$$\mathrm{op}(x_j) = -i\hbar \frac{\partial}{\partial y_j} + M_{x_j} = -\frac{i}{2\pi N} \frac{\partial}{\partial y_j} + M_{y_j}$$

$$\mathrm{op}(y_j) = i\hbar \frac{\partial}{\partial x_j} = \frac{i}{2\pi N} \frac{\partial}{\partial x_j},$$

where M_{x_j} is the operator of multiplication by the variable x_j. According to the Weyl quantization scheme,

$$\mathrm{op}\left(e^{2\pi i(\mathbf{p}^T \mathbf{x} + \mathbf{q}^T \mathbf{y})}\right) = e^{2\pi i[\mathbf{p}^T \mathrm{op}(\mathbf{x}) + \mathbf{q}^T \mathrm{op}(\mathbf{y})]},$$

where $\mathrm{op}(\mathbf{x}) = (\mathrm{op}(x_1), \ldots, \mathrm{op}(x_g))$ and $\mathrm{op}(\mathbf{y}) = (\mathrm{op}(y_1), \ldots, \mathrm{op}(y_g))$.

Theorem 4.11. *The quantized exponentials act on the sections* $s_{\boldsymbol{\mu}}$, $\boldsymbol{\mu} \in \mathbb{Z}_N^g$, *defined by (4.23), by the formula*

$$op\left(e^{2\pi i(\mathbf{p}^T \mathbf{x} + \mathbf{q}^T \mathbf{y})}\right) s_{\boldsymbol{\mu}} = e^{-\frac{2\pi i}{N} \boldsymbol{\mu}^T \mathbf{q} - \frac{\pi i}{N} \mathbf{p}^T \mathbf{q}} s_{\boldsymbol{\mu} + \mathbf{p}}.$$

Proof. Introduce the function

$$\psi_0(\mathbf{y}) = \begin{cases} 1 & \text{if } \mathbf{y} = \mathbf{0}, \\ 0 & \text{otherwise.} \end{cases}$$

Then we can write

$$s_{\boldsymbol{\mu}} = \psi_0\left(\mathbf{y} + \frac{\boldsymbol{\mu}}{N}\right) e^{-2\pi i N \mathbf{x}^T \mathbf{y}}.$$

Applying (2.24) and using the fact that $s_\mu \neq 0$ only when $\mathbf{y} = -\frac{\boldsymbol{\mu}}{N}$ we obtain

$$\begin{aligned}
&\operatorname{op}\left(e^{2\pi i(\mathbf{p}^T\mathbf{x}+\mathbf{q}^T\mathbf{y})}\right) s_{\boldsymbol{\mu}} = \operatorname{op}\left(e^{2\pi i(\mathbf{p}^T\mathbf{x}+\mathbf{q}^T\mathbf{y})}\right)\psi_0\left(\mathbf{y}+\frac{\boldsymbol{\mu}}{N}\right)e^{-2\pi i\boldsymbol{\mu}^T\mathbf{x}} \\
&= e^{2\pi i\mathbf{q}^T(-\frac{\boldsymbol{\mu}}{N})-\frac{\pi i}{N}\mathbf{p}^T\mathbf{q}}\psi_0\left(\mathbf{y}+\frac{\boldsymbol{\mu}}{N}+h\mathbf{p}\right)e^{-2\pi i N\mathbf{x}^T\mathbf{y}} \\
&= e^{-\frac{2\pi i}{N}\mathbf{q}^T\boldsymbol{\mu}-\frac{\pi i}{N}\mathbf{p}^T\mathbf{q}}\psi_0\left(\mathbf{y}+\frac{\boldsymbol{\mu}+\mathbf{p}}{N}\right)e^{-2\pi i N\mathbf{x}^T\mathbf{y}} \\
&= e^{-\frac{2\pi i}{N}\boldsymbol{\mu}^T\mathbf{q}-\frac{\pi i}{N}\mathbf{p}^T\mathbf{q}}s_{\boldsymbol{\mu}+\mathbf{p}}.
\end{aligned}$$

□

Because the sections $s_{\boldsymbol{\mu}}$ form an orthonormal basis, the formula deduced in Theorem 4.11 implies that quantized exponentials are unitary operators. The same formula yields the following multiplication rule for quantized exponentials

$$\begin{aligned}
&\operatorname{op}\left(e^{2\pi i(\mathbf{p}^T\mathbf{x}+\mathbf{q}^T\mathbf{y})}\right)\operatorname{op}\left(e^{2\pi i(\mathbf{p}'^T\mathbf{x}+\mathbf{q}'^T\mathbf{y})}\right) \\
&\quad = e^{\frac{\pi i}{N}(\mathbf{p}^T\mathbf{q}'-\mathbf{p}'^T\mathbf{q})}\operatorname{op}\left(e^{2\pi i(\mathbf{p}+\mathbf{p}')^T\mathbf{x}+(\mathbf{q}+\mathbf{q}')^T\mathbf{y})}\right).
\end{aligned}$$

We notice that by introducing quantizations of some constant functions we can obtain a group of unitary operators. Specifically, the quantization of a constant function of the form $e^{\frac{\pi i}{N}k}$, with $k \in \mathbb{Z}$, is the operator of multiplication by $e^{\frac{\pi i}{N}k}$, and so the operator associated to the function $e^{2\pi i(\mathbf{p}^T\mathbf{x}+\mathbf{q}^T\mathbf{y})+\frac{\pi i}{N}k}$ acts on the sections $s_{\boldsymbol{\mu}}$ by

$$\operatorname{op}\left(e^{2\pi i(\mathbf{p}^T\mathbf{x}+\mathbf{q}^T\mathbf{y})+\frac{\pi i}{N}k}\right)s_{\boldsymbol{\mu}} = e^{-\frac{2\pi i}{N}\boldsymbol{\mu}^T\mathbf{q}-\frac{\pi i}{N}\mathbf{p}^T\mathbf{q}+\frac{\pi i}{N}k}s_{\boldsymbol{\mu}+\mathbf{p}}. \tag{4.24}$$

Proposition 4.5. *The set of unitary operators of the form*

$$op(e^{2\pi i(\mathbf{p}^T\mathbf{x}+\mathbf{q}^T\mathbf{y})+\frac{i\pi}{N}k}), \quad \mathbf{p} \textit{ and } \mathbf{q} \in \mathbb{Z}^g, \quad k \in \mathbb{Z}$$

is a group.

Proof. Using (4.24) we obtain the multiplication rule

$$\mathrm{op}\left(e^{2\pi i(\mathbf{p}^T\mathbf{x}+\mathbf{q}^T\mathbf{y})+\frac{\pi i}{N}k}\right)\mathrm{op}\left(e^{2\pi i(\mathbf{p}'^T\mathbf{x}+\mathbf{q}'^T\mathbf{y})+\frac{\pi i}{N}k'}\right)$$
$$=\mathrm{op}\left(e^{2\pi i(\mathbf{p}+\mathbf{p}')^T\mathbf{x}+(\mathbf{q}+\mathbf{q}')^T\mathbf{y})+\frac{\pi i}{N}(k+k'+\mathbf{p}^T\mathbf{q}'-\mathbf{p}'^T\mathbf{q})}\right),$$

which shows that the product of two such operators is an operator of the same type.

The identity operator arises by setting the entries of $\mathbf{p}, \mathbf{q}$, and also k equal to zero, while

$$\mathrm{op}\left(e^{2\pi i(\mathbf{p}^T\mathbf{x}+\mathbf{q}^T\mathbf{y})+\frac{\pi i}{N}k}\right)^{-1}=\mathrm{op}\left(e^{2\pi i(-\mathbf{p}^T\mathbf{x}-\mathbf{q}^T\mathbf{y})-\frac{\pi i}{N}k}\right). \qquad \square$$

We want to understand the structure of this group. Focusing just on the exponents in the above multiplication formula, we recognize the multiplication rule for the **Heisenberg group with integer entries**. This group is

$$\mathbf{H}(\mathbb{Z}^g)=\{(\mathbf{p},\mathbf{q},k)\,|\,\mathbf{p},\mathbf{q}\in\mathbb{Z}^g, k\in\mathbb{Z}\},$$

with multiplication

$$(\mathbf{p},\mathbf{q},k)(\mathbf{p}',\mathbf{q}',k')=(\mathbf{p}+\mathbf{p}',\mathbf{q}+\mathbf{q}',k+k'+\mathbf{p}^T\mathbf{q}'-\mathbf{p}'^T\mathbf{q}).$$

The map

$$(\mathbf{p},\mathbf{q},k)\mapsto\mathrm{op}\left(e^{2\pi i(\mathbf{p}^T\mathbf{x}+\mathbf{q}^T\mathbf{y})+\frac{\pi i}{N}k}\right)$$

defines a representation of $\mathbf{H}(\mathbb{Z}^g)$ on the Hilbert space of states. To make this representation faithful, we factor it by its kernel.

Lemma 4.3. *The set of elements in $\mathbf{H}(\mathbb{Z}^g)$ that act as identity operators is the normal subgroup consisting of elements of the form $(\mathbf{p},\mathbf{q},k)$ with the coordinates of the vectors $\mathbf{p}$ and $\mathbf{q}$ being multiples of N, and k being a multiple of $2N$.*

Proof. For $(\mathbf{p},\mathbf{q},k)$ to act as the identity operator, we should have

$$e^{-\frac{2\pi i}{N}\boldsymbol{\mu}^T\mathbf{q}-\frac{\pi i}{N}\mathbf{p}^T\mathbf{q}}s_{\boldsymbol{\mu}+\mathbf{p}}=s_{\boldsymbol{\mu}}$$

for all $\boldsymbol{\mu}\in\{0,1,\ldots,N-1\}^g$. Consequently, $\mathbf{p}$ should be in $N\mathbb{Z}^g$. Then $\mathbf{p}^T\mathbf{q}$ is a multiple of N, so the coefficient $e^{-\frac{2\pi i}{N}\boldsymbol{\mu}^T\mathbf{q}-\frac{\pi i}{N}\mathbf{p}^T\mathbf{q}+\frac{\pi i}{N}k}$ equals $\pm e^{-\frac{2\pi i}{N}\boldsymbol{\mu}^T\mathbf{q}+\frac{\pi i}{N}k}$. This coefficient should be equal to 1. For $\boldsymbol{\mu}=(0,0,\ldots,0)$ this implies that $-\mathbf{p}^T\mathbf{q}+k$ should be an even multiple of N. Next, by varying $\boldsymbol{\mu}$ we conclude that each entry of $\mathbf{q}$ is a multiple of N. Because N is even, it follows that $\mathbf{p}^T\mathbf{q}$ is an even multiple of N, and consequently k is an even multiple of N. Thus any element in the kernel of the representation must belong to $N\mathbb{Z}^{2g}\times(2N)\mathbb{Z}$. It is easy to see that any element of this form is in the kernel. $\square$

Corollary 4.2. *The subgroup of the elements acting as the identity consists of all elements of the form* $(\mathbf{p},\mathbf{q},k)^N$ *with* k *even.*

Remark 4.2. In obtaining the description of the subgroup that acts as the identity on theta functions given by Lemma 4.3 it was essential that N is **even.**

It follows from Lemma 4.3 that the group of quantized exponentials is isomorphic to the quotient of $\mathbf{H}(\mathbb{Z}^g)$ by the subgroup consisting of elements of the form $(\mathbf{p},\mathbf{q},k)$, with the coordinates of $\mathbf{p}$ and $\mathbf{q}$ being multiples of N and k being a multiple of $2N$. This quotient group is itself a finite Heisenberg group. Let us recall the general definition of finite Heisenberg groups.

Definition 4.7. A finite Heisenberg group H is a central extension

$$0 \to \mathbb{Z}_m \to H \to K \to 0$$

where K is a finite abelian group and the commutator pairing

$$K \times K \to \mathbb{Z}_m, \quad (k,k') \mapsto [\tilde{k},\tilde{k}'],$$

where $\tilde{k}$, and $\tilde{k}'$ are arbitrary lifts of k and k' to H, identifies K with the group of homomorphisms from K to $\mathbb{Z}_m$.

Proposition 4.6. *The group of quantized exponentials is isomorphic to the finite Heisenberg group*

$$\{(\mathbf{p},\mathbf{q},k)\,|\,\mathbf{p},\mathbf{q} \in \mathbb{Z}_N^g, k \in \mathbb{Z}_{2N}\}$$

with the multiplication rule

$$(\mathbf{p},\mathbf{q},k)(\mathbf{p}',\mathbf{q}',k') = (\mathbf{p}+\mathbf{p}',\mathbf{q}+\mathbf{q}',k+k'+2\mathbf{p}^T\mathbf{q}').$$

The isomorphism is induced by the map $F:\mathbf{H}(\mathbb{Z}^g)\to\mathbb{Z}_N^{2g}\times\mathbb{Z}_{2N}$,

$$F(\mathbf{p},\mathbf{q},k) = (\mathbf{p}\ \mathit{mod}\ N, \mathbf{q}\ \mathit{mod}\ N, k+\mathbf{p}^T\mathbf{q}\ \mathit{mod}\ 2N).$$

Proof. The following computation shows that F is a group homomorphism

$$\begin{aligned} F((\mathbf{p},\mathbf{q},k)(\mathbf{p}',\mathbf{q}',k')) &= F((\mathbf{p}+\mathbf{p}',\mathbf{q}+\mathbf{q}',k+k'+\mathbf{p}^T\mathbf{q}'-\mathbf{p}'^T\mathbf{q}))\\ &= (\mathbf{p}+\mathbf{p}',\mathbf{q}+\mathbf{q}',k+k'+\mathbf{p}^T\mathbf{q}'-\mathbf{p}'^T\mathbf{q}+(\mathbf{p}+\mathbf{p}')^T(\mathbf{q}+\mathbf{q}'))\\ &= (\mathbf{p}+\mathbf{p}',\mathbf{q}+\mathbf{q}',k+k'+\mathbf{p}^T\mathbf{q}+\mathbf{p}'^T\mathbf{q}'+2\mathbf{p}^T\mathbf{q}')\\ &= F((\mathbf{p},\mathbf{q},k))F((\mathbf{p}',\mathbf{q}',k')). \end{aligned}$$

An element $(\mathbf{p}, \mathbf{q}, k)$ is in the kernel of this map if and only if all coordinates of the vectors $\mathbf{p}$ and $\mathbf{q}$ are multiples of N, and $k + \mathbf{p}^T\mathbf{q}$ is a multiple of $2N$. Because N is even (recall our initial choice of Planck's constant), if $\mathbf{p}$ and $\mathbf{q}$ are in $N\mathbb{Z}^g$, then $\mathbf{p}^T\mathbf{q}$ is a multiple of $2N$. Hence the last condition translates to k being itself a multiple of $2N$. Using Lemma 4.3 we conclude that F factors to the desired isomorphism. □

We denote the **finite Heisenberg group** of quantized exponentials by $\mathbf{H}(\mathbb{Z}_N^g)$, and from now on, whenever we mention the finite Heisenberg group we refer to this group. Also, to simplify the notation, we let

$$\exp(\mathbf{p}^T\mathbf{P} + \mathbf{q}^T\mathbf{Q} + kE) := \mathrm{op}\left(e^{2\pi i(\mathbf{p}^T\mathbf{x}+\mathbf{q}^T\mathbf{y})+\frac{\pi i}{N}k}\right).$$

Here $\mathbf{P}$ and $\mathbf{Q}$ are the momentum and position operators, and $E = \frac{i\pi}{N}Id$, and the notation reflects the Weyl quantization scheme. Note also that $\exp(\mathbf{p}^T\mathbf{P}+\mathbf{q}^T\mathbf{Q}+kE)$ is the image of $(\mathbf{p}, \mathbf{q}, k) \in \mathbf{H}(\mathbb{Z}^g)$ through the quotient map. The action of $\mathbf{H}(\mathbb{Z}_N^g)$ on the Hilbert space of the quantization will be called the **Schrödinger representation of the finite Heisenberg group**, by analogy with the Schrödinger representation of the Heisenberg group with real entries from §2.1.2.

One should stress out the noncommutation relation in $\mathbf{H}(\mathbb{Z}_N^g)$:

$$\begin{aligned}&\exp(\mathbf{p}^T\mathbf{P} + \mathbf{q}^T\mathbf{Q})\exp(\mathbf{p}'^T\mathbf{P} + \mathbf{q}'^T\mathbf{Q})\\ &= e^{\frac{2\pi i}{N}(\mathbf{p}^T\mathbf{q}'-\mathbf{p}'^T\mathbf{q})}\exp(\mathbf{p}'^T\mathbf{P} + \mathbf{q}'^T\mathbf{Q})\exp(\mathbf{p}^T\mathbf{P} + \mathbf{q}^T\mathbf{Q}).\end{aligned} \tag{4.25}$$

The multiplication formula in $\mathbf{H}(\mathbb{Z}_N^g)$ can either be seen as descending from the multiplication rule in $\mathbf{H}(\mathbb{Z}_N^g)$, or as given by Proposition 4.6. We prefer the first point of view, because it has a topological flavor which we now explain.

Recall the discussion from §4.2.1 in which we established the homeomorphism

$$H_1(\Sigma_g, \mathbb{R})/H_1(\Sigma_g, \mathbb{Z}) \to \mathcal{J}(\Sigma_g).$$

It shows that each element of the canonical basis corresponds to one of the $2g$ dimensions of the Jacobian variety. The map

$$\sum_j (p_j a_j + q_j b_j) \mapsto e^{2\pi i \sum_j (p_j x_j + q_j y_j)}$$

defines an isomorphism

$$H_1(\Sigma_g, \mathbb{Z}) \to \left\{e^{2\pi i(\mathbf{p}^T\mathbf{x}+\mathbf{q}^T\mathbf{y})} \mid \mathbf{p}, \mathbf{q} \in \mathbb{Z}^g\right\}, \tag{4.26}$$

between the first homology group with integer coefficients and the group of exponential functions on the Jacobian variety. Using this identification, we deduce that the Heisenberg group with integer entries $\mathbf{H}(\mathbb{Z}^g)$ is a $\mathbb{Z}$-extension of $H_1(\Sigma_g, \mathbb{Z}) = \mathbb{Z}^g \times \mathbb{Z}^g$ by the cocycle

$$c_{\mathbf{H}(\mathbb{Z}^g)}((\mathbf{p}, \mathbf{q}), (\mathbf{p}', \mathbf{q}')) = \mathbf{p}^T\mathbf{q}' - \mathbf{p}'^T\mathbf{q}.$$

Note that

$$c_{\mathbf{H}(\mathbb{Z}^g)}((\mathbf{p}, \mathbf{q}), (\mathbf{p}', \mathbf{q}')) = (\mathbf{p}, \mathbf{q}) \cdot (\mathbf{p}', \mathbf{q}'),$$

namely the algebraic intersection number of

$$\sum_j (p_j a_j + q_j b_j) \text{ and } \sum_j (p_j' a_j + q_j' b_j).$$

We conclude that $\mathbf{H}(\mathbb{Z}^g)$ is the $\mathbb{Z}$-extension of $H_1(\Sigma_g, \mathbb{Z})$ by the cocycle defined by the intersection form.

The finite dimensional Stone-von Neumann theorem

Like in the case of the Heisenberg group with real entries (see Theorem 2.1), a Stone-von Neumann Theorem holds in this case as well.

Theorem 4.12. *(Stone-von Neumann) The Schrödinger representation of* $\mathbf{H}(\mathbb{Z}_N^g)$ *is the* unique *irreducible unitary representation of this group with the property that* $\exp(kE)$ *acts as* $e^{\frac{\pi i}{N}k}I$ *for all* $k \in \mathbb{Z}$.

Proof. Let $X_j = \exp(P_j)$, $Y_j = \exp(Q_j)$, $j = 1, 2, \ldots, g$, $Z = \exp(E)$. Then $X_jY_j = Z^2Y_jX_j$, $X_jY_k = Y_kX_j$ if $j \neq k$, $X_jX_k = X_kX_j$, $Y_jY_k = Y_kY_j$, $ZX_j = X_jZ$, $ZY_j = Y_jZ$, for all i, j, and $X_j^N = Y_j^N = Z^{2N} = Id$ for all j. Because $Y_1, Y_2, \ldots, Y_g$ commute pairwise, they have a common eigenvector v. And because $Y_j^N = Id$ for all j, the eigenvalues $\lambda_1, \lambda_2, \ldots, \lambda_g$ of v with respect to the $Y_1, Y_2, \ldots, Y_g$ are roots of unity. The equalities

$$Y_jX_jv = e^{-\frac{2\pi i}{N}}X_jY_j = e^{-\frac{2\pi i}{N}}\lambda_jX_jv,$$
$$Y_jX_kv = X_kY_jv = \lambda_jX_kv, \quad \text{if } j \neq k$$

show that by applying X_j's repeatedly gives rise to an eigenvector v_0 of the commuting system $Y_1, Y_2, \ldots, Y_g$ whose eigenvalues are all equal to 1. The irreducible representation is spanned by the vectors $X_1^{n_1}X_2^{n_2}\cdots X_g^{n_g}v_0$, $n_i \in \{0, 1, \ldots, N-1\}$. Any such vector is an eigenvector of the system $Y_1, Y_2, \ldots, Y_g$, with eigenvalues respectively $e^{\frac{2\pi i}{N}n_1}$, $e^{\frac{2\pi i}{N}n_2}, \ldots,$ $e^{\frac{2\pi i}{N}n_g}$. So these vectors are linearly independent and form a basis of the irreducible representation. It is not hard to see that the action of $\mathbf{H}(\mathbb{Z}_N^g)$ on the vector space spanned by these vectors is the Schrödinger representation. $\square$

Remark 4.3. While we do not need this fact, we should notice that if we drop the condition that $\exp(kE)$ acts as multiplication by $e^{\frac{\pi i}{N}k}I$, then other irreducible representations exist. In fact, each $(\boldsymbol{\alpha}, \boldsymbol{\beta}, \gamma) \in \mathbb{Z}_d^g \times \mathbb{Z}_d^g \times \mathbb{Z}_{2N}$, where $d = \gcd(N, \gamma)$ gives rise to a distinct irreducible representation. Such a representation can be constructed by choosing a vector v, setting

$$X_j v = e^{\frac{2\pi i}{d}\alpha_j} v, \quad Zv = e^{\frac{\pi i}{N}\gamma} v,$$

and imposing $Y_j^{N/d} v = e^{\frac{2\pi i}{d}\beta_j} v$.

4.3 Theta functions via quantum mechanics

4.3.1 *Theta functions from the geometric quantization of the Jacobian variety in a Kähler polarization*

In this section we quantize the Jacobian variety in a Kähler polarization. The discussion parallels somewhat that of [Manoliu (1997)]. The variable is $\mathbf{z} = \mathbf{x} + i\mathbf{y}$, where $\mathbf{x}$ are the "momenta" and $\mathbf{y}$ are the "positions". So we are performing holomorphic quantization in what we called the momentum representation from §2.2.3.

We use the same **Planck's constant**

$$h = \frac{1}{N}, \quad N \text{ even},$$

so that the symplectic form is an integral class in cohomology. This time the prequantization line bundle is a **holomorphic** line bundle $\mathcal{L}$ with curvature

$$\frac{2\pi}{h}\omega = \frac{1}{\hbar}\omega = \pi i N (d\mathbf{z})^T \wedge Y^{-1} d\mathbf{z}.$$

Note that the first Chern class of $\mathcal{L}$ is $(1/h)\omega$, which lies in $H^2(\mathcal{J}(\Sigma_g), \mathbb{Z})$.

We try to understand again $\mathcal{L}$ by pulling it back to $\mathbb{C}^g \times \mathbb{C}$ by the quotient map

$$\mathbb{C}^g \times \mathbb{C} \to \mathcal{L}.$$

The pullback being necessarily topologically trivial, $\mathcal{L}$ is defined by a cocycle

$$\Lambda : \mathbb{C}^g \times \mathbf{\Lambda}(I, \Pi) \to \mathbb{C}$$

that tells how fibers are identified. This means that $(\mathbf{z}, \zeta)$ is identified with $(\mathbf{z}', \zeta')$ if and only if

$$\mathbf{z}' = \mathbf{z} + \boldsymbol{\lambda}, \quad \zeta' = \Lambda(\mathbf{z}, \boldsymbol{\lambda})\zeta.$$

Consistency (namely the fact that $(\mathbf{z},\zeta)\sim(\mathbf{z}',\zeta')$ is an equivalence relation) forces Λ to satisfy the cocycle condition

$$\Lambda(\mathbf{z},\boldsymbol{\lambda})\Lambda(\mathbf{z}+\boldsymbol{\lambda},\boldsymbol{\mu})=\Lambda(\mathbf{z},\boldsymbol{\lambda}+\boldsymbol{\mu}),\ \text{ for all } \mathbf{z}\in\mathbb{C}^g,\boldsymbol{\lambda},\boldsymbol{\mu}\in\boldsymbol{\Lambda}(I,\Pi).$$

In order for the line bundle to be holomorphic, the cocycle Λ should be holomorphic in $\mathbf{z}$.

Consider a Hermitian metric on $\mathcal{L}$ compatible with the curvature $(1/h)\omega$, and let us pull it back to the trivial line bundle $\mathbb{C}^g\times\mathbb{C}$. It is then given by a function $h:\mathbb{C}^g\to(0,\infty)$ such that the length of the vector w in the fiber $\mathbb{C}$ over $\mathbf{z}$ is $h(\mathbf{z})\|w\|^2$.

The condition that the quotient map is a fiber-wise isometry translates to

$$h(\mathbf{z})=|\Lambda(\mathbf{z},\boldsymbol{\lambda})|h(\mathbf{z}+\boldsymbol{\lambda}). \tag{4.27}$$

Recall that in §2.3.2 we deduced as a consequence of Lemma 2.3 that h should satisfy the differential equation

$$\bar{\partial}\partial\ln h=(-i/\hbar)\omega. \tag{4.28}$$

While h uniquely determines ω, the converse is not true. We will make a choice for h that produces a line bundle whose holomorphic sections are the canonical theta functions.

Proposition 4.7. *The function*

$$h(\mathbf{z})=e^{-\frac{\pi N}{2}(\mathbf{z}-\bar{\mathbf{z}})^TY^{-1}(\mathbf{z}-\bar{\mathbf{z}})}$$

satisfies the equation (4.28).

Proof. We denote by $\mathbf{e}_j$ the **standard basis vector**, this time of $\mathbb{C}^g$, whose kth coordinate is δ_{jk}, $j,k=1,2,\ldots,g$. We compute

$$\begin{aligned}\bar{\partial}\ln h(\mathbf{z})&=-\frac{\pi N}{2}\sum_{j=1}^g[-\mathbf{e}_j^TY^{-1}(\mathbf{z}-\bar{\mathbf{z}})-(\mathbf{z}-\bar{\mathbf{z}})^TY^{-1}\mathbf{e}_j]d\bar{z}_j\\&=\pi N\sum_{j=1}^g[\mathbf{e}_j^TY^{-1}(\mathbf{z}-\bar{\mathbf{z}})]d\bar{z}_j,\end{aligned}$$

where for the last equality we used the fact that Y^{-1} is symmetric. Next,

$$\partial\bar{\partial}\ln h(\mathbf{z})=\pi N\sum_{j=1}^g\sum_{k=1}^g(\mathbf{e}_j^TY^{-1}\mathbf{e}_k)dz_k\wedge d\bar{z}_j.$$

It is not hard to recognize this as

$$\pi N(d\mathbf{z})^T\wedge Y^{-1}d\bar{\mathbf{z}}=(-i/\hbar)\omega.$$

□

Switching to the $(\mathbf{x},\mathbf{y})$ coordinates, we can rewrite the formula for h as

$$h(\mathbf{x},\mathbf{y}) = e^{-2\pi N \mathbf{y}^T Y \mathbf{y}}.$$

Using equation (4.27), we obtain

$$\begin{aligned}|\Lambda(\mathbf{z},\boldsymbol{\lambda})| &= \frac{e^{-\frac{\pi N}{2}(\mathbf{z}-\bar{\mathbf{z}})^T Y^{-1}(\mathbf{z}-\bar{\mathbf{z}})}}{e^{-\frac{\pi N}{2}(\mathbf{z}+\boldsymbol{\lambda}-\bar{\mathbf{z}}-\bar{\boldsymbol{\lambda}})^T Y^{-1}(\mathbf{z}+\boldsymbol{\lambda}-\bar{\mathbf{z}}-\bar{\boldsymbol{\lambda}})}} \\ &= e^{-\frac{\pi N}{2}(\boldsymbol{\lambda}-\bar{\boldsymbol{\lambda}})^T Y^{-1}(\mathbf{z}-\bar{\mathbf{z}})-\frac{\pi N}{2}(\mathbf{z}-\bar{\mathbf{z}})^T Y^{-1}(\boldsymbol{\lambda}-\bar{\boldsymbol{\lambda}})+\frac{\pi N}{2}(\boldsymbol{\lambda}-\bar{\boldsymbol{\lambda}})^T Y^{-1}(\boldsymbol{\lambda}-\bar{\boldsymbol{\lambda}})} \\ &= e^{-\pi N(\boldsymbol{\lambda}-\bar{\boldsymbol{\lambda}})^T Y^{-1}(\mathbf{z}-\bar{\mathbf{z}})+\frac{\pi N}{2}(\boldsymbol{\lambda}-\bar{\boldsymbol{\lambda}})^T Y^{-1}(\boldsymbol{\lambda}-\bar{\boldsymbol{\lambda}})}.\end{aligned}$$

Let us substitute the period vectors for $\boldsymbol{\lambda}$. Note that for $j = 1, 2, \ldots, g$, $(\boldsymbol{\lambda}_{g+j} - \bar{\boldsymbol{\lambda}}_{g+j})^T Y^{-1} = 2i\mathbf{e}_j^T$. We thus have

$$|\Lambda(\mathbf{z},\boldsymbol{\lambda}_j)| = 1, \quad |\Lambda(\mathbf{z},\boldsymbol{\lambda}_{g+j})| = e^{-2\pi i N(z_j-\bar{z}_j)-2\pi N \operatorname{Im}\pi_{jj}}, \quad j = 1, 2, \ldots, g.$$

Here we used the fact that the jth entry of λ_{g+j} is π_{jj}. There are many possible Λ with these prescribed absolute values; we make the simplest choice:

$$\Lambda(\mathbf{z},\boldsymbol{\lambda}_j) = 1, \quad \Lambda(\mathbf{z},\boldsymbol{\lambda}_{g+j}) = e^{-2\pi i N z_j - \pi N \pi_{jj}}, \quad j = 1, 2, \ldots, g. \tag{4.29}$$

We stress out that other choices exist, which amount to tensoring $\mathcal{L}$ by a flat line bundle.

The cocycle Λ now defines a line bundle $\mathcal{L}$ over the Jacobian variety.

Next, we introduce the Kähler polarization

$$\frac{\partial}{\partial \bar{z}_1}, \frac{\partial}{\partial \bar{z}_2}, \ldots, \frac{\partial}{\partial \bar{z}_g}.$$

The considerations of §2.3.2 apply to show that the covariantly constant sections in the direction of this polarization are the holomorphic sections. It is customary to pull them back and view them as functions f on $\mathbb{C}^g$ that satisfy the twisted double periodicity conditions

$$\begin{aligned} f(\mathbf{z}+\boldsymbol{\lambda}_j) &= f(\mathbf{z}) \\ f(\mathbf{z}+\boldsymbol{\lambda}_{g+j}) &= e^{-2\pi i N z_j - \pi i N \pi_{jj}} f(\mathbf{z}), \end{aligned} \tag{4.30}$$

for $j = 1, 2, \ldots, g$.

Definition 4.8. The holomorphic sections of the line bundle $\mathcal{L}$ defined by the cocycle (4.29) are called **theta functions**.

These are **Riemann's theta functions**, and in the normalization of this book, they are usually referred to as **canonical theta functions**.

We will switch freely between the point of view in which theta functions are sections of $\mathcal{L}$ and the one in which they are holomorphic functions on $\mathbb{C}^g$

satisfying the periodicity conditions (4.30). We denote the space of theta functions by $\boldsymbol{\Theta}_N^{\Pi}(\Sigma_g)$.

The Hermitian metric given by h also defines the inner product on $\boldsymbol{\Theta}_N^{\Pi}(\Sigma_g)$, up to multiplication by a positive constant. We make a choice of this constant so as to favor a nice formula for an orthonormal basis. For this reason, we define the inner product to be

$$\langle f,g\rangle = (2n)^{g/2}\det(Y)^{1/2}\int_D f(\mathbf{z})\overline{g(\mathbf{z})}e^{\frac{\pi N}{2}(\mathbf{z}-\bar{\mathbf{z}})^T Y^{-1}(\mathbf{z}-\bar{\mathbf{z}})}dm,$$

where D is a fundamental domain of the lattice $\boldsymbol{\Lambda}(I,\Pi)$, and dm is the Lebesgue measure $d\mathbf{x}d\mathbf{y}$ on this domain. In the real coordinates $(\mathbf{x},\mathbf{y})$ the inner product is given by

$$\langle f,g\rangle = (2N)^{g/2}\det(Y)^{1/2}\int_{[0,1]^{2g}} f(\mathbf{x},\mathbf{y})\overline{g(\mathbf{x},\mathbf{y})}e^{-2\pi N\mathbf{y}^T Y\mathbf{y}}d\mathbf{x}d\mathbf{y}. \quad (4.31)$$

Theorem 4.13. *An orthonormal basis for the space of theta functions is given by*

$$\theta_{\boldsymbol{\mu}}^{\Pi}(\mathbf{z}) = \sum_{\mathbf{n}\in\mathbb{Z}^g} e^{2\pi i N\left[\frac{1}{2}\left(\frac{\boldsymbol{\mu}}{N}+\mathbf{n}\right)^T\Pi\left(\frac{\boldsymbol{\mu}}{N}+\mathbf{n}\right)+\left(\frac{\boldsymbol{\mu}}{N}+\mathbf{n}\right)^T\mathbf{z}\right]}, \quad \boldsymbol{\mu}\in\{0,1\ldots,N-1\}^g.$$

Definition 4.9. The theta functions defined in Theorem 4.13 are called **theta series**.

Proof. Let us first explain why each of the series from the statement converges absolutely on compact sets, and hence defines a holomorphic function. The general term of the series can be written as

$$e^{-\pi\mathbf{n}^T Y\mathbf{n}+\mathbf{a}^T\mathbf{n}+ir(\mathbf{n})},$$

where $\mathbf{a}\in\mathbb{C}^n$ does not depend on $\mathbf{n}$ and $r(\mathbf{n})$ is a real valued function. So the absolute value of the term is

$$e^{-\pi\mathbf{n}^T Y\mathbf{n}+\mathbf{a}^T\mathbf{n}}.$$

As explained during the construction of the Jacobian variety, Y is a positive definite matrix with real entries. The eigenvalues of Y are positive, and if λ is the smallest of them, then $Y-\lambda I$ is semi-positive definite, so

$$\mathbf{n}^T Y\mathbf{n}\geq\mathbf{n}^T\lambda I\mathbf{n}=\lambda\|\mathbf{n}\|^2.$$

Hence the absolute value of the term of the series is bounded from above by

$$e^{-\lambda\|\mathbf{n}\|^2+\|\mathbf{a}\|\cdot\|\mathbf{n}\|}.$$

The inequality $e^x > x+1$ implies that for large n,

$$e^{-\lambda\|\mathbf{n}\|^2+\|\mathbf{a}\|\cdot\|\mathbf{n}\|} \leq \frac{1}{\lambda\|\mathbf{n}\|^2 - \|\mathbf{a}\|\cdot\|\mathbf{n}\|}.$$

The series

$$\sum_{\mathbf{n}\in\mathbb{Z}^n} \frac{1}{\lambda\|\mathbf{n}\|^2 - \|\mathbf{a}\|\cdot\|\mathbf{n}\|}$$

is convergent because the sum of the reciprocals of squares converges, and hence the theta series are absolutely convergent. The number a depends linearly on the imaginary part of z. Hence when z varies in a compact set, all these series are bounded from above by a constant. This proves the uniform convergence on compact sets.

The computations

$$\begin{aligned}\theta_{\boldsymbol{\mu}}^{\Pi}(\mathbf{z}+\boldsymbol{\lambda}_j) = \theta_{\boldsymbol{\mu}}^{\Pi}(\mathbf{z}+\mathbf{e}_j) &= \sum_{\mathbf{n}\in\mathbb{Z}^g} e^{2\pi i N\left[\frac{1}{2}\left(\frac{\boldsymbol{\mu}}{N}+\mathbf{n}\right)^T\Pi\left(\frac{\boldsymbol{\mu}}{N}+\mathbf{n}\right)+\left(\frac{\boldsymbol{\mu}}{N}+\mathbf{n}\right)^T(\mathbf{z}+\mathbf{e}_j)\right]}\\
&= \sum_{\mathbf{n}\in\mathbb{Z}^g} e^{2\pi i(\mu_j+Nn_j)} e^{2\pi i N\left[\frac{1}{2}\left(\frac{\boldsymbol{\mu}}{N}+\mathbf{n}\right)^T\Pi\left(\frac{\boldsymbol{\mu}}{N}+\mathbf{n}\right)+\left(\frac{\boldsymbol{\mu}}{N}+\mathbf{n}\right)^T\mathbf{z}\right]}\\
&= \sum_{\mathbf{n}\in\mathbb{Z}^g} e^{2\pi i N\left[\frac{1}{2}\left(\frac{\boldsymbol{\mu}}{N}+\mathbf{n}\right)^T\Pi\left(\frac{\boldsymbol{\mu}}{N}+\mathbf{n}\right)+\left(\frac{\boldsymbol{\mu}}{N}+\mathbf{n}\right)^T\mathbf{z}\right]}\end{aligned}$$

and

$$\begin{aligned}\theta_{\boldsymbol{\mu}}^{\Pi}(\mathbf{z}+\boldsymbol{\lambda}_{g+j}) &= \sum_{\mathbf{n}\in\mathbb{Z}^g} e^{2\pi i N\left[\frac{1}{2}\left(\frac{\boldsymbol{\mu}}{N}+\mathbf{n}\right)^T\Pi\left(\frac{\boldsymbol{\mu}}{N}+\mathbf{n}\right)+\left(\frac{\boldsymbol{\mu}}{N}+\mathbf{n}\right)^T(\mathbf{z}+\boldsymbol{\lambda}_{g+j})\right]}\\
&= \sum_{\mathbf{n}\in\mathbb{Z}^g} e^{2\pi i N\left[\frac{1}{2}\left(\frac{\boldsymbol{\mu}}{N}+\mathbf{n}\right)^T\Pi\left(\frac{\boldsymbol{\mu}}{N}+\mathbf{n}\right)+\left(\frac{\boldsymbol{\mu}}{N}+\mathbf{n}\right)^T\boldsymbol{\lambda}_{g+j}+\left(\frac{\boldsymbol{\mu}}{N}+\mathbf{n}\right)^T\mathbf{z}\right]}\\
&= \sum_{\mathbf{n}\in\mathbb{Z}^g} e^{2\pi i N\left[\frac{1}{2}\left(\frac{\boldsymbol{\mu}}{N}+\mathbf{n}\right)^T\Pi\left(\frac{\boldsymbol{\mu}}{N}+\mathbf{n}\right)+\frac{1}{2}\left(\frac{\boldsymbol{\mu}}{N}+\mathbf{n}\right)^T\Pi\mathbf{e}_j+\frac{1}{2}\mathbf{e}_j\Pi\left(\frac{\boldsymbol{\mu}}{N}+\mathbf{n}\right)^T+\left(\frac{\boldsymbol{\mu}}{N}+\mathbf{n}\right)^T\mathbf{z}\right]}\\
&= \sum_{\mathbf{n}\in\mathbb{Z}^g} e^{2\pi i N\left[\frac{1}{2}\left(\frac{\boldsymbol{\mu}}{N}+\mathbf{n}+\mathbf{e}_j\right)^T\Pi\left(\frac{\boldsymbol{\mu}}{N}+\mathbf{n}+\mathbf{e}_j\right)-\frac{1}{2}\mathbf{e}_j\Pi\mathbf{e}_j+\left(\frac{\boldsymbol{\mu}}{N}+\mathbf{n}+\mathbf{e}_j\right)^T\mathbf{z}-z_j\right]}\\
&= \sum_{\mathbf{n}\in\mathbb{Z}^g} e^{2\pi i N\left[\frac{1}{2}\left(\frac{\boldsymbol{\mu}}{N}+\mathbf{n}\right)^T\Pi\left(\frac{\boldsymbol{\mu}}{N}+\mathbf{n}\right)-\frac{1}{2}\mathbf{e}_j\Pi\mathbf{e}_j+\left(\frac{\boldsymbol{\mu}}{N}+\mathbf{n}\right)^T\mathbf{z}-z_j\right]}\\
&= e^{-2\pi i N z_j-\pi i N\pi_{jj}}\theta_{\boldsymbol{\mu}}^{\Pi}(\mathbf{z})\end{aligned}$$

show that $\theta_{\boldsymbol{\mu}}^{\Pi}(\mathbf{z})$ are theta functions.

The proof that these theta series form an orthonormal set in $\boldsymbol{\Theta}_N^{\Pi}(\Sigma_g)$ is postponed until the next section, where it will be a consequence of a more general computation from the second proof to Theorem 4.14.

To show that they are a spanning set for the space of theta functions, let f be a theta function. The conditions $f(\mathbf{z}+\boldsymbol{\lambda}_j)=f(\mathbf{z})$, $j=1,2,\ldots,g$, show that f is periodic in each variable z_j, so it admits a Fourier series expansion

$$f(z)=\sum_{\mathbf{m}\in\mathbb{Z}^g} a_{\mathbf{m}}e^{2\pi i\mathbf{m}^T\mathbf{z}}.$$

The condition

$$f(\mathbf{z}+\boldsymbol{\lambda}_{g+j})=e^{-2\pi iNz_j-\pi iN\pi_{jj}}f(\mathbf{z})$$

translates to

$$\sum_{\mathbf{m}\in\mathbb{Z}^g} a_{\mathbf{m}}e^{2\pi i\mathbf{m}^T\boldsymbol{\lambda}_{g+j}}e^{2\pi i\mathbf{m}^T\mathbf{z}}=\sum_{\mathbf{m}\in\mathbb{Z}^g} a_{\mathbf{m}}e^{-\pi iN\pi_{jj}}e^{2\pi i(\mathbf{m}+N\mathbf{e}_j)^T\mathbf{z}}.$$

Equating the Fourier coefficients we obtain

$$a_{\mathbf{m}}=e^{-2\pi i\mathbf{m}^T\boldsymbol{\lambda}-\pi iN\pi_{jj}}a_{\mathbf{m}-N\mathbf{e}_j}.$$

It follows that all Fourier coefficients are determined by $a_{\mathbf{m}}$, $\mathbf{m}\in\{0,1,2,\ldots,N-1\}^g$. Hence $\boldsymbol{\Theta}_N^{\Pi}(\Sigma_g)$ has dimension at most N^g, and since we found an orthonormal set with N^g elements, that set must be an orthonormal basis. □

From this moment on we agree to take the indices $\boldsymbol{\mu}$ of the theta series $\theta_{\boldsymbol{\mu}}^{\Pi}$ modulo N, thus to consider these indices as elements of $\mathbb{Z}_N^g$.

4.3.2 *The action of the finite Heisenberg group on theta functions*

In [Weil (1964)] André Weil examined translations in the variables of theta functions and discovered an action of a finite Heisenberg group on $\boldsymbol{\Theta}_N^{\Pi}(\Sigma_g)$, which is the same Heisenberg group that we introduced in the previous section. We place Weil's work in a quantum mechanical perspective and obtain the action of the finite Heisenberg group from the quantization of exponentials. This is the complex variable version of the Schrödinger representation from §4.2.3.

Because the Jacobian variety $\mathcal{J}(\Sigma_g)$ is the quotient of $\mathbb{C}^g$ by the discrete action of a group, we can perform **equivariant** Weyl quantization. This means that we only quantize the functions that lift from the Jacobian variety and we let them act only on theta functions (with the caveat that the inner product is the integral over just the fundamental domain). The existence of the action of the finite Heisenberg group is then a direct consequence of the next result.

Theorem 4.14. *The Weyl quantization in the Kähler polarization of the exponential functions is given by*

$$op\left(e^{2\pi i(\mathbf{p}^T\mathbf{x}+\mathbf{q}^T\mathbf{y})}\right)\theta_{\boldsymbol{\mu}}^{\Pi}(\mathbf{z}) = e^{-\frac{\pi i}{N}\mathbf{p}^T\mathbf{q}-\frac{2\pi i}{N}\boldsymbol{\mu}^T\mathbf{q}}\theta_{\boldsymbol{\mu}+\mathbf{p}}^{\Pi}(\mathbf{z}), \quad \boldsymbol{\mu}\in\mathbb{Z}_N^g.$$

We present two different proofs.

Proof. The **first proof** is basically Weil's (see [Weil (1964)], [Mumford (1983)]) phrased in quantum mechanical language. It is based on the formula

$$\mathrm{op}\left(e^{2\pi i(\mathbf{p}^T\mathbf{x}+\mathbf{q}^T\mathbf{y})}\right)\phi(\mathbf{z}) = e^{2\pi i\mathbf{p}^T\mathbf{z}-\pi i h\mathbf{p}^T(\mathbf{q}-i\mathbf{p})}\phi(\mathbf{z}-h(\mathbf{q}-i\mathbf{p}))$$

which was derived in §2.3.3. Given that in our case $\mathbf{z}$ is equal to $\mathbf{x}+\Pi\mathbf{y}$ (instead of $\mathbf{x}+i\mathbf{y}$), Π plays the role of i and so the formula reads

$$\mathrm{op}\left(e^{2\pi i(\mathbf{p}^T\mathbf{x}+\mathbf{q}^T\mathbf{y})}\right)\phi(\mathbf{z}) = e^{2\pi i\mathbf{p}^T\mathbf{z}-\pi i h\mathbf{p}^T(\mathbf{q}-\Pi\mathbf{p})}\phi(\mathbf{z}-h(\mathbf{q}-\Pi\mathbf{p})).$$

It suffices to check the cases $\mathbf{p}=\mathbf{0}$ and $\mathbf{q}=\mathbf{0}$, the general case will then follow from the Baker-Campbell-Hausdorff formula.

We compute

$$\begin{aligned}
\mathrm{op}\left(e^{2\pi i\mathbf{q}^T\mathbf{y}}\right)\theta_{\boldsymbol{\mu}}^{\Pi}(\mathbf{z}) &= \theta_{\mu}(\mathbf{z}-h\mathbf{q}) = \theta_{\boldsymbol{\mu}}^{\Pi}\left(\mathbf{z}-\frac{\mathbf{q}}{N}\right)\\
&= \sum_{\mathbf{n}\in\mathbb{Z}^g} e^{2\pi i N\left[\frac{1}{2}\left(\frac{\boldsymbol{\mu}}{N}+\mathbf{n}\right)^T\Pi\left(\frac{\boldsymbol{\mu}}{N}+\mathbf{n}\right)+\left(\frac{\boldsymbol{\mu}}{N}+\mathbf{n}\right)^T\left(\mathbf{z}-\frac{\mathbf{q}}{N}\right)\right]}\\
&= \sum_{\mathbf{n}\in\mathbb{Z}^g} e^{2\pi i N\left[\frac{1}{2}\left(\frac{\boldsymbol{\mu}}{N}+\mathbf{n}\right)^T\Pi\left(\frac{\boldsymbol{\mu}}{N}+\mathbf{n}\right)+\left(\frac{\boldsymbol{\mu}}{N}+\mathbf{n}\right)^T\mathbf{z}\right]-\frac{2\pi i}{N}\boldsymbol{\mu}^T\mathbf{q}-2\pi i\mathbf{n}^T\mathbf{q}}\\
&= e^{-2\pi i\boldsymbol{\mu}^T\mathbf{q}}\sum_{\mathbf{n}\in\mathbb{Z}^g} e^{2\pi i N\left[\frac{1}{2}\left(\frac{\boldsymbol{\mu}}{N}+\mathbf{n}\right)^T\Pi\left(\frac{\boldsymbol{\mu}}{N}+\mathbf{n}\right)+\left(\frac{\boldsymbol{\mu}}{N}+\mathbf{n}\right)^T\mathbf{z}\right]} = e^{-2\pi i\boldsymbol{\mu}^T\mathbf{q}}\theta_{\boldsymbol{\mu}}^{\Pi}.
\end{aligned}$$

Also

$$\begin{aligned}
&\mathrm{op}\left(e^{2\pi i\mathbf{p}^T\mathbf{x}}\right)\theta_{\boldsymbol{\mu}}^{\Pi}(\mathbf{z})\\
&\quad= e^{2\pi i\mathbf{p}^T\mathbf{z}+\frac{\pi i}{N}\mathbf{p}^T\Pi\mathbf{p}}\sum_{\mathbf{n}\in\mathbb{Z}^g} e^{2\pi i N\left[\frac{1}{2}\left(\frac{\boldsymbol{\mu}}{N}+\mathbf{n}\right)^T\Pi\left(\frac{\boldsymbol{\mu}}{N}+\mathbf{n}\right)+\left(\frac{\boldsymbol{\mu}}{N}+\mathbf{n}\right)^T\left(\mathbf{z}+\frac{\Pi\mathbf{p}}{N}\right)\right]}\\
&\quad= \sum_{\mathbf{n}\in\mathbb{Z}^g} e^{2\pi i N\left[\frac{1}{2}\left(\frac{\boldsymbol{\mu}}{N}+\mathbf{n}\right)^T\Pi\left(\frac{\boldsymbol{\mu}}{N}+\mathbf{n}\right)+\left(\frac{\boldsymbol{\mu}}{N}+\mathbf{n}\right)^T\Pi\frac{\mathbf{p}}{N}+\frac{\pi}{N^2}\mathbf{p}^T\Pi\mathbf{p}\right]}\\
&\qquad\times e^{2\pi i N\left[\left(\frac{\boldsymbol{\mu}}{N}+\mathbf{n}\right)\mathbf{z}+\left(\frac{\mathbf{p}}{N}\right)^T\mathbf{z}\right]}\\
&\quad= \sum_{\mathbf{n}\in\mathbb{Z}^g} e^{2\pi i N\left[\frac{1}{2}\left(\frac{\boldsymbol{\mu}+\mathbf{p}}{N}+\mathbf{n}\right)^T\Pi\left(\frac{\boldsymbol{\mu}+\mathbf{p}}{N}+\mathbf{n}\right)+\left(\frac{\boldsymbol{\mu}+\mathbf{p}}{N}+\mathbf{n}\right)^T\mathbf{z}\right]},
\end{aligned}$$

and here we recognize the formula for the theta series $\theta_{\boldsymbol{\mu}+\mathbf{p}}^{\Pi}(\mathbf{z})$. The theorem is proved. □

The second proof is based on Theorem 2.5, and is taken from [Gelca and Uribe (2014)]. The operator $\mathrm{op}(f)$ associated to a function f on $\mathcal{J}(\Sigma_g)$ is the Toeplitz operator with symbol $e^{\frac{-\hbar\Delta}{4}}f$. So before proceeding with the proof, we need to clarify the definition of the Laplace operator on the Jacobian variety. We denote this operator Δ_Π, in order to emphasize its dependence on the period matrix. Recall the notation X and Y for the real and imaginary parts of Π.

On a general Riemannian manifold M, this is the Laplace-Beltrami operator. In local coordinates this operator is given by

$$\Delta f = \frac{1}{\sqrt{\det(\mathbf{g})}}\frac{\partial}{\partial x^j}\left(\mathbf{g}^{jk}\sqrt{\det(\mathbf{g})}\frac{\partial f}{\partial x^k}\right),$$

where $\mathbf{g} = (\mathbf{g}_{jk})$ is the metric and $\mathbf{g}^{-1} = (\mathbf{g}^{jk})$. If the manifold is Kähler, and if the symplectic form is given in holomorphic coordinates by

$$\omega = \frac{i}{2}\sum_{j,k}\mathbf{h}_{jk}\, dz_j \wedge d\bar{z}_k,$$

then the Laplace-Beltrami operator is

$$\Delta = 4\sum_{j,k}\mathbf{h}^{jk}\,\frac{\partial^2}{\partial z_j \partial \bar{z}_k},$$

where $(\mathbf{h}^{jk}) = (\mathbf{h}_{jk})^{-1}$.

But we should point out that on a general manifold there is no such thing as Weyl quantization (at least at the time of writing this book). Nevertheless there is a Weyl quantization in our situation and it can be defined again using the Weyl symbol $e^{-\hbar\Delta_\Pi/4}$. In the coordinates $z_j, \bar{z}_j$, $j = 1, 2, \ldots, g$, one computes that $(\mathbf{h}_{jk})^{-1} = Y^{-1}$ and therefore $(\mathbf{h}^{jk}) = Y$. We obtain the formula for the Laplace-Beltrami operator Δ_Π on the Jacobian to be

$$\sum_{j,k=1}^{g} Y_{jk}\left[(I_g + iY^{-1}X)\nabla_{\mathbf{x}} - iY^{-1}\nabla_{\mathbf{y}}\right]_j\left[(I_g - iY^{-1}X)\nabla_{\mathbf{x}} + iY^{-1}\nabla_{\mathbf{y}}\right]_k.$$

In this formula

$$\nabla_{\mathbf{x}} = \left(\frac{\partial}{\partial x_1}, \frac{\partial}{\partial x_2}, \ldots, \frac{\partial}{\partial x_g}\right) \text{ and } \nabla_{\mathbf{y}} = \left(\frac{\partial}{\partial y_1}, \frac{\partial}{\partial y_2}, \ldots, \frac{\partial}{\partial y_g}\right),$$

and the subindices j, k are the corresponding components of those vectors and the entry of the matrix of Y. We can simplify this formula as follows.

Lemma 4.4. *The Laplace-Beltrami operator in the coordinates* $(\mathbf{x}, \mathbf{y})$ *is given by*

$$\Delta_\Pi = \sum_{j,k=1}^{g}(XY^{-1}X + Y)_{jk}\,\frac{\partial^2}{\partial x_j \partial x_k} - 2(XY^{-1})_{jk}\frac{\partial^2}{\partial x_j \partial y_k} + (Y^{-1})_{jk}\frac{\partial^2}{\partial y_j \partial y_k}.$$

Proof. We compute

$$\sum_{j,k=1}^{g} Y_{jk}\left[(I_g+iY^{-1}X)\nabla_{\mathbf{x}}-iY^{-1}\nabla_{\mathbf{y}}\right]_j\left[(I_g-iY^{-1}X)\nabla_{\mathbf{x}}+iY^{-1}\nabla_{\mathbf{y}}\right]_k$$

$$=\sum_{j,k} Y_{jk}\left[\frac{\partial}{\partial x_j}+i\sum_{l_1,l_2}(Y^{-1})_{jl_1}X_{l_1l_2}\frac{\partial}{\partial x_{l_2}}-i\sum_{l_3}(Y^{-1})_{jl_3}\frac{\partial}{\partial y_{l_3}}\right]$$

$$\times\left[\frac{\partial}{\partial x_j}-i\sum_{l_1,l_2}(Y^{-1})_{kl_4}X_{l_4l_5}\frac{\partial}{\partial x_{l_2}}+i\sum_{l_6}(Y^{-1})_{jl_6}\frac{\partial}{\partial y_{l_6}}\right].$$

Multiplying out we obtain that this is equal to

$$\sum_{j,k} Y_{jk}\frac{\partial^2}{\partial x_k x_j}+\sum_{j,k,l_1,l_2,l_4,l_5} Y_{jk}(Y^{-1})_{jl_1}X_{l_1l_2}(Y^{-1})_{kl_4}X_{l_4l_5}\frac{\partial^2}{\partial x_{l_2}x_{l_5}}$$

$$-\sum_{j,k,l_1,l_2,l_6} Y_{jk}(Y^{-1})_{jl_1}X_{l_1l_2}(Y^{-1})_{kl_6}\frac{\partial^2}{\partial x_{l_2}\partial y_{l_6}}$$

$$-\sum_{j,k,l_3,l_4,l_5} Y_{jk}(Y^{-1})_{jl_3}(Y^{-1})_{kl_4}X_{l_4l_5}\frac{\partial^2}{\partial x_{l_5}\partial y_{l_3}}$$

$$+\sum_{j,k,l_3,l_6} Y_{jk}(Y^{-1})_{jl_3}(Y^{-1})_{kl_6}.$$

Because $\sum_k Y_{jk}(Y^{-1})_{kl}=\delta_{jl}$, we can simplify this to

$$\sum_{j,k} Y_{jk}\frac{\partial^2}{\partial x_k x_j}+\sum_{j,l_1,l_2,l_5}(Y^{-1})_{jl_1}X_{l_1l_2}X_{jl_5}\frac{\partial^2}{\partial x_{l_2}x_{l_5}}$$

$$-\sum_{j,l_1,l_2}(Y^{-1})_{jl_1}X_{l_1l_2}\frac{\partial^2}{\partial x_{l_2}\partial y_j}-\sum_{j,l_3,l_5}(Y^{-1})_{jl_3}X_{jl_5}\frac{\partial^2}{\partial x_{l_5}\partial y_{l_3}}$$

$$+\sum_{j,l_3}(Y^{-1})_{jl_3}\frac{\partial^2}{\partial y_{l_3}\partial y_j}.$$

Using the fact that X and Y are symmetric and renaming the indices we obtain the formula from the statement. □

Lemma 4.5. *The Laplace-Beltrami operator acts on exponential functions by the formula*

$$\Delta_{\Pi}e^{2\pi i(\mathbf{p}^T\mathbf{x}+\mathbf{q}^T\mathbf{y})}=-(2\pi)^2\mathcal{R}e^{2\pi i(\mathbf{p}^T\mathbf{x}+\mathbf{q}^T\mathbf{y})},$$

where

$$\mathcal{R}=\mathbf{q}^TY^{-1}\mathbf{q}-2\mathbf{q}^TY^{-1}X\mathbf{p}+\mathbf{p}^T(XY^{-1}X+Y)\mathbf{p}.$$

Proof. Using Lemma 4.4 we compute

$$\Delta_\Pi e^{2\pi i(\mathbf{p}^T\mathbf{x}+\mathbf{q}^T\mathbf{y})} = \sum_{j,k=1}^{g} (XY^{-1}X+Y)_{jk} \frac{\partial^2}{\partial x_j \partial x_k} e^{2\pi i(\mathbf{p}^T\mathbf{x}+\mathbf{q}^T\mathbf{y})}$$
$$-2(XY^{-1})_{jk}\frac{\partial^2}{\partial x_j \partial y_k} e^{2\pi i(\mathbf{p}^T\mathbf{x}+\mathbf{q}^T\mathbf{y})} + (Y^{-1})_{jk}\frac{\partial^2}{\partial y_j \partial y_k} e^{2\pi i(\mathbf{p}^T\mathbf{x}+\mathbf{q}^T\mathbf{y})}.$$

This is further equal to

$$-(2\pi)^2 \sum_{j,k=1}^{g} (XY^{-1}X+Y)_{jk}\, p_j p_k e^{2\pi i(\mathbf{p}^T\mathbf{x}+\mathbf{q}^T\mathbf{y})}$$
$$-2(XY^{-1})_{jk} p_j q_k e^{2\pi i(\mathbf{p}^T\mathbf{x}+\mathbf{q}^T\mathbf{y})} + (Y^{-1})_{jk} q_j q_k e^{2\pi i(\mathbf{p}^T\mathbf{x}+\mathbf{q}^T\mathbf{y})}$$
$$= -(2\pi)^2[\mathbf{q}^T Y^{-1}\mathbf{q} - 2\mathbf{p}^T XY^{-1}\mathbf{q} + \mathbf{p}^T(XY^{-1}X+Y)\mathbf{p}]e^{2\pi i(\mathbf{p}^T\mathbf{x}+\mathbf{q}^T\mathbf{y})}.$$

Using the fact that $\mathbf{p}^T XY^{-1}\mathbf{q} = \mathbf{q}^T Y^{-1} X\mathbf{p}$, which is true because X and Y are symmetric, we obtain the formula from the statement. □

We can now present the **second proof** to Theorem 4.14.

Proof. Let us introduce some useful notation local to the proof:

(1) $e(t) := \exp(2\pi i N t)$.
(2) For $\mathbf{n} \in \mathbb{Z}^g$ and $\boldsymbol{\mu} \in \{0, 1, \ldots N-1\}^g$, $\mathbf{n}_{\boldsymbol{\mu}} := \mathbf{n} + \frac{\boldsymbol{\mu}}{N}$.
(3) $Q(\mathbf{n}_{\boldsymbol{\mu}}) := \frac{1}{2}(\mathbf{n}_{\boldsymbol{\mu}}^T \Pi \mathbf{n}_{\boldsymbol{\mu}})$.
(4) $E_{\mathbf{p},\mathbf{q}}(\mathbf{x},\mathbf{y}) = e^{2\pi i(\mathbf{p}^T\mathbf{x}+\mathbf{q}^T\mathbf{y})} = e(\frac{1}{N}(\mathbf{p}^T\mathbf{x}+\mathbf{q}^T\mathbf{y}))$.

With these notations, in the $(\mathbf{x},\mathbf{y})$ coordinates

$$\theta_{\boldsymbol{\mu}}^{\Pi}(\mathbf{x},\mathbf{y}) = \sum_{\mathbf{n}\in\mathbb{Z}^g} e(Q(\mathbf{n}_{\boldsymbol{\mu}}))\, e(\mathbf{n}_{\boldsymbol{\mu}}^T(\mathbf{x}+\Pi\mathbf{y})).$$

We first compute the matrix coefficients of the Toeplitz operator with symbol $E_{\mathbf{p},\mathbf{q}}$, that is

$$\langle E_{\mathbf{p},\mathbf{q}}\theta_{\boldsymbol{\mu}}^{\Pi}, \theta_{\boldsymbol{\nu}}^{\Pi}\rangle$$
$$= (2N)^{g/2}\det(Y)^{1/2}\int_{[0,1]^{2g}} E_{\mathbf{p},\mathbf{q}}(\mathbf{x},\mathbf{y})\theta_{\boldsymbol{\mu}}^{\Pi}(\mathbf{x},\mathbf{y})\overline{\theta_{\boldsymbol{\nu}}^{\Pi}(\mathbf{x},\mathbf{y})}\, e^{-2\pi N\mathbf{y}^T Y\mathbf{y}}\, d\mathbf{x}d\mathbf{y}.$$

A calculation shows that

$$E_{\mathbf{p},\mathbf{q}}(\mathbf{x},\mathbf{y})\theta_{\boldsymbol{\mu}}^{\Pi}(\mathbf{x},\mathbf{y})\overline{\theta_{\boldsymbol{\nu}}^{\Pi}(\mathbf{x},\mathbf{y})}$$
$$= \sum_{\mathbf{m},\mathbf{n}\in\mathbb{Z}^g} e\Big[Q(\mathbf{n}_{\boldsymbol{\mu}}) - \overline{Q(\mathbf{m}_{\boldsymbol{\nu}})} + (\mathbf{n}_{\boldsymbol{\mu}+\mathbf{p}} - \mathbf{m}_{\boldsymbol{\nu}})^T\mathbf{x} + (\frac{\mathbf{q}^T}{N} + \mathbf{n}_{\boldsymbol{\mu}}^T\Pi - \mathbf{m}_{\boldsymbol{\nu}}^T\overline{\Pi})\mathbf{y}\Big].$$

The integral over $\mathbf{x} \in [0,1]^g$ of the $(\mathbf{m}, \mathbf{n})$ term will be non-zero if and only if

$$N\Big(\mathbf{n}_{\boldsymbol{\mu}+\mathbf{p}} - \mathbf{m}_{\boldsymbol{\nu}}\Big) = \boldsymbol{\mu} + \mathbf{p} - \boldsymbol{\nu} + N(\mathbf{n} - \mathbf{m}) = 0,$$

in which case the integral will be equal to one. Therefore $\langle E_{p,q}\theta^{\Pi}_{\boldsymbol{\mu}}\,,\,\theta^{\Pi}_{\boldsymbol{\nu}}\rangle = 0$ unless

$$[\boldsymbol{\nu}] = [\boldsymbol{\mu} + \mathbf{p}],$$

where the brackets represent equivalence classes in $\mathbb{Z}^g_N$. This shows that the Toeplitz operator with multiplier $E_{\mathbf{p},\mathbf{q}}$ maps $\theta^{\Pi}_{\boldsymbol{\mu}}$ to a scalar times $\theta_{\boldsymbol{\mu}+\mathbf{p}}$. We now compute the scalar.

Taking $\boldsymbol{\mu}$ in the fundamental domain $\{0, 1, \cdots, N-1\}^g$ for $\mathbb{Z}^g_N$, there is a unique representative, $\boldsymbol{\nu}$, of $[\boldsymbol{\mu} + \mathbf{p}]$ in the same domain. This $\boldsymbol{\nu}$ is of the form

$$\boldsymbol{\nu} = \boldsymbol{\mu} + \mathbf{p} + N\boldsymbol{\kappa}$$

for a unique $\boldsymbol{\kappa} \in \mathbb{Z}^g$. With respect to the previous notation, $\boldsymbol{\kappa} = \mathbf{n} - \mathbf{m}$. It follows that

$$\langle E_{\mathbf{p},\mathbf{q}}\theta^{\Pi}_{\boldsymbol{\mu}}\,,\,\theta^{\Pi}_{\boldsymbol{\nu}}\rangle = (2N)^{g/2}\det(Y)^{1/2}\sum_{\mathbf{n}\in\mathbb{Z}^g}\int_{[0,1]^g} e\Big[Q(\mathbf{n}_{\boldsymbol{\mu}}) - \overline{Q(\mathbf{m}_{\boldsymbol{\nu}})}$$
$$+\,(\frac{\mathbf{q}^T}{N} + \mathbf{n}^T_{\boldsymbol{\mu}}\Pi - \mathbf{m}^T_{\boldsymbol{\nu}}\overline{\Pi})\mathbf{y} + i\mathbf{y}^T Y\mathbf{y}\Big]\,d\mathbf{y},$$

where $\mathbf{m} = \mathbf{n} - \boldsymbol{\kappa}$ in the $\mathbf{n}$th term.

Using that $\mathbf{m}_{\boldsymbol{\nu}} = \mathbf{n}_{\boldsymbol{\mu}} + \frac{1}{N}\mathbf{p}$, one gets:

$$Q(\mathbf{n}_{\boldsymbol{\mu}}) - \overline{Q(\mathbf{m}_{\boldsymbol{\nu}})} = i\mathbf{n}^T_{\boldsymbol{\mu}}Y\mathbf{n}_{\boldsymbol{\mu}} - \frac{1}{N}\mathbf{p}^T\overline{\Pi}\mathbf{n}_{\boldsymbol{\mu}} - \frac{1}{N^2}\overline{Q(\mathbf{p})}$$

and

$$\mathbf{n}^T_{\boldsymbol{\mu}}\Pi - \mathbf{m}^T_{\boldsymbol{\nu}}\overline{\Pi} = 2i\mathbf{n}^T_{\boldsymbol{\mu}}Y - \frac{1}{N}\mathbf{p}^T\overline{\Pi},$$

and so we can write

$$\langle E_{\mathbf{p},\mathbf{q}}\theta^{\Pi}_{\boldsymbol{\mu}}\,,\,\theta^{\Pi}_{\boldsymbol{\nu}}\rangle = (2N)^{g/2}\det(Y)^{1/2}\,e\Big[-\frac{1}{N^2}\overline{Q(\mathbf{p})}\Big]\sum_{\mathbf{n}\in\mathbb{Z}^g}\int_{[0,1]^g} e\Big[i\mathbf{n}^T_{\boldsymbol{\mu}}Y\mathbf{n}_{\boldsymbol{\mu}}$$
$$-\frac{1}{N}\mathbf{p}^T\overline{\Pi}\mathbf{n}_{\boldsymbol{\mu}} + (\frac{1}{N}\mathbf{q}^T + 2i\mathbf{n}^T_{\boldsymbol{\mu}}Y - \frac{1}{N}\mathbf{p}^T\overline{\Pi})\mathbf{y} + i\mathbf{y}^T Y\mathbf{y}\Big]d\mathbf{y}.$$

Making the change of variables $\mathbf{w} := \mathbf{y} + \mathbf{n}_{\boldsymbol{\mu}}$ in the summand $\mathbf{n}$, the argument of the function e can be seen to be equal to

$$i\mathbf{w}^T Y\mathbf{w} + \frac{1}{N}(\mathbf{q}^T - \mathbf{p}^T\overline{\Pi})\mathbf{w} - \frac{1}{N}\mathbf{q}^T\mathbf{n}_{\boldsymbol{\mu}}.$$

Note that

$$e\Big(\frac{1}{N}\mathbf{q}^T\mathbf{n}_{\boldsymbol{\mu}}\Big) = e^{-2\pi i\mathbf{q}^T\boldsymbol{\mu}/N}.$$

The summation over $\mathbf{n}$ of all those integrals can be turned into a single integral over $\mathbb{R}^g$, where $\mathbb{R}^g$ is seen as the translates of $[0,1]^g$ by the elements of the lattice $\mathbb{Z}^g$. Then $\langle E_{\mathbf{p},\mathbf{q}}\theta^{\Pi}_{\boldsymbol{\mu}}\,,\,\theta^{\Pi}_{\boldsymbol{\nu}}\rangle$ is equal to

$$(2N)^{g/2}\det(Y)^{1/2}\,e\Big[-\tfrac{1}{N^2}\overline{Q(\mathbf{p})}\Big]e^{-2\pi i\mathbf{q}^T\boldsymbol{\mu}/N}\int_{\mathbb{R}^g}e^{-2\pi N\mathbf{w}^TY\mathbf{w}+2\pi i\left(\mathbf{q}^T-\mathbf{p}^T\overline{\Pi}\right)\mathbf{w}}\,d\mathbf{w}.$$

Using the formula for Gaussian integrals:

$$\int_{\mathbb{R}^g}e^{-\mathbf{x}^TA\mathbf{x}+\mathbf{b}^T\mathbf{x}}\,d\mathbf{x} = \Big(\frac{\pi^g}{\det A}\Big)^{1/2}e^{\frac{1}{4}\mathbf{b}^TA^{-1}\mathbf{b}}$$

we obtain that it is further equal to

$$\Big(\frac{1}{2N}\Big)^{g/2}\det(Y)^{-1/2}\,e^{-\frac{\pi}{2N}(\mathbf{q}^T-\mathbf{p}^T\overline{\Pi})Y^{-1}(\mathbf{q}-\overline{\Pi}\mathbf{p})}.$$

Hence

$$\langle E_{\mathbf{p},\mathbf{q}}\theta^{\Pi}_{\boldsymbol{\mu}}\,,\,\theta^{\Pi}_{\boldsymbol{\nu}}\rangle = e^{-\frac{\pi i}{N}\mathbf{p}^T\overline{\Pi}\mathbf{p}}\,e^{-2\pi i\mathbf{q}^T\boldsymbol{\mu}/N}\,e^{-\frac{\pi}{2N}(\mathbf{q}^T-\mathbf{p}^T\overline{\Pi})Y^{-1}(\mathbf{q}-\overline{\Pi}\mathbf{p})}.$$

The exponent on the right-hand side is $(-\pi/N)$ times

$$\begin{aligned}
&2i\mathbf{q}^T\boldsymbol{\mu}+i\mathbf{p}^T(X-iY)\mathbf{p}+\frac{1}{2}\Big([\mathbf{q}^T-\mathbf{p}^T(X-iY)]Y^{-1}[\mathbf{q}-(X-iY)\mathbf{p}]\Big)\\
&=2i\mathbf{q}^T\boldsymbol{\mu}+i\mathbf{p}^T(X-iY)\mathbf{p}+\frac{1}{2}\Big([\mathbf{q}^TY^{-1}-\mathbf{p}^TXY^{-1}+i\mathbf{p}^T][\mathbf{q}-X\mathbf{p}+iY\mathbf{p}]\Big)\\
&=2i\mathbf{q}^T\boldsymbol{\mu}+i\mathbf{p}^T(X-iY)\mathbf{p}+\frac{1}{2}\Big(\mathbf{q}^TY^{-1}\mathbf{q}-2\mathbf{q}^TY^{-1}X\mathbf{p}+2i\mathbf{q}^T\mathbf{p}\\
&\quad+\mathbf{p}^TXY^{-1}X\mathbf{p}-2i\mathbf{p}^TX\mathbf{p}-\mathbf{p}^TY\mathbf{p}\Big)\\
&=2i\mathbf{q}^T\boldsymbol{\mu}+i\mathbf{q}^T\mathbf{p}+\frac{1}{2}\mathcal{R},
\end{aligned}$$

where $\mathcal{R}$ is the operator defined in Lemma 4.5. We thus have

$$\langle E_{\mathbf{p},\mathbf{q}}\theta^{\Pi}_{\boldsymbol{\mu}}\,,\,\theta^{\Pi}_{\boldsymbol{\nu}}\rangle = e^{-\frac{2\pi i}{N}\mathbf{q}^T\boldsymbol{\mu}-\frac{\pi i}{N}\mathbf{q}^T\mathbf{p}-\frac{\pi}{2N}\mathcal{R}}. \tag{4.32}$$

By Lemma 4.5, $\Delta_{\Pi}(E_{\mathbf{p},\mathbf{q}}) = -(2\pi)^2\mathcal{R}E_{\mathbf{p},\mathbf{q}}$,, and therefore

$$e^{-\frac{\hbar\Delta_{\Pi}}{4}}E_{\mathbf{p},\mathbf{q}} = e^{-\frac{\Delta_{\Pi}}{8\pi N}}E_{\mathbf{p},\mathbf{q}} = e^{\frac{\pi}{2N}\mathcal{R}}E_{\mathbf{p},\mathbf{q}},$$

so that, by (4.32)

$$\langle e^{-\frac{\Delta_{\Pi}}{4N}}(E_{\mathbf{p},\mathbf{q}})\theta^{\Pi}_{\boldsymbol{\mu}}\,,\,\theta^{\Pi}_{\boldsymbol{\nu}}\rangle = e^{-\frac{2\pi i}{N}\mathbf{q}^T\boldsymbol{\mu}-\frac{\pi i}{N}\mathbf{q}^T\mathbf{p}}, \tag{4.33}$$

if $\boldsymbol{\nu} = \boldsymbol{\mu}+\mathbf{p}+N\boldsymbol{\kappa}$ and 0 otherwise. The theorem is proved. □

Corollary 4.3. *The theta series* $\theta_{\boldsymbol{\mu}}^{\Pi}$, $\boldsymbol{\mu} \in \mathbb{Z}_N^g$, *form an orthonormal set.*

Proof. Substitute $\mathbf{p} = \mathbf{q} = \mathbf{0}$ in (4.33). In this case

$$E_{\mathbf{pq}} = E_{\mathbf{0},\mathbf{0}} = I,$$

and hence

$$e^{\frac{\hbar\Delta_\Pi}{4}} E_{\mathbf{0},\mathbf{0}} = I.$$

□

As a corollary of this theorem, there is a representation of the finite Heisenberg group $\mathbf{H}(\mathbb{Z}_N^g)$ on $\boldsymbol{\Theta}_N^{\Pi}(\Sigma_g)$, given by

$$\exp(\mathbf{p}^T\mathbf{P} + \mathbf{q}^T\mathbf{Q} + kE)\theta_{\boldsymbol{\mu}}^{\Pi}(\mathbf{z}) = e^{-\frac{\pi i}{N}\mathbf{p}^T\mathbf{q} - \frac{2\pi i}{N}\boldsymbol{\mu}^T\mathbf{q} + \frac{\pi i}{N}k}\theta_{\boldsymbol{\mu}+\mathbf{p}}^{\Pi}(\mathbf{z}), \quad (4.34)$$

$\boldsymbol{\mu} \in \mathbb{Z}_N^g$. We will refer to this representation as the **Schrödinger representation of the finite Heisenberg group on the space of theta functions**. Comparing this representation with the one given in §4.2.3 we notice that the two are unitary equivalent. The unitary equivalence is a Segal-Bargmann transform.

4.3.3 *The Segal-Bargmann transform on the Jacobian variety*

In this section we give an explicit description of the Segal-Bargmann transform on the Jacobian variety. This is the unitary map which identifies the quantization model of the Jacobian variety in the real polarization from §4.2 with that in the Kähler polarization described above.

In view of Theorems 4.11 and 4.14, and of the fact that both the $s_{\boldsymbol{\mu}}$'s and the $\theta_{\boldsymbol{\mu}}^{\Pi}$'s are orthonormal bases, the Segal-Bargmann transform is the unitary map defined by

$$s_{\boldsymbol{\mu}} \mapsto \theta_{\boldsymbol{\mu}}^{\Pi}.$$

Identify $\mathbb{C}^g = \mathbb{R}^g \times \mathbb{R}^g$ by $\mathbf{z} = \mathbf{x} + \Pi\mathbf{y}$. Because of the periodicity conditions, theta functions are completely determined by their values for $(\mathbf{x}, \mathbf{y}) \in [0,1]^{2g}$. As such $\boldsymbol{\Theta}_N^{\Pi}(\Sigma_g)$ can be embedded isometrically in the space

$$L^2\big(\mathcal{J}(\Sigma_g), e^{-2\pi N\mathbf{y}^T Y\mathbf{y}}d\mathbf{x}d\mathbf{y}\big) = L^2\big([0,1]^{2g}, e^{-2\pi N\mathbf{y}^T Y\mathbf{y}}d\mathbf{x}d\mathbf{y}\big),$$

whose inner product is defined by (4.31).

Consider the orthogonal projection onto the space of theta functions

$$\pi : L^2\big(\mathcal{J}(\Sigma_g), e^{-2\pi N\mathbf{y}^T Y\mathbf{y}}d\mathbf{x}d\mathbf{y}\big) \to \boldsymbol{\Theta}_N^{\Pi}(\Sigma_g).$$

The reproducing kernel of $\boldsymbol{\Theta}_N^\Pi(\Sigma_g)$ is

$$K(\mathbf{z},\mathbf{w}) = \sum_{\boldsymbol{\nu}} \theta_{\boldsymbol{\nu}}(\mathbf{z})\overline{\theta_{\boldsymbol{\nu}}(\mathbf{w})},$$

thus the projection operator is given explicitly by the formula

$$\pi(f)(\mathbf{z}) = \int_{[0,1]^{2g}} K(\mathbf{z},\mathbf{w})f(\mathbf{w})e^{-2\pi N\mathbf{y}_\mathbf{w}^T Y\mathbf{y}_\mathbf{w}}d\mathbf{x}_\mathbf{w}d\mathbf{y}_\mathbf{w}.$$

Here, to distinguish between the complex variables $\mathbf{z}$ and $\mathbf{w}$, we make the convention that $\mathbf{z} = \mathbf{x}_\mathbf{z} + \Pi\mathbf{y}_\mathbf{z}$ and $\mathbf{w} = \mathbf{x}_\mathbf{w} + \Pi\mathbf{y}_\mathbf{w}$.

Pull back the sections $s_{\boldsymbol{\mu}}$, $\boldsymbol{\mu} \in \mathbb{Z}_N^g$, of the line bundle defined in §4.2.2 to $[0,1]^{2g}$ and extend the projection operator π to them.

Proposition 4.8. *For every* $\boldsymbol{\mu} \in \mathbb{Z}_N^g$,

$$\pi(s_{\boldsymbol{\mu}}) = (2N)^{g/2}(\det(Y))^{1/2}e^{-3i\frac{\pi}{N}\boldsymbol{\mu}^T X\boldsymbol{\mu} - \frac{5\pi}{N}\boldsymbol{\mu}^T Y\boldsymbol{\mu}}\,\theta_{-\boldsymbol{\mu}}^\Pi.$$

Proof. We compute

$$\pi(s_{\boldsymbol{\mu}}) = (2N)^{g/2}(\det(Y))^{1/2}\int_{\mathcal{J}(\Sigma_g)} K(\mathbf{z},\bar{\mathbf{w}})s_{\boldsymbol{\mu}}(\mathbf{w})e^{-2\pi N\mathbf{y}_\mathbf{w}^T Y\mathbf{y}_\mathbf{w}}d\mathbf{x}_\mathbf{w}d\mathbf{y}_\mathbf{w}$$

$$= (2N)^{g/2}(\det(Y))^{1/2}\int_{\mathcal{J}(\Sigma_g)} \sum_{\boldsymbol{\nu}} \theta_{\boldsymbol{\nu}}(\mathbf{z})\overline{\theta_{\boldsymbol{\nu}}(\mathbf{w})}s_{\boldsymbol{\mu}}(\mathbf{w})e^{-2\pi N\mathbf{y}_\mathbf{w}^T Y\mathbf{y}_\mathbf{w}}d\mathbf{x}_\mathbf{w}d\mathbf{y}_\mathbf{w}$$

$$= (2N)^{g/2}(\det(Y))^{1/2}e^{-\frac{2\pi}{N}\boldsymbol{\mu}^T Y\boldsymbol{\mu}}\sum_{\boldsymbol{\nu}}\theta_{\boldsymbol{\nu}}(\mathbf{z})$$

$$\times \int_{[0,1]^g} \overline{\theta_{\boldsymbol{\nu}}^\Pi\left(\mathbf{x}_\mathbf{w} - \Pi\frac{\boldsymbol{\mu}}{N}\right)}e^{-2\pi i\left(\frac{\boldsymbol{\mu}}{N}\right)^T\mathbf{x}_\mathbf{w}}d\mathbf{x}_\mathbf{w}.$$

The integral equals

$$\sum_{\mathbf{n}\in\mathbb{Z}^N} e^{-2\pi iN\left[\frac{1}{2}\left(\frac{\boldsymbol{\nu}}{N}+\mathbf{n}\right)^T\overline{\Pi}\left(\frac{\boldsymbol{\nu}}{N}+\mathbf{n}\right)-\left(\frac{\boldsymbol{\nu}}{N}+\mathbf{n}\right)^T\overline{\Pi}\frac{\boldsymbol{\mu}}{N}\right]}\int_{[0,1]^g} e^{-2\pi iN\left[\left(\frac{\boldsymbol{\mu}+\boldsymbol{\nu}}{N}+\mathbf{n}\right)^T\mathbf{x}_\mathbf{w}\right]}d\mathbf{x}_\mathbf{w}.$$

This is equal to zero unless $\boldsymbol{\nu} = -\boldsymbol{\mu}$ (and $\mathbf{n} = 0$), in which case we obtain

$$\pi(s_{\boldsymbol{\mu}}) = (2N)^{g/2}(\det(Y))^{1/2}e^{-3i\frac{\pi}{N}\boldsymbol{\mu}^T X\boldsymbol{\mu} - \frac{5\pi}{N}\boldsymbol{\mu}^T Y\boldsymbol{\mu}}\,\theta_{-\boldsymbol{\mu}},$$

as desired. □

Proposition 4.9. *The Segal-Bargmann transform which establishes the unitary equivalence between the Weyl quantization in the real polarization of the Jacobian defined in §4.2.3 and the Weyl quantization in the Kähler polarization of the Jacobian defined in §4.3.2 is given by*

$$[\mathcal{T}_{SB}(s)](z) = \int_{[0,1]^{2g}} K_{SB}(\mathbf{z},\mathbf{w})s(w)e^{-2\pi N\mathbf{y}_\mathbf{w}^T Y\mathbf{y}_\mathbf{w}}d\mathbf{x}_\mathbf{w}d\mathbf{y}_\mathbf{w},$$

where

$$K_{SB}(\mathbf{z},\mathbf{w}) = (2N)^{g/2}(\det(Y))^{1/2}\sum_{\boldsymbol{\nu}\in\mathbb{Z}_N^g} e^{3i\frac{\pi}{N}\boldsymbol{\mu}^T X\boldsymbol{\mu} + \frac{5\pi}{N}\boldsymbol{\mu}^T Y\boldsymbol{\mu}}\theta_{-\boldsymbol{\nu}}^\Pi(\mathbf{z})\overline{\theta_{\boldsymbol{\nu}}^\Pi(\mathbf{w})}.$$

Proof. It is not hard to see that

$$\mathcal{T}_{SB} = FD\pi,$$

where π is the projection operator onto the space of theta functions, D is the diagonal operator

$$D(\theta^{\Pi}_{\mu}) = (2N)^{g/2}(\det(Y))^{1/2}e^{3i\frac{\pi}{N}\mu^T X\mu + \frac{5\pi}{N}\mu^T Y\mu}\theta^{\Pi}_{\mu},$$

and F maps θ^{Π}_{μ} to $\theta^{\Pi}_{-\mu}$. By Proposition 4.8,

$$\mathcal{T}_{SB}(s_{\mu}) = \theta^{\Pi}_{\mu}.$$

Hence the conclusion. □

From this moment on we will concentrate only on the case of the quantization in the Kähler polarization, hence on theta functions.

4.3.4 *The algebra of linear operators on the space of theta functions and the quantum torus*

We want to describe the algebra of linear operators on the space of theta functions.

The group algebra of the Heisenberg group

Let us recall the definition of the group algebra of a group.

Definition 4.10. Given a group G, its group algebra $\mathbb{C}[G]$ the complex vector space with basis the elements of G endowed with the multiplication

$$\left(\sum_{g\in G}\alpha_g g\right)\left(\sum_{g'\in G}\beta_{g'}g'\right) = \sum_{g,g'\in G}\alpha_g\beta_{g'}gg',$$

where in each sum there is a finite number of nonzero coefficients α_g and $\beta_{g'}$.

Two cases are of interest to us, the one where $G = \mathbf{H}(\mathbb{Z}^g)$ and the one where $G = \mathbf{H}(\mathbb{Z}^g_N)$.

In the first case, let us consider the algebra $\mathbb{C}_t[\mathbf{U}^{\pm 1}, \mathbf{V}^{\pm 1}]$ of Laurent polynomials in the variables $t, U_1, U_2, \ldots, U_g, V_1, V_2, \ldots, V_g$, subject to the noncommutation relations

$$U_jV_k = t^{2\delta_{jk}}V_kU_j, \quad V_jV_k = V_kV_j, \quad U_jU_k = U_kU_j,$$
$$tU_j = U_jt, \quad tV_j = V_jt,$$

for all $j, k = 1, 2, \ldots, g$.

Proposition 4.10. *The group algebra* $\mathbb{C}[\mathbf{H}(\mathbb{Z}^g)]$ *is isomorphic to* $\mathbb{C}_t[\mathbf{U}^{\pm 1}, \mathbf{V}^{\pm 1}]$, *with the isomorphism defined by*

$$(\mathbf{p}, \mathbf{q}, k) \to t^{k-\mathbf{p}^T\mathbf{q}} U_1^{p_1} U_2^{p_2} \cdots U_g^{p_g} V_1^{q_1} V_2^{q_2} \cdots V_g^{p_g},$$

for all $\mathbf{p}, \mathbf{q} \in \mathbb{Z}^g$ *and* $k \in \mathbb{Z}$.

The finite Heisenberg group and the quantum torus at a root of unity

If we factor $\mathbb{C}_t[\mathbf{U}^{\pm 1}, \mathbf{V}^{\pm 1}]$ by the additional conditions

$$U^N = V^N = 1 \text{ and } t = e^{\frac{\pi i}{N}},$$

we obtain an algebra $\widetilde{\mathbb{C}_t}[\mathbf{U}^{\pm 1}, \mathbf{V}^{\pm 1}]$, which is dense in the **quantum torus at a root of unity**. Let us point out that the quantum torus, also known as the **noncommutative torus** is a fundamental example in noncommutative geometry (see for example [Connes (1994)], [Rieffel (1989)], [Manin (2001)]). It is a C^*-algebra spanned by non-commutative exponentials, which is a deformation quantization of the C^∞-functions on the torus. It is defined by a $*$-product of exponentials, which in our situation is

$$e^{2\pi i(\mathbf{p}^T\mathbf{x}+\mathbf{q}^T\mathbf{y})} * e^{2\pi i(\mathbf{p}'^T\mathbf{x}+\mathbf{q}'^T\mathbf{y})} = e^{\pi i h(\mathbf{p}^T\mathbf{q}'-\mathbf{q}^T\mathbf{p}')} e^{2\pi i((\mathbf{p}+\mathbf{p}')^T\mathbf{x}+(\mathbf{q}+\mathbf{q}')^T\mathbf{y})},$$

with $h = 1/N$ (so that $e^{\pi i h}$ is a root of unity, whence the name). We can think of the algebra $\widetilde{\mathbb{C}_t}[\mathbf{U}^{\pm 1}, \mathbf{V}^{\pm 1}]$ as the algebra of noncommutative trigonometric polynomials.

An easy consequence of Proposition 4.10 is the following result.

Proposition 4.11. *The algebra* $\widetilde{\mathbb{C}_t}[\mathbf{U}^{\pm 1}, \mathbf{V}^{\pm 1}]$ *is the quotient of the group algebra* $\mathbb{C}[\mathbf{H}(\mathbb{Z}_N^g)]$ *by the relation* $(0, 0, 1) = e^{\frac{\pi i}{N}}$.

We denote by $L(\mathbf{\Theta}_N^\Pi(\Sigma_g))$ the algebra of linear operators on the space of theta functions.

Proposition 4.12. *As a vector space,* $L(\mathbf{\Theta}_N^\Pi(\Sigma_g))$ *has a basis given by the operators*

$$op\left(e^{2\pi i(\mathbf{p}^T\mathbf{x}+\mathbf{q}^T\mathbf{y})}\right), \quad \mathbf{p}, \mathbf{q} \in \{0, 1, \ldots, N-1\}^g.$$

Proof. For simplicity, we will show that the operators

$$e^{\frac{\pi i}{N}\mathbf{p}^T\mathbf{q}} \mathrm{op}\left(e^{2\pi i(\mathbf{p}^T\mathbf{x}+\mathbf{q}^T\mathbf{y})}\right), \quad \mathbf{p}, \mathbf{q} \in \{0, 1, \ldots, N-1\}^g,$$

form a basis. Denote by $M_{\mathbf{p},\mathbf{q}}$ the respective matrices of these operators in the basis $\theta_{\boldsymbol{\mu}}^{\Pi}$, $\boldsymbol{\mu} \in \mathbb{Z}_N^g$. For a fixed $\mathbf{p}$, the nonzero entries of the matrices $M_{\mathbf{p},\mathbf{q}}$, $\mathbf{q} \in \{0, 1, \ldots, N-1\}^g$, are precisely those in the slots $(\mathbf{m}, \mathbf{m}+\mathbf{p})$, with $\mathbf{m} \in \{0, 1, \ldots, N-1\}^g$ (here $\mathbf{m}+\mathbf{p}$ is taken modulo N). If we vary $\mathbf{m}$ and $\mathbf{q}$ and arrange these nonzero entries in a matrix, we obtain the gth power of a Vandermonde matrix, which is therefore nonsingular. We conclude that for fixed $\mathbf{p}$, the matrices $M_{\mathbf{p},\mathbf{q}}$, $\mathbf{q} \in \{0, 1, \ldots, N-1\}^g$ form a basis for the vector space of matrices with nonzero entries in the slots of the form $(\mathbf{m}, \mathbf{m}+\mathbf{p})$. Varying $\mathbf{p}$, we obtain the desired conclusion. □

As a corollary we obtain the following description of the algebra of linear operators.

Proposition 4.13. *The algebra $L(\boldsymbol{\Theta}_N^{\Pi}(\Sigma_g))$ of linear operators on the space of theta functions is isomorphic to the algebra obtained by factoring $\mathbb{C}[\mathbf{H}(\mathbb{Z}_N^g)]$ by the relation $(0, 0, 1) = e^{\frac{i\pi}{N}}$, with the isomorphism defined by*

$$\exp(\mathbf{p}^T\mathbf{P} + \mathbf{q}^T\mathbf{Q} + kE) \mapsto op\left(e^{2\pi i(\mathbf{p}^T\mathbf{x}+\mathbf{q}^T\mathbf{Q})+\frac{\pi i}{N}k}\right).$$

Corollary 4.4. *The algebra $L(\boldsymbol{\Theta}_N^{\Pi}(\Sigma_g))$ is isomorphic to the subalgebra of trigonometric polynomials in the quantum (i.e. noncommutative) torus $\widetilde{\mathbb{C}_t}[\mathbf{U}^{\pm 1}, \mathbf{V}^{\pm 1}]$.*

We should point out that the relationship between the finite Heisenberg group associated to an abelian variety and the quantum torus was first noticed by Manin in [Manin (2001)].

4.3.5 *The action of the mapping class group on theta functions*

There is a projective representation of the mapping class group of the surface Σ_g on the space of theta functions $\boldsymbol{\Theta}_N^{\Pi}(\Sigma_g)$ which dates back to the works of Jacobi. This representation can be introduced in two different ways:

- from the action of the mapping class group on the operators while keeping the surface fixed, or
- from the action of the mapping class group on the complex structure.

That the two constructions yield the same projective representation is a consequence of the Stone-von Neumann theorem.

The action of the mapping class group of the surface on operators and its projective representation on theta functions

The projective representation of the mapping class group is the finite dimensional analogue of the representation defined in §2.5. Before we start, it is important to point out that the choice of the integer N to be **even** makes the entire symplectic group $Sp(2g, \mathbb{Z})$ act on theta functions, so we need not worry about any parity restrictions (see [Mumford (1983)] for a detailed discussion of the case $N = 1$ where parity restrictions apply).

Let h be an element of the mapping class group of the closed genus g surface Σ_g. It induces a linear automorphism

$$h_* : H_1(\Sigma_g, \mathbb{R}) \to H_1(\Sigma_g, \mathbb{R}).$$

The matrix of h_* has integer entries, determinant 1, and satisfies $h_* J_0 {h_*}^T = J_0$, where

$$J_0 = \begin{pmatrix} 0 & I_g \\ -I_g & 0 \end{pmatrix}$$

is the intersection form in $H_1(\Sigma_g, \mathbb{R})$, represented as a matrix in a canonical basis. The intersection form is bilinear, anti-symmetric, and nondegenerate on $H_1(\Sigma_g, \mathbb{R})$, so $H_1(\Sigma_g, \mathbb{R})$ endowed with this form is a symplectic vector space. In this framework, the map h_* is a **symplectic linear automorphism**. We decompose h_* into $g \times g$ blocks as

$$h_* = \begin{pmatrix} A & B \\ C & D \end{pmatrix}. \tag{4.35}$$

Proposition 4.14. *There is an action of the mapping class group $MCG(\Sigma_g)$ on $\mathbf{H}(\mathbb{Z}_N^g)$, in which an element h acts on $\mathbf{H}(\mathbb{Z}_N^g)$ by*
$h \cdot \exp(\mathbf{p}^T \mathbf{P} + \mathbf{q}^T \mathbf{Q} + kE) = \exp[(A\mathbf{p} + B\mathbf{q})^T \mathbf{P} + (C\mathbf{p} + D\mathbf{q})^T \mathbf{Q} + kE]$,
where A, B, C, D are defined by (4.35).

Proof. Consider the action of $MCG(\Sigma_g)$ on $H_1(\Sigma_g, \mathbb{Z})$ and extend it to an action on $\mathbf{H}(\mathbb{Z}_N^g)$ by

$$h \cdot (\mathbf{p}, \mathbf{q}, k) = (A\mathbf{p} + B\mathbf{q}, C\mathbf{p} + D\mathbf{q}, k).$$

Because

$$\begin{aligned} h \cdot \left[(\mathbf{p}, \mathbf{q}, k)^N\right] &= h \cdot (N\mathbf{p}, N\mathbf{q}, Nk) = (N(A\mathbf{p} + B\mathbf{q}), N(C\mathbf{p} + D\mathbf{q}), Nk) \\ &= \left[h \cdot (\mathbf{p}, \mathbf{q}, k)\right]^N, \end{aligned}$$

we deduce from Corollary 4.2 that h maps the kernel of the quotient map $\mathbf{H}(\mathbb{Z}^g) \to \mathbf{H}(\mathbb{Z}_N^g)$ to itself. It follows that the action of the mapping class group on $\mathbf{H}(\mathbb{Z}^g)$ factors to an action on $\mathbf{H}(\mathbb{Z}_g^N)$. □

Remark 4.4. For the existence of this action it is essential that the integer N is even.

We stress out that we define this action in a different way than it is usually done. When compared with other authors we use the transpose inverse of the matrix. So instead of letting h_* act by its inverse on the variable, we act by h_* on the coefficients $(\mathbf{p}, \mathbf{q})$ of the exponential function. The reason we make this choice is because it has a nice topological description which will be described in Chapter 5.

Theorem 4.15. *There is a projective unitary representation $h \mapsto \mathcal{F}(h)$ of $MCG(\Sigma_g)$ on $\mathbf{\Theta}_N^{\Pi}(\Sigma_g)$, such that for each h*

$$\mathcal{F}(h)\exp(\mathbf{p}^T\mathbf{P} + \mathbf{q}^T\mathbf{q} + kE)\mathcal{F}(h)^{-1} = h\cdot\exp(\mathbf{p}^T\mathbf{P} + \mathbf{q}^T\mathbf{q} + kE). \quad (4.36)$$

Proof. I. The map $\mathbf{H}(\mathbb{Z}_N^g) \mapsto L(\mathbf{\Theta}_N^{\Pi}(\Sigma_g))$ defined by $u \mapsto h \cdot u$ is a representation of $\mathbf{H}(\mathbb{Z}_N^g)$ (in which u acts as $h \cdot u$). This representation is irreducible, since it is basically the standard representation with the operators renamed, and it satisfies the conditions of the Stone-von Neumann Theorem (Theorem 4.12). Because of that theorem, this representation is unitary equivalent to the Schrödinger representation. Hence there is a unitary operator

$$\mathcal{F}(h) : \mathbf{\Theta}_N^{\Pi}(\Sigma_g) \to \mathbf{\Theta}_N^{\Pi}(\Sigma_g)$$

satisfying (4.36).

Because the Schrödinger representation is irreducible, Schur's Lemma implies that the map $\mathcal{F}(h)$ is unique up to multiplication by a constant of absolute value equal to 1. If $h, h' \in \text{MCG}(\Sigma_g)$, then $\mathcal{F}(h')\mathcal{F}(h)$ satisfies the exact Egorov identity (4.36) for the element $h' \circ h$ of the mapping class group, and hence $\mathcal{F}(h' \circ h)$ and $\mathcal{F}(h')\mathcal{F}(h)$ differ by a constant factor of absolute value equal to 1. We deduce that the map

$$\text{MCG}(\Sigma_g) \to \text{Aut}(\mathbf{\Theta}_N^{\Pi}(\Sigma_g)), \quad h \mapsto \mathcal{F}(h)$$

is a projective representation of the mapping class group.

II. There is a different reason we can give for the existence of the map $\mathcal{F}(h)$. The action of h can be extended linearly to $L(\mathbf{\Theta}_N^{\Pi}(\Sigma_g))$, and as such it becomes an automorphism of this algebra that keeps invariant the center $\mathbb{C}I$ (because it maps $(0, 0, 1)$ to itself). Any such automorphism is inner, which proves the existence of $\mathcal{F}(h)$ satisfying (4.36). □

In §5.4.3 we will give a topological description of $\mathcal{F}(h)$ and then in §7.4 we will use topological tools to resolve the projective ambiguity of the representation $h \mapsto \mathcal{F}(h)$.

The notation for the unitary map $\mathcal{F}(h)$ is motivated by the fact, to be explained later, that it is a Fourier transform. We recognize condition (4.36) to be an **exact Egorov identity**.

The representation of the mapping class group obtained from the change of the complex structure

The constructions of the space of theta functions and of the Schrödinger representation on it depend on

- the choice of the complex structure on the Riemann surface Σ_g,
- the choice of the canonical basis a_j, b_j, $j = 1, 2, \ldots, g$.

Each choice gives rise to a particular periodicity matrix (I, Π), and hence to a lattice $\mathbf{\Lambda}(I, \Pi)$, and consequently produces a certain abelian variety in the guise of the Jacobian variety of the surface.

We can change the complex structure on the surface and its canonical basis by acting on the surface by a homeomorphism

$$h : \Sigma_g \to \Sigma_g.$$

Identifying $\mathcal{J}(\Sigma_g)$ with $H_1(\Sigma_g, \mathbb{R})/H_1(\Sigma_g, \mathbb{Z})$, we deduce that h_* induces a symplectomorphism $\tilde{h}$ of $\mathcal{J}(\Sigma_g)$. The map $h \to \tilde{h}$ induces an action of the mapping class group of Σ_g on the Jacobian variety. This action can be described explicitly as follows. Using the decomposition of h_* into blocks (4.35) we have

$$a_j' = \sum_k a_{jk} a_k + \sum_k c_{jk} b_k \text{ and } b_j' = \sum_k b_{jk} a_k + \sum_k d_{jk} b_k.$$

Consequently the lattice whose generators are the column vectors of the period matrix (I, Π) is mapped to the lattice whose generators are the column vectors of $(A^T + C^T\Pi, D^T + B^T\Pi)$. This is not a period matrix because the left half of it is not the identity matrix. We can turn it into a period matrix by making the change of variable

$$\mathbf{z}' = (A^T + C^T\Pi)^{-1}\mathbf{z},$$

when the lattice becomes is generated by the column vectors of (I, Π'), where $\Pi' = (A^T + C^T\Pi)^{-1}(B^T + D^T\Pi)$. Note that the periodicity matrix depends only on the equivalence class of h inside the mapping class group, so we can treat h as an element of $MCG(\Sigma_g)$ instead of the much finer group of homeomorphisms.

Now we have two Jacobian varieties: one with variable $\mathbf{z}$ and periodicity matrix Π, and one with variable $\mathbf{z}'$ and periodicity matrix Π'. As algebraic varieties they are isomorphic, and for this reason one can expect that the theta functions on one of them are related to those on the other. This is indeed true, and the multiplication of a theta function in the variable $\mathbf{z}$ by a certain scaling holomorphic function turns it into a theta function in the variable $\mathbf{z}' = (A^T + C^T\Pi)^{-1}\mathbf{z}$. This is a difficult result, its proof is beyond the scope of this book, and it can be found, in the case $N = 1$ and g arbitrary in [Mumford (1983)] and for $g = 1$ and N arbitrary in [Kač and Peterson (1984)]. We will only illustrate two particular situations for $g = 1$ in the next section.

Theorems 4.13 and 4.14 imply that the Weyl quantizations of the Jacobian varieties defined by two periodicity matrices (I, Π) and (I, Π') are unitary equivalent in a canonical way, via the unitary isomorphism

$$\mathcal{U}_{\Pi,\Pi'} : \boldsymbol{\Theta}_N^{\Pi'}(\Sigma_g) \to \boldsymbol{\Theta}_N^{\Pi}(\Sigma_g),$$

defined by

$$\mathcal{U}_{\Pi,\Pi'}\left(\theta_{\boldsymbol{\mu}}^{\Pi'}\right) = \theta_{\boldsymbol{\mu}}^{\Pi}, \quad \boldsymbol{\mu} \in \mathbb{Z}_N^g.$$

So if we change the variable $\mathbf{z} \mapsto \mathbf{z}'$ (combined with the change of the periodicity matrix), add the holomorphic coefficient, then identify the theta functions in $\mathbf{z}'$ with those in $\mathbf{z}$ via $\mathcal{U}_{\Pi,\Pi'}$ we obtain an action of the mapping class group on theta functions. This is the same as the one defined above via the Stone-von Neumann theorem.

4.4 Theta functions on the Jacobian variety of the torus

To give the reader a better understanding of the general situation, and because we will use this particular case to illustrate our results in the future, let us discuss the example when the Riemann surface is the torus Σ_1.

4.4.1 *The theta functions and the action of the Heisenberg group*

In this case the Siegel upper half-space is easy to specify.

Proposition 4.15. *For every $\tau \in \mathbb{C}$ with $\operatorname{Im}\tau > 0$, there is a complex structure on the torus, a canonical basis a_1, b_1 and a holomorphic 1-form ζ_1 such that*

$$\int_{a_1} \zeta_1 = 1, \quad \int_{b_1} \zeta_1 = \tau.$$

Proof. Let the complex torus be $\mathbb{C}/\mathbb{Z}+\mathbb{Z}\tau$ and let the canonical basis be defined by the images of the lines $\mathbb{R}$ and $\mathbb{R}\tau$. The holomorphic form $\zeta_1 = dz$ satisfies the required identities. □

It is important to note that the Jacobian variety associated to the complex torus $\mathbb{C}/\mathbb{Z}+\mathbb{Z}\tau$ is precisely the same complex torus. The integration on paths of the form dz defines the holomorphic isomorphism of these tori.

Thus in the genus 1 case, the Siegel upper half-space is the upper half-plane $\operatorname{Im}\tau > 0$. The periodicity conditions for theta functions are

$$f(z+1) = f(z), \quad f(z+\tau) = e^{-2\pi i N z - \pi i N \tau} f(z).$$

The inner product on the space of theta functions is

$$\langle f, g\rangle = (2N)^{1/2} (\operatorname{Im}\tau)^{1/2} \int_0^1 \int_0^1 f(x,y)\overline{g(x,y)} e^{-2\pi N (\operatorname{Im}\tau) y^2} dx dy \quad (4.37)$$

where the real variables x, y are defined by $z = x + \tau y$.

The theta series, which define an orthonormal basis of the space of theta functions, are

$$\theta_j^\tau(z) = \sum_{n\in\mathbb{Z}} e^{\pi i N \tau \left(\frac{j}{N}+n\right)^2 + 2\pi i N \left(\frac{j}{N}+n\right) z}, \quad j = 0, 1, \ldots, N-1$$

and the finite Heisenberg group acts on theta functions by

$$\exp(pP + qQ + kE)\theta_j^\tau(z) = e^{-\frac{\pi i}{N}pq - \frac{2\pi i}{N}jq + \frac{\pi i}{N}k}\theta_{j+p}^\tau(z).$$

Recall Proposition 3.8 from §3.3.2, where it was shown that the mapping class group of the torus is the modular group $SL(2,\mathbb{Z})$ and where the generators S and T of this group were exhibited. These maps gave rise in $H_1(\Sigma_1, \mathbb{R})$ to the linear maps

$$S_* = \begin{pmatrix} 0 & -1 \\ 1 & 0 \end{pmatrix} \text{ and } T_* = \begin{pmatrix} 1 & 0 \\ 1 & 1 \end{pmatrix},$$

which define symplectomorphisms of the Jacobian variety.

4.4.2 *The action of the S map*

Let us compute the action of S on theta functions. For this we have

$$S \cdot \exp(pP + qQ) = \exp(-qP + pQ),$$

so the exact Egorov condition can be written as

$$\mathcal{F}(S)\exp(pP + qQ) = \exp(-qP + pQ)\mathcal{F}(S).$$

Let the matrix of $\mathcal{F}(S)$ be (s_{lk}). The action of the left-hand side on $\theta_j^\tau(z)$ is

$$\sum_l s_{l,j+p} e^{-\frac{\pi i}{N}pq - \frac{2\pi i}{N}jq}\theta_l^\tau(z),$$

while the action of the right-hand side is

$$\sum_l s_{l,j} e^{\frac{\pi i}{N}pq - \frac{2\pi i}{N}lp}\theta_{l-q}^\tau(z) = \sum_l s_{l+q,j} e^{-\frac{\pi i}{N}pq - \frac{2\pi i}{N}lp}\theta_l^\tau(z).$$

Setting the two equal we obtain

$$s_{l,j+p} e^{\frac{2\pi i}{N}lp} = s_{l+q,j} e^{\frac{2\pi i}{N}jq}.$$

Varying l, j, p, q we can compute using this formula s_{kl} from $s_{0,0}$, and we deduce that, up to a multiplication by a constant we can choose $s_{kl} = e^{-\frac{2\pi i}{N}kl}$. Imposing $\mathcal{F}(S)$ to be unitary, we obtain that its matrix is, up to multiplication by a complex number of absolute value 1,

$$\mathcal{F}(S) = \frac{1}{\sqrt{N}} \begin{pmatrix} 1 & 1 & 1 \cdots & 1 \\ 1 & e^{-\frac{2\pi i}{N}} & e^{-\frac{4\pi i}{N}} \cdots & e^{\frac{2(N-1)\pi i}{N}} \\ 1 & e^{-\frac{4\pi i}{N}} & e^{-\frac{8\pi i}{N}} \cdots & e^{-\frac{4(N-1)\pi i}{N}} \\ \cdots & \cdots & \cdots \cdots & \cdots \\ 1 & e^{-\frac{2(N-1)\pi i}{N}} & e^{-\frac{4(N-1)\pi i}{N}} \cdots & e^{-\frac{(N-1)^2\pi i}{N}} \end{pmatrix},$$

which is the discrete Fourier transform.

Now let us turn to the other point of view, where we change the complex variable and the periodicity matrix. For S this is done as

$$z' = \frac{z}{\tau}, \tau' = -\frac{1}{\tau}.$$

Now we have the following "functional equation" found by Carl Gustav Jacob Jacobi.

Proposition 4.16. *(Jacobi) The following identity holds*

$$e^{-\frac{\pi i N}{\tau}z^2}\theta_j^{-\frac{1}{\tau}}\left(\frac{z}{\tau}\right) = (-i\tau)^{\frac{1}{2}} N^{-\frac{1}{2}} \sum_{k=0}^{N-1} e^{-\frac{2\pi i}{N}jk}\theta_k^\tau(z),$$

where the branch of the function $\tau \mapsto (-i\tau)^{1/2}$ is chosen so that it is equal to 1 when $\tau = i$.

Proof. First let us check that

$$f(z) = e^{-\frac{\pi i N}{\tau}z^2}\theta_j^{-\frac{1}{\tau}}\left(\frac{z}{\tau}\right)$$

is a theta function for the period lattice $\mathbb{Z}+\mathbb{Z}\tau$. Using the formula for the theta series, we can write

$$f(z)=\sum_{n\in\mathbb{Z}} e^{-\frac{\pi i N}{\tau}\left(z-\frac{j}{N}-n\right)^2}.$$

From this formula it is immediate that $f(z+1)=f(z)$. For the other period, we compute

$$\begin{aligned}
f(z+\tau) &= \sum_{n\in\mathbb{Z}} e^{-\frac{\pi i N}{\tau}\left(z+\tau-\frac{j}{N}-n\right)^2} \\
&= \sum_{n\in\mathbb{Z}} e^{-\frac{\pi i N}{\tau}\left(z-\frac{j}{N}-n\right)^2-2\pi i N z-2\pi i j-2\pi i N n-\pi i N\tau} \\
&= e^{-2\pi i N z-\pi i N\tau}\sum_{n\in\mathbb{Z}} e^{-\frac{\pi i N}{\tau}\left(z-\frac{j}{N}-n\right)^2} = e^{-2\pi i N z-\pi i N\tau} f(z).
\end{aligned}$$

It follows that $f(z)\in\Theta_N^\tau(\Sigma_1)$, and consequently

$$f(z)=\sum_{k=0}^{N-1} c_k\theta_k^\tau(z),$$

for some coefficients c_k. To determine the coefficient c_k we compute the inner product $\langle f(z),\theta_k^\tau(z)\rangle$. This is an exercise in computing with Gaussian integrals. Here are the details.

To simplify things, we compute instead

$$(2N)^{1/2}(\operatorname{Im}\tau)^{-1/2}\left\langle f(z),\theta_k^\tau(z)\right\rangle,$$

namely we ignore the normalization factor in the integral (4.37). Then this is equal to

$$\begin{aligned}
&\int_0^1\int_0^1 \sum_{n\in\mathbb{Z}} e^{-\frac{\pi i N}{\tau}(x+\tau y)^2-\frac{\pi i N}{\tau}\left(\frac{j}{N}+n\right)^2+\frac{2\pi i N}{\tau}\left(\frac{j}{N}+n\right)(x+\tau y)} \\
&\quad\times\sum_{m\in\mathbb{Z}} e^{-i\pi N\bar\tau\left(\frac{k}{N}+m\right)-2\pi i N\left(\frac{k}{N}+m\right)(x+\bar\tau y)}e^{-2\pi N\operatorname{Im}\tau y^2}\,dxdy.
\end{aligned}$$

Switching summations, we obtain that this is equal to

$$\begin{aligned}
&\sum_{n,m\in\mathbb{Z}}\int_0^1\int_0^1 e^{-\frac{\pi i N}{\tau}(x+\tau y)^2-\frac{\pi i N}{\tau}\left(\frac{j}{N}+n\right)^2+\frac{2\pi i N}{\tau}\left(\frac{j}{N}+n\right)(x+\tau y)} \\
&\quad\times e^{-i\pi N\bar\tau\left(\frac{k}{N}+m\right)-2\pi i N\left(\frac{k}{N}+m\right)(x+\bar\tau y)}e^{-2\pi N\operatorname{Im}\tau y^2}\,dxdy.
\end{aligned}$$

Group all into one exponential function, and look at the exponent. After factoring out a $-\frac{\pi i N}{\tau}$, this exponent becomes

$$x^2 + 2\tau xy + \tau^2 + \left(\frac{j}{N} + n\right)^2 - 2\left(\frac{j}{N} + n\right)x - 2\tau\left(\frac{j}{N} + n\right)y$$
$$+\tau\bar{\tau}\left(\frac{k}{N} + m\right)^2 + 2\tau\left(\frac{k}{N} + m\right)x + 2\tau\bar{\tau}\left(\frac{k}{N} + m\right)y - 2i\tau \operatorname{Im}\tau y^2.$$

Completing squares we obtain that this is equal to

$$\left[x + \tau y - \left(\frac{j}{N} + n\right) + \tau\left(\frac{k}{N} + m\right)\right]^2 + (\tau\bar{\tau} - \tau^2)\left(y + \frac{k}{N}\right)^2$$
$$+2\tau\left(\frac{j}{N} + n\right)\left(\frac{k}{N} + m\right).$$

Returning to the computation of the inner product, note that

$$e^{-\frac{\pi i N}{\tau}2\tau\left(\frac{j}{N}+n\right)\left(\frac{k}{N}+m\right)} = e^{-\frac{2\pi i}{N}jk - 2\pi i(jm+kn) - 2\pi i N mn} = e^{-\frac{2\pi i}{N}jk},$$

because N, j, k, m, n are integers. Also note that in the first square there is an $x - n$, while in the second there is an $y + m$, and as m and n range over $\mathbb{Z}$, the sum of integrals becomes one single integral over $\mathbb{R} \times \mathbb{R}$. This integral is

$$e^{-\frac{2\pi i}{N}jk}\int_{\mathbb{R}}\left(\int_{\mathbb{R}} e^{-\frac{\pi i N}{\tau}\left(x+\tau y-\frac{j}{N}-\tau\frac{k}{N}\right)^2}dx\right)e^{-2\pi N \operatorname{Im}\tau\left(y+\frac{k}{N}\right)^2}dy.$$

After the substitution $\xi = x + \tau y - \frac{j}{N} - \tau\frac{k}{N}$, the inside integral becomes a standard Gaussian, and it is equal to $(-i\tau)^{1/2}N^{-1/2}$. Hence we obtain

$$e^{-\frac{2\pi i}{N}jk}(-i\tau)^{1/2}N^{-1/2}\int_{\mathbb{R}} e^{-2\pi N \operatorname{Im}\tau\left(y+\frac{k}{N}\right)^2}dy,$$

and the integral is again a Gaussian, after the substitution $\eta = y + \frac{k}{N}$. It is equal to $(2N)^{-1/2}(\operatorname{Im}\tau)^{-1/2}$, and this exactly the coefficient that we ignored in front of the inner product. We conclude that

$$\langle f(z), \theta_k^\tau(z)\rangle = (-i\tau)^{1/2}N^{-1/2}e^{-\frac{2\pi i}{N}jk},$$

and the formula is proved. □

4.4.3 *The action of the T map*

Now let us turn to the action of T on theta functions. We have

$$T \cdot \exp(pP + qQ) = \exp(pP + (p+q)Q),$$

so the exact Egorov identity is

$$\mathcal{F}(T)\exp(pP+qQ)=\exp(pP+(p+q)Q)\mathcal{F}(T).$$

Setting the matrix of $\mathcal{F}(T)$ to be (t_{lk}), the action of the left-hand side on $\theta_j^\tau(z)$ is

$$\sum_l t_{l,j+p}e^{-\frac{\pi i}{N}pq-\frac{2\pi i}{N}jq}\theta_l^\tau(z),$$

while the action of the right-hand side is

$$\sum_l t_{lj}e^{-\frac{\pi i}{N}pq-2\pi il(p+q)}\theta_{l+p}^\tau(z)=\sum_l t_{l-p,j}e^{-\frac{\pi i}{N}pq-\frac{2\pi i}{N}(l-p)(q+p)}\theta_l^\tau(z).$$

Hence

$$t_{l,j+p}e^{-\frac{2\pi i}{N}jq}=t_{l-p,j}e^{-\frac{2\pi i}{N}(l-p)(q+p)}.$$

Setting $p=0$, we obtain $t_{lj}e^{-\frac{2\pi i}{N}jq}=t_{lj}e^{-\frac{2\pi i}{N}lq}$ for all q, which can only happen if $t_{lj}=0$ when $l\neq j$. For $p=1$, $l=j+1$ we obtain $t_{j+1,j+1}=e^{-\frac{2\pi i}{N}j}t_{jj}$. We can set $t_{0,0}=1$, in which case $t_{jj}=e^{-\frac{2\pi i}{N}j^2}$, $j\geq 1$. We can think of this as a discrete Fourier transform over one point, a point of view that will become more transparent once we pass to the representation theoretic framework in the next chapter.

If we change the complex structure and the periodicity matrix by T we have the transformation

$$z'=z,\quad \tau'=\tau+1,$$

and an easy exercise shows that

$$\theta_j^{\tau+1}(z)=e^{-\frac{2\pi i}{N}j^2}\theta_j^\tau(z).$$

Conclusions to the fourth chapter

In this chapter we introduced the space of Riemann's theta functions on the Jacobian variety of a Riemann surface, and then defined the action of the finite Heisenberg group and of the mapping class group on it. This is done as follows:

1. For an integer $g\geq 1$ we consider a closed orientable genus g surface Σ_g. We endow Σ_g with a complex structure that makes it a Riemann surface. To the surface we associate the **Jacobian variety** $\mathcal{J}(\Sigma_g)$, using the lattice of periods of g linearly independent holomorphic 1-forms evaluated on a canonical basis of $H_1(\Sigma_g,\mathbb{Z})$. $\mathcal{J}(\Sigma_g)$ has a natural symplectic structure and can be viewed as the phase space of a classical

mechanical system. We set Planck's constant to be the reciprocal of an even integer, $h = 1/N$, and using the procedure of geometric quantization we associate to $\mathcal{J}(\Sigma_g)$ a space of states. This turns out to be the space of Riemann's theta functions $\mathbf{\Theta}_N^{\Pi}(\Sigma_g)$.

2. The first homology group of Σ_g can be identified with the group of exponential functions on the Jacobian variety. We apply the method of **equivariant Weyl quantization** to quantize these exponential functions in order to produce unitary operators. These operators give rise to a group, the **finite Heisenberg group** $\mathbf{H}(\mathbb{Z}_N^g)$. The operators act on theta functions by translations, in a manner first discovered by André Weil.
3. The mapping class group, $\mathrm{MCG}(\Sigma_g)$, also known as the modular group, gives rise to a group of symplectomorphisms of $\mathcal{J}(\Sigma_g)$. Via a Stone–von Neumann Theorem we deduced the existence of a projective representation of this group on the space of theta functions. This is the **representation of the mapping class group** of the surface on the space of theta functions, which is related to the action of the finite Heisenberg group by an **exact Egorov identity**.

Chapter 5

From theta functions to knots

This chapter connects the theory of theta functions to low dimensional topology. The bridge is representation theory. The main idea belongs to the author and Alejandro Uribe, and was first explained in [Gelca and Uribe (2014)]. It is different from Witten's quantum field theoretical approach.

5.1 Theta functions in the representation theoretical setting

The situation from §2.4 repeats in the case of the finite Heisenberg group. The presence of a Stone-von Neumann Theorem (Theorem 4.12) for the finite Heisenberg group $\mathbf{H}(\mathbb{Z}_N^g)$ suggests that one should be able to recover the Schrödinger representation from the group itself. This is done, like in Chapter 2, via the **induced representation**.

5.1.1 *Induced representations for finite groups*

In this paragraph we recall the construction of the induced representation for **finite groups** (see [Fulton and Harris (1991)] for more details). The group we have in mind is $\mathbf{H}(\mathbb{Z}_N^g)$. There are two equivalent ways in which the induced representation can be introduced, both of which we describe below. All groups are represented on complex vector spaces.

1. Here is the first point of view. Let G be a finite group. Assume that we are given a subgroup H of G together with a representation

$$\rho : H \to \mathrm{Aut}(W).$$

Let σ be a left coset of G modulo H, namely $\sigma \in G/H$. We define formally the space σW to be a copy of W, with its elements written formally as σw,

$w \in W$. The vector space on which G is represented is

$$\mathrm{Ind}_H^G \rho = \bigoplus_{\sigma \in G/H} \sigma W.$$

W itself is identified with the subspace of $\mathrm{Ind}_G^H \rho$ corresponding to the coset of the identity element.

To describe the action of G on $\mathrm{Ind}_H^G \rho$ we pick one representative g_σ for each coset σ and write

$$\mathrm{Ind}_H^G \rho = \bigoplus_{\sigma \in G/H} g_\sigma W.$$

For $g \in G$ and $\sigma \in G/H$, there is $h(g, \sigma) \in H$ such that

$$g g_\sigma = g_{g\sigma} h(g, \sigma),$$

The action of g on $\mathrm{Ind}_H^G \rho$ is defined on the vectors σw by

$$g \cdot \sigma w = g \cdot g_\sigma w = g_{g\sigma}[\rho(h(g, \sigma))w], \tag{5.1}$$

and then extended by linearity to the whole $\mathrm{Ind}_H^G \rho$. This is the **induced representation**.

Proposition 5.1. *The action defined by (5.1) does not depend on the choice of the representatives of the cosets. Moreover, this action restricts to an action of H that keeps W invariant, and the restriction to W coincides with ρ.*

Proof. Let g'_σ, $\sigma \in G/H$ be another set of representatives for the cosets. Then for each σ there is h_σ such that $g'_\sigma = g_\sigma h_\sigma$. If $g g'_\sigma = g'_{g\sigma} h'(g, \sigma)$, then

$$g g_\sigma h_\sigma = g_{g\sigma} h_{g\sigma} h'(g, \sigma)$$

so

$$g_{g\sigma} h(g, \sigma) h_\sigma = g_{g\sigma} h_{g\sigma} h'(g, \sigma).$$

We obtain

$$h'(g, \sigma) = h_{g\sigma}^{-1} h(g, \sigma) h_\sigma.$$

Using the action (5.1) defined by the g_σ's, we compute the action of g on $g'_\sigma w$ as

$$\begin{aligned} g \cdot g'_\sigma w &= g \cdot g_\sigma h_\sigma w = g \cdot g_\sigma[\rho(h_\sigma)w] = g_{g\sigma}[\rho(h(g, \sigma))\rho(h_\sigma)w] \\ &= g_{g\sigma} h_{g\sigma}[\rho(h_{g\sigma} h(g, \sigma) h_\sigma)w] = g'_{g\sigma}\rho(h'(g, \sigma))w. \end{aligned}$$

But this is what we obtain if we defined (5.1) using the representatives g'_σ. This shows that the action does not depend on the choice of representatives, so the induced representation is well defined.

For the second part, let $h \in H$. By choosing the representative of the coset H to be the identity element e we obtain

$$h \cdot Hw = h \cdot ew = (he)w = \rho(h)w,$$

which shows that when restricted to H the action keeps W invariant and restricts to ρ on W. □

2. The second point of view is that the induced representation can be realized as a space of functions on G that transform equivariantly under the action of the subgroup H. This approach can be generalized, under certain assumptions, to infinite groups, and was followed in §2.4. We let

$$\mathrm{Ind}_H^G \rho = \{\psi : G \to W \mid \psi(gh) = \rho(h)^{-1}\psi(g)\}.$$

The group G is represented on $\mathrm{Ind}_H^G \rho$ by

$$g_0 \cdot \psi(g) = \psi(g_0^{-1} g). \tag{5.2}$$

The next result shows that the two definitions give rise to equivalent representations.

Theorem 5.1. *The two constructions of the induced representation are equivalent to the representation of G on*

$$\mathbb{C}[G] \otimes_{\mathbb{C}[H]} W$$

induced by the left regular action of G on $\mathbb{C}[G]$.

Proof. The representation of G given in the statement of the theorem coincides actually with the first model we gave for the induced representation. This can be seen as follows. Given representatives g_σ of the left cosets of G modulo H, we have

$$\mathbb{C}[G] \otimes_{\mathbb{C}[H]} W = \bigoplus_{\sigma \in G/H} g_\sigma \otimes W.$$

The left regular action on G induces the action defined in (5.1).

For the second model, note that $\mathbb{C}[G] \otimes W$ is the space of W-valued functions on G, which we denote by $F(G, W)$. A function can be represented as

$$\sum_g g \otimes \psi(g).$$

Let us show that tensoring over $\mathbb{C}[H]$ amounts to introducing the equivariance relation $\psi(gh) = \rho(h)^{-1}\psi(g)$. Denote by $F^H(G,W)$ the space of functions satisfying this relation.

Consider the map

$$\pi_H : F(G,W) \mapsto F(G,W),$$

defined by

$$\pi_H\left(\sum_{g\in G} g\otimes\psi(g)\right) = \frac{1}{|H|}\sum_{g\in G}\sum_{h\in H} gh\otimes\rho(h)^{-1}\psi(gh).$$

Then π_H is a projection onto $F^H(G,W)$. In fact it is an orthogonal projection with respect to the counting measure. Let us analyze its kernel.

Given an element of the form $g\otimes w - gh\otimes\rho(h)^{-1}w$ we have

$$\begin{aligned}
&\pi_H\left(g\otimes w - gh\otimes\rho(h)^{-1}w\right)\\
&= \frac{1}{|H|}\left(\sum_{h\in H} gh\otimes\rho(h)^{-1}w - \sum_{h'\in H} ghh'\otimes\rho(h')^{-1}\rho(h)^{-1}w\right)\\
&= \frac{1}{|H|}\left(\sum_{h\in H} gh\otimes\rho(h)^{-1}w - \sum_{h'\in H} g(hh')\otimes\rho(hh')^{-1}w\right)\\
&= \frac{1}{|H|}\left(\sum_{h\in H} gh\otimes\rho(h)^{-1}w - \sum_{h\in H} g(h)\otimes\rho(h)^{-1}w\right) = 0.
\end{aligned}$$

So every element of this form is in the kernel.

On the other hand, modulo elements of the form $g\otimes w - gh\otimes\rho(h)^{-1}w$, every $g\otimes w\in G$ is equivalent to $gh\otimes\rho(h)^{-1}$ for every $h\in H$. Summing over all such h and dividing by $|H|$, we deduce that g is equivalent to

$$\frac{1}{|H|}\sum_{h\in H} gh\otimes\rho(h)^{-1}w.$$

The latter is in the image of π_H. We conclude that the kernel is spanned by elements of the form $g\otimes w - gh\otimes\rho(h)^{-1}w$, and this is precisely the kernel of the quotient map

$$\mathbb{C}[G]\otimes W \to \mathbb{C}[G]\otimes_{\mathbb{C}[H]} W.$$

This proves that the vector space of the second construction is $\mathbb{C}[G]\otimes_{\mathbb{C}[H]}W$.

Finally, an element g_0 acts by the left regular action as

$$g_0\sum_g g\otimes\psi(g) = \sum_g g_0g\otimes\psi(g) = \sum_g g\otimes\psi(gg_0^{-1}),$$

and this is the action from (5.2). This proves the equivalence of the two descriptions of the induced representation. □

5.1.2 *The Schrödinger representation of the finite Heisenberg group as an induced representation*

We now explain how the Schrödinger representation arises as an induced representation. Start with the Riemann surface Σ_g endowed with the canonical basis

$$a_1, a_2, \ldots, a_g, b_1, b_2, \ldots b_g$$

of its first homology group. Using this basis, $H_1(\Sigma_g, \mathbb{Z})$ can be identified with $\mathbb{Z}^{2g}$.

Now we bring into play the polarization. In §4.2.2, the real polarization used in the quantization was spanned by the vector fields

$$\frac{\partial}{\partial x_1}, \frac{\partial}{\partial x_2}, \ldots, \frac{\partial}{\partial x_g}. \tag{5.3}$$

To the real polarization we associate a Kähler polarization as in Example 2.8. This is

$$\frac{\partial}{\partial x_1} + i\frac{\partial}{\partial y_1}, \frac{\partial}{\partial x_2} + i\frac{\partial}{\partial y_2}, \ldots, \frac{\partial}{\partial x_g} + i\frac{\partial}{\partial y_g}. \tag{5.4}$$

The directions of the vector fields (5.3) in the Jacobian variety are specified by $a_1, a_2, \ldots, a_g \in H_1(\Sigma_g, \mathbb{Z})$. These elements form a Lagrangian subspace of $H_1(\Sigma_g, \mathbb{R})$. We could denote this subspace by $\mathbf{L}$, as we did in §2.4, but we don't do it in order to keep formulas simple. Instead, we recall from §2.4 the functions f such that the vector fields X_f belong to the polarization (5.3). These are the functions in the variables $y_1, y_2, \ldots, y_g$, which can be expanded in Fourier series in the exponentials of these variables. The y-coordinates are specified by

$$b_1, b_2, \ldots, b_g \tag{5.5}$$

and we think of linear combinations of the b_j's with integer coefficients as the same linear combinations of the coordinates y_j. By exponentiation the latter functions become the exponentials on the Jacobian torus that depend only on $y_1, y_2, \ldots, y_g$. To emphasize the role of $b_1, b_2, \ldots, b_g$ in this construction, we let $\mathbf{L}$ be the subgroup of $H_1(\Sigma_g, \mathbb{Z})$ generated by $b_1, b_2, \ldots, b_g$.

We now have three objects that correspond to each other:

- the real polarization (5.3),
- the Kähler polarization (5.4), and
- one "half" of a canonical basis (5.5).

Recall from §4.2.3 that the Heisenberg group $\mathbf{H}(\mathbb{Z}^g)$ is the $\mathbb{Z}$-extension of $H_1(\Sigma_g, \mathbb{Z})$ by the cocycle defined by the intersection form. Thus we can view $\mathbf{L}$ as an abelian subgroup of $\mathbf{H}(\mathbb{Z}^g)$; explicitly it is

$$\{\mathbf{0}\} \times \mathbf{Z}^g \times \{0\} \subset \mathbf{H}(\mathbb{Z}^g).$$

This subgroup factors to an abelian subgroup of $\mathbf{H}(\mathbb{Z}_N^g)$, which we denote by $\exp(\mathbf{L})$ (the notation is because we think of $\mathbf{L}$ as consisting of linear combinations of coordinate functions, while $\exp(\mathbf{L})$ consists of the exponentials of the quantizations of those coordinates functions). Let $\exp(\mathbf{L} + \mathbb{Z}E)$ be the subgroup of $\mathbf{H}(\mathbb{Z}_N^g)$ containing both $\exp(\mathbf{L})$ and the scalars $\exp(\mathbb{Z}E)$. Concretely

$$\exp(\mathbf{L} + \mathbb{Z}E) = \left\{\exp(\mathbf{q}^T\mathbf{Q} + kE), \mathbf{q} \in \mathbb{Z}^g, k \in \mathbb{Z}\right\}.$$

Then $\exp(\mathbf{L} + \mathbb{Z}E)$ is a maximal abelian subgroup of the finite Heisenberg group.

Being abelian, $\exp(\mathbf{L} + \mathbb{Z}E)$ has only 1-dimensional irreducible representations, and these representations are characters of the group. The Stone-von Neumann Theorem (Theorem 4.12) hints to considering the representation defined by the character

$$\chi_{\mathbf{L}} : \exp(\mathbf{L} + \mathbb{Z}E) \to \mathbb{C}, \quad \chi_{\mathbf{L}}(l + kE) = e^{\frac{\pi i}{N}k}.$$

To simplify the notation, we let

$$\mathcal{H}_{N,g}(\mathbf{L}) := \mathrm{Ind}_{\exp(\mathbf{L}+\mathbb{Z}E)}^{\mathbf{H}(\mathbb{Z}_N^g)} \chi_{\mathbf{L}}.$$

1. With the first definition of the induced representation,

$$\mathcal{H}_{N,g}(\mathbf{L}) = \mathbb{C}[\mathbf{H}(\mathbb{Z}_g^N)] \otimes_{\mathbb{C}[\exp(\mathbf{L}+\mathbb{Z}E)]} \mathbb{C}.$$

This means that $\mathcal{H}_{N,g}(\mathbf{L})$ is the quotient of $\mathbb{C}[\mathbf{H}(\mathbb{Z}_g^N)]$ by the subspace spanned by all elements of the form

$$u - \chi_{\mathbf{L}}(u')^{-1}uu'$$

with $u \in \mathbf{H}(\mathbb{Z}_N^g)$ and $u' \in \exp(\mathbf{L} + \mathbb{Z}E)$.

The left regular action of the finite Heisenberg group on its group algebra descends to an action on $\mathcal{H}_{N,g}(\mathbf{L})$. This action is the induced representation.

Let

$$\pi_{\mathbf{L}} : \mathbb{C}[\mathbf{H}(\mathbb{Z}_g^N)] \to \mathcal{H}_{N,g}(\mathbf{L})$$

be the quotient map. For all $u \in \mathbf{H}(\mathbb{Z}_N^g)$ and $u' \in \exp(\mathbf{L} + \mathbb{Z}E)$ we have

$$\pi_{\mathbf{L}}(uu') = \chi_{\mathbf{L}}(u')\pi_{\mathbf{L}}(u). \tag{5.6}$$

2. Alternatively, consider the space of functions

$$\mathrm{Funct}(\mathbf{H}(\mathbb{Z}_N^g)) = \{\psi : \mathbf{H}(\mathbb{Z}_N^g) \to \mathbb{C}\}.$$

Then $\mathcal{H}_{N,g}(\mathbf{L})$ is the subspace consisting of those functions satisfying the equivariance condition

$$\psi(uu') = \chi_{\mathbf{L}}^{-1}(u')\psi(u) \text{ for all } u \in \mathbf{H}(\mathbb{Z}_N^g),\ u' \in \exp(\mathbf{L} + \mathbb{Z}E). \quad (5.7)$$

The representation of $\mathbf{H}(\mathbb{Z}_N^g)$ is defined by

$$u_0 \cdot \psi(u) = \psi(uu_0^{-1}).$$

By Theorem 5.1, the two constructions are equivalent. The second facilitates the introduction of the Hilbert space structure on $\mathcal{H}_{N,g}(\mathbf{L})$. On $\mathrm{Funct}(\mathbf{H}(\mathbb{Z}_N^g))$ define the inner product

$$\langle \psi_1, \psi_2 \rangle = \frac{1}{2N^{2g+1}} \sum_{u \in \mathbf{H}(\mathbb{Z}_N^g)} \psi_1(u)\overline{\psi_2(u)}.$$

This induces an inner product on $\mathcal{H}_{N,g}(\mathbf{L})$ given by

$$\langle \psi_1, \psi_2 \rangle = \frac{1}{N^g} \sum_u \psi_1(u)\overline{\psi_2(u)},$$

where u ranges over a family of representatives of $\mathbf{H}(\mathbb{Z}_N^g)/\exp(\mathbf{L} + \mathbb{Z}E)$. Note that in this formula, $N^g = |\mathbf{H}(\mathbb{Z}_N^g)/\exp(\mathbf{L}+\mathbb{Z}E)|$, namely the number of cosets. The norm is well defined because the equivariance condition (5.7) implies that $\psi(u)\overline{\psi_2(u)}$ depends only on the equivalence class of u modulo $\exp(\mathbf{L} + \mathbb{Z}E)$.

Under the identification of the space of functions on $\mathbf{H}(\mathbb{Z}_N^g)$ with $\mathbb{C}[\mathbf{H}(\mathbb{Z}_N^g)]$, the quotient map $\pi_{\mathbf{L}}$ becomes the orthogonal projection

$$\pi_{\mathbf{L}} : \mathrm{Funct}(\mathbf{H}(\mathbb{Z}_N^g)) \to \mathcal{H}_{N,g}(\mathbf{L})$$

given by

$$(\pi_{\mathbf{L}}\psi)(u) = \frac{1}{2N^{g+1}} \sum_{u' \in \exp(\mathbf{L}+\mathbb{Z}E)} \chi_{\mathbf{L}}(u')\psi(uu').$$

The above formula for the inner product shows that if u ranges over a family of representatives of $\mathbf{H}(\mathbb{Z}_N^g)/\exp(\mathbf{L}+\mathbb{Z}E)$ then $\pi_{\mathbf{L}}(u)$ form an orthonormal basis of $\mathcal{H}_{N,g}(\mathbf{L})$. In particular

$$\pi_{\mathbf{L}}(\exp(\boldsymbol{\mu}^T\mathbf{P})), \quad \boldsymbol{\mu} \in \mathbb{Z}_N^g$$

is an orthonormal basis of $\mathcal{H}_{N,g}(\mathbf{L})$.

A left inverse of $\pi_{\mathbf{L}}$ is the inclusion map

$$s_{\mathbf{L}} : \mathcal{H}_{N,g}(\mathbf{L}) \to \mathrm{Funct}(\mathbf{H}(\mathbb{Z}_N^g)).$$

Turning to the first construction of the induced representation, this left inverse is the section

$$s_{\mathbf{L}} : \mathcal{H}_{N,g}(\mathbf{L}) \to \mathbb{C}[\mathbf{H}(\mathbb{Z}_N^g)],$$

given by

$$s_{\mathbf{L}}(\pi_{\mathbf{L}}(u)) = \frac{1}{2N^{g+1}} \sum_{u' \in \exp(\mathbf{L}+\mathbb{Z}E)} \chi_{\mathbf{L}}(u')^{-1} u u'. \tag{5.8}$$

We will use the map $s_{\mathbf{L}}$ for the abstract construction of the action of the mapping class group of the Riemann surface on theta functions.

We now prove the main result of this section, that the induced representation defined above coincides with the Schrödinger representation from Chapter 4.

Theorem 5.2. *There is a unitary transformation between the space of theta functions $\boldsymbol{\Theta}_N^{\Pi}(\Sigma_g)$ and $\mathcal{H}_{N,g}(\mathbf{L})$ defined by*

$$\theta_{\boldsymbol{\mu}}^{\Pi}(\mathbf{z}) \mapsto \pi_{\mathbf{L}}(\exp(\boldsymbol{\mu}^T \mathbf{P})), \quad \boldsymbol{\mu} \in \mathbb{Z}_N^g.$$

This transformation intertwines the Schrödinger representation on $\boldsymbol{\Theta}_N^{\Pi}(\Sigma_g)$ and the representation of $\mathbf{H}(\mathbb{Z}_N^g)$ on $\mathcal{H}_{N,g}(\mathbf{L})$ induced by the left regular action.

Proof. The map from the statement is an isomorphism of finite dimensional spaces. The norm of $\pi_{\mathbf{L}}(\exp(\boldsymbol{\mu}^T\mathbf{P}))$ is one, hence this map is unitary. We have

$$\exp(\mathbf{p}^T\mathbf{P})\exp(\boldsymbol{\mu}^T\mathbf{P}) = \exp((\mathbf{p}+\boldsymbol{\mu})^T)\mathbf{P})$$

and

$$\exp(\mathbf{q}^T\mathbf{Q})\exp(\mu^T\mathbf{P}) = e^{-\frac{\pi i}{N}\mathbf{q}^T\boldsymbol{\mu}}\exp(\boldsymbol{\mu}^T\mathbf{P})\exp(\mathbf{q}^T\mathbf{Q}).$$

It follows that

$$\exp(\mathbf{p}^T\mathbf{P})\pi_{\mathbf{L}}(\exp(\boldsymbol{\mu}^T\mathbf{P})) = \pi_{\mathbf{L}}((\mathbf{p}+\boldsymbol{\mu})^T\mathbf{P})$$
$$\exp(\mathbf{q}^T\mathbf{Q})\pi_{\mathbf{L}}(\exp(\boldsymbol{\mu}^T\mathbf{P})) = e^{-\frac{\pi i}{N}\mathbf{q}^T\boldsymbol{\mu}}\pi_{\mathbf{L}}(\exp(\boldsymbol{\mu}^T\mathbf{P}))$$

in agreement with the Schrödinger representation (4.34). □

Because of this result, we make the following definition.

Definition 5.1. The **abstract Schrödinger representation** of $\mathbf{H}(\mathbb{Z}_N^g)$ is the representation of this group induced by the character $\chi_{\mathbf{L}}$.

Under the identification of $\mathcal{H}_{N,g}(\mathbf{L})$ with $\boldsymbol{\Theta}_N^{\Pi}(\Sigma_g)$ provided by Theorem 5.2,

$$\begin{aligned}\pi_{\mathbf{L}}(\exp(\mathbf{p}^T\mathbf{P}+\mathbf{q}^T\mathbf{Q}+kE)) &= \pi_{\mathbf{L}}[\exp(\mathbf{p}^T\mathbf{P})\exp(\mathbf{q}^T\mathbf{Q}+(k+\mathbf{p}^T\mathbf{q})E)]\\ &= \chi_{\mathbf{L}}[\exp(\mathbf{q}^T\mathbf{Q}+(\mathbf{p}^T\mathbf{q}+k)E)]\pi_{\mathbf{L}}(\exp(\mathbf{p}^T\mathbf{P}))\\ &= e^{\frac{\pi i}{N}(\mathbf{p}^T\mathbf{q}+k)}\pi_{\mathbf{L}}(\exp(\mathbf{p}^T\mathbf{P}) = e^{\frac{\pi i}{N}(\mathbf{p}^T\mathbf{q}+k)}\theta_{\mathbf{p}}^{\Pi}(\mathbf{z}).\end{aligned}$$

Remark 5.1. It is important to point out that the construction of the induced representation depends only on the Lagrangian subspace $\mathbf{L}$ of $H_1(\Sigma_g,\mathbb{R})$. So this abstract model of the Schrödinger representation is associated to the pair $(\Sigma_g,\mathbf{L})$. Note that the dependence on the complex structure is also eliminated.

Remark 5.2. Although the induced representation depends only on $\mathbf{L}$, the identification of $\mathcal{H}_{N,g}(\mathbf{L})$ with $\boldsymbol{\Theta}_N^{\Pi}(\Sigma_g)$ given by Theorem 5.2 depends also on the Lagrangian subspace spanned by $a_1,a_2,\ldots,a_g$. So to define the Hilbert space of the quantization we only need half of the canonical basis, but to explicate this space and to do computations we need an entire canonical basis of $H_1(\Sigma_g,\mathbb{Z})$.

The Schrödinger representation can be extended linearly from a group representation of $\mathbf{H}(\mathbb{Z}_N^g)$ to an algebra representation of $\mathbb{C}[\mathbf{H}(\mathbb{Z}_N^g)]$. We obtain a description of the Schrödinger representation of $\mathbb{C}[\mathbf{H}(\mathbb{Z}_N^g)]$ as the left regular action of this algebra on a quotient of itself.

5.1.3 *The action of the mapping class group on theta functions in the representation theoretical setting*

For the abstract definition of the action of the mapping class group, we imitate the model from Chapter 2. We do not resolve the projective ambiguity here, nor in §5.4.3. We postpone this to Chapter 7, where we will associate a 4-dimensional manifold to each element of the mapping class group, and use the signature of this manifold to resolve the projective ambiguity.

The discrete Fourier transform induced by a pair of canonical bases

We introduce a finite dimensional analogue of the Fourier transform induced by a pair of Lagrangian subspaces from §2.5.1. Consider a Riemann surface Σ_g endowed with two canonical bases

$$a_1,a_2,\ldots,a_g,b_1,b_2,\ldots,b_g \text{ and } a_1',a_2',\ldots a_g',b_1',b_2',\ldots,b_g'. \tag{5.9}$$

Each gives rise to one version of the Jacobian variety, theta functions, and the finite Heisenberg group of quantized exponentials $\mathbf{H}(\mathbb{Z}_N^g)$. Let $\mathbf{L}$ and $\mathbf{L}'$ be the subgroups of $H_1(\Sigma_g, \mathbb{Z})$ generated by $b_1, b_2, \ldots, b_g$ and $b_1', b_2', \ldots, b_g'$ respectively. By analogy with the definition of the Fourier transform given in §2.5.1, we make the following definition.

Definition 5.2. The **discrete Fourier transform associated to the pair** $(\mathbf{L}, \mathbf{L}')$ is the linear operator

$$\mathcal{F}_{\mathbf{L},\mathbf{L}'} : \mathcal{H}_{N,g}(\mathbf{L}) \to \mathcal{H}_{N,g}(\mathbf{L}')$$

defined by

$$(\mathcal{F}_{\mathbf{L},\mathbf{L}'}\psi)(u) = \frac{\sqrt{n(\mathbf{L},\mathbf{L}')}}{N^{g+1}} \sum_{u' \in \exp(\mathbf{L}'+\mathbb{Z}E)} \psi(uu')\chi_{\mathbf{L}'}(u')$$

where $n(\mathbf{L}, \mathbf{L}')$ is the number of cosets in $\exp(\mathbf{L}' + \mathbb{Z}E)/\exp(\mathbf{L} \cap \mathbf{L}')$.

Examining the definitions of $\pi_{\mathbf{L}}$ and $s_{\mathbf{L}}$, we obtain:

Proposition 5.2. *The discrete Fourier transform determined by the pair* $(\mathbf{L}, \mathbf{L}')$ *is given by*

$$\mathcal{F}_{\mathbf{L},\mathbf{L}'} = \sqrt{n(\mathbf{L},\mathbf{L}')}\pi_{\mathbf{L}'} \circ s_{\mathbf{L}},$$

where $n(\mathbf{L}, \mathbf{L}')$ *is the number of cosets in* $\exp(\mathbf{L} + \mathbb{Z}E)/\exp(\mathbf{L} \cap \mathbf{L}')$.

This can be phrased as follows: to map an element $\pi_{\mathbf{L}}(u)$ by the discrete Fourier transform lift it to $\mathbf{H}(\mathbb{Z}_N^g)$ in all possible ways, average the results, then project to $\mathcal{H}_{N,g}(\mathbf{L})$.

Note that $n(\mathbf{L}, \mathbf{L}') = n(\mathbf{L}', \mathbf{L})$. Let us place the discrete Fourier transform in the framework of the first definition of the induced representation.

Proposition 5.3. *For* $u \in \mathbf{H}(\mathbb{Z}_N^g)$,

$$\mathcal{F}_{\mathbf{L},\mathbf{L}'}(\pi_{\mathbf{L}}(u)) = \frac{\sqrt{n(\mathbf{L},\mathbf{L}')}}{2N^{g+1}} \sum_{u' \in \exp(\mathbf{L}+\mathbb{Z}E)} \chi_{\mathbf{L}}(u')^{-1}\pi_{\mathbf{L}'}(uu'), \tag{5.10}$$

where $n(\mathbf{L}, \mathbf{L}')$ *is the number of cosets in* $\exp(\mathbf{L} + \mathbb{Z}E)/\exp(\mathbf{L} \cap \mathbf{L}')$.

The constant in front of the sum is chosen so as to make the discrete Fourier transform unitary, as the next result shows.

Proposition 5.4. *The map $\mathcal{F}_{\mathbf{L},\mathbf{L}'}$ is unitary and interpolates between the Schrödinger representations defined by* $\mathbf{L}$ *and* $\mathbf{L}'$.

Proof. Let us check that $\mathcal{F}_{\mathbf{L},\mathbf{L}'}$ is unitary. The operator

$$s_{\mathbf{L}} : \mathcal{H}_{N,g}(\mathbf{L}) \to \mathbb{C}[\mathbf{H}(\mathbb{Z}_N^g)]$$

is the adjoint of the operator

$$\pi_{\mathbf{L}} : \mathbb{C}[\mathbf{H}(\mathbb{Z}_N^g)] \to \mathcal{H}_{N,g}(\mathbf{L}).$$

Recall also that $n(\mathbf{L},\mathbf{L}') = n(\mathbf{L}',\mathbf{L})$.

Therefore

$$\begin{aligned}(\mathcal{F}_{\mathbf{L},\mathbf{L}'})^* &= \sqrt{n(\mathbf{L},\mathbf{L}')}(\pi_{\mathbf{L}'} \circ s_{\mathbf{L}})^* = \sqrt{n(\mathbf{L}',\mathbf{L})}s_{\mathbf{L}}^* \circ \pi_{\mathbf{L}'} \\ &= \sqrt{n(\mathbf{L}',\mathbf{L})}\pi_{\mathbf{L}} \circ s_{\mathbf{L}'} = \mathcal{F}_{\mathbf{L}',\mathbf{L}}.\end{aligned}$$

Hence in order to show that $\mathcal{F}_{\mathbf{L},\mathbf{L}'}$ is unitary, it suffices to show that

$$\mathcal{F}_{\mathbf{L}',\mathbf{L}} \circ \mathcal{F}_{\mathbf{L},\mathbf{L}'} = I.$$

At the heart of the proof lies the fact that the sum of the nth roots of unity is 1 if $n = 1$ and zero if $n > 1$, as this is how one usually checks that the standard discrete Fourier transform is unitary.

For $u \in \mathbf{H}(\mathbb{Z}_N^g)$, we compute

$$\begin{aligned}&\frac{4N^{2g+2}}{n(\mathbf{L},\mathbf{L}')}\mathcal{F}_{\mathbf{L}',\mathbf{L}} \circ \mathcal{F}_{\mathbf{L},\mathbf{L}'}(\pi_{\mathbf{L}}(u)) = \frac{2N^{g+1}}{\sqrt{n(\mathbf{L},\mathbf{L}')}}\mathcal{F}_{\mathbf{L},\mathbf{L}'}\left(\sum_{u' \in \exp(\mathbf{L}+\mathbb{Z}E)} \pi_{\mathbf{L}'}(uu')\right) \\ &= \sum_{u'' \in \exp(\mathbf{L}'+\mathbb{Z}E)} \sum_{u' \in \exp(\mathbf{L}+\mathbb{Z}E)} \chi_{\mathbf{L}}(u')^{-1}\chi_{\mathbf{L}'}(u'')^{-1}\pi_{\mathbf{L}}(uu'u'').\end{aligned}$$

In the inner sum, if $u'' \in \exp(\mathbf{L} \cap \mathbf{L}' + \mathbb{Z}E)$, then $u'u'' = u''u' \in \exp(\mathbf{L} + \mathbb{Z}E)$. Also, in this case $\chi_{\mathbf{L}'}(u'') = \chi_{\mathbf{L}}(u'')$. Therefore

$$\chi_{\mathbf{L}}(u')^{-1}\chi_{\mathbf{L}'}(u'')^{-1}\pi_{\mathbf{L}}(uu'u'') = \chi_{\mathbf{L}}(u'u'')\chi_{\mathbf{L}}(u'u'')\pi_{\mathbf{L}}(u) = \pi_{\mathbf{L}}(u).$$

When summing over u', this term is repeated $2N^{g+1}$ times, and when summing over $u'' \in \exp(\mathbf{L} \cap \mathbf{L}' + \mathbb{Z}E)$, the result is repeated

$$\frac{2N^{g+1}}{n(\mathbf{L},\mathbf{L}')} = |\exp(\mathbf{L} \cap \mathbf{L}' + \mathbb{Z}E)|$$

times. So when the sum is taken only over such u' and u'', then answer is

$$\frac{4N^{2g+2}}{n(\mathbf{L},\mathbf{L}')}\pi_{\mathbf{L}}(u).$$

On the other hand, let us choose $u'' \notin \exp(\mathbf{L} \cap \mathbf{L}' + \mathbb{Z}E)$. Set

$$\exp(\mathbf{L} + \mathbb{Z}E) = \{\exp(\mathbf{q}^T\mathbf{Q} + kE) \,|\, \mathbf{q} \in \mathbb{Z}_N^g, k \in \mathbb{Z}_{2N}\},$$

and

$$u'' = \exp(\mathbf{p}_0^T\mathbf{P} + \exp \mathbf{q}_0^T\mathbf{Q} + k_0E),$$

with $\mathbf{p}_0 \neq \mathbf{0}$. Then we can compute explicitly

$$\begin{aligned}
&\sum_{u' \in \exp(\mathbf{L}+\mathbb{Z}E)} \chi_{\mathbf{L}}(u')^{-1}\pi_{\mathbf{L}}(uu'u'') \\
&= \sum_{\mathbf{q},k} e^{-\frac{\pi i}{N}k}\pi_{\mathbf{L}}\left(u\exp(\mathbf{q}^T\mathbf{Q} + kE)\exp(\mathbf{p}_0^T\mathbf{P} + \mathbf{q}_0^T\mathbf{Q} + k_0E)\right) \\
&= \sum_{\mathbf{q},k} e^{-\frac{\pi i}{N}k}\pi_{\mathbf{L}}\left(u\exp(\mathbf{p}_0^T\mathbf{P} + \mathbf{q}_0^T\mathbf{Q} + k_0E)\exp(\mathbf{q}^T\mathbf{Q} + (k - 2\mathbf{p}_0^T\mathbf{q})E)\right).
\end{aligned}$$

Using (5.6) we find this to be equal to

$$\begin{aligned}
&\sum_{\mathbf{q},k} e^{-\frac{\pi i}{N}k}e^{\frac{\pi i}{N}(k-2\mathbf{p}_0^T\mathbf{q})}\pi_{\mathbf{L}}\left(u\exp(\mathbf{p}_0^T\mathbf{P} + \mathbf{q}_0^T\mathbf{Q} + k_0E)\right) \\
&\qquad = \left(\sum_{\mathbf{q},k} e^{\frac{\pi i}{N}\mathbf{p}_0^T\mathbf{q}}\right)\pi_{\mathbf{L}}\left(u\exp(\mathbf{p}_0^T\mathbf{P} + \mathbf{q}_0^T\mathbf{Q} + k_0E)\right) = 0,
\end{aligned}$$

because inside the parenthesis we have a sum of roots of unity. So when we sum over $u' \in \exp(\mathbf{L} + \mathbb{Z}E)$ and $u'' \notin \exp(\mathbf{L} \cap \mathbf{L}') + \mathbb{Z}E)$, we obtain zero.

We conclude that

$$(\mathcal{F}_{\mathbf{L},\mathbf{L}'})^* = \mathcal{F}_{\mathbf{L}',\mathbf{L}} = (\mathcal{F}_{\mathbf{L},\mathbf{L}'})^{-1},$$

which proves that $\mathcal{F}_{\mathbf{L},\mathbf{L}'}$ is unitary.

The Schrödinger representations induced by $\mathbf{L}$ and $\mathbf{L}'$ respectively are defined by the left regular actions of the Heisenberg group on $\mathcal{H}_{N,g}(\mathbf{L})$ and $\mathcal{H}_{N,g}(\mathbf{L}')$ respectively. The map $\mathcal{F}_{\mathbf{L},\mathbf{L}'}$ interpolates between these regular actions. The theorem is proved. □

The elimination of the redundancies

The choice of the section $s_{\mathbf{L}}$ in (5.8) can be simplified by taking just the average over $\exp(\mathbf{L} + \mathbb{Z}E)/\exp(\mathbb{Z}E) \simeq \exp(\mathbf{L})$. Hence

$$s_{\mathbf{L}}(\pi_{\mathbf{L}}(u)) = \frac{1}{N^g}\sum_{u' \in \exp(\mathbf{L})} uu'. \tag{5.11}$$

It is also possible to simplify the formula (5.10) for the discrete Fourier transform. To this end, note that if

$$u_1 \equiv u_2 \bmod [\exp(\mathbf{L} + \mathbb{Z}E) \cap \exp(\mathbf{L}' + \mathbb{Z}E)],$$

then

$$\chi_{\mathbf{L}}(u_1)^{-1}\pi_{\mathbf{L}'}(uu_1) = \chi_{\mathbf{L}}(u_2)^{-1}\pi_{\mathbf{L}'}(uu_2).$$

Thus in (5.10) it suffices to sum over equivalence classes modulo $\exp(\mathbf{L} + \mathbb{Z}E) \cap \exp(\mathbf{L}' + \mathbb{Z}E)$. We can make the identification

$$\exp(\mathbf{L} + \mathbb{Z}E)/[\exp(\mathbf{L} + \mathbb{Z}E) \cap \exp(\mathbf{L}' + \mathbb{Z}E)] = \exp(\mathbf{L})/\exp(\mathbf{L} \cap \mathbf{L}').$$

We obtain the following simplified formula

$$\mathcal{F}_{\mathbf{L},\mathbf{L}'}(\pi_{\mathbf{L}}(u)) = \frac{1}{\sqrt{n_r(\mathbf{L}, \mathbf{L}')}} \sum_{u' \in \exp(\mathbf{L})/\exp(\mathbf{L} \cap \mathbf{L}')} \pi_{\mathbf{L}'}(uu') \tag{5.12}$$

where $n_r(\mathbf{L}, \mathbf{L}')$ is the number of cosets in $\exp(\mathbf{L})/\exp(\mathbf{L} \cap \mathbf{L}')$.

The discrete Fourier transform induced by an element of the mapping class group

Let Σ_g be a closed genus g Riemann surface endowed with a canonical basis $a_1, a_2, \ldots, a_g, b_1, b_2, \ldots, b_g$ of its first homology group. Let $\mathbf{L}$ be the subgroup of $H_1(\Sigma_g, \mathbb{Z})$ generated by $b_1, b_2, \ldots, b_g$, which then gives rise to the maximal abelian subgroup $\exp(\mathbf{L} + \mathbb{Z}E)$, and hence to the abstract version of the Schrödinger representation of the finite Heisenberg group $\mathbf{H}(\mathbb{Z}_N^g)$.

Consider an element h of the mapping class group of Σ_g, and using it define the unitary operator

$$A(h) : \mathrm{Funct}(\mathbf{H}(\mathbb{Z}_N^g)) \to \mathrm{Funct}(\mathbf{H}(\mathbb{Z}_N^g)), \quad (A(h)\psi)(u) = \psi(h^{-1} \cdot u).$$

This descends to a unitary operator

$$A(h) : \mathcal{H}_{N,g}(\mathbf{L}) \to \mathcal{H}_{N,g}(h_*(\mathbf{L})).$$

Identifying $\mathrm{Funct}(\mathbf{H}(\mathbb{Z}_N^g)$ with $\mathbb{C}[\mathbf{H}(\mathbb{Z}_N^g)]$, we have

$$A(h) : \mathbb{C}[\mathbf{H}(\mathbb{Z}_N^g)] \to \mathbb{C}[\mathbf{H}(\mathbb{Z}_N^g)], \quad A(h)u = h \cdot u,$$

which descends to

$$A(h) : \mathcal{H}_{N,g}(\mathbf{L}) \to \mathcal{H}_{N,g}(h_*(\mathbf{L})), \quad A(h)\pi_{\mathbf{L}}(u) = \pi_{h_*(\mathbf{L})}(h \cdot u). \tag{5.13}$$

Definition 5.3. The **discrete Fourier transform defined by** h is

$$\mathcal{F}_{\mathbf{L}}(h) : \mathcal{H}_{N,g}(\mathbf{L}) \to \mathcal{H}_{N,g}(\mathbf{L}), \quad \mathcal{F}_{\mathbf{L}}(h) = \mathcal{F}_{h_*(\mathbf{L}),\mathbf{L}} \circ A(h).$$

Examining closely the definition we see that $\mathcal{F}_{\mathbf{L}}(h)$ acts on an element in $\mathcal{H}_{N,g}(\mathbf{L})$ in the following way: We lift this element in all possible ways to $\mathbb{C}[\mathbf{H}(\mathbb{Z}_N^g)]$, map the results by h_* (where we view $\mathbf{H}(\mathbb{Z}_N^g)$ as a $\mathbb{Z}_{2N}$-extension of $H_1(\Sigma_g, \mathbb{Z}_N)$), take the average, and then map the result to $\mathcal{H}_{N,g}(\mathbf{L})$ by $\pi_{\mathbf{L}}$.

The map $\mathcal{F}_{\mathbf{L}}(h)$ defined this way is unitary and satisfies the exact Egorov identity

$$\mathcal{F}_{\mathbf{L}}(h)u\mathcal{F}_{\mathbf{L}}(h)^{-1} = h \cdot u, \quad u \in \mathbf{H}(\mathbb{Z}_N^g). \tag{5.14}$$

Because of this fact it coincides, up to multiplication by a constant of absolute value 1, with the map $\mathcal{F}(h)$ defined in §4.3.5.

Theorem 5.3. *The discrete Fourier transform $\mathcal{F}_{\mathbf{L}}(h)$ is given by the formula*

$$\mathcal{F}_{\mathbf{L}}(h)(\pi_{\mathbf{L}}(u)) = \frac{1}{\sqrt{n_r(\mathbf{L}, h_*(\mathbf{L}))}} \sum_{u' \in \exp(\mathbf{L})/\exp(h_*(\mathbf{L}) \cap \mathbf{L})} \pi_{\mathbf{L}}(h \cdot (uu')).$$

Proof. Using (5.13) and (5.12) we can write

$$\mathcal{F}_{\mathbf{L}}(h)(\pi_{\mathbf{L}}(u)) = \mathcal{F}_{h_*(\mathbf{L}),\mathbf{L}}(A(h)(\pi_{\mathbf{L}}(u)) = \mathcal{F}_{h_*(\mathbf{L}),\mathbf{L}}(\pi_{h_*(\mathbf{L})}(h \cdot u)$$
$$= \frac{1}{\sqrt{n_r(\mathbf{L}, h_*(\mathbf{L}))}} \sum_{u'' \in \exp(h_*(\mathbf{L}))/\exp(h_*(\mathbf{L}) \cap \mathbf{L})} \pi_{\mathbf{L}}((h \cdot u)u'').$$

Writing $u'' = h \cdot u'$ with $u' \in \exp(\mathbf{L})/\exp(\mathbf{L} \cap h_*(\mathbf{L}))$ and using the fact that

$$(h \cdot u)(h \cdot u') = h \cdot (uu')$$

we obtain the formula from the statement. □

Example 5.1. We look at the case where $g = 1$, when the surface is Σ_1, the torus. We choose the canonical basis

$$a_1 = (1, 0) \text{ and } b_1 = (0, 1),$$

and consider the S-map of the torus, defined by the 90° counter-clockwise rotation of $\mathbb{R}^2$ (see §3.2.2). At the level of the first homology group,

$$S_* : H_1(\Sigma_1, \mathbb{Z}) \to H_1(\Sigma_1, \mathbb{Z}), \quad S_*(a_1) = b_1 \text{ and } S_*(b_1) = -a_1,$$

meaning that

$$S_* = \begin{pmatrix} 0 & -1 \\ 1 & 0 \end{pmatrix}.$$

For $\mathbf{L} = \mathbb{Z}(0,1)$ we have $S_*(\mathbf{L}) = \mathbb{Z}(1,0)$, and so

$$\exp(\mathbf{L} + \mathbb{Z}E) = \{(0,q,k) \mid q \in \mathbb{Z}_N, k \in \mathbb{Z}_{2N}\},$$
$$\exp(S_*(\mathbf{L}) + \mathbb{Z}E) = \{(p,0,k) \mid p \in \mathbb{Z}_N, k \in \mathbb{Z}_{2N}\},$$
$$\exp(\mathbf{L} \cap S_*(\mathbf{L})) = \{(0,0,0)\}.$$

Using (5.3) for $\mathcal{F}_S$, we compute

$$\mathcal{F}_{\mathbf{L}}(S)(\pi_{\mathbf{L}}(\exp(\mu P))) = \frac{1}{\sqrt{N}} \sum_{q\in\mathbb{Z}_N} \pi_{\mathbf{L}}[S \cdot (\exp(\mu P)\exp(qQ))]$$
$$= \frac{1}{\sqrt{N}} \sum_{q\in\mathbb{Z}_N} \pi_{\mathbf{L}}[S \cdot \exp(\mu P + qQ + \mu q E)]$$
$$= \frac{1}{\sqrt{N}} \sum_{q\in\mathbb{Z}_N} \pi_{\mathbf{L}}[\exp(-qP + \mu Q + \mu q E)]$$
$$= \frac{1}{\sqrt{N}} \sum_{q\in\mathbb{Z}_N} \pi_{\mathbf{L}}[\exp(-qP)\exp(\mu Q + 2\mu q E)]$$
$$= \frac{1}{\sqrt{N}} \sum_{q\in\mathbb{Z}_N} \chi_{\mathbf{L}}(\exp(\mu Q + 2\mu q E))\pi_{\mathbf{L}}(\exp(-qP))$$
$$= \frac{1}{\sqrt{N}} \sum_{q\in\mathbb{Z}_N} e^{\frac{2\pi i}{N}\mu q}\pi_{\mathbf{L}}(\exp(-qP)) = \frac{1}{\sqrt{N}} \sum_{q\in\mathbb{Z}_N} e^{-\frac{2\pi i}{N}\mu q}\pi_{\mathbf{L}}(\exp(qP)).$$

In other words

$$\mathcal{F}_{\mathbf{L}}(S)(\theta_j^{\Pi}) = \sqrt{h} \sum_{q\in\mathbb{Z}_N} e^{-2\pi i h j q}\theta_q^{\Pi}, \tag{5.15}$$

where $h = 1/N$ is Planck's constant. If we identify $\Theta_N^{\Pi}(\Sigma_1)$ with $L^2(\mathbb{Z}_N)$ by mapping $\theta_j^{\Pi}(z)$ to the characteristic function $\chi_{\{j\}}$ of the singleton $\{j\}$, then on $L^2(\mathbb{Z}_N)$,

$$([\mathcal{F}_{\mathbf{L}}(S)](f))(j) = \sqrt{h} \sum_{k\in\mathbb{Z}_N} e^{-2\pi i h k j} f(k).$$

We recognize the **standard discrete Fourier transform.**

Example 5.2. Also on the torus Σ_1, we examine the map T. For $a_1 = (1,0)$ and $b_1 = (0,1)$, we have $T(a_1) = (1,1)$ and $T(b_1) = (0,1)$. At the level of homology,

$$T_* : H_1(\Sigma_1, \mathbb{Z}) \to H_1(\Sigma_1, \mathbb{Z}), \quad T_*(a_1) = a_1 + b_1, \quad T_*(b_1) = b_1$$

that is

$$T_* = \begin{pmatrix} 1 & 0 \\ 1 & 1 \end{pmatrix}.$$

The Lagrangian subspace $\mathbf{L} = \mathbb{Z}(0,1)$ is mapped to itself. Consequently $\mathbf{L} = \mathbf{L} \cap T_*(\mathbf{L})$ and so the sum in the definition of the discrete Fourier transform has only one term. We can choose as the representative of the unique equivalence class in the quotient $\mathbf{L}/\mathbf{L} \cap T_*(\mathbf{L})$ the identity element. We compute

$$\begin{aligned}\mathcal{F}_{\mathbf{L}}(T)[\pi_{\mathbf{L}}(\exp(\mu P))] &= \pi_{\mathbf{L}}(T \cdot \exp(\mu P)) \\ &= \pi_{\mathbf{L}}[\exp(\mu P + \mu Q)] = \pi_{\mathbf{L}}[\exp(\mu P)\exp(\mu Q - \mu^2 E)] \\ &= \chi_{\mathbf{L}}(\exp(\mu Q - \mu^2 E))\pi_{\mathbf{L}}[\exp(\mu P)] = e^{-\frac{\pi i}{N}\mu^2}\pi_{\mathbf{L}}[\exp(\mu P)].\end{aligned}$$

In other words,

$$\mathcal{F}_{\mathbf{L}}(T)(\theta_j^{\Pi}) = e^{-\pi i h j^2}\theta_j^{\Pi},$$

or, on $L^2(\mathbb{Z}_N)$,

$$[(\mathcal{F}_{\mathbf{L}}(T))(f)](j) = e^{-\pi i h j^2} f(j).$$

This can be interpreted as a Fourier transform over one point. We deduce that, in genus 1, the theta series are the eigenvectors of the discrete Fourier transform associated to Dehn twist along b_1.

The same computation can be done in genus g to prove the following result.

Proposition 5.5. *Let $a_1, a_2, \ldots, a_g, b_1, b_2, \ldots, b_g$ be a canonical basis, and let $T_1, T_2, \ldots, T_g$ be the Dehn twists along the curves $b_1, b_2, \ldots, b_g$ respectively. Then the operators $\mathcal{F}_{\mathbf{L}}(T_1), \mathcal{F}_{\mathbf{L}}(T_2), \ldots, \mathcal{F}_{\mathbf{L}}(T_g)$ are simultaneously diagonalizable in the basis consisting of the theta series $\theta_{\boldsymbol{\mu}}^{\Pi}$, $\boldsymbol{\mu} \in \mathbb{Z}_N^g$, associated to the canonical basis. Moreover, the eigenvalue of the operator $\mathcal{F}_{\mathbf{L}}(T_j)$ corresponding to the eigenvector $\theta_{\boldsymbol{\mu}}^{\Pi}$ is $e^{-\frac{\pi i}{N}\mu_j^2}$.*

5.2 A heuristical explanation

The Heisenberg group with integer entries $\mathbf{H}(\mathbb{Z}^g)$, the finite Heisenberg group $\mathbf{H}(\mathbb{Z}_N^g)$, the factorization map $\pi_{\mathbf{L}} : \mathbb{C}[\mathbf{H}(\mathbb{Z}_N^g)] \to \mathcal{H}_{N,g}(\mathbf{L})$, the Schrödinger representation of the finite Heisenberg group, the discrete Fourier transform determined by a pair of canonical bases, and the projective representation of the mapping class group on $\mathcal{H}_{N,g}(\mathbf{L})$ can be given topological interpretations, which we are about to present.

5.2.1 *From theta functions to curves*

The Heisenberg group with integer entries as a group of curves

We focus first on the Heisenberg group with integer entries, $\mathbf{H}(\mathbb{Z}^g)$. As explained, this group arises when quantizing the Jacobian variety of the Riemann surface Σ_g. We have seen in §4.2.3 that this group is a central extension of $H_1(\Sigma_g, \mathbb{Z})$ by the cocycle defined by the intersection form. Because of the identification of $H_1(\Sigma_g, \mathbb{Z})$ with the group of multicurves $\mathfrak{M}(\Sigma)$ defined in §3.2.3, we can represent an element of $\mathbf{H}(\mathbb{Z}^g)$ as (γ, k), where γ is (the equivalence class of) an oriented multicurve.

We change the notation from (γ, k) to $t^k\gamma$, where t **is an abstract variable**. In this notation, the multiplication rule in $\mathbf{H}(\mathbb{Z}^g)$ is written as

$$t^k\gamma \cdot t^{k'}\gamma' = t^{k+k'+\gamma\cdot\gamma'}\gamma\gamma',$$

where $\cdot$ is the intersection form. Here the curves γ and γ' are multiplied in $\mathrm{Mul}(\Sigma)$ by smoothing the crossings (see §3.2.3).

The choice of the canonical basis a_j, b_j, $j = 1, 2, \ldots, g$, identifies $\mathbf{H}(\mathbb{Z}^g)$ with $\mathbb{Z}^{2g+1}$, and in this notation $(\mathbf{p}, \mathbf{q}, k)$ becomes $t^k(\mathbf{p}, \mathbf{q})$, where now $(\mathbf{p}, \mathbf{q})$ is the multicurve. The multiplication rule

$$t^k(\mathbf{p}, \mathbf{q})t^{k'}(\mathbf{p}', \mathbf{q}') = t^{k+k'+(\mathbf{p},\mathbf{q})\cdot(\mathbf{p}',\mathbf{q}')}(\mathbf{p}+\mathbf{p}', \mathbf{q}+\mathbf{q}')$$

can be read either algebraically, with numbers, or topologically, with curves.

Example 5.3. In $\mathbf{H}(\mathbb{Z})$, the multiplication

$$(1, 0)(0, 1) = t(1, 1)$$

is represented using curves in Figure 5.1.

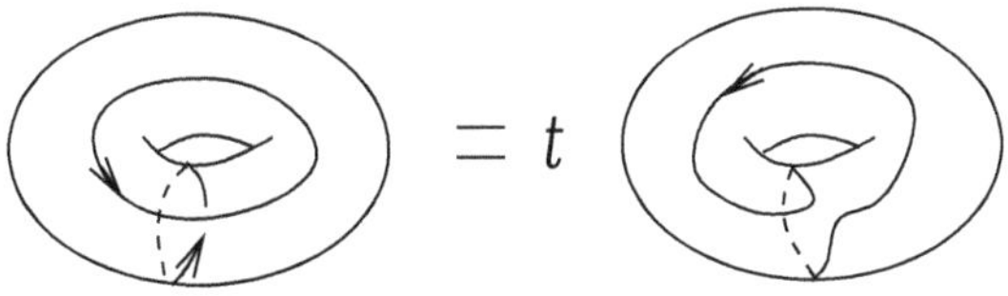

Fig. 5.1 The multiplication of $(1, 0)$ and $(0, 1)$

In order to compute the intersection form of the oriented multicurves γ and γ', we need to keep track which is the first and which is the second. This can be done by drawing γ on top of γ' and then keeping track of "over" and "under" information. A simple way to achieve this is to place both γ

and γ' in the cylinder $\Sigma_g \times [0,1]$ so that γ lies in the slice $\Sigma_g \times \{\epsilon\}$ and γ' lies in the slice $\Sigma_g \times \{\epsilon'\}$ with $\epsilon > \epsilon'$.

The elements of the group algebra $\mathbb{C}[\mathbf{H}(\mathbb{Z})]$ can be represented as

$$\sum_j c_j t^{k_j} \gamma_j,$$

or, by grouping the multicurves when they coincide, as

$$\sum_l p_l(t)\gamma_l,$$

where $p_l(t)$ are Laurent polynomials in the ring $\mathbb{C}[t, t^{-1}]$.

The finite Heisenberg group as a group of curves

We proved (Corollary 4.2) that $\mathbf{H}(\mathbb{Z}_N^g)$ is the quotient of $\mathbf{H}(\mathbb{Z}^g)$ by the subgroup consisting of elements of the form $(\mathbf{p}, \mathbf{q}, k)^N$ with k even. It is also a $\mathbb{Z}_{2N}$ extension of $H_1(\Sigma_g, \mathbb{Z}_N)$.

Note that making the element $(\mathbf{0}, \mathbf{0}, 2)^N = (\mathbf{0}, \mathbf{0}, 2N)$ be equal to the identity amounts to setting t equal to a primitive $2N$th root of unity, for example $t = e^{\frac{\pi i}{N}}$. This combined with Theorem 3.5 implies that the elements of $\mathbf{H}(\mathbb{Z}_N^g)$ can be represented as $e^{\frac{\pi i}{N}k}\gamma$, where γ is a multicurve representing an equivalence class in the group $\mathfrak{M}_N(\Sigma)$ defined in §3.2.3. Again, to keep track which multicurve is first in the product $\gamma\gamma'$, we put γ on top of γ', namely we place them in a cylinder over Σ_g with γ higher than γ'.

Factoring $\mathbb{C}[\mathbf{H}(\mathbb{Z}^g)]$ by the relation $\sim_N$ defined in §3.2.3 and setting $t = e^{\frac{\pi i}{N}}$ we obtain an algebra whose elements are of the form

$$\sum_j c_j \gamma_j,$$

where γ_j is the equivalence class in $\mathfrak{M}_N(\Sigma)$ of a multicurve and $c_j \in \mathbb{C}$, and whose multiplication is given by

$$\sum_j c_j \gamma_j \sum_k c_k \gamma'_k = \sum_{jk} c_j c_k e^{\frac{\pi i}{N}\gamma \cdot \gamma'} \gamma\gamma'.$$

In this formula, the curves γ and γ' are multiplied in $\mathrm{Mul}(\Sigma)$ by smoothing the crossings. Proposition 4.13 implies that the algebra of curves that we obtain is isomorphic to the algebra of linear operators on the space of theta functions, $L(\boldsymbol{\Theta}_N^{\Pi}(\Sigma_g))$.

Theta functions as curves

We want to interpret the factorization relation (5.6) topologically. This is probably the main "trick" of the book. Let us look at the torus Σ_1 with canonical basis $a_1 = (1,0)$, $b_1 = (0,1)$ (so $\mathbf{L} = \mathbb{Z}b_1$), and at the particular elements $u = \exp(P) \in \mathbf{H}(\mathbb{Z}_N)$ and $u' = \exp(Q) \in \exp(\mathbf{L})$. In this situation, the equality

$$\chi_{\mathbf{L}}(u')\pi_{\mathbf{L}}(u) = \pi_{\mathbf{L}}(uu')$$

reads

$$\pi_{\mathbf{L}}(\exp(Q)) = \pi_{\mathbf{L}}(\exp(P)\exp(Q)) = \pi_{\mathbf{L}}(e^{\frac{\pi i}{N}}\exp(P+Q)).$$

Representing the elements of the finite Heisenberg group as multicurves with coefficients, we rewrite this as

$$\pi_{\mathbf{L}}((1,0)) = \pi_{\mathbf{L}}(e^{\frac{\pi}{N}}(1,1)),$$

and draw it as shown in Figure 5.2.

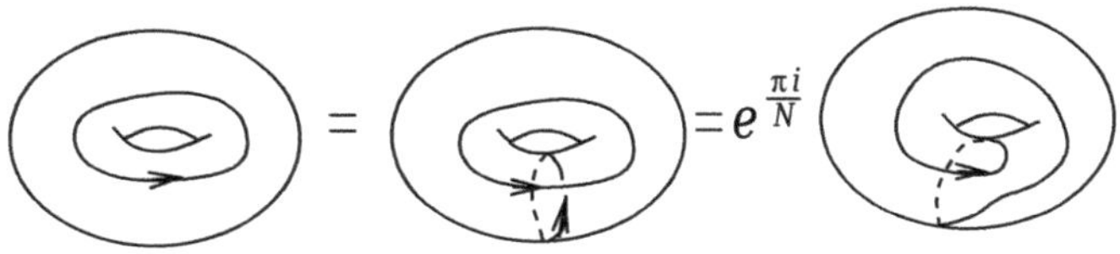

Fig. 5.2 The equivalence of $\exp(P)$ and $\exp(P+Q)$

The curves $(1,0)$ and $(1,1)$ are the same if we "fill in" the torus to produce a solid torus. How to keep track of the coefficient $e^{\frac{\pi i}{N}}$? Note that if the two curves were made out of physical rope, then when straightening the curve $(1,1)$ inside the solid torus to produce the curve $(1,0)$, it would twist around itself once. So if we replace the multicurves that represent the elements of $\mathbf{H}(\mathbb{Z}_N)$ by ribbons on the surface Σ_1, then $e^{\frac{\pi i}{N}}(1,1)$ is obtained from $(1,0)$ by transforming it by an isotopy inside the solid torus, then adding a (negative) twist and multiplying by $e^{\frac{\pi i}{N}}$.

One can examine more situations, and observe that this is always the case: if $\pi_{\mathbf{L}}(u_1) = \pi_{\mathbf{L}}(u_2)$ with $u_1, u_2 \in \mathbf{H}(\mathbb{Z}_N)$, then u_1 can be obtained from u_2 by taking an isotopy in the solid torus and introducing several twists. Each time a positive twist is introduced we multiply by $e^{-\frac{\pi i}{N}}$ and each time a negative twist is introduced we multiply by $e^{\frac{\pi i}{N}}$.

The situation repeats on an arbitrary surface Σ_g; this time we "fill in" the surface to a handlebody. Thus we arrive at the idea of representing theta functions by oriented framed multicurves in a handlebody. The original surface bounds the handlebody in such a way that the multicurves in $\exp(\mathbf{L})$ become null-homologous.

Hence the quotient map $\pi_{\mathbf{L}}$ identifies the surface Σ_g with the boundary of the handlebody H_g in such a way that the elements of $\exp(\mathbf{L})$ become null-homologous. The space of theta functions, which in the representation theoretical setting is $\mathcal{H}_{N,g}(\mathbf{L})$, can be identified with a space of oriented framed multicurves in a handlebody.

5.2.2 *The idea of a skein module*

Recall that the Schrödinger representation, extended linearly to the entire group algebra, can be viewed as the left regular action of the group algebra $\mathbb{C}[\mathbf{H}(\mathbb{Z}_N^g)]$ on a quotient of itself. The heuristical discussion from above shows that both the group algebra and its quotient, $\mathcal{H}_{N,g}(\mathbf{L})$, can be described as spaces of curves, in the cylinder over a surface respectively in a handlebody.

We will formalize these ideas in the language of skein modules introduced by Jozef Przytycki in [Przytycki (1991)]. We do this using the skein modules associated to the linking number, which were defined by Przytycki in [Przytycki (1998)].

In §1.2.2 we explained how the Jones polynomial, as well as the linking number, can be computed using skein relations. The skein relations can be applied to knots or links in $\mathbb{R}^3$ by projecting them onto a plane and then working with the diagram of the projection. But what if we consider knots and links in arbitrary manifolds?

Our heuristical discussion prompted us to consider oriented links in the cylinder $\Sigma_g \times [0,1]$. The multiplication rule is nothing but the application of the skein relations of the linking number, introduced in Chapter 1, to the projections of the multicurves onto Σ_g. This idea of considering algebras of multicurves in the cylinder over a surface with multiplication defined by skein relations belongs to Vladimir Turaev [Turaev (1989)].

But when factoring the group algebra to obtain the space of theta functions, we had to consider framed multicurves inside a handlebody. Whatever skein relations were applied on the boundary surface, those skein relations descend to the handlebody as well. Thus the need to consider skein relations in a handlebody.

The general idea of applying skein relations to knots and links in an arbitrary oriented 3-dimensional manifold belongs to Przytycki. This is done by projecting just the crossing in question onto a local planar surface inside the manifold, and then applying the skein relations. At first glance this seems nonsensical because without a preferred direction of projection even crossings are not defined. We can fix this by considering an embedded copy of $\mathbb{R}^3$ that contains two strands of the link that look like crossing in $\mathbb{R}^3$. There are many embeddings, so the skein relation can be applied in very many ways. This is exactly the point! The many possible knots and links of the manifold can be reduced, by applying skein relations, to a small family the same way knots and links in $\mathbb{R}^3$ are reduced to the empty link.

To demonstrate the usefulness of this approach, let us draw a parallel with homology. In simplicial homology, simplices show up as geometric objects in a triangulation. In singular homology, a simplex is just a continuous map from a standard simplex to a topological vector space. This makes singular homology a more general and flexible theory. The same is true in the case of skein modules. By avoiding the need of a projection plane, one can define skein relations in arbitrary manifolds.

Skein modules are constructed by the general method that governs algebraic topology: start with a large algebraic structure produced using topological objects, factor it by algebraic relations motivated by topological properties and obtain an algebraic structure that is not too complicated and carries topological information.

In this case, the large algebraic structure is the free $\mathbb{C}[t, t^{-1}]$-module with basis all framed oriented knots and links in a 3-dimensional manifold, where t is the variable used in the skein relation. Factor this by all linear relations that arise from the skein relation (seen as linear relations between knots and links instead of linear relations between their polynomial invariants). The result is the skein module of the manifold; it is a topological invariant of the manifold, and presumably it is simple enough so that it can be computed.

Of all skein modules, the ones most studied are those related to the Jones polynomial, or rather to its close relative, the Kauffman bracket. In this book we are interested in the skein modules of the linking number.

5.3 The skein modules of the linking number

5.3.1 *The definition of skein modules*

Let M be a smooth, compact, **oriented** 3-dimensional manifold, with or without boundary.

Definition 5.4. A **framed link** in M is a smooth embedding in the interior of M of a disjoint union of finitely many annuli $S^1 \times [0,1]$.

The embedded annuli are called **link components**. We can think that the unframed link is the link obtained by choosing one boundary component for each annulus. Alternatively, a framed link is a smooth embedding of finitely many disjoint circles, each of which is equipped with a nonzero smooth normal vector field. The flow of the link defined by the vector field defines the annuli in the above definition.

If in a circumstance the link or part of it happens to lie in the boundary of M we agree to "push" it inside the interior of M by an isotopy so that the condition of the definition is fulfilled.

We consider **oriented framed links**, meaning that the link components are oriented. The orientation of a link component is an orientation of one of the circles that bound the annulus.

When M is the cylinder over a surface, we represent framed links as oriented curves with the **blackboard framing**, meaning that the annulus giving the framing is always parallel to the surface. The blackboard framing is also used when drawing link diagrams. So in all our figures we represent unframed strands; they become framed by considering regular neighborhoods in the plane of the paper.

We adapt the notions of isotopy and ambient isotopy to framed knots, as well as to framed links in M. The framed links should be viewed as disjoint unions of annuli embedded in the interior of M.

Definition 5.5. Two framed links L_1 and L_2 in M are called **ambient isotopic** if there is a continuous map $F : [0,1] \times M \to M$ such that $F|\{0\} \times M = 1_M$, $(F|\{1\} \times M)(L_1) = L_2$ and $F|\{s\} \times M$ is a homeomorphism for every $s \in [0,1]$. Two framed n-component links L_1 and L_2 in M are called **isotopic** if there is a homotopy

$$H : [0,1] \times \sqcup_{j=1}^{n}(S^1 \times [0,1]) \to M$$

between them such that $H|\{s\} \times (S^1 \times [0,1])$ is a homeomorphism onto the image.

We only allow smooth isotopies and ambient isotopies, even if we do not mention it.

Definition 5.6. A **crossing** of a link in M is an embedding of the unit ball centered at the origin

$$B_1^3 = \{(x,y,z) \in \mathbb{R}^3 \,|\, x^2 + y^2 + z^2 = 1\}$$

via an orientation preserving homeomorphism, such that the strands are in the yz-plane, except close to origin where one of the strands passes through $x > 0$ and the other through $x < 0$ as to produce the crossing when projecting on the yz-plane (see Figure 5.3).

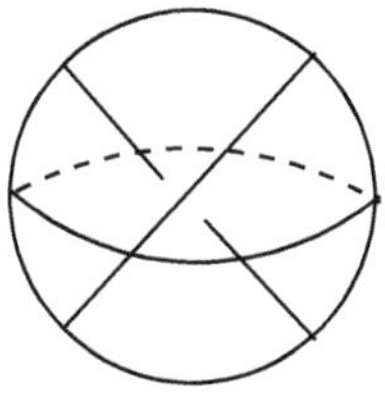

Fig. 5.3 The definition of a crossing

In the case where the link is framed, we additionally require the crossing to have the blackboard framing, meaning that inside B_1^3 the two strands have framings that are parallel to the yz-plane.

The ball is to be embedded into M via an orientation preserving embedding in any possible way that produces a crossing. In the case where M is $\mathbb{R}^3$, and we have a projection diagram for the link, the embedding can be via a linear map that maps the yz-plane to a plane parallel to the projection plane so that the crossing is an actual crossing in a diagram. But this need not be the case.

Let t be a free variable. Let $\mathbb{C}[t, t^{-1}]\text{Link}(M)$ be the free $\mathbb{C}[t, t^{-1}]$-module with basis the **ambient isotopy** classes of framed oriented links in M, including the empty link $\emptyset$. Let $\mathcal{S}(M)$ be the submodule spanned by all elements of the following two types:

(ln1) The elements depicted in Figure 5.4. Here the two terms in each skein relation depict framed links that are identical except in an embedded ball, in which they look as shown.

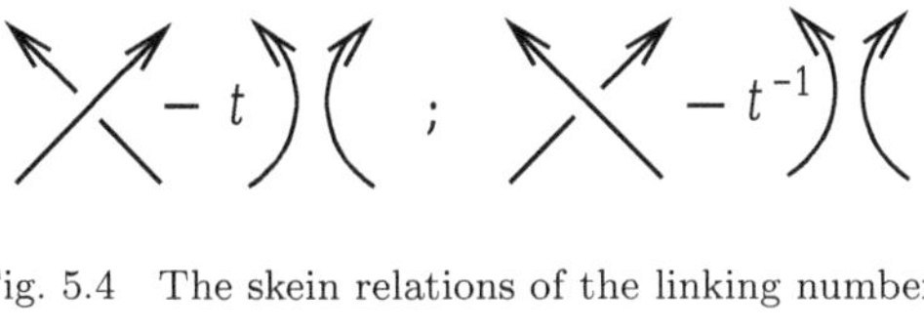

Fig. 5.4 The skein relations of the linking number

(ln2) The elements consisting of the difference between a link L in M and the link obtained by adding to L an oriented link component K that bounds an embedded disk disjoint from L and whose framing (annulus) is embedded in the disk. We represent schematically such an element as shown in Figure 5.5.

Fig. 5.5 Relation induced by adding a trivial link component

Definition 5.7. A link component K that bounds an embedded disk disjoint from the rest of the link L and whose framing (annulus) is embedded in the disk is called a **trivial framed link component** or **trivial framed knot**.

Definition 5.8. The $\mathbb{C}[t, t^{-1}]$-module

$$\mathcal{L}(M) = \mathbb{C}[t, t^{-1}]\mathrm{Link}(M)/\mathcal{S}(M)$$

is called the **linking number skein module** of M, and is denoted by $\mathcal{L}(M)$. The elements of $\mathcal{L}(M)$ are called **skeins.**

Notation: The skein which is the equivalence class modulo $\mathcal{S}(M)$ of the oriented framed link L will be denoted by $<L>$.

Rephrasing the definition, we are allowed to:

- smoothen each crossing provided that we multiply by the appropriate power of t, and
- delete the trivial link components.

The submodule $\mathcal{S}(M)$ is the submodule spanned by **skein relations**. As such, (ln1) and (ln2) can be interpreted as equivalence relations. Identifying the oriented framed links L and L' modulo $\mathcal{S}(M)$ consists of transforming L into L' by applying finitely many times the skein relations of the linking number from Figure 1.8 as well as adding and deleting trivial link components.

Remark 5.3. The name of these skein modules is motivated by the fact that the skein relations defined by Figure 5.4 are the same as those in Figure 1.8, which are used for computing the linking number.

Definition 5.9. We call $<\emptyset>$ the **empty skein.**

Lemma 5.1. *The equalities from Figure 5.6 hold in* $\mathcal{L}(M)$.

Fig. 5.6 Behaviour of skeins under Reidemeister I moves

Proof. This is an easy application of (ln1) and (ln2). □

Proposition 5.6. *Let M be either $\mathbb{R}^3$, an open or closed ball in $\mathbb{R}^3$, or the 3-dimensional sphere S^3. Then*

$$\mathcal{L}(M) \cong \mathbb{C}[t, t^{-1}].$$

Proof. The considerations in §1.2.3 show that every link L is equivalent modulo skein relations to $t^{\text{lk}(L)}\emptyset$,

$$<L>= t^{\text{lk}(L)} <\emptyset> .$$

Could it be equivalent to a skein $t^m <\emptyset>$ with $m \neq \text{lk}(L)$, so that the skein relation introduces a factorization in the coefficient ring?

The answer is negative, and here is the reason. Consider a crossing of a link L that might not arise from a link diagram. This crossing is defined by an embedding of the unit ball

$$f : B_1^3 \to \mathbb{R}^3$$

such that the crossing itself happens in the ball $B_{1/2}^3$ centered at the origin and of radius 1/2. We may assume that the embedding maps the point $0 \in B_1^3$ to $0 \in \mathbb{R}^3$. There is an ambient isotopy of $\mathbb{R}^3$ that is the identity map outside $f(B_1^3)$ such that after performing the ambient isotopy $f|B_{1/2}^3$ is the identity map. This is shown schematically in Figure 5.7. Now the crossing arises from the link diagram of the projection on the yz-plane.

If L' is the link obtained from L by applying the skein relation to this crossing, then

$$<L>= t^{\epsilon} <L'> \text{ and } \text{lk}(L) = \text{lk}(L') + \epsilon,$$

where ϵ equals $+1$ or -1.

We deduce that if $< L_1 >= t^m < L_2 >$, then the crossings used for transforming L_1 into L_2 can be made to arise from link diagrams, and so $\text{lk}(L_1) = \text{lk}(L_2) + m$. Consequently, if $< L >= t^m <\emptyset>$, then necessarily $m = \text{lk}(L)$. We conclude that the map

$$\mathcal{L}(\mathbb{R}^3) \to \mathbb{C}[t, t^{-1}], \quad \sum_j p_j(t) <L_j> \mapsto \sum_j p_j(t) t^{\text{lk}(L_j)}$$

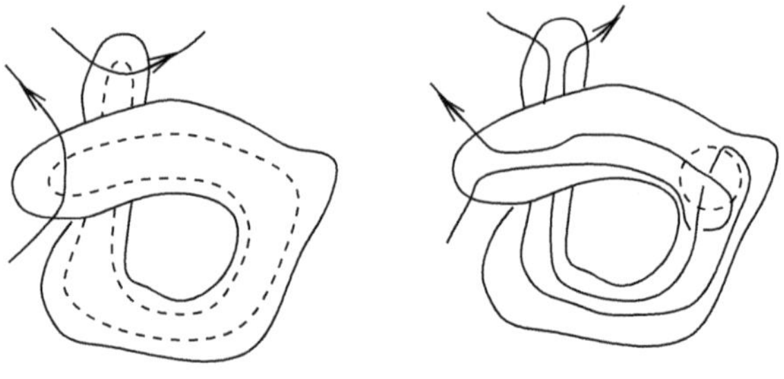

Fig. 5.7 Bringing a crossing in standard form

is a well defined module isomorphism. Hence

$$\mathcal{L}(\mathbb{R}^3) \cong \mathbb{C}[t, t^{-1}].$$

Replacing $\mathbb{R}^3$ with a closed or open ball, we obtain the same result.

Topologists prefer the 3-dimensional sphere S^3 to the 3-dimensional space, because it is compact. We have $S^3 = \mathbb{R}^3 \cup \{\infty\}$ and by considering an ambient isotopy we can make every link in S^3 and every application of the skein relation be away from ∞. Hence reducing a link to the empty link is the same in $\mathbb{R}^3$ as in S^3. We obtain

$$\mathcal{L}(S^3) \cong \mathbb{C}[t, t^{-1}],$$

and the proposition is proved. □

In the statement of the proposition we can use the equal sign because the isomorphism is canonical. The isomorphism is given by

$$\sum_j p_j(t) < L_j > \mapsto \sum_j p_j(t) t^{\mathrm{lk}(L_j)}.$$

Proposition 5.7. *Let $f : M_1 \to M_2$ be an orientation preserving smooth embedding of the smooth oriented 3-dimensional manifold M_1 in the smooth oriented manifold M_2. Then the module homomorphism*

$$f_* : \mathbb{C}[t, t^{-1}]Link(M_1) \to \mathbb{C}[t, t^{-1}]Link(M_2), \quad f_*(L) = f(L)$$

descends to a module homomorphism between $\mathcal{L}(M_1)$ and $\mathcal{L}(M_2)$. are isomorphic. If f is a homeomorphism, then the map $\mathcal{L}(M_1) \to \mathcal{L}(M_2)$ induced by it is an isomorphism.

We denote the homomorphism between $\mathcal{L}(M_1)$ and $\mathcal{L}(M_2)$ also by f_*.

Proof. The skein relations defined by (ln1) and (ln2) applied in M_1 are mapped by f to similar relations in M_2. The association $f \mapsto f_*$ is functorial. □

5.3.2 *The group algebra of the Heisenberg group as a skein algebra*

The skein algebra of a surface

Let Σ be a smooth, compact, oriented surface. The orientation of Σ induces an orientation of the cylinder $\Sigma \times [0,1]$ such that if $\mathbf{v}_1, \mathbf{v}_2$ is a positively oriented frame of Σ, then $\mathbf{v}_1, \mathbf{v}_2, \frac{\partial}{\partial t}$ is a positively oriented frame of $\Sigma \times [0,1]$. The orientation of the cylinder $\Sigma \times [0,1]$ induces orientations of its boundary components; $\Sigma \times \{1\}$ is oriented the same way as Σ, while $\Sigma \times \{0\}$ has the opposite orientation.

The identification

$$\Sigma \times [0,1] \cup \Sigma \times [0,1] \approx \Sigma \times [0,1] \tag{5.16}$$

obtained by gluing the boundary component $\Sigma \times \{0\}$ in the first cylinder to the boundary component $\Sigma \times \{1\}$ in the second cylinder by the identity map induces a multiplication for oriented framed links in $\Sigma \times [0,1]$. This can be extended $\mathbb{C}[t,t^{-1}]$-linearly to a multiplication for elements in $\mathbb{C}[t,t^{-1}]\mathrm{Link}(\Sigma \times [0,1])$. It is not hard to see that $\mathbb{C}[t,t^{-1}]\mathrm{Link}(\Sigma \times [0,1])$ endowed with this multiplication is an algebra, and that the empty link is the identity element of the multiplication.

Loosely speaking the product of two links in $\Sigma \times [0,1]$ is obtained by placing the first link on top of the second. If they are defined by link diagrams on Σ, then the product link has a diagram in which the first link is drawn on top of the second.

Lemma 5.2. *The multiplication on $\mathbb{C}[t,t^{-1}]Link(\Sigma \times [0,1])$ descends to a multiplication in $\mathcal{L}(\Sigma \times [0,1])$ which turns $\mathcal{L}(\Sigma \times [0,1])$ into an algebra with identity $\emptyset$.*

Proof. The fact that the multiplication descends to $\mathcal{L}(\Sigma \times [0,1])$ follows from the fact that if $L, L' \in \mathbb{C}[t,t^{-1}]\mathrm{Link}(\Sigma \times [0,1])$ such that $L' \in \mathcal{S}(\Sigma \times [0,1])$, then $LL', \in \mathcal{S}(\Sigma \times [0,1])$. We leave it to the reader to check that the multiplication on $\mathcal{L}(\Sigma \times [0,1])$ has all the required properties in order to turn this vector space into an algebra. □

Definition 5.10. $\mathcal{L}(\Sigma \times [0,1])$ with multiplication induced by (5.16) is called the **linking number skein algebra** of the surface Σ.

Notation: We denote the linking number skein algebra of a surface by $\mathcal{L}(\Sigma)$. However, when discussing only the module structure, we still denote it by $\mathcal{L}(\Sigma \times [0,1])$.

Intuitively two skeins are multiplied by drawing the first on top of the second and resolving all crossings.

As before, we identify a compact, orientable surface Σ, with

$$\Sigma \times \{1/2\} \subset \Sigma \times [0,1].$$

Then each multicurve in Σ gives rise to an oriented link in $\Sigma \times [0,1]$, which is framed by the blackboard framing. Hence each multicurve gives rise to a skein. If γ is the multicurve in Σ, we denote by $<\gamma>$ the corresponding skein in $\mathcal{L}(\Sigma)$.

Lemma 5.3. *If L_1 and L_2 are links in a manifold M and k is an integer such that $< L_1 >= t^k < L_2 >\in \mathcal{L}(M)$, then L_1 and L_2 are homologous in $H_1(M,\mathbb{Z})$.*

Proof. If L_1 and L_2 are ambient isotopic, they are homologous. The addition and deletion of trivial link components does not change the homology class, because these trivial link components bound disks. Also, the application of the skein relations of the linking number amounts to the smoothing of a crossing. And the smoothing does not change the cycle in $Z_1(M,\mathbb{Z})$, except for pushing the curves apart. □

Proposition 5.8. *Let Σ be a smooth compact oriented surface. Then the linking number skein module $\mathcal{L}(\Sigma \times [0,1])$ is free with basis $< \gamma >$ where γ ranges over a family of multicurves representing the homology classes of $H_1(\Sigma,\mathbb{Z})$.*

Proof. The difficulty of the proof lies in the fact that we allow other crossings besides the ones that come from projecting links on the surface. If only such particular crossings are allowed, then we are in the situation of §3.2.3 and there is little to prove. So there is a trade-off for working with general skein relations.

A first observation is that

$$H_1(\Sigma \times [0,1],\mathbb{Z}) = H_1(\Sigma,\mathbb{Z}).$$

This follows from the fact that $\Sigma = \Sigma \times \{0\}$ is a strong deformation retract of $\Sigma \times [0,1]$ with retract

$$H : [0,1] \times (\Sigma \times [0,1]), \quad H(s,x,t) = (x,ts).$$

A second observation is that, by Lemma 5.3, if $t^k L$ can be transformed into $t^m L'$ by skein relations and additions/deletions of trivial link components, then L and L' are homologous.

Consider the free $\mathbb{C}[t,t^{-1}]$ module

$$\mathbb{C}[t,t^{-1}]H_1(\Sigma,\mathbb{Z})$$

with basis the homology classes in $H_1(\Sigma,\mathbb{Z})$. We define a module homomorphism

$$F : \mathbb{C}[t,t^{-1}]\mathrm{Link}(\Sigma\times[0,1]) \to \mathbb{C}[t,t^{-1}]H_1(\Sigma,\mathbb{Z})$$

as follows. Given a link L in $\Sigma\times[0,1]$, bring it by an ambient isotopy so as to have the blackboard framing. Let

$$\pi : \Sigma\times[0,1]\to\Sigma, \quad \pi(x,t)=x$$

(remember that Σ is identified with $\Sigma\times\{1/2\}$). Consider the link diagram $\pi(L)$. Resolve all crossings of $\pi(L)$ using the skein relations to obtain $t^k r(L)$, where $k\in\mathbb{Z}$ and the link $r(L)$ has no crossings. Set

$$F(L)=t^k[r(L)]$$

where $[r(L)]$ is the homology class of $r(L)$. Extend F linearly to the whole $\mathbb{C}[t,t^{-1}]\mathrm{Link}(\Sigma_g\times[0,1])$. This is the desired homomorphism.

We claim that F is well defined. The only problem can be caused by the fact that a different ambient isotopy of L as to bring it in a blackboard framing might yield a different $F(L)$. Because in $H_1(\Sigma_g,\mathbb{Z})$, $[r(L)]=[L]$, the problem can be with the coefficient t^k.

We give k an intrinsic topological meaning. Because Σ is compact and orientable, there is an embedding

$$f : \Sigma\times[0,1]\to\mathbb{R}^3$$

via an orientation preserving diffeomorphism. Using Proposition 5.6 and Proposition 5.7 we obtain a $\mathbb{C}[t,t^{-1}]$-module homomorphism

$$f_* : \mathcal{L}(\Sigma\times[0,1])\to\mathcal{L}(\mathbb{R}^3)=\mathbb{C}[t,t^{-1}],$$

given by

$$f_*(<L>)=t^{\mathrm{lk}(L)}.$$

In particular $f_*(r(L))=t^{\mathrm{lk}(r(L))}$. Because L and $r(L)$ are equivalent modulo skein relations, the exponent k of t is given by

$$k=\mathrm{lk}(L)-\mathrm{lk}(r(L)),$$

as explained in Example 5.6.

If L and L' are ambient isotopic, then $\mathrm{lk}(L)=\mathrm{lk}(L')$. Also, in homology $[r(L)]=[r(L')]$. By Theorem 3.4, $r(L)$ can be changed into $r(L')$ by a finite number of isotopies in Σ, smoothings of crossings, and additions and deletions of trivial circles. Proposition 3.4 shows that isotopies can be replaced by ambient isotopies, and so in this case $\mathrm{lk}(r(L))$ does not change. From the proof of Theorem 3.4 it follows that we can enforce the smoothings of crossings to arise as a consequence of introducing bigons. It is an easy exercise to check that if a bigon is introduced and the crossings are smoothened then $\mathrm{lk}(r(L))$ does not change. Finally, trivial circles on the surface become trivial framed link components in $\mathbb{R}^3$, and can be deleted without changing $\mathrm{lk}(r(L))$. We conclude that k does not depend on the ambient isotopy that brings L into a link with blackboard framing, and hence F is well defined.

By taking images of links that represent homology classes, we deduce that F is onto.

Next we check that F descends to a module homomorphism

$$\hat{F} : \mathcal{L}(\Sigma_g \times [0,1]) \to \mathbb{C}[t,t^{-1}]H_1(\Sigma,\mathbb{Z}).$$

This follows from the fact that every embedded ball can be transformed by an ambient isotopy into a cylinder over a disk, so the application of the skein relation becomes part of the definition of the map F. Clearly $\hat{F}$ is onto. Let us check that it is one-to-one.

We claim that if L is a link in the blackboard framing such that $\pi(L)$ has no crossings and if γ is a multicurve representing the homology class of L, then

$$<L>=<\gamma> .$$

We may identify L with $\pi(L)$ by an ambient isotopy, thus assume that L is a multicurve on Σ_g itself. Using the proof of Theorem 3.4 combined with Proposition 3.4 we deduce that L can be transformed into γ by a succession of ambient isotopies, deletions or additions of curves, and smoothings of crossings. Moreover, the smoothings of crossings always come from eliminations of bigons, so they come in pairs of a positive and negative crossing. Hence they do not introduce additional factors of t. The claim is proved. Therefore $\hat{F}$ is an isomorphism, and the conclusion follows. □

Let Σ_g be a closed genus g surface. Endow it with a complex structure so that it becomes the Riemann surface of Chapter 4. Then all constructs of that chapter apply and we can talk about theta functions and the action of the finite Heisenberg group.

A canonical basis $a_1, a_2, \ldots, a_g, b_1, b_2, \ldots, b_g$ of $H_1(\Sigma_g, \mathbb{Z})$ gives rise to a family of skeins

$$<a_1>, <a_2>, \ldots, <a_g>, <b_1>, <b_2>, \ldots, <b_g> \in \mathcal{L}(\Sigma_g \times [0,1]).$$

Once we chose the canonical basis, we can identify a $2g$-tuple of integers, $(\mathbf{p},\mathbf{q})$, with the multicurve $(\mathbf{p},\mathbf{q})$ on Σ_g. Then $(\mathbf{p},\mathbf{q})$ defines the skein $<(\mathbf{p},\mathbf{q})>$.

Theorem 5.4. *There is an algebra isomorphism*

$$\Phi : \mathcal{L}(\Sigma_g) \to \mathbb{C}[\mathbf{H}(\mathbb{Z}^g)]$$

defined by

$$\Phi(t^k <\gamma>) = ([\gamma], k),$$

where γ ranges over all multicurves on Σ_g, with the blackboard framing, and $[\gamma]$ is the homology class of γ in $H_1(\Sigma_g, \mathbb{Z}) = \mathbb{Z}^{2g}$.

Proof. Lemma 5.3 implies that the homomorphism Φ is well defined. From Proposition 5.8 we deduce that as a module, $\mathcal{L}(\Sigma_g \times [0,1])$ is free with basis

$$<(\mathbf{p},\mathbf{q})>, \quad \mathbf{p},\mathbf{q} \in \mathbb{Z}^g.$$

The map

$$\mathbb{C}[t,t^{-1}]H_1(\Sigma_g,\mathbb{Z}) \to \mathbb{C}[\mathbf{H}(\mathbb{Z}^g)], \quad \sum_j c_j t^{k_j}(\mathbf{p}_j,\mathbf{q}_j) \mapsto \sum_j c_j(\mathbf{p}_j,\mathbf{q}_j,k_j)$$

identifies $\mathbb{C}[t,t^{-1}]H_1(\Sigma_g,\mathbb{Z})$ with $\mathbb{C}[\mathbf{H}(\mathbb{Z}^g)]$ as $\mathbb{C}$-vector spaces. In view of the proof of Proposition 5.8, this map gives rise to the map Φ, which therefore is a vector space isomorphism between the skein module $\mathcal{L}(\Sigma_g \times [0,1])$ and $\mathbb{C}[\mathbf{H}(\mathbb{Z}^g)]$.

Turning to the algebra structure, in $\mathcal{L}(\Sigma_g)$ we have

$$t^k <(\mathbf{p},\mathbf{q})> t^{k'} <(\mathbf{p}',\mathbf{q}')>= t^{k+k'+\mathbf{p}^T\mathbf{q}'-\mathbf{p}'^T\mathbf{q}} <(\mathbf{p}+\mathbf{p}',\mathbf{q}+\mathbf{q}')> .$$

This is because the difference between the number of positive crossings and the number of negative crossings of the multicurves $(\mathbf{p},\mathbf{q})$ and $(\mathbf{p}',\mathbf{q}')$ is $\mathbf{p}^T\mathbf{q}' - \mathbf{p}'^T\mathbf{q}$. We recognize the multiplication rule of the Heisenberg group $\mathbf{H}(\mathbb{Z}^g)$. □

Corollary 5.1. *The map*

$$t^k<a_1>^{p_1}<a_2>^{p_2}\cdots<a_g>^{p_g}<b_1>^{q_1}<b_2>^{q_2}\cdots<b_g>^{q_g} \mapsto t^{k-\mathbf{p}^T\mathbf{q}}(\mathbf{p},\mathbf{q}),$$

$\mathbf{p},\mathbf{q} \in \mathbb{Z}^g, k \in \mathbb{Z}$, *defines an algebra isomorphism* $\mathcal{L}(\Sigma_g) \to \mathbb{C}[\mathbf{H}(\mathbb{Z}^g)]$.

Remark 5.4. Przytycki [Przytycki (1998)] viewed $\mathcal{L}(\Sigma_g)$ as a one-parameter deformation of the group algebra $\mathbb{C}[H_1(\Sigma_g,\mathbb{Z})]$, where the deformation parameter is h and $t = e^{ih}$. Theorem 5.4 shows that this is indeed so.

The linking number skein algebra of a surface as a group algebra

Comparing the above considerations with those from §3.2.3, we realize that in the case of a surface, the skein relations allow us to define a group structure on links, similar to the group structure on $\mathfrak{M}(\Sigma)$. This is possible only for the skein algebras associated to the linking number, and not for other skein algebras, such as those associated to the Jones polynomial. For this reason we emphasize less skein groups, as compared to skein algebras. We will use the skein group of the linking number in one of the two topological interpretations of the action of the mapping class group on theta functions (§5.4.3).

Proposition 5.9. *Let Σ be a compact oriented surface. The set*

$$\mathcal{L}G(\Sigma) = \{t^k <L> \mid L \in Link(\Sigma \times [0,1]), k \in \mathbb{Z}\} \subset \mathcal{L}(\Sigma)$$

is closed under multiplication and it is a group. If $\Sigma = \Sigma_g$, a closed genus g Riemann surface, then $\mathcal{L}G(\Sigma_g)$ is isomorphic to $\mathbf{H}(\mathbb{Z}^g)$.

Proof. First, notice that the set

$$\{t^k L \mid L \in \mathrm{Link}(\Sigma \times [0,1], k \in \mathbb{Z}\} \subset \mathbb{C}[t, t^{-1}]\mathrm{Link}(\Sigma \times [0,1])$$

is closed under the application of the skein relations defined by (ln1) and (ln2). It is also closed under the operation of taking the union of two links. Hence $\mathcal{L}G(\Sigma)$ is closed under multiplication. The identity element is $<\emptyset>$. We claim that if L is a framed oriented link in $\Sigma \times [0,1]$ and k is an integer, then

$$(t^k <L>)^{-1} = t^{-k} <L^{-1}>,$$

where L^{-1} is the link L with all orientations reversed. Indeed, if we project L on Σ and resolve all crossings we obtain $t^n\gamma$, where γ is a multicurve. The same process applied to L^{-1} yields $t^{-n}\gamma^{-1}$, where γ^{-1} is obtained from the multicurve γ by reversing all orientations. Now in $\mathfrak{M}(\Sigma)$

$$\gamma\gamma^{-1} = \emptyset,$$

hence

$$t^n <\gamma> t^{-n} <\gamma^{-1}> = <\emptyset> .$$

This proves the claim. We conclude that $\mathcal{L}G(\Sigma)$ is a group.

In the case where $\Sigma = \Sigma_g$, this group is the central $\mathbb{Z}$-extension of $\mathfrak{M}(\Sigma)$ by the cocycle defined by the intersection form, hence it is $\mathbf{H}(\mathbb{Z}^g)$. □

Definition 5.11. The group $\mathcal{L}G(\Sigma)$ is called the **linking number skein group** of the surface Σ.

Corollary 5.2. *Let Σ be a surface. Then $\mathcal{L}(\Sigma)$ is the group algebra of $\mathcal{L}G(\Sigma)$.*

The action of the skein algebra of the boundary of a manifold

Let M be an oriented 3-dimensional manifold with boundary and let Σ be an oriented surface such that there is an orientation preserving homeomorphism onto the image,

$$f : \Sigma \to \partial M.$$

The image of f need not be the entire boundary of M, as the boundary of M can be disconnected or Σ can itself have boundary. The homeomorphism f induces an orientation preserving homeomorphism, which we denote also by f, from $\Sigma \times \{0\} \subset \Sigma \times [0,1]$ to ∂M. We have a homeomorphism of 3-dimensional manifolds

$$\Sigma \times [0,1] \cup_f M \approx M$$

obtained by gluing $\Sigma \times [0,1]$ to M along $f(\Sigma)$. We also have a homeomorphism

$$(\Sigma \times [0,1] \cup \Sigma \times [0,1]) \cup_f M \approx M,$$

where the two cylinders are glued by the homeomorphism between $\Sigma \times \{0\}$ on one cylinder and $\Sigma \times \{1\}$ on the other induced by the identity map on Σ. We deduce that $\mathcal{L}(\Sigma)$ acts as an algebra on $\mathcal{L}(M)$ and this action turns $\mathcal{L}(M)$ into an $\mathcal{L}(\Sigma)$-module. If $<L_0>$ is a skein in $\Sigma \times [0,1]$, viewed as a skein in a cylindrical neighborhood of the boundary of M, and $<L>$ is a skein in M, then the action is

$$<L_0><L>=<L_0 \cup L> .$$

5.3.3 *The skein module of a handlebody*

The genus g handlebody, H_g, is diffeomorphic to the cylinder over the genus 0 surface with $g+1$ boundary components, $\Sigma_{0,g+1} \times [0,1]$. We view $\Sigma_{0,g+1} \times [0,1]$ as the unit disk in $\mathbb{C}$ with g disks removed and give it the orientation of the complex plane. That induces an orientation of H_g by the rule that if $\mathbf{v}_1, \mathbf{v}_2$ is a positively oriented frame in $\Sigma_{0,g+1}$, then $\mathbf{v}_1, \mathbf{v}_2, \frac{\partial}{\partial t}$ is a positively oriented frame in H_g.

The identification of H_g with the cylinder over a disk with g holes endows $\mathcal{L}(H_g)$ with a multiplicative structure. This multiplicative structure is ambiguous, depending on the homeomorphism $H_g \to \Sigma_{0,g+1} \times [0,1]$, nevertheless we fix one choice of homeomorphism.

Consider a canonical basis $a_1, a_2, \ldots, a_g$ of the first homology of the genus g handlebody. Using the blackboard framing defined by the surface

$\Sigma_{0,g+1} \times \{1/2\}$, we can frame the curves a_j, $j = 1, 2, \ldots, g$. As such, they define framed oriented knots in H_g, which give rise to skeins in $\mathcal{L}(H_g)$. Because of the multiplicative structure of $\mathcal{L}(H_g)$, we can talk about skeins of the form $<a_1>^{k_1}<a_2>^{k_2} \cdots <a_n>^{k_n}$, $k_1, k_2, \ldots, k_n \in \mathbb{Z}$.

As a corollary of Proposition 5.8, we obtain the following result.

Proposition 5.10. *The linking number skein module $\mathcal{L}(H_g)$ of the genus g handlebody is free with basis*

$$<a_1>^{k_1}<a_2>^{k_2} \cdots <a_g>^{k_g}, \quad k_1, k_2, \ldots, k_g \in \mathbb{Z}.$$

Remark 5.5. More is true, namely that

$$<a_j> \mapsto x_j, \quad j = 1, 2, \ldots, g,$$

defines an isomorphism of algebras between $\mathcal{L}(H_g)$ and the algebra $\mathbb{C}[t, t^{-1}][x_1, x_2, \ldots, x_g]$ of polynomials in g variables with coefficients in $\mathbb{C}[t, t^{-1}]$. But we are not interested in the algebra structure of $\mathcal{L}(H_g)$.

Example 5.4. In genus 2, H_2 is homeomorphic to $\Sigma_{0,3} \times [0, 1]$. A basis of $\mathcal{L}(H_2)$ is given by $<a_1>^{k_1}<a_2>^{k_2}$, where a_1, a_2 are shown in Figure 5.8.

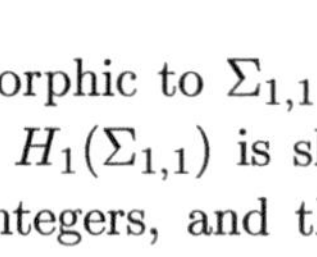

Fig. 5.8 Basis of the genus 2 handlebody

Note also that H_2 is homeomorphic to $\Sigma_{1,1} \times [0, 1]$. A set of multicurves that represent all elements of $H_1(\Sigma_{1,1})$ is shown in Figure 5.9. Here j, k range over all non-negative integers, and the box is filled with the four possibilities on the right in all possible nonequivalent ways. That these curves are a set of representatives follows from the fact that in homology they are of the form $(1, 0)^m(0, 1)^n$, where $(1, 0)$ and $(0, 1)$ are the curves that give a canonical basis of the torus. Consider the basis of $H_1(\Sigma_{1,1})$ given by the $(1, 0)$ and $(0, 1)$ curves. Then from Proposition 5.8 we deduce that the skeins defined by these curves are a basis for $\mathcal{L}(H_2)$ as a free module. This basis consists of the skeins of the form $t^{-mn} <(1, 0)>^m<(0, 1)>^n$.

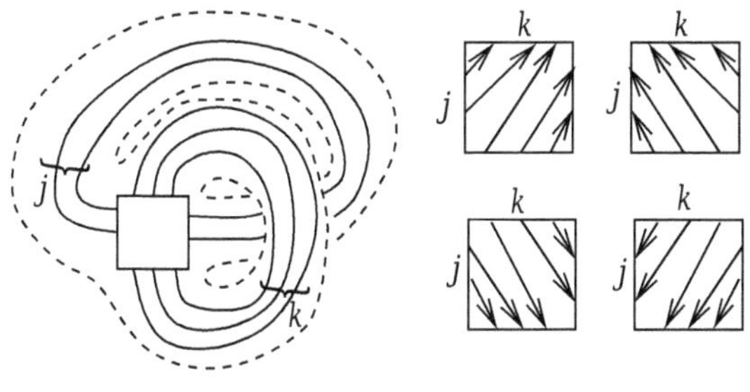

Fig. 5.9 Another basis of the genus 2 handlebody

5.4 A topological model for theta functions

5.4.1 *Reduced linking number skein modules*

The definition of the reduced linking number skein module

The interpretation of theta functions as curves, given in §5.2, leads us naturally to a reduced form of the linking number skein modules. To introduce it we need the following definition.

Definition 5.12. Let K be a framed knot in a smooth, compact, oriented 3-dimensional manifold M, defined by the embedding $f : S^1 \times [0,1] \to M$, and let n be a positive integer. The nth **parallel power** of K is the framed link obtained by restricting f to

$$S^1 \times \sqcup_{j=1}^{n} \left[\frac{j}{n+1} - \frac{1}{2n}, \frac{j}{n+1} + \frac{1}{2n} \right].$$

Intuitively, the nth parallel power of K is obtained by cutting the annulus into n annuli. We denote it by $K^{\|n}$. If the framing is viewed as a smooth vector field $\mathbf{v}$ normal to K, then $K^{\|n}$ is obtained from K by pushing K by the vectors $\frac{j}{n}\epsilon\mathbf{v}$, $j = 0, 1, \ldots, n-1$, where $\epsilon > 0$ is sufficiently small. If K is oriented, we orient the components of $K^{\|n}$ such that the orientations are parallel to that of K. We let $K^{\|0} = \emptyset$, and for $n < 0$, $K^{\|n}$ is $K^{\||n|}$ with orientations reversed.

Whenever we write $L \cup K^{\|n}$ we mean the link obtained from $L \cup K$ by taking the nth parallel power of K. The components of L are not allowed to run between the copies of K. An example of a link of the form $K_1 \cup K_2^{\|3}$ is shown in Figure 5.10.

Remark 5.6. If $K \in \Sigma \times [0,1]$, it is not always true the fact that that $< K^{\|n} >=< K >^n$. This only happens if K has the blackboard framing

Fig. 5.10 Taking the nth parallel power

and its projection onto Σ has no crossings. In the setting of §5.3.2 and §5.4, if $a_1, a_2, \ldots, a_g, b_1, b_2, \ldots, b_g$ is a canonical basis of $H_1(\Sigma_g, \mathbb{Z})$ and if $k_1, k_2, \ldots, k_g, l_1, l_2, \ldots, l_g \in \mathbb{Z}$, then

$$<a_1^{\|k_1} \cdots a_g^{\|k_g} b_1^{\|l_1} \cdots b_g^{\|l_g}>=<a_1>^{k_1} \cdots <a_g>^{k_g}<b_1>^{l_1} \cdots <b_g>^{l_g},$$

and if $a_1, a_2, \ldots, a_g$ is a canonical basis of $H_1(H_g, \mathbb{Z})$, and $k_1, k_2, \ldots, k_g \in \mathbb{Z}$, then

$$<a_1^{\|k_1} a_2^{\|k_2} \cdots a_g^{\|k_g}>=<a_1>^{k_1}<a_2>^{k_2} \cdots <a_g>^{k_g} .$$

It is now time again to fix an even positive integer N, which for us is the reciprocal of Planck's constant. Let $\mathcal{S}_N(M)$ be the submodule of $\mathcal{L}(M)$ spanned by all elements of the form

(lnr1) $(t - e^{\frac{i\pi}{N}})\sigma$, $\sigma \in \mathcal{L}(M)$,
(lnr2) $<L \cup K^{\|N}> - <L>$ for all smooth oriented framed links L and smooth oriented framed knots K in M with K disjoint from L.

Definition 5.13. The **reduced linking number skein module** of the oriented 3-dimensional manifold M is

$$\mathcal{L}_N(M) = \mathcal{L}(M)/\mathcal{S}_N(M).$$

In other words, $\mathcal{L}_N(M)$ is obtained from $\mathcal{L}(M)$ by setting $t = e^{\frac{i\pi}{N}}$ in the coefficient ring $\mathbb{C}[t, t^{-1}]$ and by deleting N parallel copies of knots whenever they occur as link components. As such, (lnr1) can be viewed as the identification of t with the primitive $2N$th root of unity, and (lnr2) as the deletion of N parallel link components. We can write these two relations shortly as

(lnr1) $t = e^{\frac{i\pi}{N}}$,
(lnr2) $<L \cup K^{\|N}>=<L>$.

The reduced linking number skein module $\mathcal{L}_N(M)$ is a $\mathbb{C}$-vector space because the variable t becomes a complex number. We still call it a module because the terminology is already established in the case of reduced Kauffman bracket skein modules, and because we want to emphasize the presence of the reduction process, from the nonreduced to the reduced module.

Remark 5.7. If K is a trivial link component, then $K^{\|N}$ consists of N trivial link components.

Lemma 5.4. *Let M be either $\mathbb{R}^3$, an open or closed 3-dimensional ball, or the 3-dimensional sphere S^3. Let L be a smooth oriented framed link in M and K a smooth oriented framed knot disjoint from L. Factor $\mathcal{L}(M)$ only by (lnr1). Then in the quotient,*

$$< L \cup K^{\|N} > = < L > .$$

Proof. Consider a crossing c of L and K. If L' is obtained from $L \cup K^{\|N}$ after smoothing the N crossings determined by c, then in the quotient

$$< L \cup K^{\|N} > = \left(t^{\pm 1}\right)^N < L' > \equiv - < L' > .$$

The skein on the right is the same regardless of whether c is a positive or negative crossing. Hence we can change the crossings between L and K at our discretion and unlink L from K. Thus we may assume that K is inside a ball that is disjoint from L. By applying an ambient isotopy, we can make sure that in some projection of $L \cup K$, K projects inside a disk disjoint from the projection of L. In this projection, every crossing of K gives rise to N^2 crossings of $K^{\|N}$. Resolving those N^2 crossings we produce a coefficient t^{N^2}, which is equal to 1, because N is even. Consequently, $K^{\|N}$ can be transformed into the empty link, which can be deleted using (ln2). This proves the result. □

This property does not hold for a general manifold M, it is only in these particular cases that the factorization relation (lnr2) is a consequence of (lnr1).

Proposition 5.11. *Let M be either $\mathbb{R}^3$, an open or closed 3-dimensional ball, or the 3-dimensional sphere S^3. Then*

$$\mathcal{L}_N(S^3) = \mathbb{C}.$$

Proof. By Lemma 5.4, relation (lnr2) is redundant in M. Using Proposition 5.6 and applying (lnr1), namely setting $t = e^{\frac{\pi i}{N}}$, we obtain the quotient map

$$\mathcal{L}(M) = \mathbb{C}[t, t^{-1}] \rightarrow \mathbb{C}.$$

Hence the conclusion. □

The reduced linking number skein module of the cylinder over a surface

Proposition 5.12. *Let Σ be a smooth compact oriented surface. Then the reduced linking number skein module $\mathcal{L}_N(\Sigma \times [0,1])$ is a vector space with basis $<\gamma>$, where γ ranges over a set of multicurves representing the homology classes of $H_1(\Sigma, \mathbb{Z}_N)$.*

Proof. Recall the maps F and $\hat{F}$ from the proof of Proposition 5.8:

$$\begin{array}{ccc} \mathbb{C}[t,t^{-1}]\mathrm{Link}(\Sigma \times [0,1]) & \xrightarrow{\ F\ } & \mathbb{C}[t,t^{-1}]H_1(\Sigma,\mathbb{Z}) \\ \downarrow & \nearrow_{\hat{F}} & \\ \mathcal{L}(\Sigma \times [0,1]) & & \end{array}$$

Let

$$\mathbb{C}H_1(\Sigma, \mathbb{Z}) \text{ and } \mathbb{C}H_1(\Sigma, \mathbb{Z}_N)$$

be the $\mathbb{C}$-vector spaces with bases $H_1(\Sigma, \mathbb{Z})$ and $H_1(\Sigma, \mathbb{Z}_N)$ respectively. Define the quotient map

$$q_1 : \mathbb{C}[t, t^{-1}]H_1(\Sigma, \mathbb{Z}) \to \mathbb{C}H_1(\Sigma, \mathbb{Z})$$

by setting $t = e^{\frac{i\pi}{N}}$, and the map

$$q_2 : \mathbb{C}H_1(\Sigma, \mathbb{Z}) \to \mathbb{C}H_1(\Sigma, \mathbb{Z}_N),$$

by extending linearly the quotient map

$$H_1(\Sigma, \mathbb{Z}) \to H_1(\Sigma, \mathbb{Z})/NH_1(\Sigma, \mathbb{Z}) = H_1(\Sigma, \mathbb{Z}_N).$$

Set

$$F_N = q_2 \circ q_1 \circ \hat{F}.$$

We have the following diagram

$$\begin{array}{ccc} \mathcal{L}(\Sigma \times [0,1]) & \xrightarrow{\ \hat{F}\ } & \mathbb{C}[t,t^{-1}]H_1(\Sigma,\mathbb{Z}) \\ \downarrow & \searrow^{F_N} & \downarrow{\scriptstyle q_2 \circ q_1} \\ \mathcal{L}_N(\Sigma \times [0,1]) & & \mathbb{C}H_1(\Sigma,\mathbb{Z}_N) \end{array}$$

The question is whether F_N factors to a homomorphism

$$\hat{F}_N : \mathcal{L}_N(\Sigma \times [0,1]) \to \mathbb{C}H_1(\Sigma, \mathbb{Z}_N).$$

The relation (lnr2) applied to a multicurve in Σ is precisely the relation by which $\mathfrak{M}(\Sigma)$ is factored in order to obtain $\mathfrak{M}_N(\Sigma)$ in Theorem 3.5. This means that the quotient map q_2 is defined by this relation. Additionally, (lnr1) defines the map q_1, so the quotient map defined by simultaneously imposing these relations is $q_2 \circ q_1$.

The only problem can arise if introducing (lnr2) before applying the homomorphism $\hat{F}$ factors any further the skein module. Note that when a curve is crossed by N parallel copies of another curve, there is no distinction between overcrossings and undercrossings. This is because once we set $t = e^{\frac{i\pi}{N}}$, when smoothing an a crossing of N parallel strands with another strand we multiply by either t^N or t^{-N}, and these are both equal to -1.

Hence if a link contains a $K^{\|N}$, then by exchanging overcrossings with undercrossings or vice-versa we can move $K^{\|N}$ (by an ambient isotopy) so that it lies inside a cylinder $\Sigma \times [0, \epsilon]$ that does not contain other link components.

We can now project $K^{\|N}$ onto the surface then resolve all crossings. We do this following the self-crossings of K. Each time we resolve the N^2 crossings of $K^{\|N}$ that arise from one self-crossing of K, we introduce a factor of $t^{\pm N^2}$, which is equal to 1 because t is a $2N$th root of unity and N is even. Then we delete the trivial circles. All that remains from $K^{\|N}$ is the Nth power of an element in $\mathrm{Mul}(\Sigma)$. This element is sent to zero by the quotient map $\mathrm{Mul}(\Sigma) \to \mathfrak{M}_N(\Sigma)$. We conclude that the map F_N does not detect the presence of the component $K^{\|N}$, so no further factorization relations are introduced. Therefore we have a well defined map

$$\hat{F}_N : \mathcal{L}_N(\Sigma \times [0,1]) \to \mathbb{C}H_1(\Sigma, \mathbb{Z}_N).$$

Every multicurve that defines a homology class in $H_1(\Sigma, \mathbb{Z}_N)$, endowed with the blackboard framing, becomes a link, and hence defines a skein. It follows that the map $\hat{F}_N$ is onto.

To prove injectivity, recall from Proposition 5.8 that modulo (ln1) and (ln2), every link L is equivalent to a multicurve in $\mathrm{Mul}(\Sigma)$. At the level of $\mathrm{Mul}(\Sigma)$, the relation (lnr2) is the same as the one defining the quotient map $\mathrm{Mul}(\Sigma) \to \mathfrak{M}_N(\Sigma)$. Hence if L is an oriented framed link and γ is a multicurve on Σ that represents the homology class of L in $H_1(\Sigma, \mathbb{Z}_N)$, then there is $k \in \mathbb{Z}$ such that in $\mathcal{L}_N(\Sigma \times [0,1])$,

$$<L>= t^k <\gamma> .$$

Hence if σ is a skein in $\mathcal{L}_N(\Sigma \times [0,1])$ and $\sum_j c_j \gamma_j$ is a linear combination of multicurves that represents the element $\hat{F}_N(\sigma) \in \mathbb{C}H_1(\Sigma, \mathbb{Z}_N)$, then $\sigma = \sum_j c_j <\gamma_j>$. This proves that $\hat{F}_N$ is one-to-one. So $\hat{F}_N$ is a vector space isomorphism between $\mathcal{L}_N(\Sigma \times [0,1])$ and $\mathbb{C}H_1(\Sigma, \mathbb{Z}_N)$ and the proposition is proved. □

The result below illustrates two particular cases.

Proposition 5.13. *(a) Let Σ_g be an oriented closed genus g surface and let $a_1, a_2, \ldots, a_g, b_1, b_2, \ldots, b_g$ be a canonical basis of $H_1(\Sigma_g, \mathbb{Z})$, which is used to define the multicurves $(\mathbf{p}, \mathbf{q})$, $\mathbf{p}, \mathbf{q} \in \mathbb{Z}^g$. Then the reduced linking number skein module $\mathcal{L}_N(\Sigma_g \times [0,1])$ is a finite dimensional vector space with basis*

$$<(\mathbf{p}, \mathbf{q})>, \quad \mathbf{p}, \mathbf{q} \in \{0, 1, 2, \ldots, N-1\}^g.$$

(b) Let H_g be an oriented genus g handlebody, and let $a_1, a_2, \ldots, a_g$ be a canonical basis of $H_1(H_g, \mathbb{Z})$. Then the reduced linking number skein module $\mathcal{L}_N(H_g)$ is a finite dimensional vector space with basis

$$<a_1^{\| m_1}><a_2^{\| m_2}> \cdots <a_g^{\| m_g}>, \quad m_1, m_2, \ldots, m_g \in \mathbb{Z}_N.$$

5.4.2 *The Schrödinger representation in the topological perspective*

Reduced linking number skein algebras

Like before, the identification

$$\Sigma \times [0,1] \cup \Sigma \times [0,1] \approx \Sigma \times [0,1]$$

obtained by gluing cylinders defines a multiplication of skeins, which gives rise to an algebra structure on $\mathcal{L}_N(\Sigma \times [0,1])$.

Definition 5.14. The algebra defined by endowing the reduced linking number skein module of the cylinder over a surface Σ with the multiplication defined by the gluing of two cylinders is called the **reduced linking number skein algebra** of the surface, and is denoted by $\mathcal{L}_N(\Sigma)$.

If M is a 3-dimensional manifold with boundary, then the identification

$$\Sigma \times [0,1] \cup M \to M,$$

obtained by gluing the bottom of $\Sigma \times [0,1]$ by an orientation preserving homeomorphism to (all or part of) the boundary of M, determines an $\mathcal{L}_N(\Sigma)$-module structure on $\mathcal{L}_N(M)$.

Let Σ_g be a genus g closed Riemann surface, endowed with a canonical basis of $H_1(\Sigma_g, \mathbb{Z})$,

$$a_1, a_2, \ldots, a_g, b_1, b_2, \ldots, b_g.$$

Using the canonical basis construct the Jacobian variety $\mathcal{J}(\Sigma_g)$, which is quantized using a Kähler polarization associated to the basis elements

$b_1, b_2, \ldots, b_g$ as in §5.1.2. The result is the space of theta functions $\mathbf{\Theta}_N^{\Pi}(\Sigma_g)$ on which act quantized exponentials. The quantized exponentials form the finite Heisenberg group $\mathbf{H}(\mathbb{Z}_N^g)$ and their action on $\mathbf{\Theta}_N^{\Pi}(\Sigma_g)$ is the Schrödinger representation. The linear combinations of quantized exponentials form $L(\mathbf{\Theta}_N^{\Pi}(\Sigma_g))$, the algebra of linear operators on the space of theta functions.

Theorem 5.5. *The algebra isomorphism Φ defined in Theorem 5.4 factors to an algebra isomorphism*

$$\Phi_N : \mathcal{L}_N(\Sigma_g) \to L(\mathbf{\Theta}_N^{\Pi}(\Sigma_g)).$$

Proof. Using the canonical basis, we define the multicurves $(\mathbf{p}, \mathbf{q})$ on Σ_g, $\mathbf{p}, \mathbf{q} \in \mathbb{Z}^g$. By Proposition 5.13 (a) $\mathcal{L}_N(\Sigma_g \times [0,1])$, as a vector space, has the basis

$$<(\mathbf{p}, \mathbf{q})>, \quad \mathbf{p}, \mathbf{q} \in \{0, 1, 2, \ldots, N-1\}^g.$$

Combining this with Proposition 4.12, we deduce that, as vector spaces, $\mathcal{L}_N(\Sigma_g \times [0,1])$ and $L(\mathbf{\Theta}_N^{\Pi}(\Sigma_g))$ are isomorphic, with the isomorphism given by

$$<(\mathbf{p}, \mathbf{q})> \mapsto \operatorname{op}\left(e^{2\pi i(\mathbf{p}^T\mathbf{x} + \mathbf{q}^T\mathbf{y})}\right), \quad \mathbf{p}, \mathbf{q} \in \{0, 1, \ldots, N-1\}^g.$$

Also,

$$<(\mathbf{p}, \mathbf{q})><(\mathbf{p}', \mathbf{q}')> = t^{\mathbf{p}^T\mathbf{q}' - \mathbf{p}'^T\mathbf{q}} <(\mathbf{p} + \mathbf{p}', \mathbf{q} + \mathbf{q}')>,$$

which corresponds to the multiplication rule for quantized exponentials. □

Recall the discussion from §4.3.4. In view of Corollary 4.4, we now have a combinatorial description of the subalgebra $\widetilde{\mathbb{C}_t}[\mathbf{U}^{\pm 1}, \mathbf{V}^{\pm 1}]$ of the $2g$-dimensional quantum torus as the reduced linking number skein algebra $\mathcal{L}_N(\Sigma_g)$. The $*$-multiplication of quantized exponentials is given by a skein relation.

In the case of reduced linking number skein algebras, there is a result similar to Proposition 5.9.

Definition 5.15. We call $\mathcal{L}G_N(\Sigma)$ the **reduced linking number skein group** of Σ.

Proposition 5.14. *Let Σ_g be a closed genus g Riemann surface. The set*

$$\mathcal{L}G_N(\Sigma) = \{t^k <L> \mid L \in \mathit{Link}(\Sigma \times [0,1]), k \in \mathbb{Z}\} \subset \mathcal{L}_N(\Sigma)$$

is closed under multiplication, and is a group. If $\Sigma = \Sigma_g$, a closed genus g Riemann surface, then $\mathcal{L}G_N(\Sigma)$ is isomorphic to $\mathbf{H}(\mathbb{Z}_N^g)$.

The Schrödinger representation

We now turn to the action of $\mathcal{L}_N(\Sigma_g)$ on $\mathcal{L}_N(H_g)$. We need a few more results of low-dimensional topology.

Lemma 5.5. *There is an orientation preserving diffeomorphism $f : \Sigma_g \to \partial H_g$ such that $f(b_j)$ is a simple closed curve that bounds an embedded disk, for each $j = 1, 2, \ldots, g$. Moreover, the homeomorphism depends only on the Lagrangian subspace $\mathbf{L}$ spanned by $b_1, b_2, \ldots, b_g$ in $H_1(\Sigma_g, \mathbb{R})$.*

To clarify the statement, the homeomorphism depends only on $\mathbf{L}$ means that if two half-bases span the same $\mathbf{L}$ then the homeomorphisms f are the same up to a composition with a homeomorphism of ∂H_g that extends to the whole H_g.

Proof. Consider the cylinder $\Sigma_g \times [0, 1]$ and place the curves b_j in $\Sigma_g \times \{0\}$. Consider a regular neighborhood of the union of these curves in $\Sigma_g \times \{0\}$, which is a disjoint union of annuli. Glue to each of these annuli a 2-handle $B^2 \times B^1$ along $(\partial B^2) \times B^1 = S^1 \times [0, 1]$. The resulting 3-dimensional manifold has two boundary components: $\Sigma_g \times \{1\}$ and a sphere. Glue a ball to this sphere along its boundary.

We claim that the result is H_g. Indeed, using the Classification of Surfaces, we may assume that $b_1, b_2, \ldots, b_g$ are in the standard position from Figure 3.7. Then this procedure consists of filling in the handlebody inside.

For the second part, note that if $b_1', b_2', \ldots, b_g'$ were another basis spanning the same $\mathbf{L}$ as $b_1, b_2, \ldots, b_g$, then, for each j, $f(b_j')$ is null-homologous and hence null-homotopic. The homotopy is a disk in H_g with boundary $f(b_j')$. Bringing this disk in general position, we may assume that it is immersed and its multiple points are away from the boundary. Now we apply the following result.

Theorem 5.6. *(Dehn's Lemma) Suppose that M is a 3-dimensional manifold and that $j : B^2 \to M$ is a map such that $f|\partial B^2$ is an embedding and $j^{-1}(j(\partial B^2)) = \partial B^2$. Then there is an embedding $k : B^2 \to M$ such that $k|\partial B^2 = j|\partial B^2$.*

This result was first formulated by Max Dehn in 1910, and was given the first complete proof by C.D. Papakyriakopoulos in 1956. A proof can be found in [Hempel (1976)] or [Jaco (1997)].

Using Dehn's Lemma and pushing the interiors of the embedded disks inside H_g we conclude that $f(b_j')$, $j = 1, 2, \ldots, g$, bound disks with interiors in the interior of H_g. Since the b'_j are not null-homologous in $H_1(\Sigma_g, \mathbb{R})$ and not homologous to each other, it follows that H_g is obtained from $\Sigma_g \times [0,1]$ by the same procedure of adding 2-handles and a ball, but with 2-handles added along b'_j, $j = 1, 2, \ldots, g$. So the very same handlebody is produced when working with $b_1, b_2, \ldots, b_g$ as when working with $b'_1, b'_2, \ldots, b'_g$. □

Lemma 5.6. *Let $f : \Sigma_g \to \partial H_g$ be an orientation preserving homeomorphism such that $f(b_j)$ is a simple closed curve that bounds an embedded disk. Then there is a realization of H_g as the cylinder $\Sigma_{0,g+1} \times [0,1]$ which turns $f(a_1), f(a_2), \ldots, f(a_g)$ into a canonical basis for $H_1(H_g, \mathbb{Z})$.*

Proof. To simplify notation, for a curve γ we denote $f(\gamma)$ also by γ (thus viewing the surface Σ_g as being the boundary of the handlebody).

Realize H_g as a cylinder such that the disks that b_j bound are vertical. Let $a'_1, a'_2, \ldots, a'_g$ be a canonical basis for H_g defined by this cylinder, so that

$$a'_j \cdot b_k = \delta_{jk}, \quad j, k = 1, 2, \ldots, g.$$

Then

$$a_j = a'_j + m_j b_j, \quad j = 1, 2, \ldots, g,$$

for some $m_j \in \mathbb{Z}$.

Consider the composition of Dehn twists

$$T_{b_1}^{-m_1} T_{b_2}^{-m_2} \cdots T_{b_g}^{-m_g},$$

which maps a_j to a'_j for each j. This composition extends to a homeomorphism of H_g onto itself. Composing the homeomorphism that realizes H_g as a cylinder with this one yields a realization of H_g as a cylinder in which $a_1, a_2, \ldots, a_g$ form a canonical basis. □

The homeomorphism f defines a gluing of $\Sigma_g \times [0,1]$ to H_g which further induces a left action of $\mathcal{L}(\Sigma_g)$ on $\mathcal{L}(H_g)$ and a left action of $\mathcal{L}_N(\Sigma_g)$ on $\mathcal{L}_N(H_g)$.

We are now in position to phrase the main result of this chapter, which establishes the topological version of the Schrödinger representation.

Theorem 5.7. *(a) There is a linear isomorphism*

$$\Psi_N : \mathcal{L}_N(H_g) \to \mathbf{\Theta}_N^{\Pi}(\Sigma_g),$$

defined by

$$\Psi_N(<\gamma>) = \theta_{[\gamma]}^{\Pi}(\mathbf{z}),$$

where γ ranges among all multicurves in $\Sigma_{0,g+1}$ and $[\gamma]$ is the homology class of γ in $H_1(H_g, \mathbb{Z}_N^g) = \mathbb{Z}_N^g$.
(b) If we identify $\mathcal{L}_N(\Sigma_g)$ with $L(\mathbf{\Theta}_N^\Pi(\Sigma_g))$ via the isomorphism defined in Theorem 5.5 then Ψ_N intertwines the left action of $\mathcal{L}_N(\Sigma_g)$ on $\mathcal{L}_N(H_g)$ and the Schrödinger representation of $L(\mathbf{\Theta}_N^\Pi(\Sigma_g))$ on $\mathbf{\Theta}_N^\Pi(\Sigma_g)$.

Proof. By Proposition 5.13, a basis of the vector space $\mathcal{L}_N(H_g)$ is given by $<\gamma>$, where the multicurve γ ranges in a set of representatives of $H_1(H_g, \mathbb{Z}_N)$. Moreover, in the proof of Proposition 5.12 we have seen that if $[\gamma_1] = [\gamma_2]$ in $H_1(H_g, \mathbb{Z}_N)$, then $<\gamma_1>=<\gamma_2>$ in $\mathcal{L}_N(H_g)$. Because $H_1(H_g, \mathbb{Z}_N) = \mathbb{Z}_N^g$, and $\theta_{\boldsymbol{\mu}}^\Pi(\mathbf{z})$, $\boldsymbol{\mu} \in \mathbb{Z}_N^g$ is a basis for $\mathbf{\Theta}_N^\Pi(\Sigma_g)$, Ψ_N defines an isomorphism of the vector spaces $\mathcal{L}_N(H_g)$ and $\mathbf{\Theta}_N^\Pi(\Sigma_g)$. This proves (a).

Consider a multicurve in Σ_g:

$$(\mathbf{p}, \mathbf{q}), \text{ with } \mathbf{p}, \mathbf{q} \in \{0, 1, \ldots, N-1\}^g.$$

Consider also a multicurve in $\Sigma_{0,g+1}$:

$$\gamma = a_1^{||\mu_1} a_2^{||\mu_2} \cdots a_g^{||\mu_g}, \quad \mu_1, \mu_2, \ldots, \mu_g \in \{0, 1, \ldots, N-1\}.$$

We compute

$$\begin{aligned} &<(\mathbf{p},\mathbf{q})><\gamma>=<(\mathbf{p},\mathbf{q})\cup\gamma>=<(\mathbf{p},\mathbf{q})(\boldsymbol{\mu},\mathbf{0})><\emptyset> \\ &= t^{\mathbf{p}^T\mathbf{0}-\boldsymbol{\mu}^T\mathbf{q}} <(\mathbf{p}+\boldsymbol{\mu},\mathbf{q})><\emptyset>= t^{-\boldsymbol{\mu}^T\mathbf{q}} <(\mathbf{p}+\boldsymbol{\mu},\mathbf{q})><\emptyset> \\ &= t^{-\boldsymbol{\mu}^T\mathbf{q}} t^{-(\mathbf{p}+\boldsymbol{\mu})^T\mathbf{q}} <(\mathbf{p}+\boldsymbol{\mu},\mathbf{0})(\mathbf{0},\mathbf{q})><\emptyset> \\ &= t^{-\boldsymbol{\mu}^T\mathbf{q}} t^{-(\mathbf{p}+\boldsymbol{\mu})^T\mathbf{q}} <(\mathbf{p}+\boldsymbol{\mu},\mathbf{0})><(\mathbf{0},\mathbf{q})><\emptyset>. \end{aligned}$$

Because acting by $<(\mathbf{0},\mathbf{q})>$ on $<\emptyset>$ amounts to adding several trivial link components, which can be deleted, this is further equal to

$$\begin{aligned} &t^{-\mathbf{p}^T\mathbf{q}-2\boldsymbol{\mu}^T\mathbf{q}} <(\boldsymbol{\mu}+\mathbf{p},\mathbf{0})><\emptyset> \\ &= t^{-\mathbf{p}^T\mathbf{q}-2\boldsymbol{\mu}^T\mathbf{q}} <a_1^{||\mu_1+p_1} a_2^{||\mu_2+p_2} \cdots a_g^{||\mu_g+p_g}>. \end{aligned}$$

We thus obtain

$$\begin{aligned} &<(\mathbf{p},\mathbf{q})><a_1^{||\mu_1} a_2^{||\mu_2} \cdots a_g^{||\mu_g}> \\ &= t^{-\mathbf{p}^T\mathbf{q}-2\boldsymbol{\mu}^T\mathbf{q}} <a_1^{||\mu_1+p_1} a_2^{||\mu_2+p_2} \cdots a_g^{||\mu_g+p_g}>. \end{aligned} \tag{5.17}$$

The multicurve $a_1^{||\mu_1} a_2^{||\mu_2} \cdots a_g^{||\mu_g}$ corresponds to the theta function $\theta_{\boldsymbol{\mu}}^\Pi(\mathbf{z})$, while the multicurve $a_1^{||\mu_1+p_1} a_2^{||\mu_2+p_2} \cdots a_g^{||\mu_g+p_g}$ corresponds to the theta function $\theta_{\boldsymbol{\mu}+\mathbf{p}}^\Pi(\mathbf{z})$ (here the indices are taken modulo N). Also $t = e^{\frac{\pi i}{N}}$. We recognize in (5.17) the formula for the Schrödinger representation (4.34):

$$\exp(\mathbf{p}^T\mathbf{P} + \mathbf{q}^T\mathbf{Q})\theta_{\boldsymbol{\mu}}^\Pi(\mathbf{z}) = e^{-\frac{\pi i}{N}\mathbf{p}^T\mathbf{q} - \frac{2\pi i}{N}\boldsymbol{\mu}^T\mathbf{q}} \theta_{\boldsymbol{\mu}+\mathbf{p}}^\Pi(\mathbf{z}).$$

This proves (b). □

Remark 5.8. Every skein in $\mathcal{L}_N(H_g)$ can be represented as $\sum_j c_j < \gamma_j >$, where γ_j are multicurves in $\Sigma_{0,g+1}$. Let $\boldsymbol{\mu}_j$ be the homology class of γ_j in $H_1(H_g, \mathbb{Z}_N) = \mathbb{Z}_N^g$. Then the isomorphism Ψ is given explicitly by

$$\Psi\left(\sum_j c_j < \gamma_j >\right) = \sum_j c_j \theta_{\boldsymbol{\mu}_j}.$$

The heuristical discussion from §5.2.1 implies that the map

$$f_* : \mathcal{L}_N(\Sigma \times [0,1]) \to \mathcal{L}_N(H_g)$$

induced by homeomorphism f is the topological analogue of the projection operator

$$\pi_{\mathbf{L}} : \mathbb{C}[\mathbf{H}(\mathbb{Z}_N^g)] \to \mathcal{H}_{N,g}(\mathbf{L}).$$

The information defining f consists of two parts:

1. The condition that $f(b_j)$ bound disks, for $j = 1, 2, \ldots, g$, is the topological equivalent to

   $$\pi_{\mathbf{L}}(uu') = \pi_{\mathbf{L}}(u)$$

   for $u \in \mathbf{H}(\mathbb{Z}_N^g)$, $u' \in \exp(\mathbf{L})$. This condition is therefore equivalent to the choice of the polarization when defining the Hilbert space of the quantization. If

   $$f_1, f_2 : \Sigma_g \to \partial H_g$$

   are two homeomorphisms satisfying this condition, then there is a homeomorphism $f : H_g \to H_g$ such that

   $$f_2 = f \circ f_1.$$

   The homeomorphism f maps $\mathcal{L}_N(H_g)$ isomorphically onto itself, which allows us to identify the Schrödinger representation defined by f_1 with that defined by f_2.
2. The condition $f(a_j) = a_j$, $j = 1, 2, \ldots, g$, is the topological equivalent of explicating the theta series. It is the choice that permits explicit computations.

Example 5.5. To the theta series $\theta_0^\tau(z), \theta_1^\tau(z), \ldots, \theta_{N-1}^\tau(z)$ of the torus Σ_1 correspond the skeins shown in Figure 5.11.

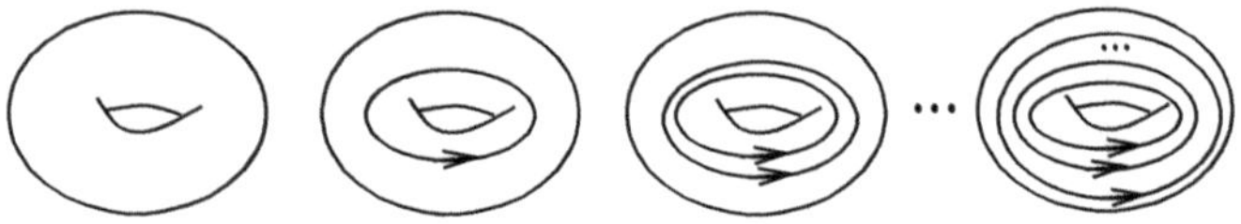

Fig. 5.11 Theta series of the torus

Example 5.6. We illustrate the topological version of the Schrödinger representation, in the case of the genus 1 surface, with the pictorial check of the formula

$$\exp(P + 2Q)\theta_1^{\mathcal{T}}(z) = t^{-1\cdot 2 - 2\cdot 2\cdot 1}\theta_{1+1}^{\mathcal{T}}(z) = t^{-6}\theta_2^{\mathcal{T}}(z).$$

The theta series $\theta_1^{\mathcal{T}}(z)$ is represented in $H_1 = \Sigma_{0,2} \times [0,1]$ by the curve a_1. The operator $\exp(P + 2Q)$ is represented by the curve $(1,2)$ on the torus $\Sigma_1 = \partial H_1$ (slightly pushed inside so as to lie in the interior).

Figure 5.12 shows the product of the curve $(1,2)$ on the torus, which represents $\exp(P+2Q)$, and the curve a_1 in the solid torus, which represents $\theta_1^{\mathcal{T}}(\mathbf{z})$. On the left we show the framed curves as they sit in the solid torus, so that $(1,2)$ has the blackboard framing of the boundary torus. On the right we show the curves as projected on the annulus $\Sigma_{0,2}$, with the convention that they have the blackboard framing of the annulus, and hence of the plane of the paper.

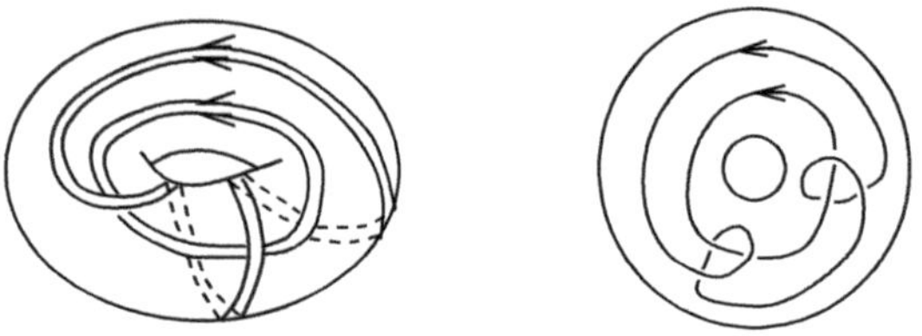

Fig. 5.12 Diagram describing the Schrödinger representation

Examining the diagram on the right, we notice that all 6 crossings have negative sign. We resolve them using the skein relations of the linking number (Figure 5.4) obtaining the curves on the left side of Figure 5.13. Then we delete the trivial link components (Figure 5.5) to obtain the curves from the right side of Figure 5.13. We recognize the final skein to be $t^{-6}a_1^{\|2}$, which is the diagrammatic form of $t^{-6}\theta_2^{\mathcal{T}}(z)$.

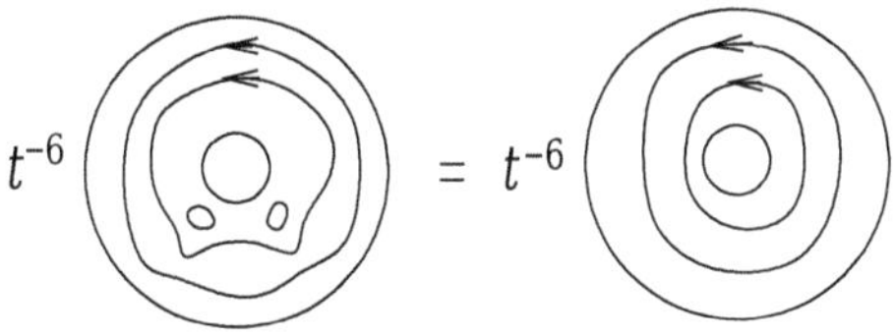

Fig. 5.13 Diagram computing the Schrödinger representation

5.4.3 *The action of the mapping class group on theta functions in the topological perspective*

Now we turn our attention to the discrete Fourier transform, and translate in topological language the formula (5.12) for the discrete Fourier transform induced by a pair $(\mathbf{L}, \mathbf{L}')$ as well as the formula from from Theorem 5.3 for the discrete Fourier transform induced by an element of the mapping class group.

Once again, let Σ_g be a closed genus g Riemann surface endowed with a canonical basis $a_1, a_2, \ldots, a_g, b_1, b_2, \ldots, b_g$ of $H_1(\Sigma_g, \mathbb{Z})$. The complex structure and the canonical basis give rise to the Jacobian variety, to theta functions, and to the action of the finite Heisenberg group.

Let h be an element of the mapping class group, $\mathrm{MCG}(\Sigma_g)$. This element induces the linear map $h_* : H_1(\Sigma_g, \mathbb{Z}) \to H_1(\Sigma_g, \mathbb{Z})$, which in the canonical basis has a matrix

$$h_* = \begin{pmatrix} A & B \\ C & D \end{pmatrix}.$$

The element h acts on $\mathbf{H}(\mathbb{Z}_N^g)$ by

$$h \cdot \exp(\mathbf{p}^T\mathbf{P} + \mathbf{q}^T\mathbf{q} + kE) = \exp[(A\mathbf{p} + B\mathbf{q})^T\mathbf{P} + (C\mathbf{p} + D\mathbf{q})^T\mathbf{Q} + kE],$$

as in Proposition 4.14 (see §4.3.5). This action can be extended linearly to $\mathbb{C}[\mathbf{H}(\mathbb{Z}_N^g)]$. Factor $\mathbb{C}[\mathbf{H}(\mathbb{Z}_N^g)]$ by the relations

$$\exp(kE) = e^{\frac{\pi i}{N}k}, \quad k \in \mathbb{Z},$$

to obtain $L(\mathbf{\Theta}_N^\Pi(\Sigma_g))$. Then h induces a linear map acting on $L(\mathbf{\Theta}_N^\Pi(\Sigma_g))$.

Using Theorem 5.5, we identify $L(\mathbf{\Theta}_N^\Pi(\Sigma_g))$ with $\mathcal{L}_N(\Sigma_g)$. We thus obtain an action of h on $\mathcal{L}_N(\Sigma_g)$.

Proposition 5.15. *The action of $MCG(\Sigma_g)$ on $\mathbf{H}(\mathbb{Z}_N^g)$ defined in Proposition 4.14 gives rise to the action of the mapping class group on $\mathcal{L}_N(\Sigma_g)$ given by*

$$h\left(\sum_j c_j < L_j >\right) = \sum_j c_j < (h \times Id)(L_j) >,$$

where $Id : [0,1] \to [0,1]$ *is the identity map.*

Proof. Note that $h \times Id$ is a homeomorphism of $\Sigma_g \times [0,1]$ onto itself. If

$$H : (\Sigma_g \times [0,1]) \times [0,1] \to \Sigma_g \times [0,1]$$

is an ambient isotopy, then $(h \times Id) \circ H$ is also an ambient isotopy. Also, if f is an embedding of the unit ball into $\Sigma_g \times [0,1]$ that defines a crossing of L, then $(h \times Id) \circ f$ is an embedding of the unit ball that defines a crossing of $h(L)$. A trivial link component of L is mapped by $h \times Id$ to a trivial link component of $(h \times Id)(L)$. And N parallel copies of a link component are mapped to N parallel copies of a link component. It follows that the map $h : \mathcal{L}_N(\Sigma_g) \to \mathcal{L}_N(\Sigma_g)$ from the statement is well defined.

To show that it coincides with the action of h on operators, it suffices to check it on a vector space basis of $\mathcal{L}_N(\Sigma_g)$. We choose the basis consisting of the multicurves $(\mathbf{p}, \mathbf{q}) \in \{0, 1, \ldots, N-1\}$. And indeed, the image of the multicurve $(\mathbf{p}, \mathbf{q})$ through h is the multicurve

$$(A\mathbf{p} + B\mathbf{q}, C\mathbf{p} + D\mathbf{q}).$$

The result is proved. □

Notation: We denote the action of h on skeins in $\mathcal{L}_N(\Sigma_g)$ shortly by

$$\sigma \mapsto h(\sigma).$$

Let H_g be the genus g handlebody endowed with a canonical basis $a_1, a_2, \ldots, a_g$ of the first homology group. To the canonical basis of $H_1(\Sigma_g, \mathbb{Z})$ and the canonical basis of $H_1(H_g, \mathbb{Z})$, we associate an orientation preserving homeomorphism

$$f : \Sigma_g \to \partial H_g$$

such that

$$f(a_j) = a_j \text{ and } f(b_j) \text{ bound disks}, j = 1, 2, \ldots, g.$$

The homeomorphism f defines the Schrödinger representation (see §5.4.2).

Consider the tubular neighborhood N_0 of ∂H_g which is the image of $\Sigma_g \times [0,1]$. Given a link $L \in H_g$, there is an ambient isotopy

$$H : H_g \times [0,1] \to H_g$$

such that $H(\cdot, 1)$ embeds L into N_0. Intuitively, H represents one way of pushing L towards the boundary, i.e. of lifting L to a regular neighborhood of the boundary. Map $H(L,1)$ to $\Sigma_g \times [0,1]$ by f^{-1}. The skein

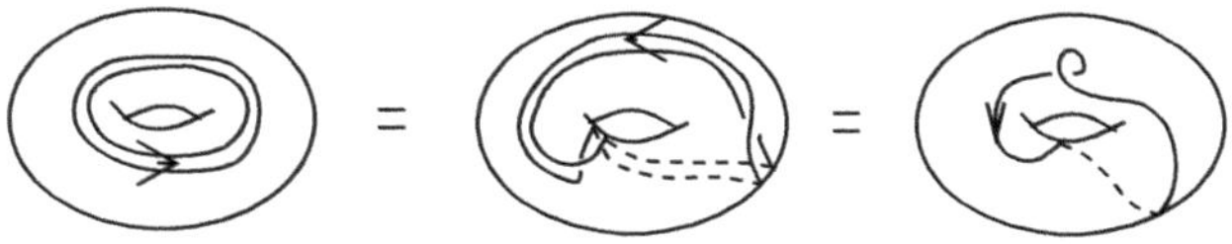

Fig. 5.14 Pushing $\theta_1^\tau(z)$ to the boundary torus

$<f^{-1}(H(L,1))>$ is an element of the reduced linking number skein group $\mathcal{L}G_N(\Sigma_g) \simeq \mathbf{H}(\mathbb{Z}_N^g)$.

Example 5.7. One possible way of pushing the oriented framed curve representing $\theta_1^\tau(z)$ to the boundary torus is depicted in Figure 5.14. The result is the curve $t(1,1)$.

There are many different ways of choosing the ambient isotopy H, i.e. there are many ways of pushing L towards the boundary. But because the group $\mathcal{L}G_N(\Sigma_g)$ is finite, there are only finitely many distinct skeins that can be produced from L by lifting L to a regular neighborhood of the boundary and viewing it as a reduced skein, and then mapping the result to $\Sigma_g \times [0,1]$.

Let

$$\mathbf{s}_f(L) = \frac{1}{n(H)} \sum_H <f^{-1}(H(L,1))>,$$

where the sum is taken over a family of ambient isotopies that produce all possible distinct skeins in $\mathcal{L}G_N(\Sigma_g)$, and $n(H)$ is the number of such isotopies.

Lemma 5.7. *If* $<L_1>=<L_2>$ *then* $\mathbf{s}_f(L_1) = \mathbf{s}_f(L_2)$.

Proof. Assume that L_2 is ambient isotopic to L_1 via an ambient isotopy $H_{2,1}$. Then

$$\begin{aligned}\mathbf{s}_f(L_1) &= \frac{1}{n(H)} \sum_H <f^{-1}(H(L_1,1))> \\ &= \frac{1}{n(H)} \sum_H <f^{-1}(H \circ H_{2,1}(L_2,1))> \\ &= \frac{1}{n(H')} \sum_{H'} <f^{-1}(H'(L_2,1))>= \mathbf{s}_f(L_2).\end{aligned}$$

In other words, $H_{2,1}$ establishes a one-to-one correspondence between the terms in the sum associated to L_1 and the terms in the sum associated to L_2.

The relations (ln1), (ln2), (lnr1), and (lnr2) in H_g are mapped via $f^{-1}\circ H(\cdot,1)$ to the same relations in $\Sigma_g\times[0,1]$. This proves the lemma. □

Based on this result, for each skein of the form $<L>$ with L a link in H_g, we can define

$$\hat{\mathbf{s}}_f(<L>)=\mathbf{s}_f(L).$$

Extend $\hat{\mathbf{s}}_f$ by linearity to a vector space homomorphism

$$\hat{\mathbf{s}}_f:\mathcal{L}(H_g)\to\mathcal{L}(\Sigma_g).$$

For simplicity, we image Σ_g in the standar position so that $\Sigma_g=\partial H_g$ and so f is the inclusion.

Proposition 5.16. *Under the identification of $\mathcal{L}_N(H_g)$ with $\boldsymbol{\Theta}_N^{\Pi}(\Sigma_g)$, the map $\hat{\mathbf{s}}_f$ coincides with the map $s_{\mathbf{L}}$.*

Proof. Consider a basis element $\pi_{\mathbf{L}}(\exp(\boldsymbol{\mu}^T\mathbf{P}))$. As a skein, it is of the form

$$<f((\boldsymbol{\mu},\mathbf{0}))>.$$

Consider also an element $\exp(\mathbf{q}^T\mathbf{Q})\in\exp(\mathbf{L})$, $\mathbf{q}\in\{0,1,\ldots,N-1\}^g$. Then

$$\exp(\boldsymbol{\mu}^T\mathbf{P})\exp(\mathbf{q}^T\mathbf{Q})=t^{\boldsymbol{\mu}^T\mathbf{q}}\exp(\boldsymbol{\mu}^T\mathbf{P}+\mathbf{q}^T\mathbf{Q})$$

which as a skein is

$$t^{\boldsymbol{\mu}^T\mathbf{q}}<(\boldsymbol{\mu},\mathbf{q})>.$$

Using for $s_{\mathbf{L}}$ the simplified formula (5.11), we have

$$\begin{aligned}s_{\mathbf{L}}(\pi_{\mathbf{L}}(\exp(\boldsymbol{\mu}^T\mathbf{P}))&=\frac{1}{N^g}\sum_{\mathbf{q}\in\{0,1,\ldots,N-1\}^g}(\boldsymbol{\mu}^T\mathbf{P})\exp(\mathbf{q}^T\mathbf{Q})\\&=\frac{1}{N^g}\sum_{\mathbf{q}\in\{0,1,\ldots,N-1\}^g}t^{\boldsymbol{\mu}^T\mathbf{q}}\exp(\boldsymbol{\mu}^T\mathbf{P}+\mathbf{q}^T\mathbf{Q}).\end{aligned}$$

Or, in the language of skein modules

$$s_{\mathbf{L}}(<f((\boldsymbol{\mu},\mathbf{0}))>)=\frac{1}{N^g}\sum_{\mathbf{q}'\in\{0,1,\ldots,N-1\}^g}t^{\mathbf{p}^T\mathbf{q}'}<(\boldsymbol{\mu},\mathbf{q})>.$$

A close look at the skeins

$$t^{\boldsymbol{\mu}^T\mathbf{q}}<(\boldsymbol{\mu},\mathbf{q})>,\quad\mathbf{q}\in\mathbb{Z}_N^g$$

shows that they represent all possible ways of pushing $<f((\boldsymbol{\mu},\mathbf{0}))>$ to the boundary and then mapping by the inverse of f to the cylinder over the surface. The coefficient $t^{\boldsymbol{\mu}^T\mathbf{q}}$ accounts for changing the framed curve on the boundary to have the blackboard framing (see Example 5.7). □

In view of this result we can give the map $\mathbf{s}_f$ a description that is easier to work with.

Corollary 5.3. *Let L be a link in H_g, isotoped to a regular neighborhood N_0 of the boundary of ∂H_g. Let $r : H_g \times [0,1] \to (H_g \backslash N_0)$ be a deformation retraction. Then*

$$s_f(L) = \frac{1}{N}\sum f^{-1}(L + rf(L_0)),$$

where L_0 runs through a family of links representing the elements of $\exp(\mathbf{L}) \subset \mathcal{L}G_N(\Sigma_g)$.

In words, this means that we push in some way L close to the boundary of the torus, then insert close to the core of the torus links that span the kernel of $\pi_{\mathbf{L}}$ and then take the average.

This topological reformulation of $s_{\mathbf{L}}$ will now allow us to place the discrete Fourier transforms $\mathcal{F}_{\mathbf{L},\mathbf{L}'}$ and $\mathcal{F}_{\mathbf{L}}(h)$ in a topological perspective.

Consider the handlebody H_g endowed with the canonical basis of its first homology group $a_1, a_2, \ldots, a_g$. Let Σ_g be a Riemann surface and $a_1, a_2, \ldots, a_g, b_1, b_2, \ldots, b_g$ and $a'_1, a'_2, \ldots, a'_g, b'_1, b'_2, \ldots, b'_g$ be two canonical bases of $H_1(\Sigma_g, \mathbb{Z})$. Let $\mathbf{L}$ and $\mathbf{L}'$ be the subgroups of $H_1(\Sigma_g)$ generated by $b_1, b_2, \ldots, b_g$ and $b'_1, b'_2, \ldots, b'_g$ respectively. Choose homeomorphisms

$$f, f' : \Sigma_g \to \partial H_g$$

such that

$$f(a_j) = a_j \text{ and } f(b_j) \text{ bound disks}, j = 1, 2, \ldots, g,$$
$$f'(a'_j) = a_j \text{ and } f'(b'_j) \text{ bound disks}, j = 1, 2, \ldots, g.$$

The homeomorphisms f and f' define two different models of the Schrödinger representation as actions of the reduced linking algebra of Σ_g on the reduced linking number skein module of H_g. The two models are identified by the discrete Fourier transform $\mathcal{F}_{\mathbf{L},\mathbf{L}'}$. As a corollary of Proposition 5.16, we have the following result.

Theorem 5.8. *Up to multiplication by a constant, the discrete Fourier transform $\mathcal{F}_{\mathbf{L},\mathbf{L}'}$ is given by*

$$\mathcal{F}_{\mathbf{L},\mathbf{L}'}(\sigma) = f'(\hat{s}_f(\sigma)), \quad \sigma \in \mathcal{L}_N(H_g).$$

In other words, $\mathcal{F}_{\mathbf{L},\mathbf{L}'}$ acts on a skein $< L >$ defined by a link L in the following way: lift L in all non-equivalent ways to a regular neighborhood of the boundary of H_g, map the results to $\Sigma_g \times [0,1]$ using f^{-1}, then take the arithmetic mean, and map the result to H_g by f'.

We now turn to the discrete Fourier transform determined by an element of the mapping class group. Let H_g be the genus g handlebody with the canonical basis of the first homology group $a_1, a_2, \ldots, a_g$ and Σ_g the genus g surface with the canonical basis of the first homology group $a_1, a_2, \ldots, a_g, b_1, b_2, \ldots, b_g$. Let also $h \in \mathrm{MCG}(\Sigma_g)$.

The subgroup $\mathbf{L}$ of $H_1(\Sigma_g, \mathbb{Z})$ generated by $b_1, b_2, \ldots, b_g$ defines the abstract version of the Schrödinger representation and the discrete Fourier transform $\mathcal{F}_{\mathbf{L}}(h)$.

There are two actions of $\mathcal{L}_N(\Sigma_g)$ on $\mathcal{L}_N(H_g)$, one which is the Schrödinger representation, and one in which the skein σ acts the way $h(\sigma)$ acts in the Schrödinger representation. By the Stone-von Neumann theorem, the two representations are unitary equivalent; they are related by the isomorphism $\mathcal{F}_{\mathbf{L}}(h)$. In topological framework, the **exact Egorov identity** reads

$$\mathcal{F}_{\mathbf{L}}(h)\sigma\mathcal{F}_{\mathbf{L}}(h)^{-1} = h(\sigma), \quad \sigma \in \mathcal{L}_N(\Sigma_g). \tag{5.18}$$

Again, as a corollary of Proposition 5.16 we have the following result.

Theorem 5.9. *Up to multiplication by a constant, the discrete Fourier transform $\mathcal{F}_{\mathbf{L}}(h)$ is equal to the map*

$$\sigma \mapsto f(h(\hat{\mathbf{s}}_f(\sigma))), \quad \sigma \in \mathcal{L}_N(H_g).$$

In other words, up to multiplication by a constant, the discrete Fourier transform of a skein of the form $<L>$, with L a framed oriented link in H_g, is obtained by lifting the link L in all possible nonequivalent ways (as a reduced skein) to a regular neighborhood of the boundary, mapping the results by h, taking the arithmetic mean of the images, and then pushing the result inside H_g.

One can also check that the map defined this way satisfies the exact Egorov identity (5.18) with the action of h on skeins in $\mathcal{L}_N(\Sigma)$, hence by Schur's lemma, it is a scalar multiple of $\mathcal{F}_{\mathbf{L}}(h)$.

Example 5.8. We will exemplify the topological version of the discrete Fourier transform by showing how the S-map on the torus acts on the theta function $\theta_1^\tau(z)$. Recall the definition of the S-map from §3.3.2.

Choose the canonical basis $a_1 = (1, 0)$, $b_1 = (0, 1)$, with $\mathbf{L} = \mathbb{Z}b_1$. The theta function $\theta_1^\tau(z)$ is represented in the solid torus by the curve shown in Figure 5.15, with the blackboard framing of the paper.

There are N linearly independent skeins that are liftings of this curve to the boundary. They arise by pushing this curve close to the boundary

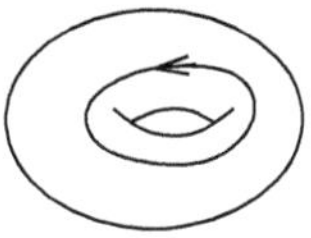

Fig. 5.15 Skein representing $\theta_1^\tau(z)$

torus, then inserting the multicurves $(0,j)$, $j=0,1,\ldots,N-1$, close to the core of the solid torus. In other words, the nonequivalent results of lifting the curve $(1,0)$ to the boundary are

$$(1,0)(0,j) = t^j(1,j), \quad j=0,1,\ldots,N-1.$$

These are represented in Figure 5.16. The curves are shown in the blackboard framing of the boundary torus. Note that the powers of t show up in the process of actually pushing the curve of $\theta_1^\tau(z)$ to the boundary since we have to twist it in order to obtain the blackboard framing of the boundary torus (see Example 5.7). Here the last curve spins $N-1$ times around the torus in the $(0,1)$ direction.

Fig. 5.16 Liftings of $\theta_1^\tau(z)$ to the boundary torus

The S-map sends these curves to the ones in Figure 5.17, which, after being pushed inside the solid torus, become the ones from Figure 5.18.

Fig. 5.17 Computation of $\mathcal{F}_S\theta_1^\tau(z)$

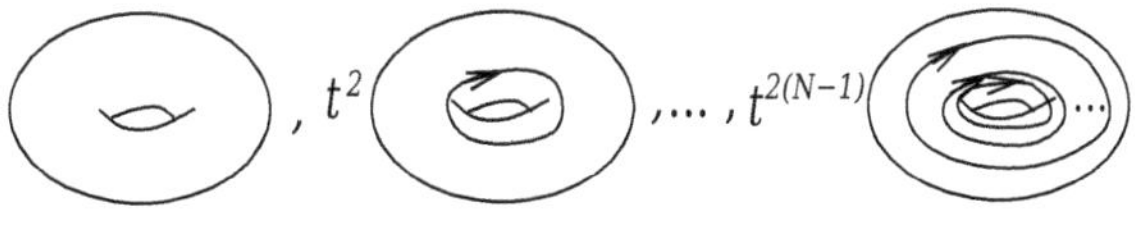

Fig. 5.18 Computation of $\mathcal{F}_S\theta_1^\tau(z)$

Note that in each skein the arrow points the opposite way as for $\theta_j^{\mathcal{T}}(z)$. Using the identity

$$<L \cup \gamma^{\| N}> = <L>,$$

and the fact that two parallel curves of opposite orientation cancel each other in the skein module, we can replace j parallel strands by $N-j$ parallel strands with opposite orientation. Also, $\theta_0^{\mathcal{T}}(z) = \theta_N^{\mathcal{T}}(z)$. Hence the skeins depicted in Figure 5.18 are

$$t^{2j}\theta_{N-j}^{\mathcal{T}}(z), \quad j = 1, \ldots, N.$$

Taking their arithmetic mean, we obtain

$$\frac{1}{N}\sum_{j=0}^{N-1} t^{2j}\theta_{N-j}^{\mathcal{T}}(z) = \frac{1}{N}\sum_{j=0}^{N-1} e^{\frac{2\pi i j}{N}}\theta_{N-j}^{\mathcal{T}}(z) = \frac{1}{N}\sum_{j=0}^{N-1} e^{-\frac{2\pi i j}{N}}\theta_j^{\mathcal{T}}(z).$$

This is equal, up to multiplication by a constant, to the standard discrete Fourier transform (5.15) of $\theta_1^{\mathcal{T}}(z)$. In fact it is equal to

$$N^{1/2}\mathcal{F}_{\mathbf{L}}(S)\theta_1^{\mathcal{T}}(z).$$

Example 5.9. Start in the same setting as in Example 5.8. Recall the Dehn twist $T = T_{b_1}$ along the curve $b_1 = (0,1)$ on the torus. Let us compute the action of $\mathcal{F}_{\mathbf{L}}(T)$ on $\theta_1^{\mathcal{T}}(z)$. As in Example 5.8, $\theta_1^{\mathcal{T}}(z)$ is represented by the framed oriented curve in Figure 5.15, and the distinct liftings of this curve to the boundary are those from Figure 5.16.

Applying T we transform these into the framed oriented curves from Figure 5.19. Here the last curve spins N times around the torus in the $(0,1)$ direction.

Fig. 5.19 Computation of $\mathcal{F}_T\theta_1^{\mathcal{T}}(z)$

Pushing these curves inside the solid torus, and taking the arithmetic mean we obtain

$$\sum_{j=0}^{N-1} t^{-j-1}t^j\theta_1^{\mathcal{T}}(z) = t^{-1}\theta_1^{\mathcal{T}}(z).$$

This is equal to

$$\mathcal{F}_T\theta_1^{\mathcal{T}}(z),$$

by Proposition 5.5.

Conclusions to the fifth chapter

In this chapter we derived a low dimensional model for theta functions, the action of the finite Heisenberg group, and the representation of the mapping class group (also known as the modular group) using the skein modules associated to the linking number. More precisely

1. we **identified the space of theta functions** on the Jacobian variety $\mathcal{J}(\Sigma_g)$ with the **reduced linking number skein module** of the handlebody H_g bounded by Σ_g,
2. we **identified the group algebra of the finite Heisenberg group** $\mathbf{H}(\mathbb{Z}_N^g)$ with the **reduced linking number skein algebra** of the surface Σ_g, and described the action of the finite Heisenberg group on theta functions as the action of the reduced linking number skein algebra of Σ_g on the reduced linking number skein module of the handlebody H_g,
3. we described the image of a theta function under the **discrete Fourier transform** defined by an element h of the mapping class group as being obtained by pushing the skein in H_g that represents the theta function in all possible ways to the boundary, mapping the images by h, taking the average and pushing the result back inside H_g.

Chapter 6

Some results about 3- and 4-dimensional manifolds

The next step in our study of theta functions is a more detailed analysis of the discrete Fourier transform defined by an element of the mapping class group. We will be able to relate this discrete Fourier transform to 3- and 4-dimensional topology. To enpower the reader with the necessary background, we devote this chapter to reviewing the necessary facts from low-dimensional topology. For more details the reader can consult [Rolfsen (2003)].

All manifolds are **orientable**, and for an oriented manifold M, $-M$ denotes the manifold with orientation reversed. To distinguish easily between 3- and 4-dimensional manifolds, we use for the latter the boldface script. S^n and B^n denote the n-dimensional sphere and ball respectively.

6.1 3-dimensional manifolds obtained from Heegaard decompositions and surgery

In this section we explain the two standard ways of presenting compact, orientable, 3-dimensional manifolds without boundary.

6.1.1 *The Heegaard decompositions of a 3-dimensional manifold*

Let M be a compact, oriented, 3-dimensional manifold without boundary.

Definition 6.1. A **Heegaard decomposition** of M is a triple (H_g, H_g', h), where H_g and H_g' are handlebodies of the same genus and $h : \partial H_g \to \partial H_g'$ is a homeomorphism such that M is obtained by gluing H_g and H_g' along their boundaries by h:

$$M = H_g \cup_h H_g'.$$

Theorem 6.1. *Every closed orientable 3-dimensional manifold without boundary admits a Heegaard decomposition.*

Proof. As mentioned in Chapter 3, M can be triangulated (for the proof see [Moise (1952)]). This means that M can be decomposed into tetrahedra such that any two tetrahedra share a face, an edge, or a vertex. Let M^1 be the 1-skeleton of M, which is the graph formed by the vertices and edges. We want to construct a regular neighborhood of M^1 in M. Since we only assume a piece-wise linear structure on M, we rely on this structure for the definition of the regular neighborhood. We use the barycentric subdivision of the triangulation.

Recall that the barycentric subdivision is obtained by dividing each edge by its midpoint into 2 edges, each face by its medians into 6 faces, and each tetrahedron by the planes that join vertices with medians of the opposite face into 24 tetrahedra.

Consider the second barycentric subdivision (i.e. the barycentric subdivision of the barycentric subdivision), and define the regular neighborhood N of M^1 to be the union of the (closed) tetrahedra of the subdivision that intersect M^1. We want to show that N is a handlebody.

Define the neighborhood N_v of a vertex v and the neighborhood N_e of an edge e. It is not hard to see that, for each vertex v, N_v is topologically a ball. These balls are glued to each other along edges, and the gluing is by inserting N_e. Let us see how this gluing is performed. For an edge e, the neighborhood N_e can be embedded in $\mathbb{R}^3$, so we can visualize it like in Figure 6.1, where it is drawn in one dimension less. Note that the two balls

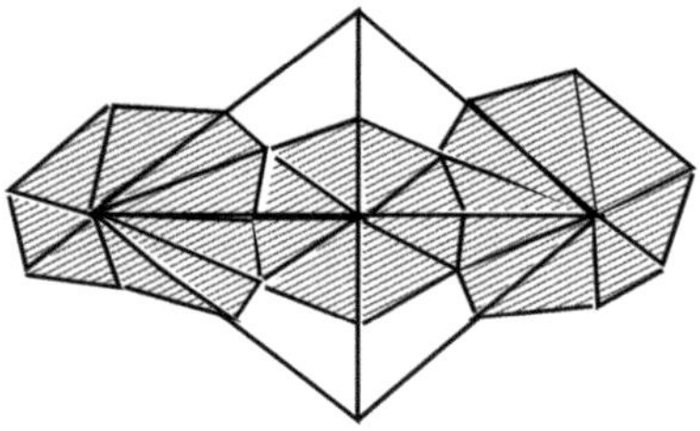

Fig. 6.1 Neighborhood of an edge

are joined by a $[0,1] \times B^2$, hence N is obtained by joining finitely many balls by disjoint 1-handles $[0,1] \times B^2$. The result is a handlebody.

Let $\widehat{M}^1$ be the dual in the second barycentric subdivision of the 1-skeleton M^1. It consists of the union of all vertices and edges that do not lie on a simplex that meets M^1. Let $\widehat{N}$ be regular neighborhood of $\widehat{M}^1$ (in the second barycentric subdivision). For the same reason, $\widehat{N}$ is a handlebody. Moreover, M is obtained by gluing N and $\widehat{N}$ along their boundaries. This is a Heegaard decomposition of M. □

Remark 6.1. By smoothing at vertices and along edges, we can make ∂H_g and $\partial H'_g$ smooth, and we can turn h into a diffeomorphism.

6.1.2 *3-dimensional manifolds obtained from surgery*

The mapping cylinder of a homeomorphism

Handlebodies are simple objects, so almost all information in the Heegaard decomposition

$$M = H_g \cup_h H'_g$$

is carried by the gluing homeomorphism h. There is one choice, h_{S^3}, which yields $M = S^3$. We can see this as follows. S^3 is made out of two balls glued along a 2-dimensional sphere. This is the Heegaard decomposition of S^3 using genus 0 handlebodies. Carve a 1-handle out of the first ball and add it to the second. The two balls now become solid tori, and thus we obtain the Heegaard decomposition of the sphere with genus 1 handlebodies. We can repeat this procedure of adding handles g times to obtain a Heegaard decomposition of S^3 with handlebodies of any genus g. This decomposition is sketched in Figure 6.2.

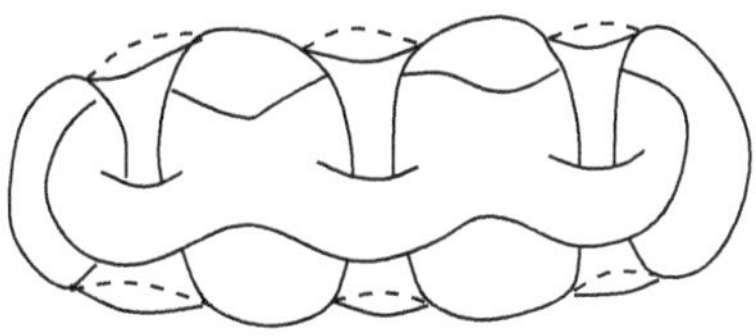

Fig. 6.2 Heegaard decomposition of the 3-dimensional sphere

For every other manifold M we need to compose h_{S^3} with a homeomorphism h of $\Sigma_g = \partial H_g$. A way to produce M is to start with

$$S^3 = H_g \cup_{h_{S^3}} H'_g,$$

and then insert between H_g and H'_g the mapping cylinder of h.

Definition 6.2. Let X and Y be topological spaces and let $f : X \to Y$ be a continuous function. The **mapping cylinder** I_f of f is

$$I_f = (X \times [0,1] \sqcup Y)/\sim$$

where $\sim$ identifies $(x, 1)$ with $f(x)$ for every $x \in X$.

In our situation, because h is a homeomorphism, I_h is homeomorphic to $\Sigma_g \times [0,1]$. But the homeomorphism

$$F : \Sigma_g \times [0,1] \to I_h$$

is not the identity map. Instead

$$F|\Sigma_g \times \{0\} = Id \text{ and } F|\Sigma_g \times \{1\} = h,$$

where Id is the identity map. We call $F(\Sigma_g \times \{0\})$ the **bottom** of the mapping cylinder and $F(\Sigma_g \times \{1\})$ the **top** of the mapping cylinder. The map F identifies both the top and the bottom with Σ_g.

We orient I_h so that

$$\partial I_h = \Sigma_g \times \{1\} \cup (-\Sigma_g \times \{0\}),$$

meaning that the orientation of the top agrees with that of Σ_g while the orientation of the bottom disagrees.

If $h = h' \circ h$, then $I_h = I_{h'} \cup I_h$, where the top of I_h is glued to the bottom of $I_{h'}$ via the identity map on Σ_g.

We want to give an explicit description of I_h as a 3-dimensional manifold. Recall that, by Theorem 3.9, h is a composition of Dehn twists along nonseparating simple closed curves. For this reason we will concentrate first on Dehn twists.

Let γ be an oriented nonseparating simple closed curve on Σ_g, defining the Dehn twist T_γ. Let A be an open annulus which is a regular neighborhood of γ in Σ_g. Choose $\epsilon < 1/2$, and consider the open solid torus

$$N_\gamma = A \times \left(\frac{1}{2} - \epsilon, \frac{1}{2} + \epsilon\right),$$

which is an open regular neighborhood of $\gamma \times \{1/2\}$ in $\Sigma_g \times [0,1]$ (see Figure 6.3).

We claim that there is a homeomorphism

$$F_0 : (\Sigma_g \times [0,1]) \backslash N_\gamma \to (\Sigma_g \times [0,1]) \backslash N_\gamma$$

such that $F_0|\Sigma_g \times \{0\} = Id$ and $F_0|\Sigma_g \times \{1\} = T_\gamma$. Indeed, the twist T_γ can be made to happen entirely inside A, so that $T_\gamma|\Sigma_g \backslash A = Id$. We can define

$$F_0 = T_\gamma \times Id \text{ on } A \times \left[\frac{1}{2} + \epsilon, 1\right]$$

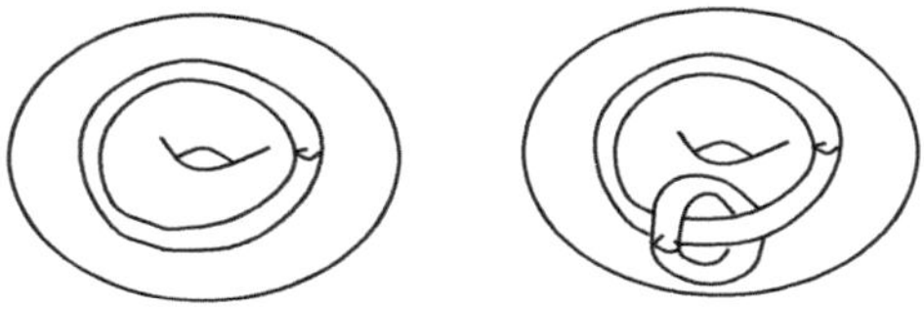

Fig. 6.3 $\Sigma_g \times [0,1]$ and $\Sigma_g \times [0,1]\backslash N_\gamma$

and let it be equal to the identity elsewhere.

The boundary ∂N_γ is a torus $S^1 \times S^1$, of which $\gamma \times [\frac{1}{2}+\epsilon]$ is $(1,0)$. When twisting $A \times [\frac{1}{2}+\epsilon, 1]$ by $T_\gamma \times Id$, the curve $(0,1)$ on this torus becomes $(1,1)$, while the curve $(1,0)$ does not change. This means that on the boundary of N_γ, F_0 is the twist about γ. We tried to sketch this in Figure 6.4. The claim is proved.

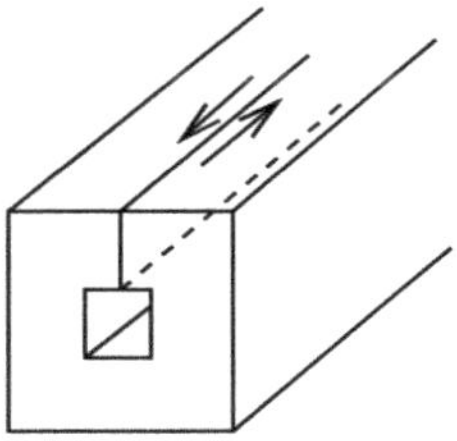

Fig. 6.4 Twisting $\Sigma_g \times [0,1]\backslash N_\gamma$

Because T_γ is a nontrivial element of the mapping class group of the torus, F_0 cannot be extended over N_γ. But we can obtain an extension by "twisting" N_γ. Glue N_γ back to $(\Sigma_g \times [0,1])\backslash N_\gamma$ along their boundaries in such a way that the $(0,1)$ curve is identified with $(1,1)$ while $(1,0)$ is identified with $(1,0)$. We claim that the resulting 3-dimensional manifold is I_{T_γ}. Indeed, the identity map on N_γ extends F_0 to a homeomorphism

$$F : \Sigma_g \times [0,1] \to I_{T_\gamma},$$

which satisfies $F|\Sigma_g \times \{0\} = Id$ and $F|\Sigma_g \times \{1\} = T_\gamma$.

In general, the homeomorphism h can be written as a product of twists, say $h = T_1 T_2 \cdots T_n$, by Theorem 3.9. Then

$$I_h = I_{T_1} \cup I_{T_2} \cup \cdots \cup I_{T_n}.$$

Thus I_h can be obtained by removing several disjoint solid tori from $\Sigma_g \times [0,1]$ and then gluing them back by some homeomorphisms of their boundaries.

3-dimensional manifolds via surgery

Let us examine the procedure of obtaining the mapping cylinder of a homeomorphism more closely and in a more general situation.

Definition 6.3. The procedure of removing a solid torus from a 3-dimensional manifold M and gluing it back by a homeomorphism of its boundary is called **Dehn surgery** or, in short, **surgery**.

The Dehn surgery used for the mapping cylinder is of a particular type, and as such it can be encoded by a framed link. Let us explain.

Let M be the manifold on which we perform surgery, and let N be one of the solid tori to be removed and then glued back. If the gluing homeomorphism is changed by an isotopy, the resulting 3-dimensional manifold does not change. So the result of surgery depends on the equivalence class of the gluing homeomorphism in $\mathrm{MCG}(\Sigma_1)$. The images of the $(1,0)$ and $(0,1)$ contain all the information.

The images c_1 and c_2 of $(1,0)$ and $(0,1)$ must intersect at one point and the intersection must have positive algebraic intersection number. We claim that c_1 is irrelevant for the result of the surgery. Choose c_1 to be any curve that intersects c_2 positively at one point. Its preimage intersects $(0,1)$ positively at exactly one point, so it must be of the form $(1,n)$, $n \in \mathbb{Z}$. Change the solid torus by a homeomorphism (n full twists) so that this curve becomes $(1,0)$; this does not change c_2. Now $(1,0)$ is mapped to c_1 while $(0,1)$ is still mapped to c_2, as desired.

In the case of the mapping cylinder, the curve $(0,1)$ on the boundary of N_γ is glued to the curve $(1,1)$ curve on boundary component of $(\Sigma_g \times [0,1])\backslash N_\gamma$. Both the torus $N_\gamma \subset \Sigma_g \times [0,1]$ and the $(1,1)$-curve can be encoded by a framed knot K in $\Sigma_g \times [0,1]$. This framed knot is an embedded annulus in N_γ that has one boundary component equal to γ and the other equal to the $(1,1)$-curve on the boundary.

Conversely, if we are given a framed knot K, namely an embedded annulus, it defines the surgery torus as a regular neighborhood of one of its boundary components. Choose this solid torus so that it does not intersect the other boundary component. Then the annulus intersects the boundary of the torus along a simple closed curve, which is the image of the curve $(0,1)$ of the torus attached by surgery. This is shown in Figure 6.5 for a knot with the blackboard framing. In this figure, the solid torus (depicted on the right) is glued to the knot complement in such a way that the dotted lines are identified.

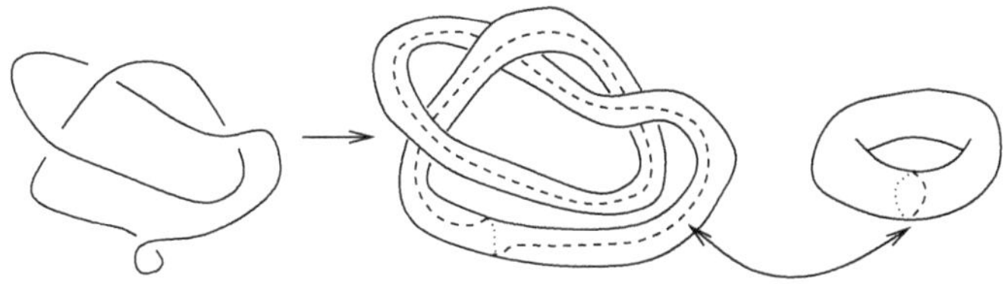

Fig. 6.5 Surgery on a framed knot

Definition 6.4. If an oriented 3-dimensional manifold M' is obtained by performing surgery on a manifold M, which surgery is encoded by a framed link L in M, we say that M' is **obtained by performing surgery on** $L \subset M$.

As a corollary of the above discussion, we obtain the following result.

Theorem 6.2. *The mapping cylinder of a homeomorphism can be obtained from* $\Sigma_g \times [0,1]$ *by performing surgery on a framed link in* $\Sigma_g \times [0,1]$.

We should point out that there are surgery descriptions of the mapping cylinder that do not come from writing the homeomorphism as a product of Dehn twists.

Example 6.1. Let b_1 be the element of the canonical basis depicted in Figure 3.7 from §3.1.3. The framed knot which encodes T_{b_1} is depicted in Figure 6.6. The drawing should be interpreted as follows. From the cylinder over the surface we present only the top, $\Sigma_g \times \{1\}$. To define the surgery knot, start with an annulus A that is a regular neighborhood of the curve b_1 on Σ_g. Embed the surface Σ_g in in the cylinder $\Sigma_g \times [0,1]$ as $\Sigma_g \times \{1/2\}$. Then A becomes a framed knot in $\Sigma_g \times [0,1]$. Modify this knot by a positive twist with respect to the blackboard framing of the knot (this is obtained by adding the kink in the figure). The result is the surgery knot, which in the figure is presented very close to the surface $\Sigma_g \times \{1\}$.

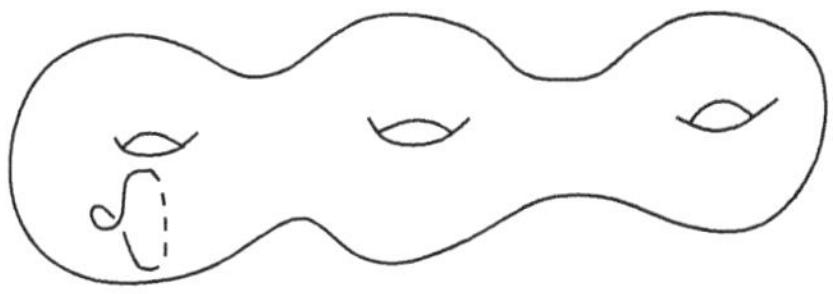

Fig. 6.6 Surgery knot of the Dehn twist

We can now state the main result.

Theorem 6.3. *(Lickorish-Wallace) Every compact, orientable, 3-dimensional manifold without boundary can be obtained by surgery on a framed link in S^3.*

Proof. Let M be the manifold. By Theorem 6.1 there is a Heegaard decomposition

$$M = H_g \cup_{h_0} H'_g.$$

To obtain M we can start with the Heegaard decomposition of the 3-dimensional sphere,

$$S^3 = H_g \cup_{h_{S^3}} H'_g,$$

then insert between H_g and H'_g the mapping cylinder of the homeomorphism h such that $h_0 = h_{S^3} \circ h$. We obtain

$$M = H_g \cup_{Id} I_h \cup_{h_{S^3}} H'_g.$$

The cylinder I_h can be obtained by starting with a regular neighborhood of ∂H_g in S^3, and then performing surgery on a framed link. Consequently, M is obtained by performing surgery on a framed link in S^3. □

Definition 6.5. If a manifold M is obtained by performing surgery on a framed link L in S^3, then L is called a **surgery diagram** for M.[1]

Notation: We denote the 3-dimensional manifold obtained by performing surgery on the framed link L by M_L.

Remark 6.2. Every framed link in S^3 yields by surgery a 3-dimensional manifold, and the link might not necessarily arise from the mapping cylinder of the gluing homeomorphism in some Heegaard decomposition.

Example 6.2. Surgery on each of the framed knots in Figure 6.7 yields S^3. The two framed knots are not the same, they do not coincide under an ambient isotopy of S^3. They are obtained from the trivial knot by adding a positive twist in the first case and a negative twist in the second. They are referred to as the **trivial knots with framing** $+1$ **and** -1 respectively.

[1]It should be pointed out that in 3-dimensional topology there is a more general type of surgery specified by a link, namely surgery with rational coefficients. Within that framework, the particular surgery considered in this book is known as surgery with integer coefficients.

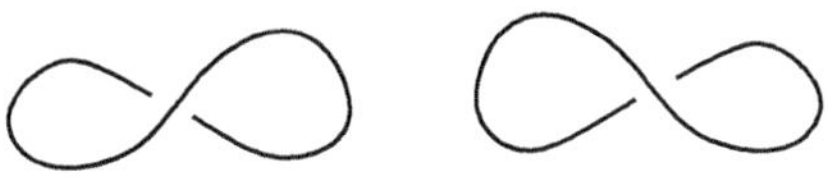

Fig. 6.7 Surgery knots that give rise to S^3

That surgeries on these framed knots yield S^3 is obvious for topologists. For mathematicians who prefer formulas, this simple situation allows explicit calculations. We explain the knot on the left of Figure 6.7.

The 3-dimensional sphere can be represented as

$$S^3 = \{(z_1, z_2) \in \mathbb{C}^2 \mid |z_1|^2 + |z_2|^2 = 2\}.$$

It has a Heegaard decomposition into the solid tori

$$S_1 = \{(z_1, z_2) \mid |z_1| \leq |z_2|\} \text{ and } S_2 = \{(z_1, z_2) \mid |z_1| \geq |z_2|\}.$$

They are glued along the torus

$$S^1 \times S^1 = \{(e^{2\pi is}, e^{2\pi it}) \mid s, t \in [0, 1)\}.$$

We can make S_2 be the surgery torus, which is removed from S^3 and then glued back in. The gluing defined by the knot in question is so that the curve $t \mapsto (1, e^{2\pi it})$, $t \in [0, 1]$, on the boundary of S^2 is identified with the curve $t \mapsto (e^{2\pi it}, e^{2\pi it})$, $t \in [0, 1]$, on the boundary of S_1.

If the second curve were $t \mapsto (1, e^{2\pi it})$, $t \in [0, 1]$, then S_2 would return to its original position, and the result would be S^3. We modify S_1 by an homeomorphism so that this is the case. The homeomorphism is the twist $f : S_1 \to S_1$ given by

$$f(re^{2\pi is}, \sqrt{2 - r^2}e^{2\pi it}) = (re^{2\pi i(s-t)}, \sqrt{2 - r^2}e^{2\pi it}), \quad r \in [0, 1], s, t \in [0, 1).$$

Note that f extends to the solid torus S_1 the Dehn twist on its boundary about the curve $s \mapsto (e^{-2\pi is}, 1)$. The homeomorphism f now extends to a homeomorphism between the result of the surgery and S^3, so surgery yields S^3.

In general, surgery on a link consisting only of trivial components, each with framing ± 1, is S^3.

Example 6.3. Surgery on the trivial framed knot (Figure 6.8) gives rise to $S^1 \times S^2$.

Here is why. In the explicit coordinates of the previous example, identify S_1 with S_2 via the homeomorphism $f : \mathbb{C}^2 \to \mathbb{C}^2$, $(z_1, z_2) \mapsto (z_2, z_1)$. Then surgery glues one copy of S_2 to another copy by the identity homeomorphism of the boundary. The two copies of S_2 are homeomorphic to the

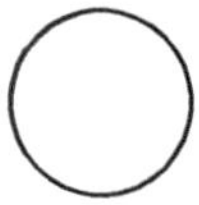

Fig. 6.8 Surgery knot that gives rise to $S^1 \times S^2$

cartesian product of S^1 and the disk, so let the first be $S^1 \times B^2_+$ and let the second be $S^1 \times B^2_-$.

Let f_+ be a homeomorphism from the upper hemisphere S^2_+ of S^2 to the disk B^2_+, and let f_- be a homeomorphism from the lower hemisphere S^2_- of S^2 to B^2_-, so that f_+ and f_- coincide on the common circle. Then

$$f : S^1 \times S^2 \to S^1 \times B^2_+ \cup S^1 \times B^2_-,\ f(e^{2\pi it}, x) = \begin{cases} f_+(e^{2\pi it}, x) & \text{if } x \in S^2_+ \\ f_-(e^{2\pi it}, x) & \text{if } x \in S^2_- \end{cases}$$

is a homeomorphism between the result of the surgery and $S^1 \times S^2$.

Definition 6.6. The **connected sum** of two oriented n-dimensional manifolds is the n-dimensional manifold obtained by removing an open ball from the interior of each manifold and identifying the boundaries of those balls by an orientation reversing homeomorphism.

Notation: We denote the connected sum of M_1 and M_2 by $M_1 \# M_2$.

Example 6.4. The genus g surface is the connected sum of g tori.

Proposition 6.1. *If there is an embedded 2-dimensional sphere in S^3 that separates the framed links L_1 and L_2, then*

$$M_{L_1 \cup L_2} = M_{L_1} \# M_{L_2}.$$

Example 6.5. Surgery on the trivial link with n components (Figure 6.9) yields

$$\#_{k=1}^n S^1 \times S^2,$$

the connected sum of n copies of $S^1 \times S^2$.

Fig. 6.9 Trivial link with n components

6.2 The interplay between 3-dimensional and 4-dimensional topology

6.2.1 *3-dimensional manifolds are boundaries of 4-dimensional handlebodies*

We start by drawing a parallel between the construction of surfaces by adding handles to S^2 and the construction of 3-dimensional manifolds by performing surgery on S^3. Recall the definition of a handle from §3.3.3. A handle is just a ball, it becomes a 0-, 1-, 2-, etc. handle depending on how you attach it. Here we are interested only in **3-dimensional 1-handles** and **4-dimensional 2-handles**. A 3-dimensional 1-handle is a $B^1 \times B^2$ attached along $\partial B^1 \times B^2$. A 4-dimensional 2-handle is a $B^2 \times B^2$ attached along $\partial B^2 \times B^2$.

We can also interpret the construction of a surface by adding "handles" to S^2, these are not quite the same type of handles that we had in mind above or when talking about Morse theory, nevertheless by abuse of language we might sometimes refer to these as handles, although in truth they are boundaries of 1-handles.

(a) The process of adding a "handle" to S^2 is shown in Figure 6.10. We remove two disks from S^2, and in the remaining holes we glue $[0,1] \times S^1$ along $\{0,1\} \times S^1$. Interpret $\{0,1\}$ as the 0-dimensional sphere S^0, which is the boundary of the 1-dimensional unit ball B^1, with the latter identified with $[0,1]$. So we remove from S^2 an $S^0 \times B^2$, producing a manifold with boundary $S^0 \times S^1$, and to this boundary we glue $B^1 \times S^1$ along $\partial B^1 \times S^1$.

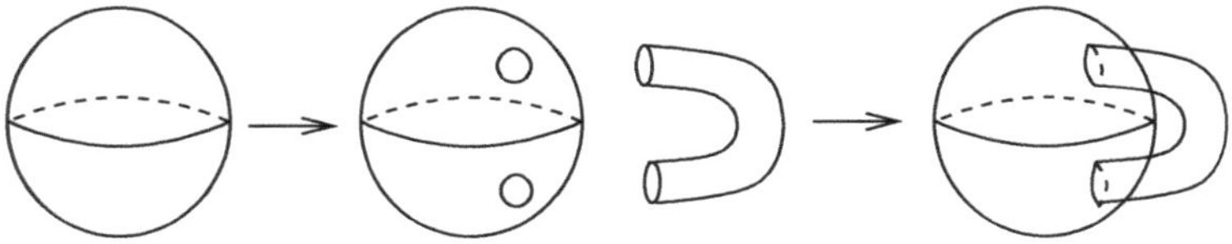

Fig. 6.10 Adding a handle to S^2

(b) By analogy, when performing surgery along a link component, we remove from S^3 an $S^1 \times B^2$, obtaining a manifold with boundary $S^1 \times S^1$. To this we glue $B^2 \times S^1$ along $\partial B^2 \times S^1$.

Closed orientable surfaces are boundaries of handlebodies. These handlebodies are obtained by adding 3-dimensional 1-handles. Let us draw a parallel between adding a 1-handle to a 3-dimensional ball versus adding a 2-handle to a 4-dimensional ball.

(a) To add a 1-handle, start with the 3-dimensional ball, B^3. Choose two disjoint disks on its boundary, which we identify with $S^0 \times B^2$. Take a 3-dimensional 1-handle $B^1 \times B^2$, whose boundary is decomposed into $B^1 \times \partial B^2 = B^1 \times S^1$ and $\partial B^1 \times B^2 = S^0 \times B^2$. Glue the handle to B^3 such that $\partial B^1 \times B^2$ is glued to the two disks on the boundary of B^3. The boundary of the result is the sphere with one handle discussed above.
(b) To add a 4-dimensional 2-handle to the 4-dimensional ball B^4, consider an $S^1 \times B^2$ on the boundary $\partial B^4 = S^3$. The handle that we attach is $B^2 \times B^2$, whose boundary is the union of $B^2 \times \partial B^2 = B^2 \times S^1$ and $\partial B^2 \times B^2 = S^1 \times B^2$. The handle is glued by identifying $\partial B^2 \times B^2$ with the $S^1 \times B^2$ in ∂B^4.

Proposition 6.2. *Any surgery on a component of a framed link in S^3 arises as the boundary of a 2-handle addition to B^4.*

Proof. Denote by S_1 the solid torus to be removed from S^3 when doing surgery. The framing of the link component determines a curve γ to which the $(0,1)$-curve of the attached torus is glued in the process of surgery. If we identify S_1 with $S^1 \times B^2$, then $\gamma = (1,n)$ for some $n \in \mathbb{Z}$.

The boundary of the 2-handle $B^2 \times B^2$ consists of the solid tori $B^2 \times \partial B^2$ and $\partial B^2 \times B^2$, which can be both modeled by $S^1 \times B^2$. In this model, the two are glued (to produce the boundary $\partial(B^2 \times B^2)$) of the 2-handle) along their boundary tori in such a way that the curves $(1,0)$ and $(0,1)$ on the first are identified with the curves $(0,1)$ and $(1,0)$ on the second. When surgery is performed, the solid torus $B^2 \times \partial B^2$ is glued so that the curve $(0,1)$ on its boundary is identified with $\gamma \in \partial S_1$. This means that the curve $(1,0)$ on the boundary of $\partial B^2 \times B^2$ is identified with $\gamma = (1,n)$. But there is a homeomorphism of the solid tori $\partial B^2 \times B^2$ and S_1 that maps $(1,0)$ to $(1,n)$, because the Dehn twist about $(0,1)$ can be extended to the solid torus. Hence the gluing homeomorphism of the surgery in S^3 can be extended to a 4-dimensional handle attachment, as claimed. □

An easy application of the Seifert-van Kampen Theorem implies that the fundamental group of the resulting manifold is trivial. We thus obtained:

Proposition 6.3. *Every closed, orientable, 3-dimensional manifold is the boundary of an orientable simply connected 4-dimensional manifold.*

Notation: We use the boldface notation $\mathbf{M}_L$ for the 4-dimensional manifold obtained by adding handles as specified by the framed link L. So $M_L = \partial\mathbf{M}_L$. The manifold $\mathbf{M}_L$ is sometimes called a **4-dimensional handlebody**.

When performing surgery we can start with a smooth structure on S^3 and make sure that the original link is smooth. Also after attaching the handles we can smoothen the corners, thus obtaining a smooth 4-dimensional manifold. We therefore have a stronger result.

Proposition 6.4. *Every closed, orientable, smooth 3-dimensional manifold is the boundary of a smooth, orientable, simply connected 4-dimensional manifold.*

As a rule, in the examples that follow the upper index specifies a dimension, while a lower index and all other symbols are used for enumeration. So B_3^2 denotes a 2-dimensional ball and $S_0^{3\prime}$ denotes a 3-dimensional sphere.

Example 6.6. The result of adding a 2-handle to B^4 as specified by the trivial framed knot from Figure 6.8 is $B^2 \times S^2$, the trivial disk bundle over S^2. This is because we can decompose S^2 into a union of two disks B_1^2 and B_2^2 such that B^4 is $B^2 \times B_1^2$ and the handle $B^2 \times B^2$ is $B^2 \times B_2^2$.

Example 6.7. If we add a 2-handle to B^4 along the first, respectively the second trivial knot components from Figure 6.11, we obtain the complex projective plane $\mathbb{C}P^2$ and $-\mathbb{C}P^2$ respectively, with a 4-dimensional ball removed.

Fig. 6.11 Handlebody descriptions of $\mathbb{C}P^2 \backslash B^4$ and $-\mathbb{C}P^2 \backslash B^4$.

To see this, recall that the **complex projective plane** $\mathbb{C}P^2$ is obtained by gluing three copies C_1, C_2, C_3 of $\mathbb{C} \times \mathbb{C}$ by the maps

$$c_{12} : C_1 \to C_2, \quad c_{12}(z_1, z_2) = (1/z_1, z_2/z_1),$$
$$c_{23} : C_2 \to C_3, \quad c_{23}(z_1, z_2) = (z_1/z_2, 1/z_2),$$
$$c_{31} : C_3 \to C_1, \quad c_{31}(z_1, z_2) = (z_2/z_1, 1/z_1),$$

where the gluing takes place on the domains of these maps. We can make the overlaps to be solely over manifolds of one dimension less, the same way that we obtain $\mathbb{C}P^1$ by gluing two unit disks along their boundaries. Then $\mathbb{C}P^2$ is obtained by gluing three copies D_1, D_2, D_3 of $B^2 \times B^2$ by the homeomorphisms

$$\begin{aligned} f_{12} &: \partial B^2 \times B^2 \to \partial B^2 \times B^2, \quad f_{12}(z_1, z_2) = (1/z_1, z_2/z_1), \\ f_{23} &: B^2 \times \partial B^2 \to B^2 \times \partial B^2, \quad f_{23}(z_1, z_2) = (z_1/z_2, 1/z_2) \\ f_{31} &: \partial B^2 \times B^2 \to B^2 \times \partial B^2, \quad f_{31}(z_1, z_2) = (z_2/z_1, 1/z_1). \end{aligned}$$

Here we identified B^2 with the unit disk in $\mathbb{C}$.

The gluing of D_1 and D_2 by f_{12} is the gluing of a handle to B^4 as specified by the framed trivial knot on the left of Figure 6.7. This is because f_{12} extends the homeomorphism on the boundary

$$(e^{2\pi is}, e^{2\pi it}) \mapsto (e^{-2\pi is}, e^{2\pi i(t-s)}).$$

The latter maps the curve $(1,0)$ to $(1,1)$ and the curve $(0,1)$ to $(0,1)$, so it is the surgery homeomorphism of the specified knot. We conclude that the result of surgery is $\mathbb{C}P^2\backslash D_3$, namely the projective space with a 4-dimensional ball removed.

The trivial knot on the right side of Figure 6.11, which has framing -1 yields also $\mathbb{C}P^2$, however one should conjugate everywhere in the second coordinate, and so one should reverse the orientations of all manifolds. Hence the handle addition specified by the trivial knot with framing -1 yields $\mathbb{C}P^2$, but with opposite orientation.

For further application, we want to show that the manifold $\mathbb{C}P^2\backslash B^4$ obtained here can also be identified with a B^2-bundle over S^2. The local charts are again B^4 and the handle $B^2 \times B^2$, where the ball B^4 is seen as $B^2 \times B^2$ in such a way that

$$S^3 = \partial B^4 = \partial B^2 \times B^2 \cup B^2 \times \partial B^2$$

is a Heegaard decomposition of S^3 into the solid torus to which the handle is glued and its complement.

Now we have two disk bundles $B_1^{2'} \times B_1^2$ and $B_2^{2'} \times B_2^2$ that are glued to each other so that $\partial B_1^{2'} \times B_1^2$ and $\partial B_2^{2'} \times B_2^2$ are identified via a homeomorphism. This homeomorphism is a twist of the solid torus that takes $(1,0)$ to $(1,1)$ and it can be performed so as to preserve the fiber structure of $S^1 \times B^2$. Hence this homeomorphism defines a disk bundle over the sphere.

Example 6.8. If the framed link L is obtained from the framed link L' by adding a trivial framed link component with framing ± 1, namely one of the two trivial link components from Figure 6.7, then $\mathbf{M}_L$ is obtained by taking the connected sum of $\mathbf{M}_{L'}$ with the complex projective plane $\pm\mathbb{C}P^2$.

Let us explain. If L_0 is the trivial link component, then by the previous example $\mathbf{M}_{L_0}$ is $\mathbb{C}P^2$ with a B^4 removed. The manifold $\mathbf{M}_L$ is obtained from $\mathbf{M}_{L'}$ and $\mathbf{M}_{L_0}$ by

- removing a ball B_1^4 from $\mathbf{M}_{L'}$ such that $\partial B_1^4 \cap \partial \mathbf{M}_{L'}$ is a 3-dimensional ball B_1^3,
- removing a ball B_2^4 from $\mathbf{M}_{L_0}$ such that $\partial B_2^4 \cap \partial \mathbf{M}_{L_0}$ is a 3-dimensional ball B_2^3,
- gluing $\mathbf{M}_{L'} \backslash B_1^4$ to $\mathbf{M}_{L_0} \backslash B_2^4$ by an orientation reversing homeomorphism $\partial B_1^4 \backslash B_1^3 \to \partial B_2^4 \backslash B_2^3$, which homeomorphism restricts to a homeomorphism $\partial B_1^3 \to \partial B_2^3$.

This is shown schematically in Figure 6.12. On the boundary,

$$M_L = M_{L'} \# M_{L_0} = M_{L'} \# S^3 \equiv M_{L'}.$$

So the boundary does not change. Thus we can ignore what happens on the boundary, and build $\mathbf{M}_L$ as the connected sum of $\mathbf{M}_{L'}$ and $\mathbf{M}_{L_0}$, then on the boundary cap $M_{L_0} = S^3$ with a ball. Hence $\mathbf{M}_L$ is obtained by taking the connected sum of $\mathbf{M}_{L'}$ with $\mathbb{C}P^2$, respectively $-\mathbb{C}P^2$, as claimed.

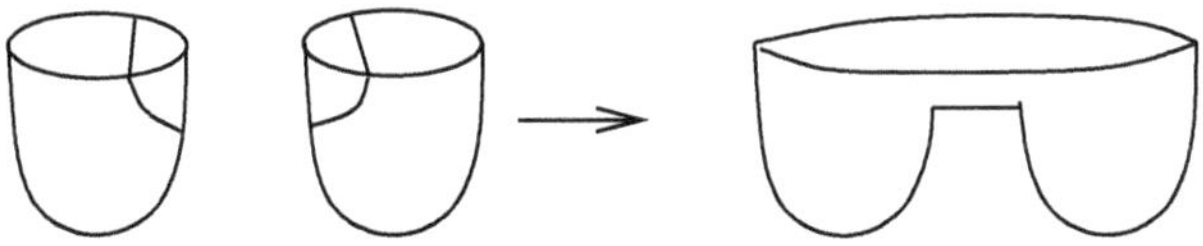

Fig. 6.12 Connected sum of two surgeries

Example 6.9. Let L be a framed link that can be separated by an embedded S^2 from the Hopf link , depicted in Figure 6.13 (here the framing is the blackboard framing). Consider the framed link $L' = L \cup$ . Then

$$\mathbf{M}_{L'} = \mathbf{M}_L \#(S^2 \times S^2).$$

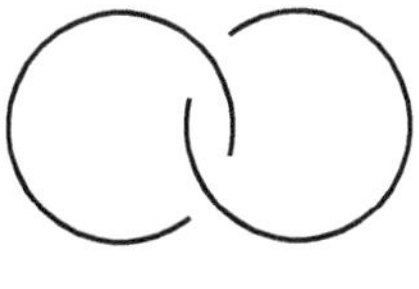

Fig. 6.13 Hopf link

Because of what we explained in the previous example, in order to prove this it suffices to show that the result of adding 2-handles to B^4 as specified by ⚭ is $S^2 \times S^2$ minus a ball.

Let us first examine $M_{⚭} = \partial \mathbf{M}_{⚭}$. View the 3-dimensional sphere as

$$S^3 = \{(z_1, z_2) \in \mathbb{C}^2 \,|\, |z_1|^2 + |z_2|^2 = 2\},$$

and the two components of the Hopf link as

$$\{(z_1, 0) \,|\, |z_1| = \sqrt{2}\} \text{ and } \{(0, z_2) \,|\, |z_2| = \sqrt{2}\}.$$

Regular neighborhoods of the two components are the solid tori

$$S_1 = \{(z_1, w) \,|\, |z_1| + |w|^2 = 2, \quad |w| \leq r\}$$

and

$$S_2 = \{(w, z_2) \,|\, |w|^2 + |z_2|^2 = 2, \quad |w| \leq r\},$$

where $0 < r < 1$. The manifold $S^3 \backslash (S_1 \cup S_2)$ is the cylinder over the torus: $S^1 \times S^1 \times (0,1)$. As $r \to 1$, the cylinder approaches the torus $\{(z_1, z_2) \,|\, |z_1| = |z_2| = 1\}$. From this we deduce that $M_{⚭}$ is the result of gluing two solid tori $S^1 \times B^2$ so that the curves $(1,0)$ and $(0,1)$ on the boundary of the first solid torus are mapped respectively to $(0,1)$ and $(1,0)$ on the second solid torus. Consequently the result of the surgery is $M_{⚭} = S^3$.

Next we examine what happens at the 4-dimensional level. In view of the commentary at the end of Example 6.6, the manifold $M_{⚭}$ is the result of a procedure known to topologists as **plumbing**. Plumbing is performed by starting with two disk bundles over surfaces, in our case over 2-dimensional spheres,

$$B_j^2 \hookrightarrow E_j \to S_j^2, \quad j = 1, 2.$$

Then choose a disk in each sphere, $B_j^{2'} \subset S_j^2$, and identify $B_1^2 \times B_1^{2'}$ with $B_2^2 \times B_2^{2'}$ by switching base and fiber directions $(x, y') \mapsto (y', x)$.

In our case the disk bundles are trivial: $E_j = S_j^2 \times B_j^2$. Let the common 4-dimensional ball be

$$B^4 = \{(z_1, z_2) \in \mathbb{C}^2 \,|\, |z_1|^2 + |z_2|^2 \leq 2\}.$$

Then because we switch the base and fiber directions, z_1 and z_2 are switched, so the two handles are attached along the solid tori S_1 and S_2. By Example 6.6, the attachment of a handle defined by the trivial knot is a trivial disk bundle. Thus $\mathbf{M}$ is the result of plumbing two trivial disk bundles.

To understand why the result is $S^2 \times S^2$ with a ball missing, a good idea is to picture this in a lower dimension, namely when "plumbing" two annuli $S_1^1 \times B_1^1$ and $S_2^1 \times B_2^1$. In this case we identify two rectangles $B_1^{1'} \times B_1^1$ and $B_2^{1'} \times B_2^1$ by switching coordinates, as shown in Figure 6.14, and the result is a torus without a disk. What is missing for this to be a torus $S^1 \times S^1$ is the disk $(S_1^1\backslash B_1^1) \times (S_2^1\backslash B_2^2)$.

Fig. 6.14 "Plumbing" two copies of $B^1 \times S^1$

Returning to the 4-dimensional situation, the picture is the same. The part that is missing in order for $\mathbf{M}$ to be $S^2 \times S^2$ is $(S_1^2\backslash B_1^{2'}) \times (S^2\backslash B_2^{2'})$.

Example 6.10. Let L be a framed link in S^3 that is separated by an embedded S^2 from the framed link  depicted in Figure 6.15 (with the blackboard framing). Consider the framed link

$$L' = L \cup$$ .

We want to show that

$$\mathbf{M}_{L'} = \mathbf{M}_L \# S^2 \tilde{\times} S^2.$$

where $S^2 \tilde{\times} S^2$ is an S^2-bundle over S^2. We will prove in §6.2.2 that this bundle is nontrivial, meaning that it is not equivalent as a bundle to the cartesian product $S^2 \times S^2$. It is not even homeomorphic as a manifold to $S^2 \times S^2$.

Fig. 6.15 Surgery that yields $(S^2 \tilde{\times} S^2)\backslash B^4$

It suffices to show that surgery on the link from Figure 6.15 yields an S^2-bundle over S^2 with a ball missing. Like in Example 6.9, we are plumbing the trivial bundle $S^2 \times B^2$ (corresponding to the link component on the left, as explained in Example 6.6) with the nontrivial B^2 bundle over S^2 from Example 6.7 (corresponding to the link component on the right).

Repeating the argument from Example 6.9, we see that the boundary M $= \partial\mathbf{M}$ is obtained by gluing two solid tori so that $(1,0)$ is identified with $(0,1)$ and $(0,1)$ is identified with $(1,1)$. As the second torus can be twisted so that $(1,1)$ becomes $(1,0)$, we deduce that M $= S^3$.

We glue to $\mathbf{M}$ a 4-dimensional ball B_3^4 along S^3 and and show that the new manifold $\mathbf{M}$ $\cup B_3^4$ is an S^2-bundle over S^2. To this end write $B_3^4 = B_3^2 \times B_3^{2'}$ so that $\partial B_3^2 \times B_3^{2'} \cup B_3^2 \times \partial B_3^{2'}$ is the Heegaard decomposition of S^3 given by the solid tori mentioned above.

The picture now looks as follows: There is the original ball B_0^4 to which two 2-handles, $B_1^4 = B_1^2 \times B_1^{2'}$ and $B_2^4 = B_2^2 \times B_2^{2'}$ are attached. To the boundary of this manifold we attach the ball $B_3^4 = B_3^2 \times B_3^{2'}$. If we glue together B_4^0 and B_1^4 we obtain the trivial disk bundle over the sphere, which we view as a trivial sphere bundle over the disk. If we glue B_2^4 and B_3^4, then, fiberwise, disks are capped with disks so as to produce spheres, and the result is a sphere bundle over the disk.

The manifold $\mathbf{M}$ $\cup B_3^4$ is obtained by gluing these two sphere bundles. The gluing is performed over $S^1 \times S^2$ by identifying $B_0^{2'} \times \partial B_0^2 \cup B_1^2 \times \partial B_1^{2'}$ with $B_2^2 \times \partial B_2^{2'} \cup B_3^2 \times \partial B_3^{2'}$. The identification in the second term of the union preserves the fiber structure because it is the identity map. The identification in the first term is by a homeomorphism of the solid torus that maps $(1,0)$ to $(1,n)$, and this homeomorphism can be isotoped to one that preserves the fiber structure. Hence the result is an S^2-bundle over S^2.

6.2.2 *The signature of a 4-dimensional manifold*

The definition of the signature

Let us recall the definition of the intersection form from Chapter 3, in the particular case of smooth 4-dimensional manifolds.

Let $\mathbf{M}$ be a smooth, connected, compact, oriented, 4-dimensional manifold. Then $H_2(\mathbf{M}, \mathbb{R})$ is finite dimensional and every homology class can be represented by an oriented smooth surface without boundary, possibly disconnected, embedded in the interior of $\mathbf{M}$. This is in particular the case if $\mathbf{M}$ is a 4-dimensional handlebody like those discussed in the previous section. A general position argument in differential topology implies that given two embedded compact surfaces without boundary Σ and Σ' in $\mathbf{M}$, one can perturb them without changing their homology classes so that they intersect transversally at finitely many points.

Let P be an intersection point of Σ and Σ'. Let also $\mathbf{v}_1, \mathbf{v}_2$ be a 2-dimensional frame at P that determines the orientation of Σ and $\mathbf{v}_1', \mathbf{v}_2'$ a 2-dimensional frame at P that determines the orientation of Σ'. Then $\mathbf{v}_1, \mathbf{v}_2, \mathbf{v}_1', \mathbf{v}_2'$ is a frame in the tangent space of $\mathbf{M}$ at P. The **index** of the intersection of P is $+1$ if this frame is a positively oriented, and -1 if it is a negatively oriented. The **algebraic intersection number** $\Sigma \cdot \Sigma'$ is the sum of the indices of all intersection points of Σ and Σ'.

The algebraic intersection number depends only on the homology classes of Σ and Σ' and therefore descends to a bilinear pairing

$$\cdot : H_2(\mathbf{M}, \mathbb{R}) \times H_2(\mathbf{M}, \mathbb{R}) \to \mathbb{R},$$

which is the **intersection form**.

Remark 6.3. An alternative way to define the intersection pairing on $H_2(\mathbf{M}, \mathbb{R})$ is by using Poincaré duality. This allows us to identify $H_2(\mathbf{M}, \mathbb{R})$ with $H^2(\mathbf{M}, \mathbb{R})$. Using the same duality, we also identify and $H^4(\mathbf{M}, \mathbb{R})$ with $H_0(\mathbf{M}, \mathbb{R}) = \mathbb{R}$. Then the bilinear pairing is the cup product

$$\cup : H^2(\mathbf{M}, \mathbb{R}) \times H^2(\mathbf{M}, \mathbb{R}) \to H^4(\mathbf{M}, \mathbb{R}) = \mathbb{R}.$$

Switching to de Rham cohomology, the cup product is the wedge $(\alpha, \beta) \mapsto \alpha \wedge \beta$, or, with the identification $H^4(\mathbf{M}, \mathbb{R}) = \mathbb{R}$ given by integration,

$$(\alpha, \beta) \mapsto \int_M \alpha \wedge \beta.$$

So via Poincaré duality the intersection form in $H_2(M, \mathbb{R})$ becomes the integral of the wedge of two 2-forms in $H^2(M, \mathbb{R})$.

Proposition 6.5. *The matrix of the intersection pairing in* $H_2(\mathbf{M}, \mathbb{R})$ *in a given basis is finite dimensional and symmetric.*

Proof. Because $H_2(\mathbf{M}, \mathbb{R})$ is a finite-dimensional real vector space, the intersection pairing is given by a finite dimensional square matrix A. If Σ and Σ' represent homology classes in $H_2(\mathbf{M}, \mathbb{R})$, and if $\mathbf{v}_1, \mathbf{v}_2$ respectively $\mathbf{v}_1', \mathbf{v}_2'$ are frames giving the orientations of Σ respectively Σ' at an intersection point, then the frames $\mathbf{v}_1, \mathbf{v}_2, \mathbf{v}_1', \mathbf{v}_2'$ and $\mathbf{v}_1', \mathbf{v}_2', \mathbf{v}_1, \mathbf{v}_2$ have the same orientation. Hence

$$[\Sigma] \cdot [\Sigma'] = [\Sigma'] \cdot [\Sigma],$$

proving that the matrix is symmetric. □

We conclude that the matrix of the intersection pairing in a given basis has only real eigenvalues. We define the signature of the intersection form to be the signature of this matrix, namely the difference between the number of positive and the number of negative eigenvalues.

Definition 6.7. Let $\mathbf{M}$ be a smooth, compact, oriented, 4-dimensional manifold. The **signature** of $\mathbf{M}$, denoted $\sigma(\mathbf{M})$, is the signature of the intersection form on $H_2(\mathbf{M}, \mathbb{R})$.

We stress out that the homology groups and hence the signature can be defined for manifolds that do not have a smooth structure, and such manifolds do exist in dimension 4, we just used this particular but more intuitive definition of the intersection form and of the signature. One should observe that the signature as well as the eigenvalues of the matrix of the intersection form are topological invariants of the oriented 4-dimensional manifold $\mathbf{M}$. This is because the eigenvalues are invariant under change of basis. We will use this property in Proposition 6.6 below.

The signature of a 4-dimensional handlebody

For studying the action of the discrete Fourier transform on theta functions we only need the signature of 4-dimensional handlebodies.

Let L be an oriented smooth framed link in S^3 with link components $L_1, L_2, \ldots, L_n$. Let M_L be the 3-dimensional manifold obtained by performing surgery on L and let $\mathbf{M}_L$ be the 4-dimensional handlebody obtained by adding handles to B^4 as specified by L.

Definition 6.8. The **linking matrix** A_L of a link L is the matrix whose jk entry equals $\operatorname{lk}(L_j, L_k)$, where for $j \neq k$, $\operatorname{lk}(L_j, L_k)$ is the linking number or L_j and L_k, while for $j = k$, $\operatorname{lk}(L_j, L_j)$ is the writhe of L_j (i.e. the linking number of L_j and a parallel copy of itself).

Let $\sigma(L)$ be the signature of A_L namely the difference between its number of positive eigenvalues and its number of negative eigenvalues.

Theorem 6.4. *There is a basis of $H_2(\mathbf{M}_L, \mathbb{R})$ such that, in this basis, the matrix of the intersection pairing of $\mathbf{M}_L$ coincides with A_L. Consequently,*

$$\sigma(\mathbf{M}_L) = \sigma(L).$$

Proof. Each handle can be continuously deformed to become a 2-dimensional disk attached to ∂B^4 along its corresponding link component.

The ball B^4 can be contracted to a point. Therefore the handlebody $\mathbf{M}_L$ admits a strong deformation retraction which is a wedge of spheres (also called bouquet of spheres, this is obtained by joining several spheres at one point). The situation is depicted, in one dimension less, in Figure 6.16. So if n is the number of link components of L, then

$$H_2(\mathbf{M}_L, \mathbb{R}) = H_2(\vee_{k=1}^n S^2, \mathbb{R}) = \mathbb{R}^n.$$

Moreover, a basis for $H_2(\mathbf{M}_L, \mathbb{R})$ consists of disks attached to the link components of L and then capped inside B^4. The way we cap these disks is irrelevant, since B^4 is contractible. So we will cap these disks using Seifert surfaces of the link components.

Fig. 6.16 Turning a handlebody into a wedge of spheres

A **Seifert surface** of a knot $K \subset S^3$ is a compact connected oriented surface S_K in S^3 such that $\partial S_K = K$. Every knot admits a Seifert surface, which can be constructed as follows. Orient K and consider one of its knot diagrams. Smoothen each crossing to produce a collection of circles. Cap each circle with a disk so that the disks are disjoint, and orient the disks using the orientation of the knot K (so that when we travel along K, on the "up" side of the disk, the disk is on the left). These disks will be part of S_K.

In the location of each crossing add to S_K a band with one half-twist between two disks, so that the crossing is recovered on the boundary of S_K. The half-twisted band is connected to the disks so as to preserve orientation, so the resulting surface is oriented. The result is the Seifert surface of the knot. The construction of the Seifert surface of the figure-eight knot is shown in Figure 6.17 (the $\pm$ signs represent the orientation).

In the case of the link L, the components are oriented, and we use their orientation to construct the Seifert surface. Then for each link component L_j we have a handle, and that handle gives rise to a basis element of $H_2(\mathbf{M}_L, \mathbb{R})$ which is represented by a compact, connected, oriented surface S_j obtained by attaching a disk to a Seifert surface S_{L_j} along its boundary circle. Given two link components L_j and L_k, the surfaces S_j and S_k are

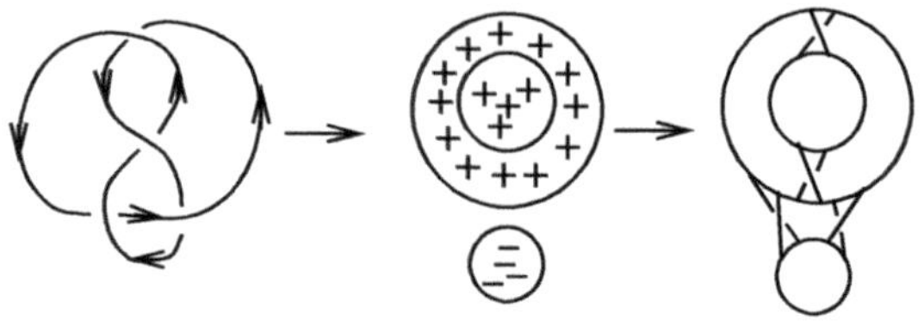

Fig. 6.17 The Seifert surface of the figure-eight knot

not in general position; they might intersect along a subsurface. But we might bring them in general position by pushing the interior of the Seifert surface S_{L_k} inside of B^4, away from the boundary $S^3 = \partial B^4$. Then S_j and S_k intersect where the Seifert surface S_{L_j} intersects L_k, and by perturbing L_k this intersection can be made finite and transverse.

The orientation of S_j at a point on L_j is given by a frame of two vectors, one of which is tangent to L_j in the direction of the orientation of L_j and the other points towards the interior of the Seifert surface. The same is true for S_k. So at an intersection point p of S_j and S_k, the frame that decides the sign of p has the fourth vector always pointing towards the interior of B^4. Hence it is the orientation in S^3 of the frame formed by the first three vectors that decides the sign of p.

For each intersection point p, carve from S_{L_j} a small disk D_p at each intersection point with L_k. Orient ∂D_p so that the disk that was carved out is on the left of it. Using Stokes' Theorem, like in the proof of Theorem 1.1, we have

$$\mathrm{lk}(L_j, L_k) = \frac{1}{4\pi} \sum_{p \in S_{L_j} \cap L_k} \int_{\partial D_p} \int_{L_k} \frac{\mathbf{r_1} - \mathbf{r_2}}{\|\mathbf{r_1} - \mathbf{r_2}\|^3} \cdot \left(\frac{d\mathbf{r_1}}{ds} \times \frac{d\mathbf{r_2}}{dt} \right) dtds.$$

And, by Remark 1.2, each term in the sum is equal to the $+1$ or -1, depending if the 3-dimensional frame discussed above is positively or negatively oriented. We conclude that the algebraic intersection number of the homology classes corresponding to L_j and L_k is precisely $\mathrm{lk}(L_j, L_k)$.

The same considerations apply when $L_j = L_k$, in which case L_k is substituted by a parallel copy of L_j. The theorem is proved. □

Corollary 6.1. *If we add to* L *a trivial link component with framing* $+1$ *or* -1, *then the signature of* $\mathbf{M}_L$ *changes by* $+1$ *respectively* -1.

The fact mentioned in the corollary can also be seen by noticing that when adding a handle to $\mathbf{M}_L$ as specified by the trivial link component with framing $+1$, or -1, we take the connected sum of $\mathbf{M}_L$ with either a positively oriented or a negatively oriented $\mathbb{C}P^2$. There is only one homology class in $H_2(\mathbb{C}P^2)$, which is represented by a surface. This surface intersects a perturbed copy of itself once, positively in the first case and negatively in the second.

Example 6.11. Recall the manifold $\mathbf{M}$ defined by the Hopf link from Figure 6.13. Then depending on the orientation of the two link components, the matrix A_L which defines the intersection pairing in $H_2(\mathbf{M}$$, \mathbb{R})$ is equal to one of the following matrices

$$\begin{pmatrix} 0 & 1 \\ 1 & 0 \end{pmatrix} \text{ or } \begin{pmatrix} 0 & -1 \\ -1 & 0 \end{pmatrix}. \tag{6.1}$$

These matrices have eigenvalues 1 and -1. We conclude that

sign($\mathbf{M}$) $= 0$.

If we glue a ball to the boundary of this manifold, its second homology group, and hence signature does not change. Thus the intersection pairing in $H_2(S^2 \times S^2, \mathbb{R})$ has eigenvalues $+1$ and -1 and

$$\operatorname{sign}(S^2 \times S^2) = 0.$$

Example 6.12. Similarly, let us consider the manifold $\mathbf{M}$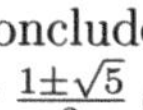 defined by the link from Figure 6.15. Depending on the orientation, the matrix A_L is

$$\begin{pmatrix} 0 & 1 \\ 1 & 1 \end{pmatrix} \text{ or } \begin{pmatrix} 0 & -1 \\ -1 & 1 \end{pmatrix}. \tag{6.2}$$

The eigenvalues are $\frac{1 \pm \sqrt{5}}{2}$ in the first situation and $\frac{-1 \pm \sqrt{5}}{2}$ in the second. Again one is positive and one is negative. As above we conclude that the intersection pairing in $H_2(S^2 \tilde{\times} S^2, \mathbb{R})$ has eigenvalues either $\frac{1 \pm \sqrt{5}}{2}$ or $\frac{-1 \pm \sqrt{5}}{2}$, depending on orientation, and

$$\operatorname{sign}(S^2 \tilde{\times} S^2) = 0.$$

Proposition 6.6. *Up to an equivalence of bundles, there are only two oriented S^2-bundles over S^2, namely $S^2 \times S^2$ and $S^2 \tilde{\times} S^2$.*

Proof. That $S^2 \times S^2$ and $S^2 \tilde{\times} S^2$ are not equivalent bundles follows from the fact that they are not homeomorphic as 4-dimensional manifolds, which is implied by the fact that the eigenvalues of their intersection forms in second homology are different (see Examples 6.11 and 6.12).

Any S^2-bundle over S^2 is obtained by gluing two S^2-bundles over B^2 along $S^1 = \partial B^2$. The two bundles over disks are trivial because the disks are contractible. Hence the information about the resulting bundle is contained in the gluing map $S^1 \to \mathrm{Diff}^+(S^2)$. Since two homotopic maps give the same bundle, the bundles are classified by $\pi_1(\mathrm{Diff}^+(S^2))$. Stephen Smale [Smale (1959)] proved that $\mathrm{Diff}^+(S^2)$ deformation retracts to $SO(3)$. Hence

$$\pi_1(\mathrm{Diff}^+(S^2)) = \pi_1(SO(3)) = \mathbb{Z}_2.$$

This shows that there are only two bundles, and these must be the ones exhibited here. □

6.3 Changing the surgery link

Example 6.2 shows that there are infinitely many framed links in S^3 that are surgery diagrams for S^3 itself, namely all framed links whose link components are (unlinked) trivial knots with framings ± 1. There are many more, and the same is true for all manifolds. In this section we describe Robion Kirby's method [Kirby (1978)] for deciding if two surgery diagrams yield the same 3-dimensional manifold.

6.3.1 *Handle slides*

Given two handles, one can slide one over the other by an isotopy (see Figure 6.18). Here the analogy between surfaces and 3-dimensional manifolds breaks. In the 2-dimensional situation, sliding a 1-handle over another does not change the description of the boundary surface as the result of adding "handles" to S^2. This is due to the fact that the process of 1-handle addition to a 3-dimensional ball is very simple. By contrast, while the sliding of a 2-handle over another does not change the 3-dimensional manifold nor the 4-dimensional manifold that it bounds, it does change the surgery link. Let us see how.

In the case of the 3-dimensional ball with 1-handles attached, a handle slides over another so that one connected component of $B^2 \times \partial B^1$ slides by an isotopy along the boundary $\partial B^2 \times B^1$ of another handle (see Figure 6.18). Let us perform the same motion in the 4-dimensional situation.

Consider L a framed oriented link in S^3 and let $\mathbf{M}_L$ be the 4-dimensional manifold obtained by attaching 2-handles to B^4 as specified by L. Consider two 2-handles $\mathbf{H}_1$ and $\mathbf{H}_2$, each of which is identified with a product of 2-

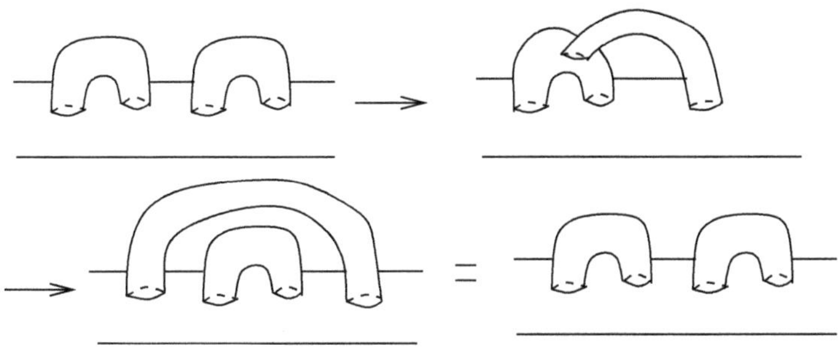

Fig. 6.18 Sliding one handle over another on a surface

dimensional disks as

$$\mathbf{H}_1 = B_1^2 \times B_1^{2'} \text{ and } \mathbf{H}_2 = B_2^2 \times B_2^{2'}.$$

Let L_1 and L_2 be the framed link components that define the addition of these handles.

We want to slide the attaching part of the boundary of $\mathbf{H}_1$, namely $B_1^2 \times \partial B_1^{2'}$, over the attaching part of the boundary of $\mathbf{H}_2$, namely $\partial B_2^2 \times B_2^{2'}$.

In ∂B^4, $B_1^2 \times \partial B_1^{2'}$ is identified with a solid torus S_1 that contains L_1 as an embedded annulus, with one boundary component of L_1 being the core of the torus and the other being a curve of the form $(1, n)$ on the boundary.

Grab S_1 at one point and pull it towards L_2 until it reaches the other handle. This is done rigorously as follows. Choose an embedded rectangle $R = [0,1] \times [0,1] \subset S^3$ that does not intersect any of the components of L except L_1 and L_2. It intersects L_1 over $[0,\epsilon] \times [0,1]$ in such a way that $\{0\} \times [0,1]$ and $\{\epsilon\} \times [0,1]$ coincide with arcs in different boundary components of L_1, and it intersects L_2 over $[1-\epsilon, 1] \times [0,1]$ such that $\{1-\epsilon\} \times [0,1]$ and $\{1\} \times [0,1]$ coincide with arcs in different boundary components of L_2 for some small $\epsilon > 0$. Delete from L_1 the portion that coincides with $[0,\epsilon] \times (\epsilon, 1-\epsilon)$ and and replace it with

$$[0, 1-2\epsilon] \times [0,\epsilon] \cup [1-3\epsilon, 1-2\epsilon] \times [\epsilon, 1-\epsilon] \cup [0, 1-\epsilon] \times [1-\epsilon, 1-2\epsilon].$$

The newly obtained link component, which we call L_1', has a part that runs very close to L_2. Note that L_1 and L_1' are ambient isotopic link components by an ambient isotopy that is supported in a neighborhood of the rectangle R.

Now slide L_1' through $\partial B_2^2 \times B_2^{2'}$. To see how this happens, identify the solid torus $\partial B_2^2 \times B_2^{2'}$ with $S^1 \times B^2$. Recall that when attaching $\mathbf{H}_2$, one of the boundary components of L_2 is identified with $\{1\} \times S^1$. Slide the part of L_1' that is close to L_2 through the handle so that the slide happens in $\{1\} \times B^2$. Then L_1' becomes the band connected sum over R of L_1 with a parallel copy of L_2 defined below.

Definition 6.9. Let K_1 and K_2 be two disjoint framed knots in S^3, viewed as embedded annuli. Let also $R = [0,1] \times [0,1]$ be a rectangle embedded in S^3 whose intersection with K_1 is $[0,\epsilon] \times [0,1]$ such that $\{0\} \times [0,1]$ and $\{\epsilon\} \times [0,1]$ coincide with arcs in different boundary components of K_1 and whose intersection with K_2 is $[1-\epsilon,1] \times [0,1]$ such that $\{1-\epsilon\} \times [0,1]$ and $\{1\} \times [0,1]$ coincide with arcs in different boundary components of K_2 (see Figure 6.19). The **band connected sum over** R of K_1 and K_2, denoted by $K_1 \#_R K_2$, is the knot obtained by deleting from $K_1 \cup K_2 \cup R$ the portion of R that coincides with $[1,0] \times (\epsilon, 1-\epsilon)$.

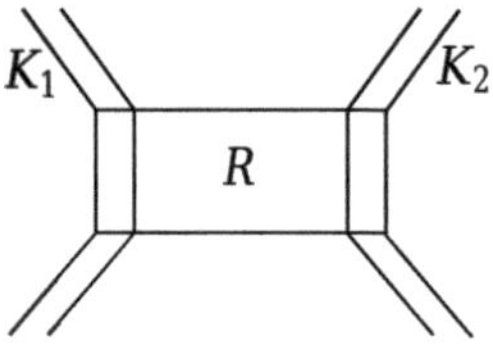

Fig. 6.19 Position of the band with respect to the framed knots

Example 6.13. An example of the band sum of two trefoil knots is shown in Figure 6.20.

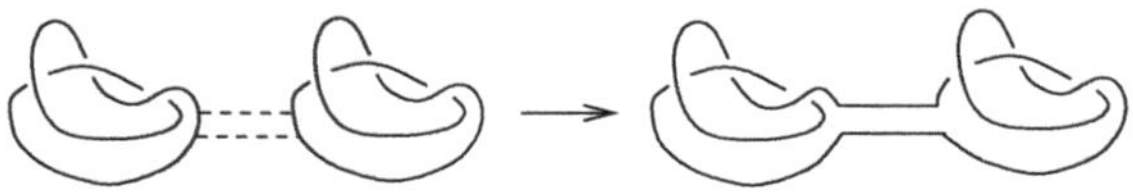

Fig. 6.20 Band sum of two knots.

Definition 6.10. Let L be a framed link and let L_1 and L_2 be two of its components. The **slide** of L_1 over L_2 is the band connected sum of L_1 with a parallel copy of L_2, determined by a band that does not intersect L_2 or any other link component.

Figure 6.21 depicts the slide of a trefoil (on the left) over a figure eight knot (on the right). Both link components are shown in the blackboard framing.

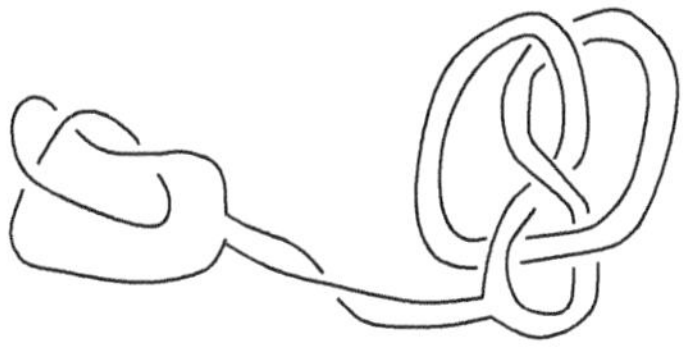

Fig. 6.21 Slide of a trefoil over a figure eight knot

Sliding of a handle over the complex projective plane

The slide of a knot over a trivial knot with framing +1 is shown in Figure 6.22. In general, we can slide simultaneously several link components by making them run parallel to each other and very close to the side $\{0\} \times [0, 1]$ of the rectangle R, and then sliding them over as if they were just one component (Figure 6.23).

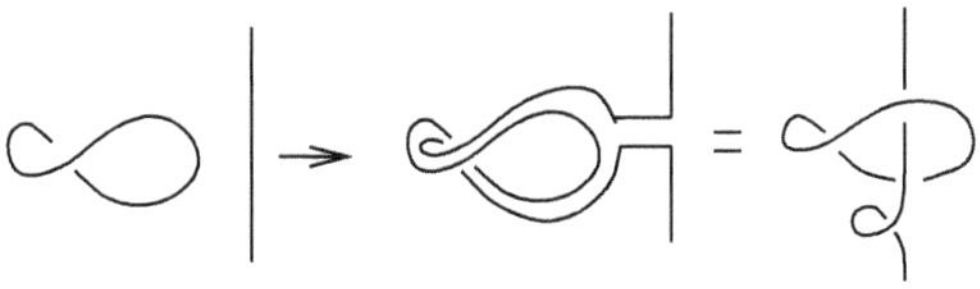

Fig. 6.22 Slide of a knot over a trivial knot with framing +1

Fig. 6.23 Sliding several link components over a trivial knot

An ambient isotopy allows us to present the result of simultaneously sliding several link components of a trivial knot with framing +1 also as shown on the left of Figure 6.24, while on the right of this figure it is shown the result of sliding over a trivial knot with framing −1. Note that by performing such a simultaneous slide we can entangle or untangle link components.

Definition 6.11. We will call the operation of sliding a link component over a trivial link component with framing ±1 a **(k0) move**.

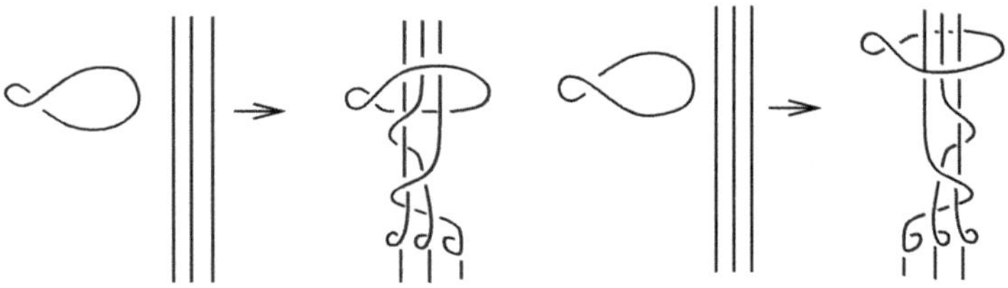

Fig. 6.24 Sliding several link components over a trivial knot

In Example 6.7 and Example 6.8, it was explained that a trivial link component with framing ± 1 corresponds to a copy of $\pm\mathbb{C}P^2$. Thus at the 4-dimensional level, Figure 6.23 and Figure 6.24 show how to slide handles over a $\pm\mathbb{C}P^2$. These handle slides can be performed back and forth.

Example 6.14. Let us revisit the manifold $\mathbf{M}$ from Example 6.9, obtained by adding handles as prescribed by the Hopf link. If we take the connected sum of this manifold with $\mathbb{C}P^2$, and perform handle slides as shown in Figure 6.25, we deduce that

$\mathbf{M}$ $$\#\mathbb{C}P^2 = (\mathbb{C}P^2\#\mathbb{C}P^2\#(-\mathbb{C}P^2))\backslash B^4.$$

Adding the missing ball we obtain that

$$(S^2 \times S^2)\#\mathbb{C}P^2 = \mathbb{C}P^2\#\mathbb{C}P^2\#(-\mathbb{C}P^2).$$

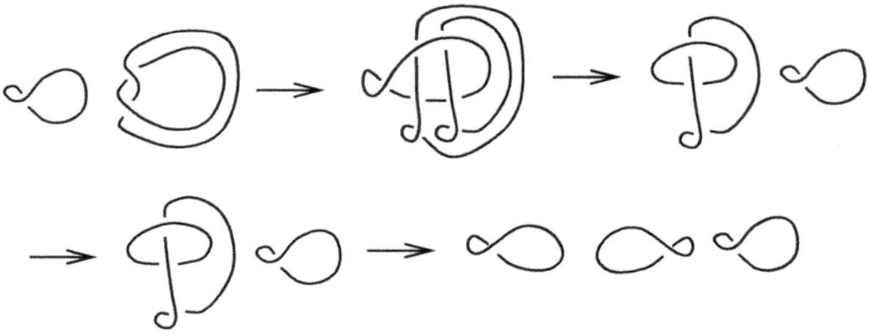

Fig. 6.25 Adding $\mathbb{C}P^2$ to $S^2 \times S^2$

Example 6.15. Similarly, for the handlebody $\mathbf{M}$ from Example 6.10 which was defined by the link in Figure 6.15, if we perform a (k0) move of the link as in Figure 6.26 we obtain a link with two components, one which is the trivial knot with framing $+1$ and one which is the trivial knot with framing -1. Hence

$\mathbf{M}$  $= (\mathbb{C}P^2\#(-\mathbb{C}P^2))\backslash B^4$

and consequently

$$S^2\tilde{\times}S^2 = \mathbb{C}P^2\#(-\mathbb{C}P^2).$$

Fig. 6.26 $S^2\tilde{\times}S^2 = \mathbb{C}P^2\#(-\mathbb{C}P^2)$

6.3.2 *Kirby's theorem*

We are now in position to state and prove the result from [Kirby (1978)] which specifies when two surgeries yield the same manifold.

Theorem 6.5. *(Kirby) Given two framed links L_1 and L_2 in S^3, there is an orientation preserving homeomorphism $M_{L_1} \to M_{L_2}$ if and only if L_1 and L_2 can be changed into one another by a sequence of the following two operations:*

(k1) *add or subtract from the link an unknotted circle with framing ± 1, which is separated from the other link components by an embedded 2-dimensional sphere in S^3;*

(k2) *given two link components, replace one of them by the result of sliding it over the other link component.*

Definition 6.12. The operations (k1) and (k2) are called Kirby moves.

Remark 6.4. The operation (k0) from the previous section is also known as Kirby move, it is in fact a particular case of the (k2) move.

Proof. From Example 6.2 and the above discussion on handle slides, it follows that if two framed links are transformed into one another by a sequence of the Kirby moves (k1) and (k2), then they give rise by surgery to homeomorphic 3-dimensional manifolds.

The proof of the converse, presented below, follows closely [Kirby (1978)]. It is based on both Morse theory and Cerf theory, which were reviewed in Chapter 3.

We start with two framed links L_1 and L_2 that yield homeomorphic 3-dimensional manifolds M_{L_1} and M_{L_2}. We may assume that the manifolds are smooth and that there is a diffeomorphism h between them. Consider smooth 4-dimensional handlebodies $\mathbf{M}_{L_1}$ and $\mathbf{M}_{L_2}$ with boundaries M_{L_1} and M_{L_2}. We will glue the 4-dimensional manifolds $\mathbf{M}_{L_1}$ and $-\mathbf{M}_{L_2}$ ($\mathbf{M}_{L_2}$ with orientation reversed) via the mapping cylinder $\mathbf{I}_h$ of h. Because h is a homeomorphism, I_h is homeomorphic to $M_{L_1} \times [1,2]$. We glue $\mathbf{I}_h$ to $\mathbf{M}_{L_1} \cup (-\mathbf{M}_{L_2})$ so that the bottom $M_{L_1} \times \{1\}$ of the mapping cylinder is identified via the identity map with $\partial \mathbf{M}_{L_1} = M_{L_1}$, and the top $M_{L_1} \times \{2\}$ of the mapping cylinder is identified via the identity map with $\partial(-\mathbf{M}_{L_2}) = -M_{L_2}$.

Let $\mathbf{N} = \mathbf{M}_{L_1} \cup I_h \cup (-\mathbf{M}_{L_2})$ be the resulting 4-dimensional manifold. Note that $\mathbf{N}$ is homeomorphic to the gluing of $\mathbf{M}_{L_1}$ and $-\mathbf{M}_{L_2}$ via h, but it is because we insert the mapping cylinder in between that we are able to put a smooth structure on $\mathbf{N}$. Changing L_1 by applying repeatedly (k1) amounts to adding to $\mathbf{M}_{L_1}$, and hence to $\mathbf{N}$, copies of $\pm\mathbb{C}P^2$. So by using (k1) several times we can make sure that $\text{sign}(\mathbf{N}) = 0$. We now use the following result [Rokhlin (1952)] (see [Melvin (1984)] for a proof).

Theorem 6.6. *(Rohlin) A smooth oriented compact 4-dimensional manifold bounds a smooth oriented compact 5-dimensional manifold if and only if its signature is equal to zero.*

Let W^5 be a smooth 5-dimensional manifold such that $\partial W^5 = \mathbf{N}$, whose existence is guaranteed by Rohlin's theorem.

The Morse theoretical part of the proof

Consider a Morse function

$$f : W^5 \to [1,2]$$

with the property that $f^{-1}(i) = \mathbf{M}_{L_i}$, $i = 1,2$, and $f|I_h \approx M_{L_1} \times [1,2]$ is the projection onto the second coordinate. Recall the Fundamental Theorem of Morse theory (Theorem 3.6), which relates critical points to handles.

By pushing the critical points of index 0 and 5 along ascending and descending manifolds respectively, we can cancel each of them, in a birth/death singularity, with another critical point. Thus we may assume that there are only critical points of index $1,2,3,4$.

Let p be a critical point of index k. Set $q = f(p)$ and choose $\epsilon > 0$ such that p is the only critical point in $f^{-1}([q-\epsilon, q+\epsilon])$. Then $f^{-1}([1, q+\epsilon])$ is obtained from $f^{-1}([1, q-\epsilon])$ by adding a 1-handle $B^k \times B^{5-k}$. View the handle addition as taking the connected sum with $S^k \times B^{5-k}$ (see Figure 6.27). Then on the boundary,

$$f^{-1}(q+\epsilon) = f^{-1}(q-\epsilon)\#(S^k \times S^{4-k}),$$

where $S^{4-k} = \partial B^{5-k}$. Because $S^k \times S^{4-k}$ is homeomorphic to $S^{4-k} \times S^k$, we can also produce the same 4-dimensional manifold $f^{-1}(q+\epsilon)$ from $f^{-1}(q-\epsilon)$ by attaching a $5-k$ handle to the 5-dimensional manifold $f^{-1}([1, q-\epsilon])$. Since we are only interested in the manifolds $\mathbf{M}_{L_1}$ and $\mathbf{M}_{L_2}$, we can replace the manifold W^5 by another in which only 2- and 3-handles show up.

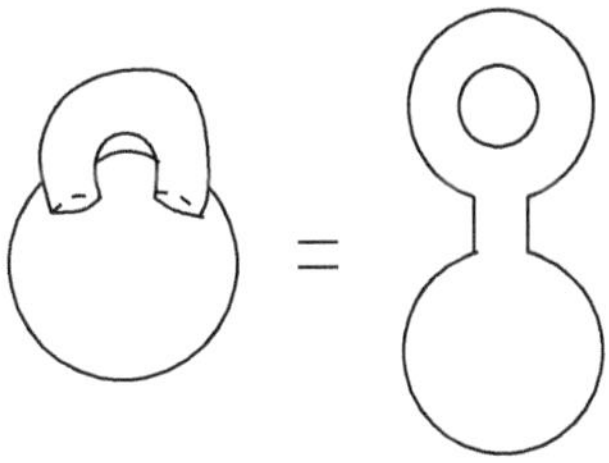

Fig. 6.27 Handle addition as a connected sum

The manifold $\mathbf{M}_{L_1}$ is simply connected. When passing over a point of index 2 or 3 we add a 2- or 3-handle to W^5. As a consequence of the Seifert-van Kampen Theorem, the level sets $f^{-1}(y)$ stay simply connected. Let us examine what happens when we add a 2-handle $B^2 \times B^3$ to a 5-dimensional manifold X with simply connected boundary. We are gluing $\partial B^2 \times B^3$ to ∂X so that ∂X changes by removing an $S^1 \times B^3$ and adding a $B^2 \times \partial B^3 = B^2 \times S^2$. In ∂X, $\partial B^2 = S^1$ lies in a disk. Using this disk we can cap the boundary of the handle $B^2 \times S^2$ with another $B^2 \times S^2$, to obtain an S^2-bundle over S^2.

We deduce that the addition of a 2-handle to W^5 gives rise on the boundary to taking the connected sum with an S^2-bundle over S^2. By Proposition 6.6, there are only two such bundles, $S^2 \times S^2$ and $S^2 \tilde{\times} S^2$. So when passing from one level set to another over an index 2 critical point, we take the connected sum with $S^2 \times S^2$ or $S^2 \tilde{\times} S^2$. In view of Example 6.14 and Example 6.15, on the 3-dimensional manifold which is the boundary of the level set one can achieve these by performing the Kirby moves (k1) and (k2).

For critical points of index 3 we go backwards. If the 5-dimensional manifold X is obtained from the 5-dimensional manifold X' by adding a 3-handle, then X' is obtained from X by removing a 3-handle. The removal of a 3-handle can be achieved by adding a 2-handle, as shown schematically in Figure 6.28. Kirby moves are reversible, so the change on the boundary of level sets can again be achieved using Kirby moves.

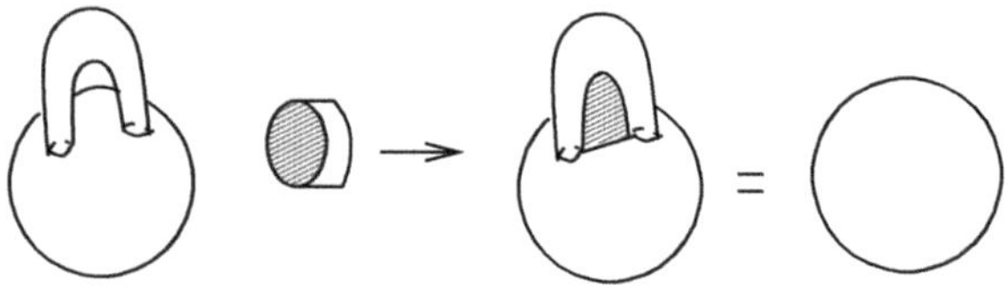

Fig. 6.28 Removal of a handle

We conclude that by performing Kirby moves we can bring M_{L_1} and M_{L_2} to a situation in which W^5 has no critical points, so it is just the cylinder over $\mathbf{M}_{L_1}$. Thus we may assume that $\mathbf{M}_{L_1}$ and $\mathbf{M}_{L_2}$ are one and the same manifold.

The Cerf theoretical part of the proof

Now we are in the situation of a 4-dimensional manifold $\mathbf{M}$ on which the links L_1 and L_2 give two different handlebody structures. For $j = 1$ or 2, the handlebody structure given by L_j is obtained by adding 2-handles $B^2 \times B^2$ to the ball B^4. Each handle can be modeled as the neighborhood of an index 2 critical point of a Morse function, and so the two handle decompositions are given by Morse functions

$$f_j : \mathbf{M} \to [-1, 1],$$

with $f_j^{-1}(-1)$ the only critical point of index 0, no critical points of index 1, and all critical points of index 2 in $f_j^{-1}((0,1))$. Then $f_j^{-1}(0) = S^3$ and $f_j^{-1}(1) = \partial\mathbf{M} = M_{L_1} = M_{L_2}$. Since there are no interesting phenomena below $f_j^{-1}(0)$, we only focus on the functions $f_j|f_j^{-1}([0,1])$, namely on

$$f_j : \mathbf{M}\backslash B^4 \to [0, 1].$$

For easy reference, we split the boundary of $\mathbf{M}$ into $\partial_-\mathbf{M} = S^3$ and $\partial_+\mathbf{M} = M_{L_1} = M_{L_2}$.

Recall the definition of the strata $\mathcal{F}^0$ and $\mathcal{F}^1$ from Chapter 3. Using Cerf's Theorem (Theorem 3.8) we connect f_1 and f_2 by a path $F : [0,1] \to \mathcal{F}^0 \cup \mathcal{F}^1$, which traverses $\mathcal{F}^1$ at finitely many points.

If $F(t) \in \mathcal{F}^0$ then $f_t = F(t)$ is a Morse function, which can have critical points of index $0, 1, 2, 3$, or 4. We will show that we can avoid critical points of index $1, 3$, and 4, and so that we only have 1 critical point of index 0 and several critical points of index 2. This situation corresponds to adding 2-handles to a ball, and hence is described by a link. Thus for all but finitely many t, $F(t)$ gives a handlebody structure described by a link. This link will not change as long as we move in $\mathcal{F}^0$. We will then show that crossing $\mathcal{F}^1$ amounts to performing Kirby moves. At this point we recall Lemma 3.6, Lemma 3.7, and Lemma 3.8 from Chapter 3.

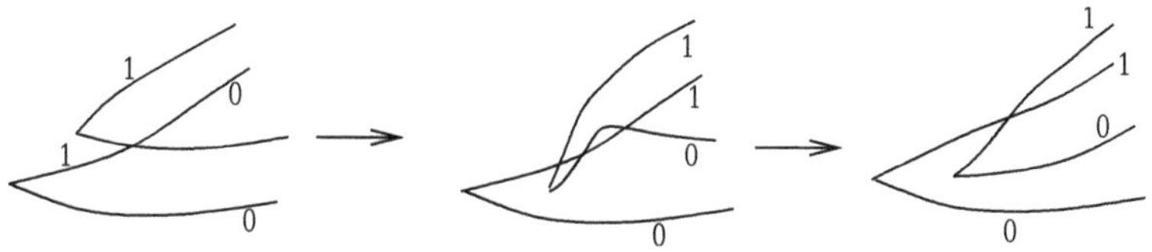

Fig. 6.29 Arranging the index 0 critical points

Step 1. We proceed to change $F(t)$ so that its new graphic has no critical points of index 0 and 4, and equals the old graphic when t is close to 0 and to 1. First, by using repeatedly the Beak Lemma and the Independent Singularities Lemma as shown in Figure 6.29, we can make a neighborhood of the index 0 critical points in the graphic to look as shown in Figure 6.30.

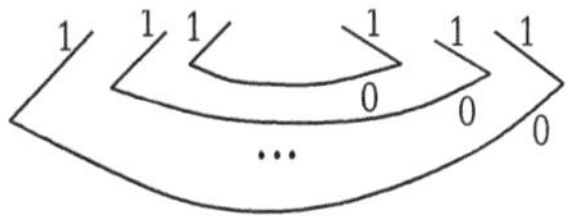

Fig. 6.30 Position of index 0 critical points

Next, we remove index 0 critical points one by one, starting with the innermost in the configuration from Figure 6.30. Using the Independent Singularities Lemma, we move all other critical points away from this configuration. We arrive at two possible configurations:

- we start with an index 1 critical point, a pair of index 0 and index 1 critical points is born, the index 0 point is canceled with the original index 1 point, and one index 1 point is left, or
- one of the index 1 points is canceled by an index 2 point.

Let us analyze the first case. The fortunate situation is where the two index 1 critical points cross so as to form a swallow tail, and then we can make the index 0 critical point disappear using the Swallow Tail Lemma. But it might happen that the two critical points cross back and forth, which would prevent the swallow tail to form. A typical situation is shown in Figure 6.31. If the descending manifold of the higher index 1 critical point does never intersect the ascending manifold of the lower index 1 critical point, there is no difficulty in removing the bigons: just move one critical point up and the other down.

The only problem is when these manifolds intersect. Each horizontal slice (which has dimension 3) is intersected by the ascending manifold in a dimension 2 set and the descending manifold in a dimension 0 set, and the two are generically disjoint. A general position argument allows us to change the path $F(t)$ in a situation in which the descending manifold of an index 1 point intersects the ascending manifold of an index 1 point only for finitely many t.

Following [Kirby (1978)], we mark this situation on the graphic by drawing a vertical segment between the curves that depict the two critical points. In the handle language, the vertical lines depict a situation where one handle is sliding over another, and we choose the path so that only finitely many handle slides occur.

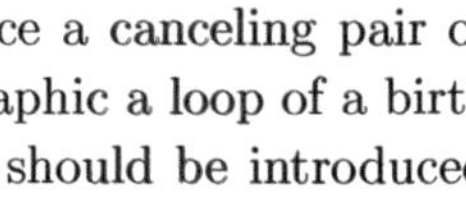

Fig. 6.31 Index 1 critical points crossing back and forth

The trick is to introduce a canceling pair of critical points of index 1 and 2, producing in the graphic a loop of a birth followed by a death. The new index 1 critical point should be introduced near the critical point of index 0 in such a way that its descending manifold intersects the ascending manifold of the index 0 critical point in exactly one point.

Now we perform the sequence of operations represented in Figure 6.32. At the first step we cancel the index 0 critical point with the newly introduced index 1 critical point. Then we apply the Triangle Lemma several times. Finally we use the Swallow Tail Lemma to get rid of the index 0 critical point.

In the second case we can simply remove the index 0 point as shown in Figure 6.33. Here on the first step we used the Independent Singularity

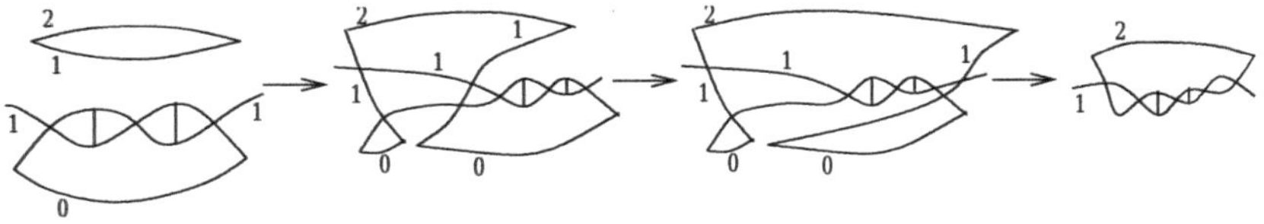

Fig. 6.32 Elimination of an index 0 critical point

Lemma, on the second step we used the Beak Lemma, and on the third we used the Swallow Tail Lemma.

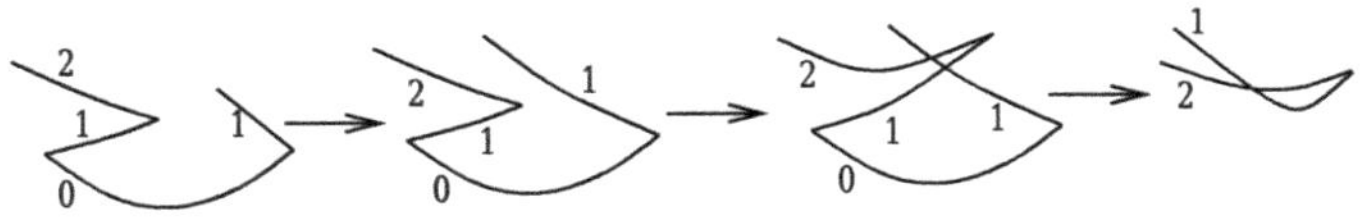

Fig. 6.33 Elimination of an index 0 critical point

The same method turned upside down eliminates all index 4 points. So now we have no 0- or 4-handles.

Step 2. We proceed to eliminate all index 1 and index 3 critical points, thus removing all 1- and 3-handles. The idea is to replace each such critical point by an index 2 critical point. We focus on index 1 critical points since the index 3 critical points can be turned into index 1 points by changing f to $-f$, and then we can eliminate them by the same procedure.

Exactly how in the previous situation the index 0 points are born and removed using index 1 points, the index 1 points are born and removed using index 2 points. By applying the Beak Lemma, we can arrange these birth/death situations like in Figure 6.34. Here the situation is more complicated than before: index 0 critical points had no descending manifolds, so any bigons that their level curves formed could be removed. But index 1 critical points do have descending manifolds, so in the graphic we could have many vertical lines between the level curves of index 1 critical points.

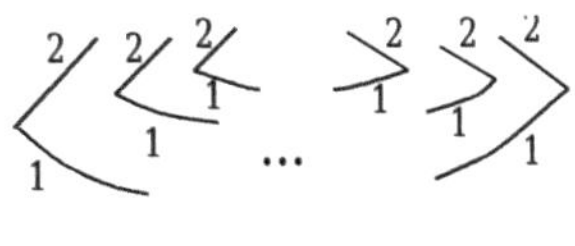

Fig. 6.34 Position of index 1 critical points

Nevertheless, we can perform the move from Figure 6.35. There we first change the birth point slightly so that the descending and ascending manifolds no longer intersect. In graphical language, we move the vertical bar to the left until it disappears. Then we use the Beak Lemma and remove a bigon to interchange the beak points (the one that was first on the right moved to the left).

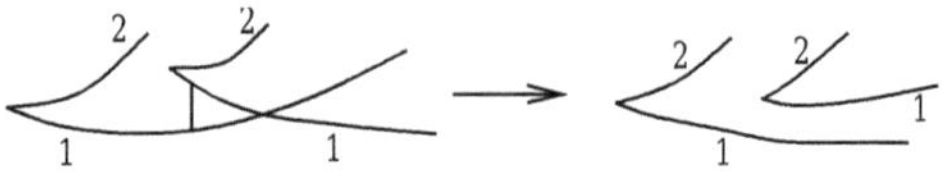

Fig. 6.35 Interchanging two beaks

After applying this move a number of times, we arrive at the configuration from Figure 6.36.

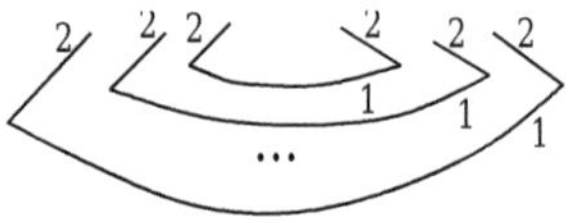

Fig. 6.36 Position of index 1 critical points

Let us concentrate on the lowest index 1 critical point, p_{low}, and let us look first only at the region of the graphic between the birth and death point. Denote by D_t the descending 1-dimensional manifold of p_{low} at time t. Because below p_{low} there are no index 1 or index 0 critical points, using a general position argument we can make sure that D_t avoids all ascending manifolds of critical points, and thus ends on $\partial_-\mathbf{M} = S^3$. So D_t is an arc with boundary points in $\partial_-\mathbf{M}$. Connect the end points of D_t by an arc E_t in a regular neighborhood of $\partial_-\mathbf{M}$ to obtain a circle S_t^1.

Fix t and consider a triangulation of $\mathbf{M}$ so that S_t^1 is part of the 1-skeleton. The handlebody $\mathbf{M}$ is simply connected (Proposition 6.4), so the tangent bundle $T\mathbf{M}$ of $\mathbf{M}$ has a unique trivialization over the 1-skeleton compatible with the orientation of $\mathbf{M}$. The restriction of $T\mathbf{M}$ to S_t^1 is therefore trivial, and the trivialization of $T\mathbf{M}$ gives a trivialization of the normal bundle of S_t^1, which identifies this bundle with $S_t^1 \times B^3$. The boundary of $S_t^1 \times B^3$ is of the form $S^1 \times S^2$. Remove $S_t^1 \times B^3$ from $\mathbf{M}$ and glue in an $B^2 \times S^2$ along $\partial B^2 \times S^2 = S^1 \times S^2$. As explained in the Morse theoretical part of the proof, the $B^2 \times S^2$ corresponds to a 2-handle, so when replacing $S_t^1 \times B^3$ by $B^2 \times S^2$, we replaced an index 1 critical point p_{low} by an index 2 critical point. And in the process $\mathbf{M}$ was changed to $\mathbf{M}\#(S^2 \times S^2)$.

We want to show that the new situation corresponds to a path $F(t)$, so surgery must be continuous with respect to t. Let t_0 be the time of the birth of p_{low} and choose a reference time t_r between the birth and death and very close to t_0. Consider the index 2 critical point which cancels p_{low}. Its descending 2-manifold is a disk. Part of the boundary of this disk is D_{t_r}, and we can let the rest of its boundary be E_t. Some care needs to be taken on $\partial_-\mathbf{M}$, but a regular neighborhood of this boundary is of the form $S^3 \times [0,1]$, and we can choose E_{t_r} to lie on the top part of it, so that we can still take a tubular neighborhood of $D_{t_r} \cup E_{t_r} = S^1_{t_r}$ inside $\mathbf{M}$. When t varies back and forth from t_r, the descending manifold of the index 2 critical point varies by a smooth ambient isotopy. Thus the regular neighborhood $S^1_t \times B^3$ varies smoothly, and we can make sure that the attaching of $B^2 \times S^2$ is done so that it varies smoothly with t. Thus the surgery varies smoothly between the birth and the death point. Let us check what happens at the birth and death points of p_{low}:

The birth point. Because there are no index 4 critical points, the intersection of a level set of the function $F(t_0)$ with a descending manifold is of dimension at most 2. So by transversality we can make sure that there is $q \in \partial_-\mathbf{M}$ such that there is a ball B^3_q centered at q that does not intersect descending manifolds except those of p_{low} and of the index 2 critical point that cancels it. Then there is a column $B^3_q \times [0,1] \in \mathbf{M}$ which, at time t_0, flows to B^3_q, with $B^3_q \times \{0\} = B^3_q \in \partial_-\mathbf{M}$ and $B^3_q \times \{1\} \in \partial_+\mathbf{M}$. Because the reference point t_r where we made our first choice of S^1_t is close to t_0, we can make sure that, for $t > t_0$ and close to t_0, the $S^1_t \times B^3$ that we replace lies in this column. So at $t = t_0$, we change the surgery link by adding the Hopf link (with blackboard framing) from Figure 6.13 separated from the rest of the link by ∂B^3_q. As a corollary of the discussion in Example 6.14 we conclude that the birth point can be handled by Kirby moves.

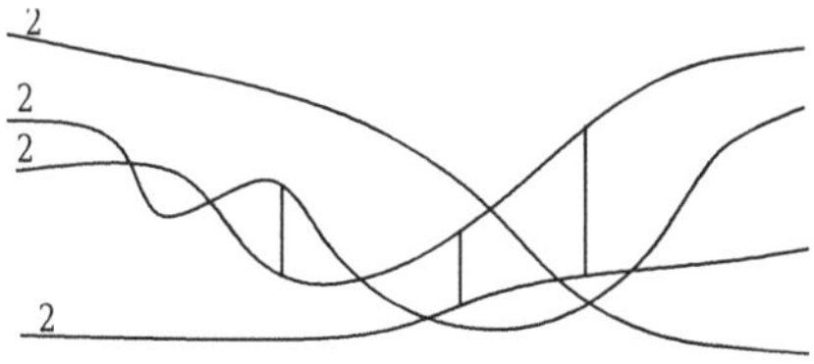

Fig. 6.37 Graphic containing index 2 critical points only

The death point. Let t_1 be the time of the death point. We want to use the same type of column for the death point. The difficulty lies in the fact that the reference point t_r where we pick our first S^1_t is close to t_0 but not to t_1, so we are not guaranteed that S_t^1 lies in such a column when t is close to t_1. This is because we are forced to vary S_t^1 smoothly. But $\partial_-\mathbf{M} = S^3$, so there is an isotopy carrying E_t, when t is closed to t_1 inside such a column. So the death point can be modeled by the removal of a the Hopf link from Figure 6.13 separated from the rest of the link by a sphere. And again by Example 6.14 this can be accomplished using Kirby moves.

Repeating this procedure we can eliminate the index 1 critical points one-by-one, and consequently we can eliminate the index 3 critical points as well.

We are now in position to complete the proof of the theorem. There are only index 2 critical points, so the graphic of $F(t)$ looks like in Figure 6.37. For all but finitely many t, $f_t = F(t)$ is a Morse function. A critical point of index 2 has a descending B^2 which intersects $S^3 = \partial_-\mathbf{M}$ along a circle - the link component of the 2-handle defined by that critical point. As t varies from 1 to 2, the evolution of the descending 2-disks yields an isotopy from the framed link L_1 to the framed link L_2, except for the following possibility: For some t, the descending 2-disk of f_t may not reach S^3 but hit instead a critical point with smaller critical value (according to our notation convention, this is depicted by a vertical line in the graphic from Figure 6.37). This happens exactly when a 2-handle is being slid over another 2-handle. This corresponds to the move (k2). The theorem is proved. □

6.4 Surgery for 3-dimensional manifolds with boundary

Next we turn our attention to 3-dimensional manifolds with boundary, and derive results analogous to those from the previous sections for this situation.

6.4.1 *A relative version of Kirby's theorem*

We will describe a relative version of Kirby's theorem. First, let us introduce the notion of a ribbon graph.

Definition 6.13. A **ribbon graph** in a 3-dimensional manifold M is an oriented surface that is the disjoint union of annuli (loops) and several connected surfaces, each of these connected surfaces being obtained by gluing finitely many rectangles (edges) to finitely many disks (vertices) where an edge $[a, b] \times [0, 1]$ is glued to the boundary of one or two disks such that $\{a\} \times [0, 1]$ and $\{b\} \times [0, 1]$ is identified each with an arc.

In plain language, a ribbon graph is a graph with fattened edges (so as to keep track of the order of the edges at a vertex and of the twistings of edges). As specified in the definition, we require ribbon graphs to be oriented surfaces, but we point out that other authors allow nonorientable surfaces as well. We are only interested in ribbon graphs with trivalent vertices. An example of such a ribbon graph, embedded in $\mathbb{R}^3$, is shown in Figure 6.38. Viewing the ribbon graph as having a deformation retract which is an actual graph, and orienting the latter, we can talk about **oriented ribbon graphs**. From now on, in our drawings we represent ribbon graphs as actual graphs, being understood that the actual surface is a regular neighborhood of the graph in the plane of the paper.

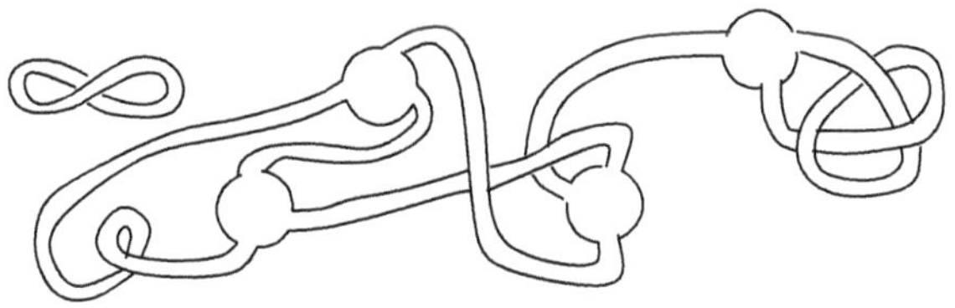

Fig. 6.38 Ribbon graph

We now introduce a particular family of ribbon graphs. For each $g \geq 1$, consider the graph in the xy-plane obtained as the union of the circles $(x - j)^2 + y^2 = 1/9$, $j = 1, 2, \ldots, g$, and the segments $j + 1/3 \leq x \leq j+2/3, y = 0$, $j = 1, 2, \ldots, g-1$. Orient each of the circles counterclockwise and of the linear segments in the positive direction, so that now the graph is oriented. We define the ribbon graph $\mathcal{G}_g$ to be a regular neighborhood of this oriented graph in the plane. The graph $\mathcal{G}_3$ is shown in Figure 6.39 (the arrows of the orientation are not shown to keep the figure simple). Note that if we consider a regular neighborhood of $\mathcal{G}_g$ in the 3-dimensional space, then we obtain the genus g handlebody in which the g oriented circles determine a canonical basis. For this reason we call $\mathcal{G}_g$ the **graph associated to the genus** g **handlebody**. The unframed graph consisting of the circles and segments will be called the **core** of $\mathcal{G}_g$.

If $g = 0$, then the graph $\mathcal{G}_0$ is just a disk and its core is a point. This is the graph associated to B^3, which is the genus 0 handlebody. There is no need to worry about framing, in this case it carries no information. If

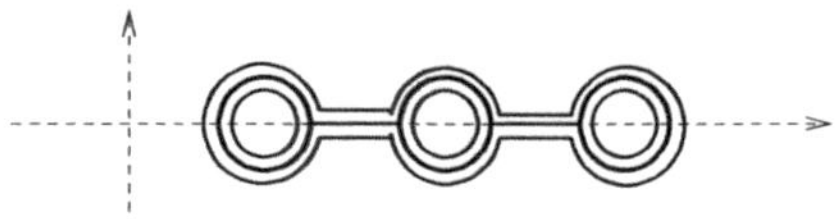

Fig. 6.39 The graph $\mathcal{G}_3$

$g = 1$, then $\mathcal{G}_1$ is the trivial framed knot.

Definition 6.14. A **handlebody graph** in a smooth 3-dimensional manifold M is a smooth embedding of the disjoint union of $\mathcal{G}_{g_1}, \mathcal{G}_{g_2}, \ldots, \mathcal{G}_{g_k}$, for some $n \geq 1$ and $g_1, g_2, \ldots, g_k \geq 0$.

The **core** of the handlebody graph is the image through the embedding of the cores of the $\mathcal{G}_j$.

If L is a framed link in S^3, and Γ is a handlebody graph in S^3 disjoint from L, then Γ survives surgery along L and gives rise to a handlebody graph in M_L. We denote this graph by Γ_L.

Given a framed knot K and a handlebody graph Γ in S^3, a **slide** of Γ over S^3 can be defined in the same way as the slide of a knot over another knot. In this case one picks an edge of Γ and then slides it over K, along a band that connects the edge to K. The result is still a handlebody graph. The slide depends both on the edge that was chosen and on the band that connects it to K.

Proposition 6.7. *Given two framed links L_1 and L_2 in S^3, and two handlebody graphs Γ, Γ' in S^3, with Γ disjoint from L_1 and Γ' disjoint from L_2, there is an orientation preserving homeomorphism $M_{L_1} \to M_{L_2}$ that maps Γ_{L_1} to Γ'_{L_2} if and only if the pairs (L_1, Γ) and (L_2, Γ') can be changed into one another by isotopies and finitely many of the following three operations:*

- **(k1)** *add or subtract from the link an unknotted circle with framing ± 1, which is separated from the other link components by an embedded 2-dimensional sphere in S^3;*
- **(k2)** *given two link components, replace one of them by the result of sliding it over the other link component;*
- **(k3)** *replace the trivalent ribbon graph by its slide over a link component.*

Proof. Kirby moves (k1) and (k2) change the manifold to one homeomorphic to it. The move (k3) arises from an isotopy of the manifold in which the handlebody graph slides through a surgery torus. Hence if the pairs can be changed into one another by the three moves, then the desired homeomorphism exists.

For the converse we need to re-examine the proof of Kirby's theorem. Let $h : M_{L_1} \to M_{L_2}$ be the homeomorphism that maps Γ_{L_1} to Γ'_{L_2}. It is possible that h does not arise from an isotopy, so we cannot simply apply Kirby's theorem to obtain the same surgery diagram and then apply (k3) repeatedly to slide one graph to the other. Instead we do the Morse theoretical part of the proof using the particular homeomorphism h that maps Γ_{L_1} to Γ'_{L_2}.

First, we make sure that the additions of $\pm\mathbb{C}P^2$ that change the signature are done so that the surgery trivial link components lie in balls that do not intersect Γ_{L_1}. It is harder to see why we can avoid the situation that the link components entangle $\Gamma_{L_1} \times \{\alpha\} \in I_h$ when we add a 5-dimensional 2-handle to W. This can nevertheless be done by using an isotopy to move away the graph from the area where we attach the handle, as sketched in Figure 6.40. We can also avoid the handle slides that come with the attachment of the $S^2 \times S^2$ and $S^2 \tilde{\times} S^2$.

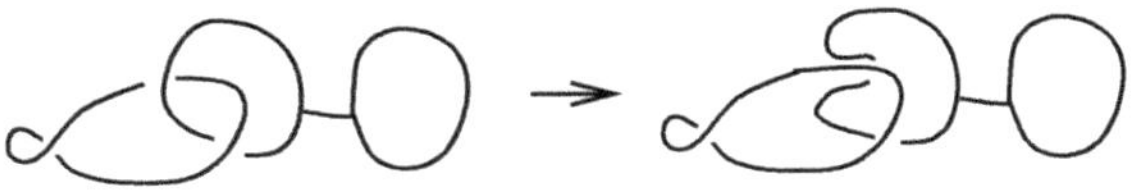

Fig. 6.40 Avoiding a handle

So at the end of the Morse theoretical part of the proof we end up with $\mathbf{M}_{L_1} \cup I_{h_0} \cup \mathbf{M}_{L_2}$, where now the mapping cylinder I_{h_0} is isotopic to the identity. Perform the isotopy, which amounts to using (k3) several times, to obtain the same 4-dimensional manifold with the same embedded graph in its boundary, but with two different Morse functions that yield two different handle-decompositions.

Next we pass to the Cerf theoretical part of the proof. All $\pm\mathbb{C}P^2$ additions and all handle-slides that appear in the process of changing one Morse function into the other can be done away from the graph. The only time when the graph might interfere with the process is when we move the Morse functions inside the stratum $\mathcal{F}^0$ of the space of Morse functions, in order to move the graph away from the events. In the end we might not arrive at the same pair (link,graph), like in the situation depicted in Figure 6.41. But again the two configurations differ by an isotopy of the 4-dimensional manifold, so at the level of surgery diagrams they can be changed one into the other by a sequence of (k3) moves. This completes the proof.

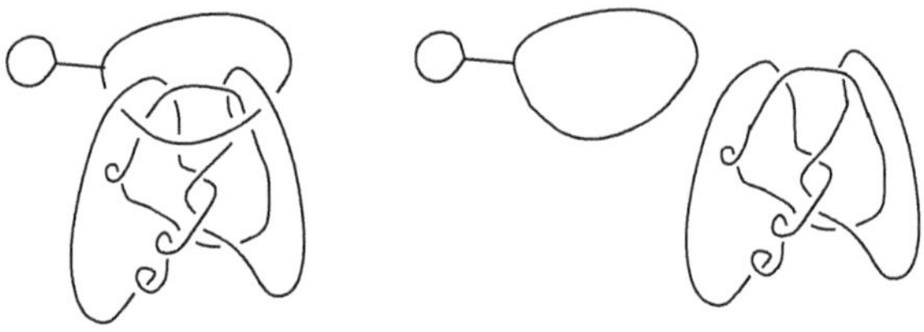

Fig. 6.41 Two different pairs (link,graph) with the same Morse function

□

It is time to turn to manifolds with boundary. The Lickorish-Wallace theorem, which states that a 3-dimensional manifolds without boundary can be obtained by surgery on a link in S^3, has an analogue for manifolds with boundary.

Proposition 6.8. *Every smooth, connected, compact, orientable 3-dimensional manifold can be obtained by performing surgery on a framed link in the complement of an open regular neighborhood of a handlebody graph in S^3.*

Proof. If M is a smooth, compact, orientable 3-dimensional manifold, then there exists a finite collection of handlebodies which, when glued to the connected components of ∂M, yield a closed manifold M'. In M', the handlebodies form a regular neighborhood of a handlebody graph. By the Lickorish-Wallace Theorem (Theorem 6.3), M' can be obtained by performing surgery on a framed link L in S^3. The surgery tori lie inside M' and they form a regular neighborhood of a link L' in M'. A general position argument allows us to push the handlebody graph off the surgery tori. Hence the handlebody graph in M' comes from a handlebody graph in S^3, and we are done. □

We conclude that every smooth, compact, orientable 3-dimensional manifold with boundary can be represented by a pair (L, Γ) consisting of a framed surgery link and a handlebody graph in S^3. We will call such a pair a **surgery diagram for the manifold with boundary** and denote the resulting manifold $M_{L,\Gamma}$. The question is when two surgery diagrams represent the same manifold. We will prove a somewhat weaker version of Kirby's theorem for manifolds with boundary, in which the boundaries are parametrized. This situation imposes rigidity on the boundary [2].

In each genus g we fix a (standard) surface Σ_g which bounds (in the standard way) the handlebody H_g.

Definition 6.15. Given a boundary component $\Sigma' \subset \partial M$ of the smooth, compact, oriented 3-dimensional manifold M, a **parametrization** of this component is an orientation reversing homeomorphism $f : \Sigma_g \to \Sigma'$ from a fixed surface Σ_g of the same genus as Σ'.

Let (L, Γ) be a pair of a framed link and handlebody graph that defines a 3-dimensional manifold with boundary. Let Γ_g be a component of the graph Γ, whose regular neighborhood $N(\Gamma_g)$ is a genus g handlebody in $S^3 \backslash L$. Let $f_0 : \mathcal{G}_g \to \Gamma_g$ be a homeomorphism that restricts to a homeomorphism of the cores. Let H_g be a regular neighborhood of $\mathcal{G}_g$ in S^3. Then f_0 extends to $f_1 : H_g \to N(\Gamma_g)$. This homeomorphism gives rise to a parametrization f of the boundary component of $M_{L,\Gamma}$ that corresponds to Γ_g by the surface $\Sigma_g = \partial H_g$.

$$\begin{array}{ccccc}
\mathcal{G}_g & \longrightarrow & H_g & \longleftarrow & \Sigma_g \\
{\scriptstyle f_0}\downarrow & & {\scriptstyle f_1}\downarrow & & {\scriptstyle f}\downarrow \\
\Gamma_g & \longrightarrow & N(\Gamma_g) & \longleftarrow & \partial N(\Gamma_g)
\end{array}$$

Doing this for all boundary components, we obtain a parametrization of $\partial M_{L,\Gamma}$ by a surface Σ which is a disjoint union of finitely many closed surfaces.

Conversely, given a parametrization of the boundary component of M we can use this parametrization to fill in the handlebody to M and hence the parametrization gives rise to a homeomorphism from one of the graphs $\mathcal{G}_g$ to the corresponding graph component of the surgery diagram of M.

Now we obtain as a direct corollary of Proposition 6.7 the following result, which is the desired relative version of Kirby's theorem.

Theorem 6.7. *The manifolds with boundary $M_{L_1,\Gamma}$ and $M_{L_2,\Gamma'}$ are homeomorphic under a homeomorphism that preserves the parametrizations of the boundaries induced by Γ and Γ' respectively if and only if the pairs (L_1, Γ) and (L_2, Γ') can be changed into one another by isotopies and finitely many of the following moves:*

[2] If we allow the boundary to change by a homeomorphism then the situation is significantly more complicated.

(k1) *add or subtract from the link an unknotted circle with framing* ± 1, *which is separated from the other link components by an embedded 2-dimensional sphere in* S^3;

(k2) *given two link components, replace one of them by the result of sliding it over the other link component;*

(k3) *replace the trivalent ribbon graph by its slide over a link component.*

One should point out that in the same way that we can pass from the 3-dimensional sphere to an arbitrary 3-dimensional manifold through surgery, one can also pass from one manifold to another.

Proposition 6.9. *Let* M *and* M' *be two smooth, compact, orientable manifolds with or without boundary. Assume moreover that* ∂M *is homeomorphic to* $\partial M'$. *Then there is a framed link* L *in* M *such that* M' *is obtained from* M *by performing surgery on* L.

Proof. It suffices to discuss the case of connected manifolds since in the general case we can treat the connected components separately. Let $\partial M \approx \partial M' \approx \sqcup_{j=1}^{n} \Sigma_{g_j}$. Since both M and M' can be obtained by performing surgery on links in $S^3 \backslash \sqcup_{j=1}^{n} H_{g_j}$, the only thing to check is that surgery is reversible, namely that one can pass from M to $S^3 \backslash \sqcup_{j=1}^{n} H_{g_j}$ via surgery on a link in M.

Assume that M is obtained from $S^3 \backslash \sqcup_{j=1}^{n} H_{g_j}$ by surgery on a link L, and let L_k be a link component. Let N_k be a regular neighborhood of L_k identified with the standard solid torus $S^1 \times B^2$ in such a way that the framing of L_k determines the curve $(1,0)$ on the boundary $S^1 \times S^1$ of the solid torus. Surgery removes N_k and replaces it by a solid torus $N_k' = S^1 \times B^2$ which is glued so that the $(0,1)$ curve on its boundary is glued to the framing curve. Also, up to a reparametrization of N_k' by $S^1 \times B^2$, the $(1,0)$ curve on $\partial N_k'$ is glued to $(0,1)$ on ∂N_k.

Identify N_k' with $S^1 \times B^2 = S^1 \times \{z \mid |z| = 1\}$ and consider the framed link component

$$L_k' = S^1 \times [0,1] \subset N_k' \subset M.$$

Then by performing surgery on $L_k' \in M$, the solid torus N_k' in M is removed and then it is replaced back by N_k. This shows that the surgery process is reversible, and the proposition is proved.

In short, if M is obtained from $S^3 \backslash \sqcup_{j=1}^{n} H_{g_j}$ by removing solid tori from $S^3 \backslash \sqcup_{j=1}^{n} H_{g_j}$ and gluing them back in a "twisted" fashion, then $S^3 \backslash_{j=1}^{n} H_{g_j}$ can be obtained from M by removing those tori and gluing them back in the original "untwisted" fashion. □

6.4.2 *Cobordisms via surgery*

The category of cobordisms

Let us introduce the **category of 3-dimensional cobordisms**. This will be needed in Chapter 7 to construct the topological quantum field theory associated to theta functions – a universal object that incorporates the theories of theta functions for all surfaces.

The **objects** of this category are parametrized surfaces, where by the parametrization of a surface Σ we mean an orientation preserving homeomorphism $h : \Sigma_{fixed} \to \Sigma$ from some fixed (parametrizing) surface Σ_{fixed}.

The **morphisms** of this category are 3-dimensional cobordisms.

Definition 6.16. A **3-dimensional cobordism** is a triple

$$(M, \partial_- M, \partial_+ M)$$

where M is an oriented, compact, 3-dimensional manifold and $\partial_- M, \partial_+ M$ are parametrized oriented closed subsurfaces which partition ∂M, and the orientation of $\partial_+ M$ agrees with that of ∂M while that of $\partial_- M$ disagrees.

We call a cobordism $(M, \partial_- M, \partial_+ M)$ **connected** if M is connected. Two 3-dimensional cobordisms $(M, \partial_- M, \partial_+ M)$ and $(M', \partial_- M', \partial_+ M')$ are called homeomorphic if there exists an orientation preserving homeomorphism $h : M \to M'$ which restricts to $\partial_\pm M$, such that the composition of $h|\partial_\pm M$ with the parametrization of $\partial_\pm M$ gives the parametrization of $\partial_\pm M'$.

Two cobordisms $(M, \partial_- M, \partial_+ M)$ and $(M', \partial_- M', \partial_+ M')$ can be composed if and only if $\partial_+ M$ and $-\partial_- M'$ are homeomorphic under a homeomorphism that preserves orientations and parametrizations. In this case, for the composition

$$(M'', \partial_- M'', \partial_+ M'') := (M', \partial_- M', \partial_+ M')(M, \partial_- M, \partial_+ M),$$

the manifold M'' is obtained by gluing M to M' by the parametrization preserving homeomorphism that identifies $\partial_+ M$ with $\partial_- M'$, $\partial_- M'' = \partial_- M$, and $\partial_+ M'' = \partial_+ M'$, both with the same parametrizations as before.

Surgery diagrams for cobordisms and homeomorphisms

Next we introduce some constructs due to Vladimir Turaev [Turaev (1994)]. The first is the representation of a cobordism by a surgery diagram.

Proposition 6.10. *Let $(M, \partial_- M, \partial_+ M)$ be a connected 3-dimensional cobordism. Then there is a surgery diagram for M such that:*
(i) the handlebody graph representing $\partial_\pm M$ lies in the yz-plane of $\mathbb{R}^3 \subset S^3$ in such a way that its "circles" are of the form $(y-j)^2 + (z-k)^2 = 1/9$ and its "edges" lie on the line $z = 1$ for $\partial_- M$ and $z = 2$ for $\partial_+ M$,
(ii) the surgery link lies entirely inside the slice $\mathbb{R}^2 \times (1,2)$.

A depiction of such a surgery diagram is given in Figure 6.42. Here, to be as general as possible, we included a genus 0 boundary component represented by a dot.

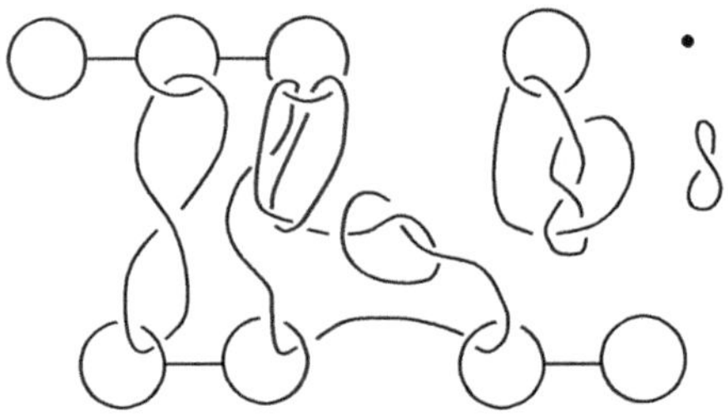

Fig. 6.42 The representation of a cobordism by a surgery diagram

Proof. Using Proposition 6.8, we represent M by a surgery diagram (L, Γ) where L is a framed link and Γ is a handlebody graph in S^3. By introducing trivial link components with framing ± 1 and performing Kirby (k0) moves, we can unknot the "circles" of the handlebody graphs and unlink them from each other.

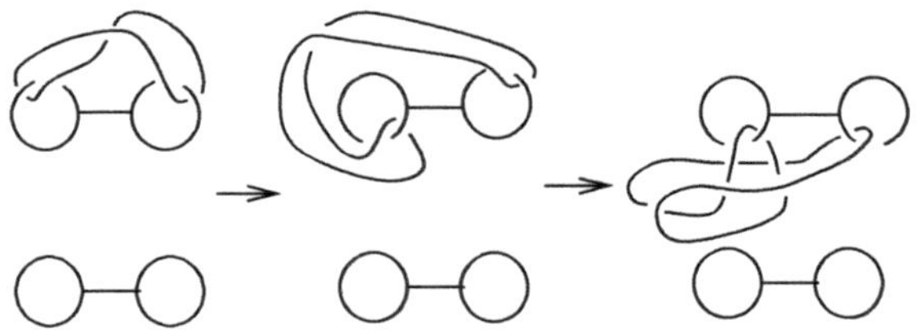

Fig. 6.43 Moving a link to the slice $\mathbb{R}^2 \times (1,2)$

Next split Γ into the handlebody graphs Γ_- and Γ_+ corresponding to $\partial_- M$ and $\partial_+ M$ and isotope Γ_- and Γ_+ to the locations required by (i). Finally isotope the link as to lie inside the slice $\mathbb{R}^2 \times (1,2)$ as suggested in Figure 6.43. □

Example 6.16. Let Σ_g be a closed genus g surface, $g \geq 1$. Topologically, the cylinder $\Sigma_g \times [0,1]$ is the complement of the graph depicted in Figure 6.44.

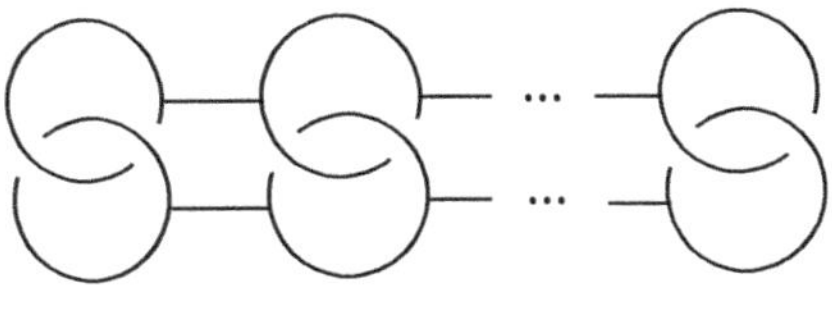

Fig. 6.44 The cylinder over a surface as the complement of a graph

The cylinder $\Sigma_g \times [0,1]$ can be turned into the cobordism

$$(\Sigma_g \times [0,1], \Sigma_g \times \{0\}, \Sigma_g \times \{1\}).$$

But the diagram from Figure 6.44 does not represent the cylinder as desired in Proposition 6.10, since the handlebody graphs are linked.

A correct surgery diagram for this cobordism is shown in Figure 6.45.

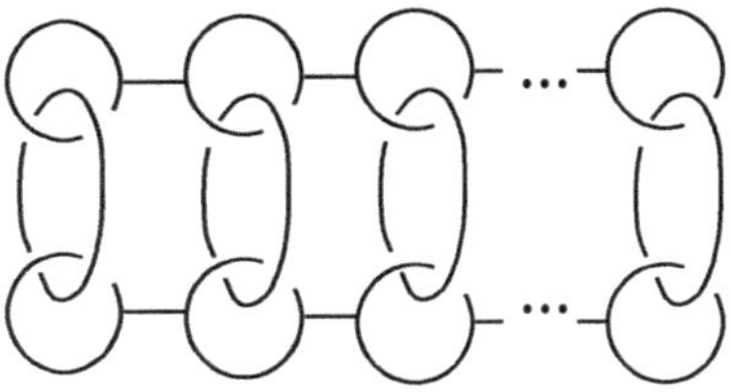

Fig. 6.45 Surgery diagram for the cylinder over a surface

To see why this is so, recall how $\#_{k=1}^{g} S^1 \times S^2$ is obtained by gluing two copies of the handlebody H_g. Remove from each handlebody a regular neighborhood of its core. We obtain two copies of $\Sigma_g \times [0,1]$, which are depicted in Figure 6.46. They are glued along the dotted lines in the figure. The result of the gluing is of course $\Sigma_g \times [0,1]$.

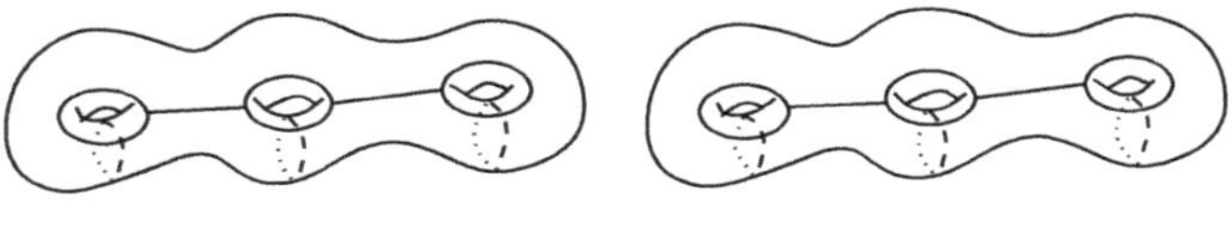

Fig. 6.46 Two copies of $\Sigma_g \times [0,1]$

Now assume that the two handlebodies are glued to each other and let us push the graph that is the core of the second handlebody to the first handlebody, as depicted in Figure 6.47. Note that the complement of the two trivalent graphs is still $\Sigma_g \times [0,1]$.

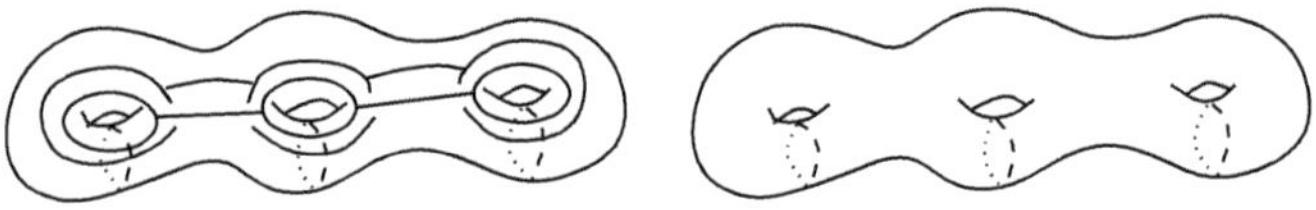

Fig. 6.47 Slide of the handlebody graph from one handlebody to the other

Next let us perform the gluing. The result is shown in Figure 6.48. To better see why this is so, the reader might wish to forget for a moment the presence of the two trivalent graphs, and recall Examples 6.3 and 6.5. Note that the surgery curves are close to the boundary of the first handlebody, and hence away from the two trivalent graphs. An isotopy (just flip one of the graphs) changes this to Figure 6.45.

Fig. 6.48 Surgery diagram for the cylinder over a surface

We will return to this explanation with more details in the proof of Theorem 6.8 below.

Example 6.17. If I_h is the mapping cylinder of a homeomorphism of Σ_g, then a (planar) surgery diagram for I_h can be obtained by adding a surgery link to the diagram from Figure 6.45. One should remark that just very few surgery links added to this diagram yield mapping cylinders of homeomorphisms.

From this moment on, by a surgery diagram for a connected 3-dimensional cobordism $(M, \partial M_-, \partial M_+)$ we will mean a triple (L, Γ_-, Γ_+) of a framed link and two handlebody graphs that satisfy properties (i) and (ii) from the statement of Proposition 6.10, such that $(L, \Gamma_- \cup \Gamma_+)$ is a surgery diagram for M and Γ_- and Γ_+ are the handlebody graphs that correspond to $\partial_- M$ respectively $\partial_+ M$.

The following result will be essential in associating a topological quantum field theory to theta functions.

Theorem 6.8. *[Turaev (1994)] If (L, Γ_-, Γ_+) and $(L', \Gamma'_-, \Gamma'_+)$ are surgery diagrams for the connected 3-dimensional cobordisms $(M, \partial_- M, \partial_+ M)$ and $(M', \partial_- M', \partial_+ M')$ respectively then a surgery diagram for the composition $(M', \partial_- M', \partial_+ M')(M, \partial_- M, \partial_+ M)$ is obtained as follows:*
(i) place the surgery diagram (L, Γ_-, Γ_+) in $\mathbb{R}^3 \subset S^3$ such that the "circles" of $\partial_- M$, are of the form $(y-j)^2 + (z-1)^2 = 1/9$ and its "edges" lie on $z = 1$ and the "circles" of $\partial_+ M$, are of the form $(y-j)^2 + (z-2)^2 = 1/9$, for and its "edges" lie on $z = 2$;
(ii) place the surgery diagram $(L', \Gamma'_-, \Gamma'_+)$ in $\mathbb{R}^3 \subset S^3$ such that the "circles" of $\partial_- M$, are of the form $(y-j)^2 + (z-2)^2 = 1/9$ and its "edges" lie on $z = 2$ and the "circles" of $\partial_+ M$, are of the form $(y-j)^2 + (z-3)^2 = 1/9$, for and its "edges" lie on $z = 3$; so the "circles" and "edges" of Γ'_- and Γ_+ that correspond under gluing overlap;
(iii) place a horizontal surgery circle in $z = 2$ around all but one of the connected components of $\Gamma'_- = \Gamma_+$ and frame it by an annulus that lies in this plane;
(iv) delete the edges of $\Gamma_+ = \Gamma'_-$ and interpret the "circles" of these overlapping graphs as surgery circles.

We stress out that Γ_+ and Γ'_- overlap entirely. An illustration of the composition is shown in Figure 6.49. Note that in this picture we can slide the horizontal circle at the very right over the horizontal circle in the middle to obtain the horizontal circle that links the first two vertical circles. We can also flip the horizontal circle in the middle so that it links the two surgery circles on the left. Thus it is irrelevant on which two of the three possible locations we place the horizontal circles.

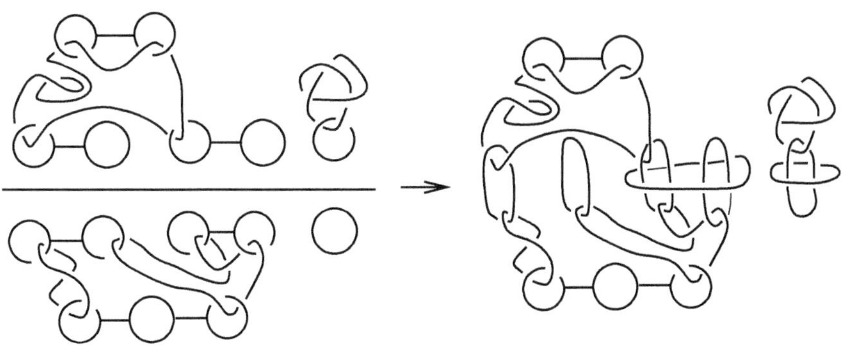

Fig. 6.49 The gluing of the two cobordisms on the left bottom and top yields the cobordism on the right

Proof. A key ingredient for understanding the proof of the theorem is Example 6.3, in which surgery along the trivial knot yields $S^1 \times S^2$. This surgery is the result of gluing two solid tori along their boundaries so that the $(1,0)$ curve is mapped to the $(1,0)$ curve and the $(0,1)$ curve is mapped to the $(0,1)$ curve. Now suppose that in the second torus we have a link which is represented by some link diagram. If we slide this link to the first solid torus by pushing it through the boundary we obtain a link in the first torus which can be represented by precisely the same diagram (see Figure 6.50 where the solid tori lie on the outside of the unknotted circles). This is because the link lies in a cylindrical neighborhood of the boundary of the second torus and we slide it to a cylindrical neighborhood of the boundary of the first torus. And in the plane of the paper the two cylindrical neighborhoods are both on the "outside", so the link diagrams are represented the same way.

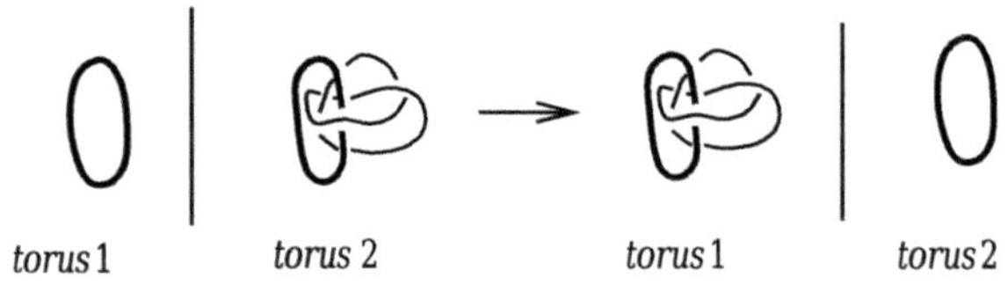

Fig. 6.50 Sliding a link from one solid torus to the other

Now we look at the general situation. First we analyze the case of two 3-dimensional cobordisms, $(M, \partial_- M, \partial_+ M)$ and $(M', \partial_- M', \partial_+ M')$, represented by the diagrams (L, Γ_-, Γ_+), $(L', \Gamma'_-, \Gamma'_+)$, such that $\partial_+ M$ and hence $\partial_- M'$ are connected. For precisely the same reason as above we can slide the diagram consisting of L' and the handlebody graph Γ'_+ from the second cobordism to precisely the same diagram inside the first cobordism (see Figure 6.51). So we can slide everything from the second cobordism to the first, leaving the second cobordism to be just a plain handlebody. Gluing this handlebody to the complement of Γ_+ in S^3 amounts to performing a surgery on trivial link with as many components as is the genus of Γ_+. Each of this links replaces the corresponding "circle" of Γ_+. This proves the theorem in the case where $\partial_+ M$ is connected.

The gluing of two cobordisms along a disconnected surface can be split into steps where at each step we glue along a connected component. The first step is the same as above, but at subsequent steps we glue the manifold to itself along two boundary components. Hence we need to understand how the surgery diagram changes under self-gluings. The simplest case

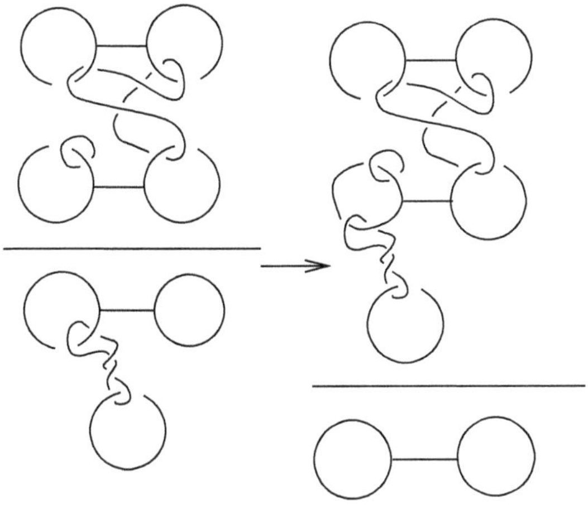

Fig. 6.51 Sliding of the cobordism diagram

comes handy, so let us consider the situation where the two surfaces are tori and hence the corresponding handlebody graph components form a pair of unknotted, unlinked circles. Cut S^3 by a sphere into to parts, each of which containing one of the tori. Each of the pieces is a solid torus with a ball removed. Now glue the two pieces along the two tori, to obtain an $S^1 \times S^2$ with two balls removed. We need to understand what happens after we identify the boundaries of the two balls.

Since the two balls can be isotoped inside a bigger ball, it suffices to understand what happens if in S^3 with two balls removed we identify the boundaries of the two balls. Turning one of the boundary spheres inside-out we see that S^3 minus two balls is the cylinder over a sphere. Gluing the ends of the cylinder yields therefore $S^1 \times S^2$.

We conclude that the result of gluing S^3 along the boundaries of two missing tori corresponding to handlebody graph components that are unknotted, unlinked circles is $(S^1 \times S^2)\#(S^1 \times S^2)$. Now if instead of just two circles we consider two graph components each with g "circles", we obtain

$$(\#_{j=1}^{g} S^1 \times S^2)\#(S^1 \times S^2),$$

i.e. the connected sum of $g+1$ $S^1 \times S^2$'s. Turning to the general case of the surgery diagram of the gluing of two cobordism, we know how to handle the first g of the $S^1 \times S^2$'s, namely exactly like in the case of the gluing of one cobordism to another along a connected surface: we delete the two graphs and replace each pair of corresponding "circles" by just one circle that runs from one graph to the other. Let us see what happens with the last $S^1 \times S^2$. Again the simplest case is illuminating. When cutting along the 2-dimensional sphere, we severe all strands that run between the two graph components. After gluing along that sphere, the strands now run in the direction of S^1, so they link with the surgery circle. Now push this surgery circle towards the g surgery circles of the other $S^1 \times S^2$'s to obtain the desired surgery diagram. The theorem is proved. □

Notation: If $(L, \Gamma_- \Gamma_+)$ and $(L', \Gamma'_-, \Gamma'_+)$ are the surgery diagrams of the two cobordisms, than the theorem tells us how to associate a surgery diagram $(L'', \Gamma_-, \Gamma'_+)$ to the composition. We denote the link L'' by $L' \circ L$.

In view of Examples 6.16 and 6.17 we have the following corollary.

Corollary 6.2. *If (L, Γ_-, Γ_+) and $(L', \Gamma'_-, \Gamma'_+)$ are graphs representing the mapping cylinders the elements h respectively h' of the mapping class group, then the mapping cylinder of $h' \circ h$ is obtained by translating the diagrams of h and h' in the plane until Γ_+ and Γ'_- overlap (in the manner imposed by the gluing), then deleting the "edges" of the overlap and interpreting the "circles" of the overlap as surgery components.*

This rule for composing mapping cylinders is shown in Figure 6.52.

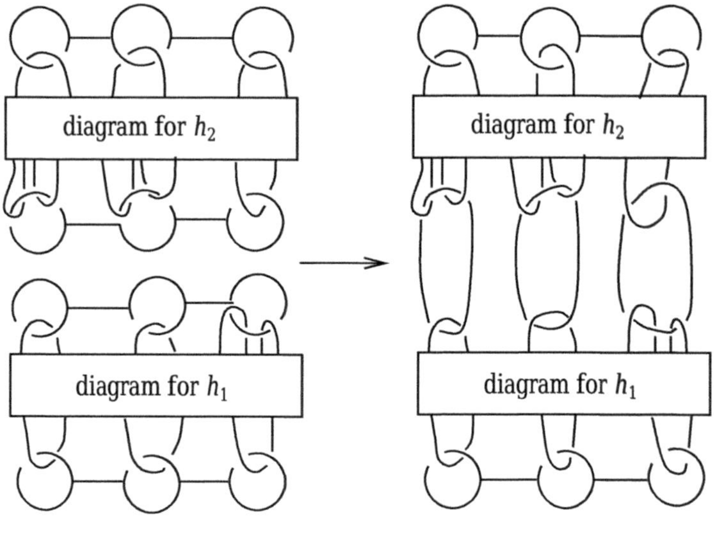

Fig. 6.52 Composition of mapping cylinders

6.5 Wall's formula for the nonadditivity of the signature of 4-dimensional manifolds

In this section we present a theorem by C.T.C. Wall describing how the signature of 4-dimensional manifolds changes under gluings with corners. Wall's Theorem will be used in Chapter 7 to resolve the projective ambiguity of the representation of the mapping class group of a Riemann surface on the space of theta functions associated to that surface.

The theorem makes use of the Maslov index, so we recommend the reader to review the definitions and results from §2.5.2.

Notation: In this section we consider only homology and cohomology with real coefficients, and we ignore the coefficient field in the notation, that is

$$H_*(M) := H_*(M, \mathbb{R}), \quad H^*(M) := H^*(M, \mathbb{R}).$$

6.5.1 *Lagrangian subspaces in the boundary of a 3-dimensional manifold*

Let M be an oriented manifold of dimension $2k$, k a positive integer. Here we are only interested in the cases $k = 1$ and $k = 2$. Recall the intersection form in $H_k(M)$, $k = 1, 2$. This bilinear form is nondegenerate if $k = 1$, but might be degenerate when $k = 2$. By analogy with symplectic forms (see §2.3.1), for a subspace $V \subset H_k(M)$ we can define the orthogonal of V as

$$V^\perp = \{\mathbf{w} \in H_k(M) \,|\, \mathbf{v} \cdot \mathbf{w} = 0 \text{ for all } \mathbf{v} \in V\}.$$

We can extend the intersection form to relative homology as well, and define orthogonals by the above formula.

If $k = 1$ and so the intersection form is symplectic, then we have the notion of a Lagrangian subspace, characterized by the space being equal to its symplectic orthogonal. For example, by Proposition 3.2, the intersection form on $H_1(\Sigma_g, \mathbb{R})$ is symplectic, and if $a_1, a_2, \ldots, a_n, b_1, b_2, \ldots, b_n$ is a canonical basis, then both $\mathrm{Span}(a_1, a_2, \ldots, a_n)$ and $\mathrm{Span}(b_1, b_2, \ldots, b_n)$ are Lagrangian subspaces of $H_1(\Sigma_g, \mathbb{R})$. Note that $\mathrm{Span}(b_1, b_2, \ldots, b_n)$ is the kernel of the map $H_1(\Sigma_g, \mathbb{R}) \to H_1(H_g, \mathbb{R})$ determined by the standard identification of Σ_g with ∂H_g. The next result shows that this is a general method for producing Lagrangian subspaces in $H^1(\Sigma_g)$.

Proposition 6.11. *Let M be a compact, oriented, smooth 3-dimensional manifold with oriented boundary, and endow $H_1(\partial M)$ with the intersection* *form (which is symplectic). Then the kernel of the map $H_1(\partial M) \to H_1(M)$ induced by the inclusion $\partial M \hookrightarrow M$ is a Lagrangian subspace of $H_1(\partial M)$.*

Proof. The proof is based on Poincaré duality and its version for manifolds with boundary, Poincaré-Lefschetz duality.

We place the map from the statement in the long exact sequence in homology of the triple $(\partial M,\ M,\ (M, \partial M))$:

$$\cdots \to H_2(M, \partial M) \to H_1(\partial M) \to H_1(M) \to H_1(M, \partial M) \to \cdots .$$

Because of Poincaré duality, this sequence can be identified, canonically, with the long-exact sequence in cohomology

$$\cdots \to H^1(M) \to H^1(\partial M) \to H^2(M, \partial M) \to H^2(M) \to \cdots .$$

We first check that the kernel of $H^1(\partial M)$ in $H^1(M)$ is isotropic with respect to the intersection form. Because of the exact sequence, this amounts to showing that the image of $H_2(M, \partial M)$ inside $H_1(\partial M)$ is isotropic. Using Poincaré duality and also Theorem 4.4, we deduce that we must check that the image of $H^1(M)$ inside $H^1(\partial M)$ is isotropic with respect to the symplectic form on $H^1(M)$ defined by

$$(\alpha, \beta) \mapsto \int_{\partial M} \alpha \wedge \beta.$$

So let α and β be 1-forms representing classes in $H^1(M)$. Using Stokes' Theorem, we compute

$$\int_{\partial M} (\alpha|\partial M) \wedge (\beta|\partial M) = \int_M d(\alpha \wedge \beta) = \int_M d\alpha \wedge \beta - \int_M \alpha \wedge d\beta.$$

The last expression is 0, because α and β, representing cohomology classes, are closed.

We are left to check that the dimension of the kernel is half the dimension of $H_1(\partial M)$. We have

$$\begin{aligned}\dim(H_1(\partial M)) &= \dim\ker(H_1(\partial M) \to H_1(M)) \\ &\quad + \dim\operatorname{coker}(H_2(M, \partial M) \to H_1(\partial M)).\end{aligned}$$

But the map

$$H_2(M, \partial M) \to H_1(\partial M)$$

is dual to the map

$$H^1(\partial M) \to H^2(M, \partial M), \tag{6.3}$$

so

$$\dim \operatorname{coker}(H_2(M,\partial M) \to H_1(\partial M)) = \dim \ker(H^1(\partial M) \to H^2(M,\partial M)).$$

By Poincaré duality the map $H^1(\partial M) \to H^2(M,\partial M)$ coincides with the map $H_1(\partial M) \to H_1(M)$. We conclude that

$$\dim(H_1(\partial M)) = 2\dim\ker(H_1(\partial M) \to H_1(M)),$$

and the lemma is proved. □

Example 6.18. By choosing half of a canonical basis in each boundary component of M, we can span a Lagrangian subspace of $H_1(\partial M)$.

Example 6.19. If $M = \Sigma_g \times [0,1]$, where Σ_g is a closed genus g surface, then

$$\operatorname{Span}([x \times \{0\}] - [x \times \{1\}]),$$

where x runs through a family of curves in Σ_g that span $H_1(\Sigma_g)$, is a Lagrangian subspace of $H_1(\partial M) = H_1(\Sigma_g) \oplus H_1(\Sigma_g)$. Recall that the bracket denotes the homology class of the curve.

6.5.2 *Wall's theorem*

The setting of Wall's theorem consists of two compact, oriented 4-dimensional manifolds that are glued by a diffeomorphism along a 3-dimensional manifold with boundary. In this book we apply the theorem to 4-dimensional handlebodies, so we can also assume that manifolds are smooth and can be triangulated so that everything that was discussed elsewhere is valid here, too.

In Wall's notation:

- $\mathbf{Y}_-$, $\mathbf{Y}_+$, and $\mathbf{Y}$ are compact oriented 4-dimensional manifolds,
- X_0, X_-, and X_+ are compact oriented 3-dimensional manifolds,
- Z is a closed oriented surface

such that

$$\begin{aligned}
&\mathbf{Y} = \mathbf{Y}_- \cup \mathbf{Y}_+,\\
&\partial\mathbf{Y}_- = X_- \cup (-X_0), \quad \partial\mathbf{Y}_+ = X_0 \cup (-X_+)\\
&\partial X_- = \partial X_0 = \partial X_+ = Z.
\end{aligned}$$

This configuration is sketched in Figure 6.53.

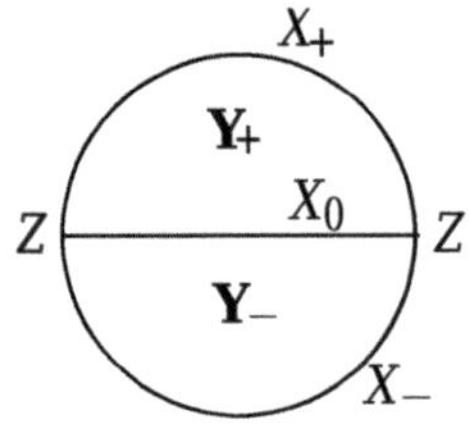

Fig. 6.53 The gluing of two 4-dimensional manifolds

The inclusion of Z in X_-, X_0, X_+ induces corresponding maps in first homology. Define

$$\mathbf{L}_- = \ker(H_1(Z) \to H_1(X_-)), \quad \mathbf{L}_0 = \ker(H_1(Z) \to H_1(X_0)),$$
$$\mathbf{L}_+ = \ker(H_1(Z) \to H_1(X_+)).$$

By Proposition 6.11, the spaces $\mathbf{L}_-$, $\mathbf{L}_0$, and $\mathbf{L}_+$ are Lagrangian subspaces of $H_1(Z, \mathbb{R})$.

Theorem 6.9. *[Wall (1969)] In the above setting*

$$\sigma(\mathbf{Y}) = \sigma(\mathbf{Y}_-) + \sigma(\mathbf{Y}_+) - \tau(\mathbf{L}_-, \mathbf{L}_0, \mathbf{L}_+),$$

where τ is the Maslov index.

Remark 6.5. This formula looks identical as the one in Wall's paper, but one should notice that the Maslov index is the negative of Wall's non-additivity function σ, and the orientations of X_-, X_0, and X_+ chosen here are opposite to those from [Wall (1969)]; they agree with the convention in [Turaev (1994)].

Proof. The proof uses various facts from algebraic topology and is taken from [Wall (1969)]. Again we work with homology with real coefficients, and in the notation we ignore the coefficient field.

First, note that in the long exact sequence in homology of the triple $\partial\mathbf{Y}, \mathbf{Y}, (\mathbf{Y}, \partial\mathbf{Y})$:

$$\cdots \to H_2(\partial\mathbf{Y}) \stackrel{i_*}{\to} H_2(\mathbf{Y}) \stackrel{p_*}{\to} H_2(\mathbf{Y}, \partial\mathbf{Y}) \to H_1(\partial\mathbf{Y}) \to \cdots,$$

the two vector spaces in the middle are dual to each other, since because of Lefschetz-Poincaré duality $H_2(\mathbf{Y}, \partial\mathbf{Y}) = H^2(\mathbf{Y})$. If we choose dual bases in $H_2(\mathbf{Y})$ and $H^2(\mathbf{Y})$, then the matrix of intersection numbers in $H_{2k}(\mathbf{Y})$ coincides with the matrix of p_*.

Let $A_{\mathbf{Y}}$ be this matrix; it defines the antisymmetric bilinear form

$$B_{\mathbf{Y}}(\mathbf{v}, \mathbf{w}) = \mathbf{v}^T A_{\mathbf{Y}} \mathbf{w}.$$

The radical of this form, which is the set of the vectors $\mathbf{w}$ such that $\mathbf{v}^T A_{\mathbf{Y}} \mathbf{w} = 0$ for all $\mathbf{v}$, is the same as the kernel of p_*, and because of exactness this is $i_*(H_2(\partial\mathbf{Y}))$.

So if we factor $H_2(\mathbf{Y})$ by $i_*(H_2(\partial\mathbf{Y}))$, then $B_{\mathbf{Y}}$ descends to a nondegenerate bilinear form $\hat{B}_{\mathbf{Y}}$ on $H_2(\mathbf{Y})/i_*(H_2(\partial\mathbf{Y}))$, whose signature equals the signature of $B_{\mathbf{Y}}$ and is therefore equal to $\sigma(\mathbf{Y})$. Note that

$$H_2(\mathbf{Y})/i_*(H_2(\partial\mathbf{Y})) = H_2(\mathbf{Y})/\ker p_* = p_*(H_2(\mathbf{Y})),$$

and so we have a splitting

$$H_2(\mathbf{Y}) = i_*(H_2(\partial\mathbf{Y})) \oplus p_*(H_2(\mathbf{Y})). \tag{6.4}$$

The same remarks apply if we replace $\mathbf{Y}$ by $\mathbf{Y}_+$ or $\mathbf{Y}_-$, and we use the same splittings

$$H_2(\mathbf{Y}_\pm) = i_*(H_2(\partial\mathbf{Y}_\pm)) \oplus p_*(H_2(\mathbf{Y}_\pm)),$$

with $\sigma(\mathbf{Y}_\pm)$ equal to the signature of $\hat{B}_{\mathbf{Y}_\pm}$ on $p_*(H_2(\mathbf{Y}_\pm))$.

The compositions of maps $\mathbf{Y}_\pm \hookrightarrow \mathbf{Y} \to \mathbf{Y}/\partial\mathbf{Y}$ allow us to map $H_2(\mathbf{Y}_-)$ and $H_2(\mathbf{Y}_+)$ inside $H_2(\mathbf{Y}, \partial\mathbf{Y})$. The images of these two spaces have intersection equal to $\{0\}$. Moreover, they are orthogonal with respect to the intersection form, namely $\Sigma_- \cdot \Sigma_+ = 0$ if $\Sigma_\pm \in H_2(\mathbf{Y}_\pm)$.

Also, because there is a relative inclusion map $(X_0, Z) \to (\mathbf{Y}, \partial\mathbf{Y})$, one can map $H_2(X_0, Z)$ to a subspace inside $H_2(\mathbf{Y}, \partial\mathbf{Y})$.

Lemma 6.1. *The subspace of $H_2(\mathbf{Y}, \partial\mathbf{Y})$ orthogonal (with respect to the intersection form) to the image of $H_2(\mathbf{Y}_-) \oplus H_2(\mathbf{Y}_+)$ is the image of $H_2(X_0, Z)$.*

Proof. First, note that $H_2(\mathbf{Y}_-)$ and $H_2(\mathbf{Y}_+)$ are indeed orthogonal to each other with respect to the intersection form.

By Poincaré-Lefschetz duality $H_2(\mathbf{Y}_\pm, \partial\mathbf{Y}_\pm) = H^2(\mathbf{Y}_\pm)$, and the latter is dual to is dual to $H_2(\mathbf{Y}_\pm)$. Because of this, the matrix of the map

$$H_2(\mathbf{Y}, \partial\mathbf{Y}) \to H_2(\mathbf{Y}, \mathbf{Y}_+ \cup \partial\mathbf{Y}) \cong H_2(\mathbf{Y}_-, \partial\mathbf{Y}_-)$$

is the intersection matrix of the homology classes in $H_2(\mathbf{Y}, \partial\mathbf{Y})$ with the homology classes in the image of $H_2(\mathbf{Y}_-)$. Similarly for $\mathbf{Y}_+$. Hence the subspace orthogonal to $H_2(\mathbf{Y}_-)$ is the kernel of this map. It follows that the subspace orthogonal to both $H_2(\mathbf{Y}_-)$ and $H_2(\mathbf{Y}_+)$ is the intersection of the two kernels, i.e. the kernel of the map

$$H_2(\mathbf{Y}, \partial\mathbf{Y}) \to H_2(\mathbf{Y}_-, \partial\mathbf{Y}_-) \oplus H_2(\mathbf{Y}_+, \partial\mathbf{Y}_+).$$

Place this in the Mayer-Vietoris sequence

$$\cdots \to H_2(\mathbf{Y}_- \cap \mathbf{Y}_+, \partial\mathbf{Y}_- \cap \partial\mathbf{Y}_+) \to H_2(\mathbf{Y}, \partial\mathbf{Y})$$
$$\to H_2(\mathbf{Y}_-, \partial\mathbf{Y}_-) \oplus H_2(\mathbf{Y}_+, \partial\mathbf{Y}_+) \to \cdots$$

and use the fact that $(\mathbf{Y}_- \cap \mathbf{Y}_+, \partial\mathbf{Y}_- \cap \partial\mathbf{Y}_+) = (X_0, Z)$ to conclude the proof of the lemma. □

Returning to the proof of the theorem, the maps

$$H_2(\mathbf{Y}_\pm) \to H_2(\mathbf{Y})$$

induced by the inclusion preserve the splitting (6.4). Hence we have maps

$$p_*(H_2(\mathbf{Y}_\pm)) \to p_*(H_2(\mathbf{Y})),$$

and these preserve the intersection numbers. Because the intersection form on $p_*(H_2(\mathbf{Y}_\pm))$ is non-singular and the two images are orthogonal to each other, both maps are injective and their images form a direct sum. Let V denote the orthogonal complement of $p_*(H_2(\mathbf{Y}_-)) \oplus p_*(H_2(\mathbf{Y}_+))$ in $p_*(H_2(\mathbf{Y}))$. Since signature is additive for direct sums, the signature of the intersection form restricted to V is

$$\sigma(\mathbf{Y}) - \sigma(\mathbf{Y}_-) - \sigma(\mathbf{Y}_+).$$

To complete the proof we compute the signature of the restriction to V of the intersection form.

We can also map $i_*(H_2(\partial\mathbf{Y}_-)) \oplus i_*(H_2(\partial\mathbf{Y}_+))$ inside $p_*(H_2(\mathbf{Y}))$, using the composition of inclusions $\partial\mathbf{Y}_\pm \hookrightarrow \mathbf{Y}_\pm \hookrightarrow \mathbf{Y}$. Because of the splitting (6.4), the image is a subspace of V. Denote this image by S.

The orthogonal $S^\perp$ of S in V with respect to the intersection form is then orthogonal to the image of

$$i_*(H_2(\partial\mathbf{Y}_-)) \oplus p_*(H_2(\mathbf{Y}_-)) \oplus i_*(H_2(\partial\mathbf{Y}_+)) \oplus p_*(H_2(\mathbf{Y}_+)),$$

namely to the image of $H_2(\mathbf{Y}_-) \oplus H_2(\mathbf{Y}_+)$. By Lemma 6.1, this is the image in $H_2(\mathbf{Y}, \partial\mathbf{Y})$ of $H_2(X_0, Z)$. More precisely

$$S^\perp = p_*(H_2(\mathbf{Y})) \cap \operatorname{im}(H_2(X_0, Z) \to H_2(\mathbf{Y}, \partial\mathbf{Y})).$$

Note that this space contains the image of $i_*(H_2(\partial\mathbf{Y}_-)) \oplus i_*(H_2(\partial\mathbf{Y}_+))$ since the inclusion $\partial\mathbf{Y}_\pm \hookrightarrow (\mathbf{Y}, \partial\mathbf{Y})$ factors through $(\partial\mathbf{Y}_\pm, X_\pm)$, and $H_2(\partial\mathbf{Y}_\pm, X_\pm) \cong H_2(X_0, Z)$. In other words $S \subset S^\perp$.

Let W be a complement of S in $S^\perp$. Then $W \cong S^\perp/S$. Because $(S^\perp)^\perp = S$, the radical of the restriction to $S^\perp$ of the intersection form is S. Consequently the intersection form is nondegenerate on $S^\perp/S$, and hence on W. It follows that the form is nondegenerate on $W^\perp$ as well and we can decompose

$$V = W \oplus W^\perp.$$

So the signature of the intersection form on V is the sum of its restrictions to W and $W^\perp$.

We have $S \subset W^\perp$ and in $W^\perp$, $S^\perp = S$ (this is because the radical of the restriction to $S^\perp$ of the intersection form is S). It follows that we can find a basis $\mathbf{v}_1, \mathbf{v}_2, \ldots, \mathbf{v}_k, \mathbf{w}_1, \mathbf{w}_2, \ldots, \mathbf{w}_k$ of $W^\perp$ such $\mathbf{v}_1, \mathbf{v}_2, \ldots, \mathbf{v}_k$ is a basis for S and the intersection form on $L^\perp$ is $\sum \pm \mathbf{v}_j \oplus \mathbf{w}_j$, for some choice of the signs $\pm$. In other words, there is a basis of $W^\perp$ such that the matrix of the intersection form is block-diagonal, with the blocks on the diagonal of the form

$$\begin{bmatrix} 0 & 1 \\ 1 & 0 \end{bmatrix} \text{ or } \begin{bmatrix} 0 & -1 \\ -1 & 0 \end{bmatrix}.$$

Each of these blocks has a positive eigenvalue and a negative eigenvalue, so the signature of the matrix is zero. Hence the signature of the intersection form restricted to $W^\perp$ is zero.

It follows that the signature of the intersection form on V is the same as the signature of the intersection form on W. We will now relate the latter to the Maslov index. To this end, we want to define an isomorphism

$$W \to \frac{\mathbf{L}_0 \cap (\mathbf{L}_- + \mathbf{L}_+)}{\mathbf{L}_0 \cap \mathbf{L}_- + \mathbf{L}_0 \cap \mathbf{L}_+}. \tag{6.5}$$

Recall that W was defined to be an additive complement of the space

$$S = \operatorname{im}(H_2(\partial \mathbf{Y}_-) \oplus H_2(\partial \mathbf{Y}_+) \to H_2(\mathbf{Y}, \partial \mathbf{Y}))$$

in

$$\begin{aligned} S^\perp &= \operatorname{im}(H_2(\mathbf{Y}) \to H_2(\mathbf{Y}, \partial \mathbf{Y})) \cap \operatorname{im}(H_2(X_0, Z) \to H_2(\mathbf{Y}, \partial \mathbf{Y})) \\ &= \operatorname{im}(H_2(X_- \cup X_0 \cup X_+) \to H_2(\mathbf{Y}, \partial \mathbf{Y})). \end{aligned}$$

The isomorphism (6.5) is defined as follows. Let x_1 be an element of $S^\perp$. Because $S \subset S^\perp$, x_1 can be lifted, non-uniquely, to $x_2 \in H_2(X_0, Z)$. Choose $x_3 = \delta x_2 \in H_1(Z)$ (here $\delta : H_2(X_0, Z) \to H_1(Z)$ is the operator that arises in the long exact sequence; geometrically it maps the cycle x_2 to its boundary). The map

$$x_1 \bmod S \mapsto x_3 \bmod (\mathbf{L}_0 \cap \mathbf{L}_- + \mathbf{L}_0 \cap \mathbf{L}_+) \tag{6.6}$$

is the desired isomorphism.

Step 1. Let us show that this definition does not depend on the chosen representative x_1 for a class in $W = S^{\perp}/S$. This amounts to showing that every element $x_1 \in S$ is mapped to 0. If

$$x_1 \in \mathrm{im}(H_2(\partial\mathbf{Y}_-) \oplus H_2(\partial\mathbf{Y}_+) \to H_2(\mathbf{Y}, \partial\mathbf{Y}))$$

then

$$x_1 \in \mathrm{im}(H_2(\partial\mathbf{Y}_-) \oplus H_2(\partial\mathbf{Y}_+) \to H_2(X_- \cup X_0 \cup X_+, \partial\mathbf{Y})).$$

Note that $H_2(X_- \cup X_0 \cup X_+, \partial\mathbf{Y}) = H_2(X_0, Z)$, so this readily defines

$$x_2 \in \mathrm{im}(H_2(\partial\mathbf{Y}_-) \oplus H_2(\partial\mathbf{Y}_+) \to H_2(X_0, Z)).$$

We want to show that

$$\delta[\mathrm{im}(H_2(\partial\mathbf{Y}_-) \oplus H_2(\partial\mathbf{Y}_+) \to H_2(X_0, Z))] = (\mathbf{L}_0 \cap \mathbf{L}_-) + (\mathbf{L}_0 \cap \mathbf{L}_+).$$

For this we examine separately

$$H_2(\partial\mathbf{Y}_\pm) \to H_2(\partial\mathbf{Y}_\pm, X_\pm) \cong H_2(X_0, Z) \xrightarrow{\delta} H_1(Z)$$

for each choice of signs. The map defined by this composition shows up in the Mayer-Vietoris sequence

$$H_2(\partial\mathbf{Y}_\pm) \to H_1(Z) \to H_1(X_\pm) \oplus H_1(X_0),$$

so the image of $H_2(\partial\mathbf{Y}_\pm)$ is equal to

$$\ker(H_1(Z) \to H_1(X_\pm) \oplus H_1(X_0)) = \mathbf{L}_\pm + \mathbf{L}_0.$$

So if $x_1 \in S$ then $x_3 \in (\mathbf{L}_- \cap \mathbf{L}_0) + (\mathbf{L}_+ \cap \mathbf{L}_0)$, showing that the map is well defined.

Step 2. Let us prove that the map is one-to-one. If x_3 is in $(\mathbf{L}_- \cap \mathbf{L}_0) + (\mathbf{L}_+ \cap \mathbf{L}_0)$ then from the exactness of the long exact sequence of the triple $(X_0, Z, (X_0, Z))$ we deduce that $x_2 \in \mathrm{im}(H_2(\partial\mathbf{Y}_-) \oplus H_2(\partial\mathbf{Y}_+) \to H_2(X_0, Z))$. This necessarily implies that $x_1 \in S$.

Step 3. Next, we show that the image of the map (6.6) coincides with $\mathbf{L}_0 \cap (\mathbf{L}_- + \mathbf{L}_+)$. First, let us show that $x_3 \in \mathbf{L}_0 \cap (\mathbf{L}_- + \mathbf{L}_+)$. When x_1 ranges in $S^{\perp}$, then the total range of values of x_2 is the set of elements in $H_2(X_0, Z)$ whose image in $H_2(\mathbf{Y}, \partial\mathbf{Y})$ is also in the image of $H_2(\mathbf{Y})$. In the long exact sequence, the image of $H_2(\mathbf{Y})$ inside $H_2(\mathbf{Y}, \partial\mathbf{Y})$ is the kernel of the δ operator. So the range of values of x_2 is the set of elements in $H_0(X, Z)$ that are mapped to 0 in $H_1(\partial\mathbf{Y})$.

Thus the range of x_3 is

$$\mathrm{im}(H_2(X_0, Z) \to H_1(Z)) \cap \ker(H_1(Z) \to H_1(\partial\mathbf{Y})).$$

The first member is $\mathbf{L}_0$. For the second use the long exact sequence of the triple $(Z, \partial\mathbf{Y}, (\partial\mathbf{Y}, Z))$. In this sequence,

$$\ker(H_1(Z) \to H_1(\partial\mathbf{Y})) = \operatorname{im}(H_2(\partial\mathbf{Y}, Z) \to H_1(Z)).$$

We have

$$\begin{aligned}\operatorname{im}(H_2(\partial\mathbf{Y}, Z) \to H_1(Z)) &= \operatorname{im}(H_2(\partial X_-, Z) \oplus H_2(\partial X_+, Z) \to H_1(Z)) \\ &= \operatorname{im}(H_2(\partial X_-, Z) \to H_1(Z)) + \operatorname{im}(H_2(\partial X_+, Z) \to H_1(Z)).\end{aligned}$$

Using the long exact sequence of the triples $Z, X_\pm, (X\pm, Z)$, we conclude that

$$\operatorname{im}(H_2(\partial X_\pm, Z) \to H_1(Z)) = \ker(H_1(Z) \to H_1(X_\pm)) = \mathbf{L}_\pm.$$

Hence $x_3 \in \mathbf{L}_0 \cap (\mathbf{L}_- + \mathbf{L}_+)$, as claimed, and moreover we obtain that the map (6.6) is onto.

Thus we have established the existence of the isomorphism (6.5). We want to determine the image of the intersection form through this isomorphism.

Let $\mathbf{l}_0 \in \mathbf{L}_0 \cap (\mathbf{L}_- + \mathbf{L}_+)$. Then there are $\mathbf{l}_\pm \in \mathbf{L}_\pm$ with

$$\mathbf{l}_- + \mathbf{l}_0 + \mathbf{l}_+ = 0.$$

Recall that

$$\begin{aligned}\mathbf{L}_\pm &= \ker(H_1(Z) \to H_1(X_\pm)) = \operatorname{im}(H_1(X_\pm, Z) \stackrel{\partial}{\to} H_1(Z)) \\ \mathbf{L}_0 &= \ker(H_1(Z) \to H_1(X_0)) = \operatorname{im}(H_1(X_0, Z) \stackrel{\partial}{\to} H_1(Z))\end{aligned}$$

(here ∂ associates to a relative cycle its boundary). We can therefore find the cycles $\zeta_0 \in Z_2(X_0, Z))$, $\zeta_\pm \in Z_2(X_\pm, Z)$, with $\mathbf{l}_0 = \partial\zeta_0$, $\zeta_\pm = \partial\zeta_\pm$, and so

$$\partial(\zeta_+ + \zeta_0 + \zeta_-) = 0 = \partial\zeta_+ + \partial\zeta_0 + \partial\zeta_- = 0.$$

This shows that the cycle $\zeta_+ + \zeta_0 + \zeta_+$ has no boundary so it represents an element in $H_2(\mathbf{Y})$. Viewing it as an element in $H_2(\mathbf{Y}, \partial\mathbf{Y})$ it represents the same homology class as ζ_0 because $\zeta_\pm$ live on the boundary of $\mathbf{Y}$. We conclude that $\zeta_+ + \zeta_0 + \zeta_-$ is a cycle in $\mathbf{Y}$ that represents the element in $S^\perp$ that is mapped to $\mathbf{l}_0$ by (6.6).

Let $\mathbf{l}_0'$, $\zeta_+', \zeta_0', \zeta_-'$ be another system of such elements. Deform ζ_+' and ζ_0 slightly inside $\mathbf{Y}_+$ and ζ_-' inside $\mathbf{Y}$. Then in $\mathbf{Y}$

$$(\zeta_+ + \zeta_0 + \zeta_-) \cap (\zeta_+' + \zeta_0' + \zeta_-') = \zeta_0 \cap \zeta_-',$$

as it can be seen sketched in Figure 6.54.

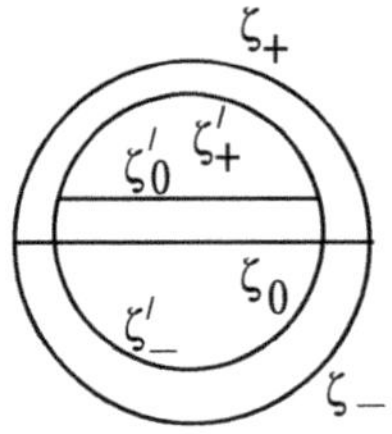

Fig. 6.54 The intersection of the cycles $\zeta_+, \zeta_0, \zeta_-, \zeta_+', \zeta_0', \zeta_-'$

Let $\mathbf{l}_0, \mathbf{l}_0'$ correspond to $x_1, x_1' \in W = S^\perp/S$. Then the intersection form evaluated on x_1 and x_1' becomes the algebraic intersection number of ζ_0 and ζ_-'. This is the same as the intersection number of ζ_0 and $\partial\zeta_-'$ in X_0 and this is further equal to the negative of the algebraic intersection number of $\partial\zeta_0$ and $\partial\zeta_-'$ in Z, because $\partial X_0 = -Z$.

Comparing with the notation from §2.5.2, $\mathbf{L}_1 = \mathbf{L}_0$, $\mathbf{L}_2 = \mathbf{L}_+$, and $\mathbf{L}_3 = \mathbf{L}_-$, while the symplectic form ω is the intersection form in $H_1(Z)$. Recall that $\mathbf{l}_0 = \partial\zeta_0$, and let $\mathbf{l}_-' = \partial\zeta_-'$. Then, by (2.31), the bilinear form used for defining the Maslov index is

$$B_0(\mathbf{l}_0, \mathbf{l}_0') = \omega(\mathbf{l}_0, \mathbf{l}_-').$$

Consequently, the signature of the intersection form on W is $-\tau(\mathbf{L}_0, \mathbf{L}_+, \mathbf{L}_-)$ (again the negative because the orientations of X_0 and Z do not agree), which is in fact equal to $-\tau(\mathbf{L}_-, \mathbf{L}_0, \mathbf{L}_+)$ by Proposition 2.8 (i). The same is true about the signature of the intersection form on V. This completes the proof. □

Corollary 6.3. *(Novikov) If the smooth, compact, oriented 4-dimensional manifolds $\mathbf{Y}_-$ and $\mathbf{Y}_+$ are glued by an orientation reversing homeomorphism between two connected components of their boundaries so as to yield the 4-dimensional manifold $\mathbf{Y}$, then*

$$\sigma(\mathbf{Y}) = \sigma(\mathbf{Y}_-) + \sigma(\mathbf{Y}_+).$$

6.6 The structure of the linking number skein module of a 3-dimensional manifold

We conclude this chapter with the computation of the linking number skein module, $\mathcal{L}(M)$, introduced in Chapter 5, for a general 3-dimensional manifold M. This is more of an aside, since from the point of view of theta functions we are not really interested in the linking number skein modules but in the **reduced** linking number skein modules. As we will see in §7.3 the latter are very simple. Nevertheless, since we might have stirred up reader's curiosity, and since the detour is not too long, we present Jozef Przytycki's explicit computation of the linking number skein module of an arbitrary compact oriented 3-dimensional manifold.

This a particular case of the more general result of Przytycki [Przytycki (1998)] which covers all possible coefficient rings. We are only concerned with the coefficient ring being $\mathbb{C}[t, t^{-1}]$.

Some preliminary notation is needed. Let M be a smooth, compact, orientable 3-dimensional manifold with or without boundary. Because a link and a union of closed surfaces in M can be perturbed so as to intersect transversally in finitely many points, we can define their algebraic intersection number. This number turns out to depend only on the homology classes of the link and the union of surfaces, and it defines a bilinear form of intersection of 1-cycles and 2-cycles

$$\phi : H_1(M, \mathbb{Z}) \times H_2(M, \mathbb{Z}) \to \mathbb{Z}.$$

Let

$$T(H_1(M, \mathbb{Z}) = \{\alpha \in H_1(M, \mathbb{Z}) \,|\, \phi(\alpha, \beta) = 0 \text{ for all } \alpha \in H_2(M, \mathbb{Z}\}.$$

Example 6.20. If M has no boundary, then $T(H_1(M, \mathbb{Z}))$ is the torsion part of $H_1(M, \mathbb{Z})$.

Example 6.21. If $M = H_g$ is the genus g handlebody, then because $H_2(H_g, \mathbb{Z}) = 0$, the pairing ϕ is identically equal to zero. So $T(H_1(H_g, \mathbb{Z}) = H_1(H_g, \mathbb{Z})$.

Example 6.22. If $M = \Sigma_g \times [0, 1]$, the pairing is still identically equal to zero. This is because $H_1(M, \mathbb{Z})$ is spanned by a canonical basis $a_1, a_2, \ldots, a_g, b_1, b_2, \ldots, b_g$, while $H_2(\Sigma_g \times [0, 1])$ is generated by the surface $\Sigma_g \times \{0\}$. We can place the canonical basis at the top of the cylinder, so that no intersections occur. Thus $T(H_1(\Sigma_g \times [0, 1], \mathbb{Z})) = H_1(\Sigma_g \times [0, 1], \mathbb{Z})$.

We define the **multiplicity** of a class $\alpha \in H_1(M, \mathbb{Z})$, denoted $mul(\alpha)$ as follows:

- If $\alpha \in T(H_1(M, \mathbb{Z}))$ set $mul(\alpha) = 0$.
- If $\alpha \notin T(H_1(M, \mathbb{Z}))$, then $\beta \mapsto \phi(\alpha, \beta) \in \mathbb{Z}$ defines a map from $H_2(M, \mathbb{Z})$ to a nonzero subring of $\mathbb{Z}$, which is necessarily of the form $n_\alpha \mathbb{Z}$, $n_\alpha > 0$. In this case set $mul(\alpha) = n_\alpha$.

Example 6.21 and Example 6.22 show that in the case of a handlebody and of a cylinder over a surface, each homology class has multiplicity zero.

Before we proceed with Przytycki's theorem, we state and prove a well known result about 2-cycles bounded by a link.

Lemma 6.2. *If the oriented link L inside the 3-dimensional manifold M is null-homologous, then there is a smooth, compact, connected, oriented surface Σ, embedded (smoothly) in M, such that $\partial\Sigma = L$.*

Proof. Of course, we can use a triangulation and remove corners and self-crossings from a 2-cycle. But, unlike the proof of the fact that 1-cycles can be represented by smooth multicurves, here more care is needed because a general position argument won't remove multiple points of 2-cycles.

Consider a triangulation of M. Because each 3-dimensional simplex is contractible, there is a homotopy that maps the link to the 1-skeleton. By eventually passing to a subdivision, we can make the homotopy be an ambient isotopy (thus removing self-intersections), and thus we can a priori assume that the link is part of the 1-skeleton. The fact that it is null-homotopic means that it bounds a 2-cycle in the triangulation. The 2-cycle is of the form $f(\Sigma)$ where Σ is an oriented surface and f is an immersion that maps $\partial\Sigma$ to L. We can choose a differentiable structure on M so that the "corners" of this surface (i.e. the regions where two triangles meet along an edge or several triangles meet at a vertex) are made smooth. This is done by the same procedure that turns a convex polyhedron into a sphere.

Finally, we have to eliminate the self-crossings. Along edges that are not part of L, an even number of triangles of Σ meet (so that the edge is canceled in homology). By passing to a finer triangulation and using an ambient isotopy of M we can make sure that along the interiors of edges only 2 or 4 triangles of $f(\Sigma)$ meet (so that all points of higher multiplicity are at vertices). So along each edge, the self-crossing of the 2-cycle is homeomorphic to the crossing of 2 planes, while at each vertex where self-crossing occurs, the two-cycle looks like the intersection of several planes that meet at one point.

Let $X = \cup_{k=0}^{3}[0, i^k] \subset \mathbb{C}$, which is a cross with center at the origin of the complex plane. We can find an $X \times S^1$ immersed in the 2-cycle as follows. Start with an edge where a self-crossing occurs. Choose a neighborhood of that edge in the 2-cycle that looks like $X \times [0, 1]$. Trace this both ways to the endpoints of the edge. Because of the structure of the self-intersection at a vertex, we can unambiguously continue with the same $X \times [0, 1]$ pattern beyond the endpoints. We continue in this manner, and we produce a path in the 1-skeleton. This path might cross itself, but because of the finiteness of the 2-skeleton, it must close up. Once we have the $X \times S^1$ immersed into the 2-cycle, we can smoothen the crossing X to two disjoint curves. This removes some self-intersection from the 2-cycle. The operation is sketched in Figure 6.55.

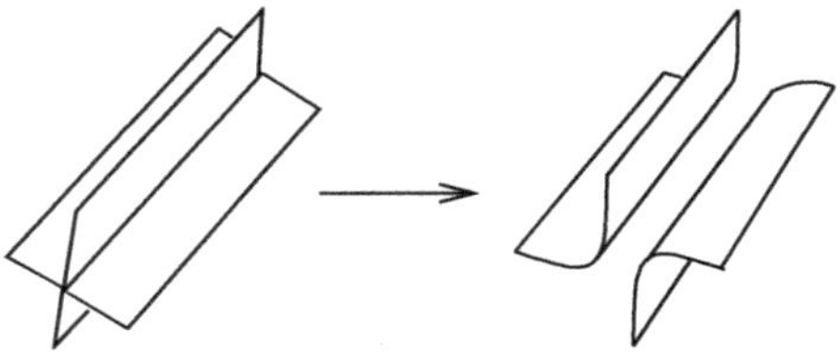

Fig. 6.55 Removing a self-intersection

Repeating this procedure finitely many times, we can transform the 2-cycle into an embedded surface whose boundary is still L.

Finally, to make the surface connected, we can add "tubes" between its connected components. This is done by choosing a path from one connected component to another, add the cylinder which is the boundary of a regular neighborhood of the path and delete the two disks around the end-points of the path. The procedure is illustrated in Figure 6.56.

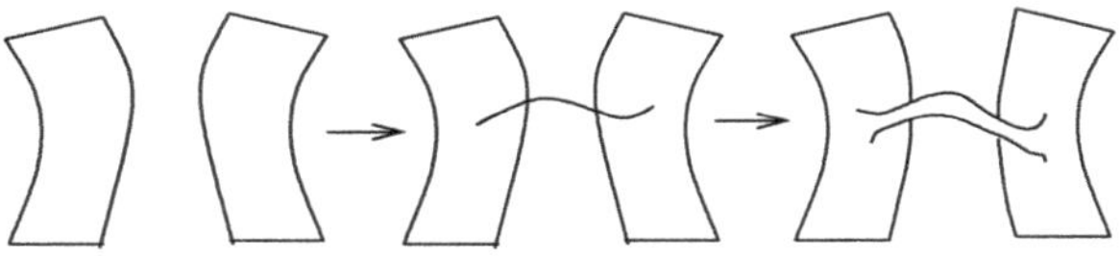

Fig. 6.56 Adding a tube

□

Now we can state and prove the main result of this section.

Theorem 6.10. *(Przytycki) The linking number skein module $\mathcal{L}(M)$ is isomorphic, as a $\mathbb{C}[t,t^{-1}]$-module, to*

$$\bigoplus_{\alpha \in H_1(M,\mathbb{Z})} \mathbb{C}[t,t^{-1}]/(t^{mul(2\alpha)} - 1).$$

Proof. Let I be the ideal generated by elements of the form $(t^{mul(\alpha)}-1)\alpha$, where α ranges over $H_1(M,\mathbb{Z})$. We see that

$$\bigoplus_{[L]\in H_1(M,\mathbb{Z})} \mathbb{C}[t,t^{-1}]/(t^{mul(2[L])}-1) = \mathbb{C}[t,t^{-1}]H_1(M,\mathbb{Z})/I.$$

We want to construct an isomorphism

$$\mathbb{C}[t,t^{-1}]H_1(M,\mathbb{Z})/I \to \mathcal{L}(M).$$

For that we need some notation and some results.

Let K and K' be two knots in M obtained by choosing two different framings of the same smooth simple closed curve γ in M. Then K and K' are embedded annuli, and to pass from K to K' one has to perform a finite number of (full) twists of the annulus of K around γ, as described in Figure 6.57. Because the manifold M is oriented, it is clear which twists are positive, which are negative (just embed a ball in $B^3 \hookrightarrow M$ by an orientation preserving diffeomorphism, and decide the sign of the twist in B^3). Count the number of twists to be performed on K to obtain K', and let $\Delta(K,K')$ be this number if the twists are positive, and the negative of this number if the twists are negative. In general, for framed links L and L' obtained by framing the same family of curves, let $\Delta(L,L')$ be the sum of Δ's for all curves from the collection. Then in $\mathcal{L}(M)$,

$$<L'> = t^{\Delta(L,L')} <L> .$$

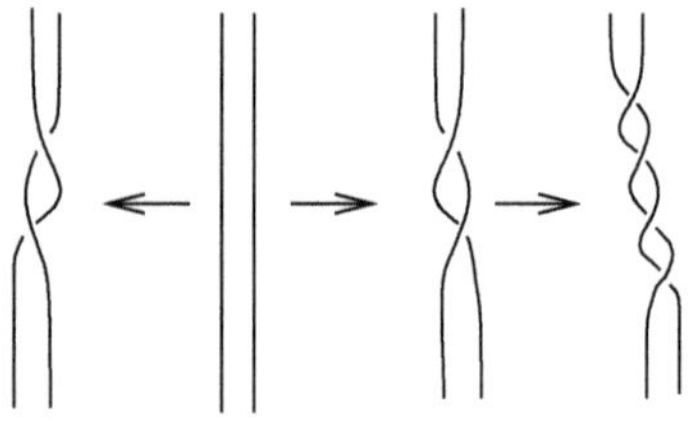

Fig. 6.57 Performing full twists on a framed knot

Let L_1 and L_2 be two framed links in M, so that the underlying unframed links bound a surface Σ in M. Part of $\partial\Sigma$ is the underlying unframed link of L_1, the other part is the underlying unframed link of $-L_2$. Endow these with the framing defined by a regular neighborhood of the boundary in Σ, and call the new framed links $\partial_1\Sigma$ respectively $\partial_2\Sigma$. Define

$$\Delta_\Sigma(L_1,L_2) = \Delta(\partial_2\Sigma, L_2) - \Delta(\partial_1\Sigma, L_1).$$

Note that $\partial_1\Sigma$ and $\partial_2\Sigma$ are curves on Σ endowed with the blackboard framing, and they are also homologous in Σ. By the discussion in §5.3.2,

$<\partial_1\Sigma>=<\partial_2\Sigma>$ in a cylindrical neighborhood of Σ and hence in M. It follows that, in $\mathcal{L}(M)$,

$$<L_2>= t^{\Delta(L_1,L_2)} <L_1> .$$

Lemma 6.3. *Let L_1 and L_2 be framed links in M representing the same homology class in $H_1(M,\mathbb{Z})$. Then*

$$\Delta_\Sigma(L_1,L_2)(mod\ 2mul([L_1]))$$

does not depend on Σ.

Proof. Let Σ_1 and Σ_2 be two surfaces embedded in M such that the underlying unframed link of $L_1 \cup (-L_2)$ bounds both Σ_1 and Σ_2. Define $\partial_j\Sigma_k$, $j,k = 1,2$ as above. Consider the 2-cycle

$$\beta = [\Sigma_1 \cup (-\Sigma_2)].$$

Because $[L_1] = [L_2]$, $\phi([L_1],\beta) = \phi([L_2],\beta)$.

Let us interpret $\phi([L_1],\beta)$ geometrically. Consider a solid torus which is a tubular neighborhood of the unframed link underlying L_1. The framing of L_1 determines a simple closed curve on the boundary of this solid torus, and $\phi([L_1],\beta)$ counts with sign the number of times this curve intersects $\Sigma_1 \cup (-\Sigma_2)$. This is the sum of the number of times it intersects Σ_1 and the number of times it intersects Σ_2, again counted with sign. The first of these numbers is the difference between the framing of L_1 and $\partial_1\Sigma_1$ and the second is the difference between the framing of L_1 and $\partial_1\Sigma_2$.

Doing the same for the L_2 boundary component, we deduce that

$$\phi([L_1],\beta) - \phi([-L_2],\beta) = \Delta_{\Sigma_1}(L_1,L_2) - \Delta_{\Sigma_2}(L_1,L_2).$$

Because $[L_1] = [L_2]$, $\phi([L_1],\beta) = \phi([L_2],\beta)$, so the first of these differences is $2\phi([L_1],\beta)$. The latter is a multiple of $2mul([L_1])$, and hence $\Delta_{\Sigma_1}(L_1,L_2) - \Delta_{\Sigma_2}(L_1,L_2)$ is a multiple of $2mul([L_1])$. The lemma is proved. □

For two homologous links, L_1 and L_2, we define

$$\delta(L_1,L_2) = \Delta_\Sigma(L_1,L_2)(\text{mod } 2mul([L_1])) \in \mathbb{Z}_{2mul(L_1,L_2)}.$$

Because of Lemma 6.2 and Lemma 6.3 $\delta(L_1,L_2)$ is well defined.

We are now in position to introduce the desired isomorphism. First, for each homology class $\alpha \in H_1(M,\mathbb{Z})$ we fix a link L_α such that $[L_\alpha] = \alpha$. Let the trivial class 0 be represented by the trivial knot. Consider the homomorphism of $\mathbb{C}[t,t^{-1}]$-modules

$$F : \mathbb{C}[t,t^{-1}]\text{Link}(M) \to \mathbb{C}[t,t^{-1}]H_1(M,\mathbb{Z})/I$$

defined by

$$F(L) = t^{\delta(L_{[L]},L)} L_{[L]}.$$

Note that $L_{[L]}$ might or might not be L depending on whether or not L is the link that represents the homology class $[L]$. Nevertheless L and $L_{[L]}$ are homologous so the exponent of t makes sense.

One should point out that, by Lemma 6.3, $\delta(L_{[L]}, L)$ is well defined modulo $2mul([L])$, so the element $t^{\delta(L_{[L]},L)} L_{[L]}$ is well defined in $\mathbb{C}[t, t^{-1}]H_1(M, \mathbb{Z})/I$. Thus F is well defined. This homomorphism factors through the skein relation (ln2) (Figure 5.5) because the trivial class is represented by the trivial knot.

Let us show that F factors through the skein relations (ln1) (Figure 5.4). Note that if L_2 is obtained from L_1 by applying the first of the skein relations in Figure 5.4, then L_1 and L_2 are homologous, so $L_{[L_1]} = L_{[L_2]}$. It follows that

$$\delta(L_{[L_2]}, L_2) - \delta(L_{[L_1]}, L_1) = \delta(L[L_1], L_2) - \delta(L_{[L_1]}, L_1) = \delta(L_1, L_2).$$

The skein relation translates to

$$<L_2>= t <L_1>,$$

so all we have to check is that

$$\delta(L_1, L_2) = 1.$$

For this we need to find a surface Σ such that $\Delta_\Sigma(L_1, L_2) = 1$. Consider the link $L_1^{\|2}$, where we take the double parallel copy of each link component. Then outside a ball B^3 containing the crossing, $L_1^{\|2}$ consists of a copy of L_1 and a copy of L_2, since outside the ball, L_1 and L_2 are the same. So outside B^3 we let Σ be the annulus that connects one parallel copy of L_1 to the other. Inside the ball we choose Σ as shown in Figure 6.58. Here L_1, L_2 and $\partial_2\Sigma$ are with the blackboard framing, but $\partial_1\Sigma$ is not. In fact $\partial_1\Sigma$ differs from the part of L_1 that crosses over by a negative half-twist, and from the part of L_1 that crosses under by another negative half-twist. In other words, $\Delta(\partial\Sigma, L_1) = -1$. So

$$\Delta_\Sigma(L_1, L_2) = \Delta(\partial_2\Sigma, L_2) - \Delta(\partial_2\Sigma, L_1) = 0 - (-1) = 1.$$

The same considerations apply to negative crossings. Hence F descends to a module homomorphism

$$\hat{F} : \mathcal{L}(M) \to \mathbb{C}[t, t^{-1}]H_1(M, \mathbb{Z})/I.$$

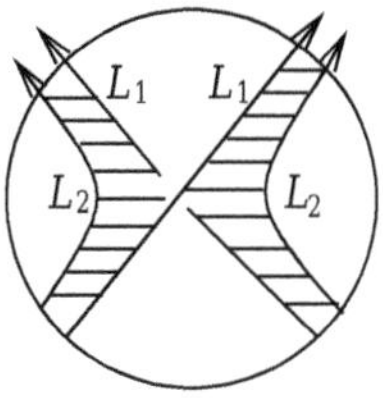

Fig. 6.58 Surface that bounds the skein relation (ln1)

To show that $\hat{F}$ is an isomorphism, we construct its inverse. First, there is a $\mathbb{C}[t, t^{-1}]$-homomorphism

$$G : \mathbb{C}[t, t^{-1}]H_1(M, \mathbb{Z}) \to \mathcal{L}(M), \quad G([L]) = <L> .$$

Let us explain why G is well defined. Assume that L is homologous to L'. Then $L \cup L'^{-1}$ is null-homologous, where L'^{-1} is L' with orientation reversed. Consider the surface Σ provided by Lemma 6.2. Embed a cylinder $\Sigma \times [0, 1]$ around Σ in M. Then, by Proposition 5.8, $< L >=< L' >$ in $\mathcal{L}(\Sigma \times [0, 1])$, and so they are mapped to equal skeins by the skein module homomorphism induced by the embedding of the cylinder in M.

The next result shows that G descends to a homomorphism

$$\hat{G} : \mathbb{C}[t, t^{-1}]H_1(M, \mathbb{Z})/I \to \mathcal{L}(M).$$

Lemma 6.4. *Let L be a framed oriented link and Σ a closed oriented surface in M. Then, in $\mathcal{L}(M)$,*

$$(t^{2\phi([L],[\Sigma])} - 1) <L>= 0.$$

Proof. Let K be a trivial framed knot that bounds a disk B^2 in Σ disjoint from L. Orient K so that its orientation agrees of that of Σ. Add K to L K to obtain $L' = L \cup K$; by the skein relation (ln2), $< L \cup K >=< L >$. Let γ be a multicurve on Σ such that each component γ_k bounds a disk B_k^2 having one intersection point with L. Orient the components of γ so that $K \cup \gamma$ bounds the surface $\Sigma \backslash (B^2 \cup (\cup_k B_k^2))$. Then K and γ are homologous in $H_1(\Sigma \backslash (B^2 \cup (\cup_k B_k^2)), \mathbb{Z})$. Using Proposition 5.8 we deduce that $<K>=<\gamma>$ in an embedded $\Sigma \backslash (B^2 \cup (\cup_k B_k^2)) \times [0, 1]$. Consequently,

$$<L>=<L \cup K>=<L \cup \gamma> .$$

We now remove the components of γ using the skein relations as shown in Figure 6.59. Note that the number of times a component of γ spins positively around L minus the number of times a component of γ spins negatively around L is precisely the algebraic intersection number of L and Σ, which is equal to $\phi([L], [\Sigma])$. The lemma is proved. $\square$

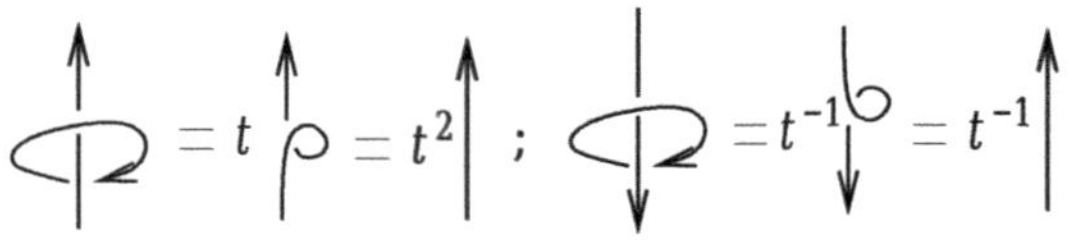

Fig. 6.59 The removal of γ

We return to the proof of the theorem. Each homology class $\alpha \notin T(H_1(M,\mathbb{Z}))$ can be represented by some link L, and we can find a closed surface Σ such that $\mathrm{Mul}(\Sigma)([L]) = \phi([L],[\Sigma])$. Hence, as a corollary of Lemma 6.4, we obtain that

$$(t^{2mul([L])} - 1) < L >= 0.$$

it follows that $\hat{G}$ is well defined.

It is not hard to check that

$$\hat{F} \circ \hat{G} = Id \text{ and } \hat{G} \circ \hat{F} = Id.$$

The theorem is proved. □

Conclusions to the sixth chapter

From the contents of this chapter the following will play an important role in the subsequent chapters:

1. The description of an element of the mapping class group of a closed genus g surface by **surgery** on a link in the cylinder over the surface.
2. The surgery **construction of closed 3-dimensional manifolds**.
3. **Kirby's theorem**, which states that two surgery links that yield the same 3-dimensional manifold can be changed one into the other by adding/deleting trivial link components with framing ± 1 and by performing slides of one link component over another.
4. The surgery description of **cobordisms** and the relative version of Kirby's theorem in this case, with a particular situation being the cobordisms that describe mapping cylinders of elements of the mapping class group.
5. **Wall's non-additivity formula**, which describes how the signature of 4-dimensional manifolds changes under gluings with corners.

Chapter 7

The discrete Fourier transform and topological quantum field theory

In this chapter we will reveal a profound relationship between the discrete Fourier transform that acts on theta functions and 3- and 4-dimensional topology. As an application, we will use 4-dimensional manifolds to resolve the projectivity of the representation of the mapping class group on theta functions. We will construct a topological quantum field theory that brings together, for the Riemann surfaces of all genera, their Hilbert spaces of theta functions, the actions of the finite Heisenberg group and of the mapping class group, as well as the discrete Fourier transforms defined by pairs of canonical bases. Moreover, we will prove that such a topological quantum field theory is unique.

7.1 The discrete Fourier transform and handle slides

After the detour through 3- and 4-dimensional topology, we return to the setting of Chapters 4 and 5. We intend to give a topological description of the discrete Fourier transform, different from the one in Chapter 5.

7.1.1 *The discrete Fourier transform as a skein*

Consider a closed genus g Riemann surface Σ_g, $g \geq 1$, endowed with a canonical basis $a_1, a_2, \ldots, a_g, b_1, b_2, \ldots, b_g$. As before, let $\mathbf{L}$ be the subgroup of $H_1(\Sigma_g, \mathbb{Z})$, generated by $b_1, b_2, \ldots, b_g$, which gives rise to a particular version of the Jacobian variety, theta functions, Schrödinger representation, and representation of the mapping class group.

Let h be an element of the mapping class group and let $\mathcal{F}_{\mathbf{L}}(h)$ be the discrete Fourier transform induced by h. Combining Proposition 4.12 with Theorem 5.5, we deduce that every linear map on the space of theta functions $\mathbf{\Theta}_N^\Pi(\Sigma_g)$ can be represented as left multiplication on $\mathcal{L}_N(H_g)$ by a skein in $\mathcal{L}_N(\Sigma_g)$. In particular this is true for the discrete Fourier transform $\mathcal{F}_{\mathbf{L}}(h)$. We want to determine the skein associated to the discrete Fourier transform.

In deriving the formula for the skein representing $\mathcal{F}_{\mathbf{L}}(h)$, the **exact Egorov identity** (4.36) in its topological incarnation (5.18) will play the key role. Recall that translated to the language of skeins, this identity reads

$$\mathcal{F}_{\mathbf{L}}(h)\sigma\mathcal{F}_{\mathbf{L}}(h)^{-1} = h(\sigma), \quad \sigma \in \mathcal{L}_N(\Sigma_g). \tag{7.1}$$

The case of the Dehn twist about b_1

Let $h = T_{b_1}$, the positive Dehn twist about a smooth oriented simple closed curve that represents b_1.

First, recall the identification between skeins and operators provided by Theorem 5.5. Assimilating the operators of the form $\exp(kE)$ with constants, we can write

$$\mathcal{F}_{\mathbf{L}}(T_{b_1}) = \sum c_{\mathbf{p},\mathbf{q}} \exp(\mathbf{p}^T\mathbf{P} + \mathbf{q}^T\mathbf{Q}).$$

Our goal is to compute the coefficients $c_{\mathbf{p},\mathbf{q}}$. Switching to skeins, we rewrite the exact Egorov identity as

$$T_{b_1}(\sigma)\mathcal{F}_{\mathbf{L}}(T_{b_1}) = \mathcal{F}_{\mathbf{L}}(T_{b_1})\sigma$$

Because $T_{b_1}(\sigma) = \sigma$ for all skeins that do not contain curves that intersect b_1, it follows that $\mathcal{F}_{\mathbf{L}}(T_{b_1})$ commutes with all such skeins. It also commutes with the multiples of b_1 (viewed as a skein with the blackboard framing). Switching back to the operator notation, we conclude that $\mathcal{F}_{\mathbf{L}}(T_{b_1})$ commutes with all operators of the form $\exp(\mathbf{p}^T\mathbf{P} + \mathbf{q}^T\mathbf{Q})$ with p_1, the first entry of $\mathbf{p}$, equal to 0. Because of the noncommutation relation (4.25), we must have

$$\mathcal{F}_{\mathbf{L}}(T_{b_1}) = \sum_{j=0}^{N-1} c_j \exp(jQ_1).$$

To compute the coefficients c_j, we use the exact Egorov identity in the particular case where $\sigma = a_1$, namely where the operator is $\exp(P_1)$. Since $T \cdot \exp(P_1) = \exp(P_1 + Q_1)$ the Egorov identity reads

$$\exp(P_1 + Q_1) \sum_{j=0}^{N-1} c_j \exp(jQ_1) = \sum_{j=0}^{N-1} c_j \exp(jQ_1) \exp(P_1).$$

We transform this further into

$$\sum_{j=0}^{N-1} c_j e^{\frac{\pi i}{N} j} \exp[P_1 + (j+1)Q_1] = \sum_{j=0}^{N-1} c_j e^{-\frac{\pi i}{N} j} \exp(P_1 + jQ_1),$$

or, taking into account that $\exp(P_1) = \exp(P_1 + NQ_1)$,

$$\sum_{j=0}^{N-1} c_{j-1} e^{\frac{\pi i}{N}(j-1)} \exp(P_1 + jQ_1) = \sum_{j=0}^{N-1} c_j e^{-\frac{\pi i}{N} j} \exp(P_1 + jQ_1),$$

where $c_{-1} = c_{N-1}$. It follows that $c_j = e^{\frac{\pi i}{N}(2j-1)} c_{j-1}$ for all j. Normalizing so that $\mathcal{F}_{\mathbf{L}}(T_{b_1})$ is a unitary map and $c_0 > 0$, we obtain $c_j = N^{-1/2} e^{\frac{\pi i}{N} j^2}$, and hence

$$\mathcal{F}_{\mathbf{L}}(T_{b_1}) = N^{-1/2} \sum_{j=0}^{N-1} e^{\frac{\pi i}{N} j^2} \exp(jQ_1).$$

In the language of skeins,

$$\mathcal{F}_{\mathbf{L}}(T_{b_1}) = N^{-1/2} \sum_{j=0}^{N-1} t^{j^2} a_1^j.$$

This is the same as

$$\mathcal{F}_{\mathbf{L}}(T_{b_1}) = N^{-1/2} \sum_{j=0}^{N-1} \gamma^j \tag{7.2}$$

where γ is obtained by adding one full positive twist to the framing of a_1 (the twist is positive in the sense that, as skeins, $\gamma = ta_1$). The curve γ is depicted in Figure 7.1. We recognize γ as the surgery curve for the Dehn twist T_{b_1}, described in Example 6.1 from §6.1. Of course T can be described by more than one surgery along a framed link in $\Sigma_g \times [0,1]$, but in the future when we refer to the **surgery curve** of the Dehn twist T_{b_1} we will always mean this particular curve, unless otherwise specified.

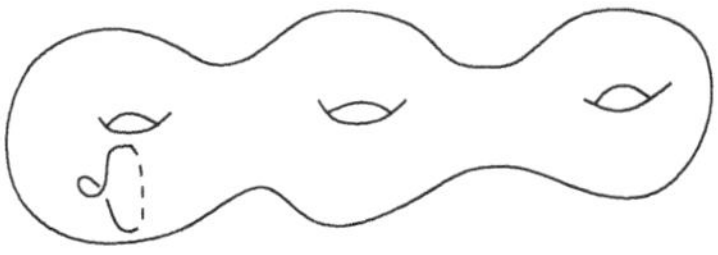

Fig. 7.1 The skein of the Fourier transform of the Dehn twist about b_1

The skein Ω

Formula (7.2) leads us to consider a skein that will play a central role in this chapter. Consider the solid torus H_1 endowed with a canonical basis of its first homology consisting of the element a_1. Under the identification $H_1 = A \times [0,1]$ (where A is the annulus $\{z \in \mathbb{C} \,|\, 1 \le |z| \le 3\}$), turn a_1 into a skein by endowing it with the blackboard framing of the cylinder.

The reduced linking number skein module of $\mathcal{L}_N(H_1)$ is, by Proposition 5.13, an N-dimensional vector space with basis

$$\emptyset, a_1, \ldots, a_1^{\| N-1}.$$

Alternately, it is the vector space of theta functions, $\boldsymbol{\Theta}_N^{\Pi}(\Sigma_1)$, with basis the theta series $\theta_0^{\tau}(z), \theta_1^{\tau}(z), \ldots, \theta_{N-1}^{\tau}(z)$, where τ is a complex number with positive imaginary part.

Definition 7.1. We define the element

$$\Omega = N^{-1/2} \sum_{j=0}^{N-1} < a_1^{\| j} > = N^{-1/2} \sum_{j=0}^{N-1} \theta_j^{\tau}(z) \tag{7.3}$$

in $\mathcal{L}_N(H_1) = \boldsymbol{\Theta}_N^{\Pi}(\Sigma_1)$.

The diagram of Ω is shown in Figure 7.2.

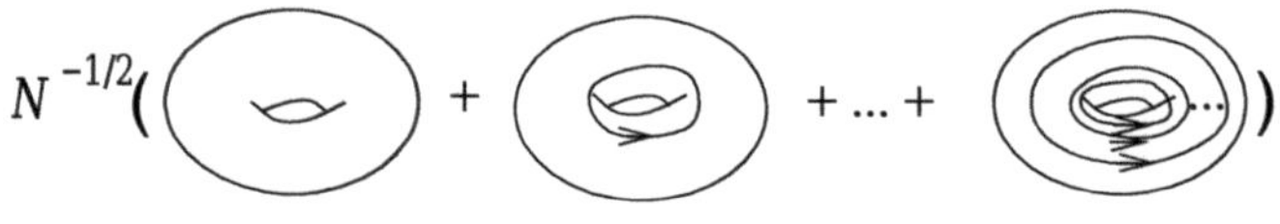

Fig. 7.2 The skein Ω

Remark 7.1. Note that

$$\Omega = S\theta_0^{\tau}(z),$$

where S is the S-map on the torus.

Now consider a framed oriented knot K in the interior of some smooth, compact, oriented manifold M. This knot has a regular neighborhood in M consisting of a solid torus N on whose boundary the framing determines a curve. Remove the interior of N from M and glue to the resulting manifold, along ∂N, the solid torus H_1 with Ω inside such that the circle $\{z \,|\, |z| = 1\} \times \{0\}$ is glued to the framing curve. This gives rise to a skein in $\mathcal{L}_N(M)$, which we denote by $\Omega(K)$, and call **the coloring of the knot K by Ω**. So we have the following definition.

Definition 7.2. The coloring of the framed oriented knot K by Ω is

$$\Omega(K) = N^{-1/2} \sum_{j=0}^{N-1} < K^{\|j} > .$$

Example 7.1. In the case where $N = 4$ and K is the figure 8 knot, the skein $\Omega(K)$ is depicted in Figure 7.3.

Fig. 7.3 The skein $\Omega(K)$ for K the figure 8 knot

We extend this to links.

Definition 7.3. If L is a framed oriented link, with link components $L_1, L_2, \ldots, L_k$, we let $\Omega(L)$, the **coloring of the link L by Ω**, be the skein in $\mathcal{L}_N(M)$ given by

$$\Omega(L) = N^{-1/2} \sum_{j_1=0}^{N-1} \sum_{j_2=0}^{N-1} \cdots \sum_{j_k=0}^{N-1} < L_1^{\|j_1} \cup L_2^{\|j_2} \cup \cdots \cup L_k^{\|j_k} > .$$

In other words, we replace each component L_j of L by $\Omega(L_j)$.

We can interpret (7.2) as saying that the skein associated to $\mathcal{F}_{\mathbf{L}}(T_{b_1})$ is the coloring of the surgery curve of the Dehn twist T_{b_1} by Ω.

Here are two straightforward properties of Ω that will be used in the sequel.

Proposition 7.1. *(i) The skein $\Omega(L)$ is independent of the orientations of the components of L.*
(ii) The skein relation from Figure 7.4 holds, where the n parallel strands point in the same direction.

Proof. (i) In $\mathcal{L}_N(H_1)$,

$$<(-a_1)^{\|j}>=<a_1^{\|N-j}>,$$

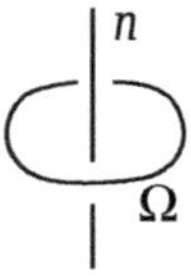

$$= \begin{cases} N^{1/2}\langle\phi\rangle & \text{if } n=0 \\ 0 & \text{if } 0<n<N \end{cases}$$

Fig. 7.4 An identity for Ω

so Ω does not change when changing the orientation of a_1. Therefore the orientation of a_1 is irrelevant.

(ii) When $n = 0$ there is nothing to prove, the trivial knot decorated by Ω is just $N^{-1/2}(1 + 1 + \cdots + 1) = N^{1/2}$.

If $n \neq 0$, then by resolving all crossings in the diagram we obtain n vertical parallel strands with the coefficient

$$N^{-1/2} \sum_{j=0}^{N-1} t^{\pm 2nj} = N^{-1/2} \cdot \frac{t^{2Nn} - 1}{t^{2n} - 1}.$$

where the signs in the exponents are either all positive, or all negative. Since t^2 is a primitive Nth root of unity, this is equal to zero. Hence the conclusion. □

The case of a product of Dehn twists

A first remark, consequence of the above discussion, is that the skein associated to the Dehn twist about b_1 does not depend on the choice of the canonical basis of the surface Σ_g. This is because the exact Egorov identity involves only skeins in the cylinder over the surface, and this identity completely determines the skein associated to the Fourier transform, regardless of which choice we make for the polarization that gives rise to the Schrödinger representation. So for the moment we can write simply $\mathcal{F}(T)$ for $\mathcal{F}_{\mathbf{L}}(T)$.

Since by Proposition 3.1 every nonseparating simple closed curve is the b_1-curve of a canonical basis, we deduce that the skein associated to the positive Dehn twist about that curve is obtained by coloring the surgery curve of the Dehn twist by Ω. It is not hard to see that the same is true for the negative Dehn twist. By Theorem 3.9 every element of the mapping class group of Σ_g is the product of Dehn twists about nonseparating curves. Thus we have the following result.

Proposition 7.2. *Let $h \in MCG(\Sigma_g)$ be obtained as a composition of Dehn twists $h = T_1T_2 \cdots T_n$ about nonseparating curves. Express each Dehn twist*

T_j by surgery on a curve γ_j as above, and consider the link $L_h = \gamma_1 \cup \gamma_2 \cup \cdots \cup \gamma_n$, which expresses h as surgery on the framed link L_h in $\Sigma_g \times [0,1]$. Then, up to multiplication by a number of absolute value 1, the discrete Fourier transform $\mathcal{F}(h) : \mathcal{L}_N(H_g) \to \mathcal{L}_N(H_g)$ is given by

$$\mathcal{F}(h)\beta = \Omega(L_h)\beta.$$

We will derive a better formulation in §7.4.

Notation: From now on we denote the discrete Fourier transform by $\mathcal{F}(h)$, ignoring $\mathbf{L}$, since the discrete Fourier transform is given by a skein which is independent of $\mathbf{L}$.

7.1.2 *The exact Egorov identity and handle slides*

Next, we give the exact Egorov identity a topological interpretation in terms of handle slides. For this we look at its skein theoretical version (7.1). We start again with some heuristics.

Example 7.2. Consider the torus Σ_1 with a choice of canonical basis for $H_1(\Sigma_g, \mathbb{Z})$. This choice allows us to represent the oriented simple closed curves on the torus by pairs of integers (p, q) like in Chapter 3. As before, we view these curves as skeins with the blackboard framing.

For the positive twist T along the curve $(0, 1)$ and the operator represented by the curve $(1, 0)$ the exact Egorov identity reads

$$\mathcal{F}(T)(1,0) = (1,1)\mathcal{F}(T).$$

In §7.1.1 we gave a description of $\mathcal{F}(T)$ as the skein in $\mathcal{L}_N(\Sigma_1)$ obtained by coloring the surgery curve of T by Ω. The exact Egorov identity then describes the commutation of two skeins in $\mathcal{L}_N(\Sigma_1)$. This commutation is shown graphically in Figure 7.5.

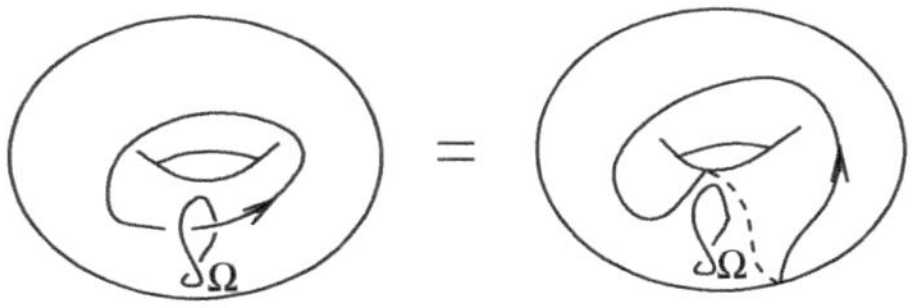

Fig. 7.5 The exact Egorov identity on the torus

The diagram on the right is the same as the one from Figure 7.6.

As such, the curve $(1, 1)$ is obtained by sliding the curve $(1, 0)$ along the surgery curve of the positive twist (recall Definition 6.10).

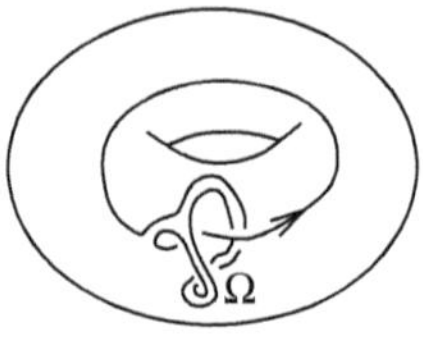

Fig. 7.6 The exact Egorov identity as the slide of one knot over another

We can adapt the above analysis to the genus 2 surface and obtain the next result. Recall the basis of $\mathcal{L}_N(\Sigma_2)$, as a vector space, given by pairs $((p_1,p_2),(q_1,q_2))$ with $p_1,p_2,q_1,q_2 \in \mathbb{Z}_N$.

Lemma 7.1. *In $\mathcal{L}_N(\Sigma_2)$, the following commutation relation holds*

$$\mathcal{F}(T_{b_1})((1,0),(0,0)) = ((1,1),(0,0))\mathcal{F}(T_{b_1}).$$

Consequently, in $\mathcal{L}_N(\Sigma_2)$, one can slide the curve $((1,0),(0,0)) \subset \Sigma_2$ over the surgery curve of T_{b_1} colored by Ω.

We are now in position to prove a handle slide invariance of skeins in arbitrary manifolds as a direct corollary of the exact Egorov identity.

Theorem 7.1. *Let M be a smooth, compact, oriented 3-dimensional manifold, L an arbitrary oriented framed link in M and K_0 and K two oriented framed knots in M disjoint from L. Then, in $\mathcal{L}_N(M)$, one has the equality*

$$<L \cup K_0 \cup \Omega(K)>=<L \cup (K_0\#K) \cup \Omega(K)>,$$

however one does the band sum $K_0\#K$.

Remark 7.2. To make sure that the reader understands, the skein $< L \cup K_0 \cup \Omega(K) >$ is $N^{-1/2}\sum_{j=0}^{N-1} <L \cup K_0 \cup K^{||j}>$.

Proof. Isotope K_0 along the embedded band that defines $K_0\#K$ to a knot K_0' that intersects K. There is an embedded punctured torus $\Sigma_{1,1}$ in M, disjoint from L, which contains $K_0' \cup K$ on its boundary, as shown in Figure 7.7 a). In fact, by looking at a neighborhood of this torus, we can find an embedded $\Sigma_{1,1}\times[0,1]$ such that $K_0'\cup K \subset \Sigma_{1,1}\times\{0\}$. The boundary of this cylinder is a genus 2 surface Σ_2, and K_0' and K_1 lie in a punctured torus of this surface and intersect at exactly one point. By pushing off K_0' to a knot isotopic to K_0 (which we identify with K_0), we see that we can place K_0 and K in an embedded $\Sigma_2 \times [0,1]$ such that $K_0 \in \Sigma_2 \times \{0\}$ and $K \in \Sigma_2 \times \{1/2\}$.

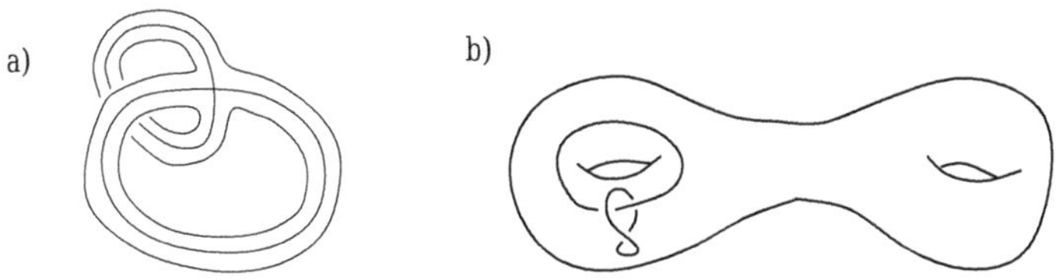

Fig. 7.7 Turning a slide into the exact Egorov identity

By performing a twist in $\Sigma_{1,1} \times [0,1]$ we can change the framing of K in such a way that K_0 and K look inside $\Sigma_2 \times [0,1]$ like in Figure 7.7 b). Then K_0 is mapped to $K_0 \# K$ in $\Sigma_2 \times \{1\}$ by the Dehn twist of Σ_2 with surgery diagram K. Hence the equality

$$K_0 \cup \Omega(K) = (K_0 \# K) \cup \Omega(K)$$

in $\Sigma_2 \times [0,1]$ is just the exact Egorov identity, which we know is true. By embedding $\Sigma_2 \times [0,1]$ in $\Sigma_{1,1} \times [0,1]$ we conclude that this equality holds in $\Sigma_{1,1} \times [0,1]$. By embedding $\Sigma_{1,1} \times [0,1]$ in M with the link L inside, we conclude that the identity from the statement holds (as a consequence of Proposition 5.7). □

Remembering our discussion from Chapter 6 on how sliding one knot over another is related to handle slides in a 4-dimensional handlebody, we see that the above theorem relates the discrete Fourier transform and the exact Egorov identity that it satisfies to handle slides.

Theorem 7.1 has three applications:

- the construction of a topological invariant for closed 3-dimensional manifolds;
- the computation of the reduced linking number skein modules of 3-dimensional manifolds;
- a topological way to resolve the projective ambiguity of the representation of the mapping class group.

We will elaborate on these three applications in the sections that follow. Then, combining them, we will construct a topological quantum field theory associated to classical theta functions. This will be done in the framework established by Sir Michael Atiyah [Atiyah (1990)], [Atiyah (1988)].

7.2 The Murakami-Ohtsuki-Okada invariant of a closed 3-dimensional manifold

7.2.1 *The construction of the invariant*

Theorem 7.1 shows that if L is a framed link in S^3, then the skein $\Omega(L)$ is invariant in $\mathcal{L}_N(S^3)$ under the Kirby move (k2) performed on L. We have proved in Chapter 5 that $\mathcal{L}_N(S^3)$ is equal to $\mathbb{C}\emptyset$. By ignoring the empty link we can therefore identify $\mathcal{L}_N(S^3)$ with $\mathbb{C}$.

Invariance under (k2) suggest the possibility of defining a numerical topological invariant that associates to a compact, connected, oriented 3-dimensional manifold M without boundary the number $\Omega(L) \in \mathcal{L}_N(S^3) = \mathbb{C}$, where L is a surgery link for M. In view of Kirby's Theorem (Theorem 6.5) we also need to check invariance under the Kirby move (k1).

Recall the unknots with framing $+1$ respectively -1 from Figure 7.8. For further reference, we will call these framed knots U_+ respectively U_-.

Fig. 7.8 The unknots U_+ and U_-

Lemma 7.2. *In any 3-dimensional manifold, the following equalities hold*

$$\Omega(U_+) = e^{\frac{\pi i}{4}} <\emptyset>, \quad \Omega(U_-) = e^{-\frac{\pi i}{4}} <\emptyset> .$$

Consequently $\Omega(U_+) \cup \Omega(U_-) =<\emptyset>$.

Proof. We have

$$\Omega(U_+) = N^{-1/2} \sum_{j=0}^{N-1} t^{j^2} <\emptyset> .$$

Because N is even,

$$\sum_{j=0}^{N-1} t^{j^2} = \sum_{j=0}^{N-1} e^{\frac{\pi i}{N} j^2} = \sum_{j=0}^{N-1} e^{\frac{\pi i}{N}(N+j)^2} = \sum_{j=N}^{2N-1} t^{j^2}.$$

Hence

$$\sum_{j=0}^{N-1} t^{j^2} = \frac{1}{2} \sum_{j=0}^{2N-1} e^{\frac{2\pi i}{2N} j^2}.$$

The last expression is a **Gauss sum**, which is equal to $e^{\frac{\pi i}{4}}N^{1/2}$ (see [Lang (1994)] page 87). This proves the first formula.

On the other hand,

$$\Omega(U_-) = N^{-1/2}\sum_{j=0}^{N-1} e^{-\frac{\pi i}{N}j^2} <\emptyset>$$

which is the complex conjugate of $\Omega(U_+)$. Hence, the second formula. □

Unfortunately the lemma shows that $\Omega(L)$ is not invariant under the Kirby move (k1) performed on L. But the Kirby move (k1) changes $\sigma(L)$ (the difference in the number of positive and of negative eigenvalues of the linking matrix) by $+1$ and -1 respectively, depending on whether we add a U_+ or a U_-. So if we incorporate $\sigma(L)$ in the definition of the invariant we obtain a true 3-dimensional manifold invariant. We have the following result.

Theorem 7.2. *Given a closed, oriented, 3-dimensional manifold M obtained as surgery on the framed link L in S^3, the number*

$$Z_N(M) = e^{-\frac{\pi i}{4}\sigma(L)}\Omega(L)$$

is a topological invariant of the manifold M.

This invariant was introduced by Hitoshi Murakami, Tomotada Ohtsuki, and Masae Okada in [Murakami *et al.* (1992)]. They did it by finding a linking number analogue of the Reshetikhin-Turaev invariant of 3-dimensional manifolds [Reshetikhin and Turaev (1991)], the latter being associated to the Jones polynomial. The above discussion shows that this invariant is associated naturally to theta functions and it that it arises from the study of the discrete Fourier transform.

Let us return to the framework of Chapter 6 and change slightly the point of view. If M is obtained by performing surgery on L, that is $M = M_L$, then L also specifies how to add handles to B^4 to obtain a 4-dimensional manifold $\mathbf{M}_L$ such that $\partial\mathbf{M}_L = M_L$. Theorem 6.4 shows that $\sigma(\mathbf{M}_L) = \sigma(L)$, so in the above formula we can substitute $\sigma(\mathbf{M}_L)$ in the exponent. We can therefore think that $\mathbf{M}_L$ endows the 3-dimensional manifold M with an integer $\sigma(\mathbf{M}_L)$.

Definition 7.4. We call the pair $(M, \sigma(\mathbf{M}_L)$ a **framed** oriented, compact, connected, 3-dimensional manifold.[1]

[1]This is because $\sigma(\mathbf{M}_L)$ can also be calculated from the framing of the tangent bundle of M (see [Walker (1991)]).

Note that every integer n is of the form $\sigma(\mathbf{M}_L)$ for some choice of L that yields the original manifold M. So all pairs (M, n) where M is a manifold and n is an integer are framed manifolds. We can therefore define an invariant of framed, compact, connected, oriented 3-dimensional manifolds without boundary by

$$Z_N(M, n) = \Omega(L)$$

where L is a link such that $\sigma(L) = \sigma(\mathbf{M}_L) = n$.

7.2.2 *The computation of the invariant*

The Murakami-Ohtsuki-Okada invariant was computed explicitly by its authors for all manifolds. We present here their work, as a detour through algebraic topology and number theory. This section is somewhat technical and can be skipped by a reader interested solely in the theory of theta functions.

First, rewrite the invariant as

$$Z_N(M) = e^{-\frac{i\pi}{4}\sigma(L)} N^{-n/2} \sum_{\boldsymbol{\nu} \in \mathbb{Z}_N^n} t^{\boldsymbol{\nu}^T A_L \boldsymbol{\nu}}, \tag{7.4}$$

where n is the number of components of the framed link L and A_L is the linking matrix of L.

Remark 7.3. In this formula the range of $\boldsymbol{\nu}$ should be $\{0, 1, \ldots, N-1\}^n$, but this can be identified with $\mathbb{Z}_N^n$ since if we add to $\boldsymbol{\nu}$ an element of $N\mathbb{Z}^n$ the exponent of t will change by a multiple of N^2. Because N is even, this is actually a multiple of $2N$, and t raised to a multiple of $2N$ is equal to 1.

The absolute value of the invariant

Let us compute $|Z_N(M)|$. To do this, we need to examine the linear transformation defined by the linking matrix

$$T_L : \mathbb{Z}_N^n \to \mathbb{Z}_N^n, \quad T_L \boldsymbol{\nu} = A_L \boldsymbol{\nu}.$$

Lemma 7.3. *$kerT_L$ is isomorphic to $H^1(M, \mathbb{Z}_N)$.*

Proof. Consider the triple of 3-dimensional manifolds (M, N_1, N_2), where N_1 is S^3 minus an open regular neighborhood of L and N_2 is the disjoint union of the closures of the n solid tori that form the regular neighborhood of L, so that $M = N_1 \cup N_2$. Then $N_1 \cap N_2$ is a disjoint union of n tori $S^1 \times S^1$. The triple (M, N_1, N_2) does not quite satisfy the conditions for the existence of a Meyer-Vietoris sequence, since the interiors of N_1 and N_2 do not cover M. Nevertheless we can enlarge N_1 and N_2 for this to happen, and the old N_1 and N_2 are deformation retracts of the new ones, while the tori in the intersections of the old N_1 and N_2 are deformation retracts of the new $N_1 \cap N_2$. So we have a Mayer-Vietoris exact sequence in cohomology for this triple

$$\cdots \xrightarrow{0} H^1(M, \mathbb{Z}_N) \to H^1(N_1, \mathbb{Z}_N) \oplus H^1(N_2, \mathbb{Z}_N) \xrightarrow{f} H^1(N_1 \cap N_2, \mathbb{Z}_N) \to \cdots.$$

Explicitly this sequence is

$$\begin{aligned} \cdots \xrightarrow{0} H^1(M, \mathbb{Z}_N) \to H^1(S^3 \backslash N_2, \mathbb{Z}_N) \oplus \oplus_{k=1}^n H^1(S^1 \times B^2, \mathbb{Z}_N) \\ \xrightarrow{f} \sum_{n=1}^{k} H^1(S^1 \times S^1, \mathbb{Z}) \to \cdots, \end{aligned}$$

and, even more explicitly, it is

$$\cdots \xrightarrow{0} H^1(M, \mathbb{Z}_N) \to \mathbb{Z}_N^n \oplus \mathbb{Z}_N^n \xrightarrow{f} \mathbb{Z}_N^{2n} \to \cdots.$$

Let us determine the matrix of the map f. Using Poincarè-Lefschetz duality, we can identify $H^1(N_1, \mathbb{Z}_N)$ with $H_2(N_1, \partial N_1, \mathbb{Z}_N)$. Then the generators of $\mathbb{Z}_N^n = H^1(N, \mathbb{Z}_N) = H_2(N_1, \partial N_1, \mathbb{Z}_N)$ can be taken to be the Seifert surfaces of the link components of L. By the same Poincaré-Lefschetz duality, $H^1(S^1 \times B^2, \mathbb{Z}_N)$ can be identified with $H_2(S^1 \times B^2, \partial(S^1 \times B^2), \mathbb{Z}_N)$ and the latter has as generator the surface $\{1\} \times B^2$. Also $H^1(S^1 \times S^1, \mathbb{Z}_N)$ is identified with $H_1(S^1 \times S^1, \mathbb{Z}_N)$. In this setting we can find the dual of f to have the matrix

$$\begin{pmatrix} Id & A_L \\ Id & 0 \end{pmatrix},$$

and hence f itself has the matrix

$$\begin{pmatrix} Id & Id \\ A_L & 0 \end{pmatrix},$$

where of course we use the fact that A_L is symmetric. From this matrix we can see that $\ker f$ is isomorphic to $\ker T_L$.

Also, by the first isomorphism theorem $\ker f$ is isomorphic with $H^1(M, \mathbb{Z}_N)$. This proves the lemma. □

For further reference, we will denote the isomorphism introduced in Lemma 7.3 by ι.

Given the discussion in Remark 7.3, we can regard the quadratic map $Q(\boldsymbol{\nu}) = \boldsymbol{\nu}^T A_L \boldsymbol{\nu}$ as a map $\mathbb{Z}_N^n \to \mathbb{Z}_{2N}$. The restriction $Q_0 = Q|\ker T_L$ is a homomorphism into $\{0, N\} \subset \mathbb{Z}_{2N}$ because

$$(\boldsymbol{\nu} + \boldsymbol{\nu}')^T A_L (\boldsymbol{\nu} + \boldsymbol{\nu}') = \boldsymbol{\nu}^T A_L \boldsymbol{\nu} + \boldsymbol{\nu}'^T A_L \boldsymbol{\nu}' + 2\boldsymbol{\nu}^T A_L \boldsymbol{\nu}',$$

and $2\boldsymbol{\nu}^T A_L \boldsymbol{\nu}'$ is a multiple of $2N$.

The next lemma, as well as the final result about the absolute value of $Z_N(M)$, can be stated elegantly in terms of the cup product $\cup$ in cohomology. For those unfamiliar with this operation, let us point out that in the case of the coefficient ring equal to $\mathbb{R}$, when the kth cohomology group is isomorphic to the kth de Rham cohomology group (see §4.1.1), the cup product is just the wedge product of differential forms:

$$H^k(M) \times H^l(M) \xrightarrow{\wedge} H^{k+l}(M).$$

But a cup product in cohomology can be defined for every ring of coefficients. We only describe the cup product for the situations where it is applied: the cup product of two elements in $H^1(M, \mathbb{Z}_N)$ and of an element in $H^1(M, \mathbb{Z}_N)$ and an element in $H^2(M, \mathbb{Z}_N)$, and for that we use Poincaré duality.

Under the identification of $H^1(M, \mathbb{Z}_N)$ with $H_2(M, \mathbb{Z}_N)$, each cohomology class can be represented by a compact oriented surface without boundary in M which is branched over a finite number of disjoint circles and the number of sheets meeting along each circle is a multiple of N. If cohomology classes α_1 and α_2 are represented by surfaces Σ_1 and Σ_2, then by bringing the surfaces in general position they intersect transversally along a finite collection of curves. Let γ be this intersection. For p a point in a connected component of γ, choose vector the vector $\mathbf{v} \in T_p\gamma$ and the vectors $\mathbf{v}_j \in T_p\Sigma_j$ such that $\mathbf{v}, \mathbf{v}_j$ is an orientation frame for Σ_j, $j = 1, 2$ and $\mathbf{v}, \mathbf{v}_1, \mathbf{v}_2$ is an orientation frame of M. Then $\mathbf{v}$ defines an orientation of the connected component of γ. With this convention γ is an oriented multicurve.

Then, the homology class $[\gamma] \in H_1(M, \mathbb{Z}_N)$ corresponds by Poincaré duality to a cohomology class in $H^2(M, \mathbb{Z}_N)$. This cohomology class is $\alpha_1 \cup \alpha_2$. Note that $\alpha_1 \cup \alpha_2 = -\alpha_2 \cup \alpha_1$.

If $\alpha \in H^1(M, \mathbb{Z}_N)$ and $\beta \in H^2(M, \mathbb{Z}_N)$, then α can be represented by an oriented branched surface Σ and β can be represented by an oriented curve γ. Brought in general position, Σ and γ intersect transversally at finitely many points. Associate to each of these points the sign $+1$ if the tangent frame giving the orientation of Σ and the tangent vector giving the orientation of γ form, in this order, an orientation frame for M, and -1 otherwise. Then the intersection of Σ and γ defines an element of $H_0(M, \mathbb{Z}_N)$, which by Poincaré duality corresponds to a cohomology class in $H^3(M, \mathbb{Z}_N)$. This is $\alpha \cup \beta$. Again $\beta \cup \alpha = -\alpha \cup \beta$.

Now we are in position to state the next result.

Lemma 7.4. *We have the following commuting diagram*

$$\begin{array}{ccc} ker T_L & \xrightarrow{\quad Q_0 \quad} & \{0, N\} \subset \mathbb{Z}_{2N} \\ \downarrow \iota & & \downarrow \times\frac{1}{2} \\ H^1(M, \mathbb{Z}_N) & \xrightarrow{\quad \Psi \quad} & \{0, \frac{N}{2}\} \subset H^3(M, \mathbb{Z}_N) = \mathbb{Z}_N \end{array}$$

where Ψ is defined by $\Psi(\alpha) = \alpha \cup \alpha \cup \alpha$. Consequently $Q_0 \equiv 0$ if and only if $\alpha \cup \alpha \cup \alpha = 0$ for all $\alpha \in H^1(M, \mathbb{Z}_N)$.

Proof. Let $\boldsymbol{\nu}$ be an element in $\ker T_L \subset \mathbb{Z}_N^n$, and set $\alpha = \iota(\boldsymbol{\nu})$. We will show that $\alpha \cup \alpha \cup \alpha = Q_0(\boldsymbol{\nu})/2$.

Let Σ be a branched surface representing the Poincaré dual of α. Let $N_1 = S^3 \backslash int(N_2)$ and $N_2 = \cup_{k=1}^n (S^1 \times B^2)_k$ be the 3-dimensional manifolds defined in Lemma 7.3, where we label the solid tori of N_2 in order to keep track of the corresponding link components. Then we can perturb Σ so that its branch locus lies in N_1.

Since $[\Sigma]$ is the Poincaré dual of $\iota(\boldsymbol{\nu})$, $\Sigma \cap \partial(S^1 \times B^2)_k$ is a union of $\tilde{\nu}_k$ circles in $\partial(S^1 \times B^2)_k$ each of which is parallel to the framing curve γ_k of the link component, where $\tilde{\nu}_k$ is a positive integer which modulo N is equal to ν_k, with $\boldsymbol{\nu}^T = (\nu_1, \nu_2, \cdots, \nu_n)$. Let $m_k = \{1\} \times \partial B^2$ be the meridian of $(S^1 \times B^2)_k$. Using the Seifert-van Kampen Theorem (which in this case leads to the Wirtinger presentation of $\pi_1(L) = \pi_1(N_1)$) together with the fact that $H_1(M, \mathbb{Z})$ is the abelianization of $\pi_1(M)$, we deduce that the meridians generate $H_1(M, \mathbb{Z})$. Using this fact, we may assume that the branching locus of Σ is a union of m_k's. Let $a_k N$ be the number of sheets of Σ meeting along m_k.

The surface Σ can be interpreted as a relative cycle in the pair formed by M and the branching locus. So $\Sigma \cap N_1$ can be interpreted as a relative cycle in the pair formed by N_1 and ∂N_1 union with the branching locus. It follows that in $H_1(N_1, \mathbb{Z})$,

$$\sum_k a_k N[m_k] = \sum_k \tilde{\nu}_k [\gamma_k].$$

Recalling the proof of Lemma 7.3, we have

$$\begin{pmatrix} [\gamma_1] \\ [\gamma_2] \\ \cdots \\ [\gamma_n] \end{pmatrix} = A_L \begin{pmatrix} [m_1] \\ [m_2] \\ \cdots \\ [m_n] \end{pmatrix},$$

with the appropriate orientations of the meridians. We obtain the following relation between the a_k and the $\tilde{\nu}_k$

$$\begin{pmatrix} a_1 N \\ a_2 N \\ \cdots \\ a_n N \end{pmatrix} = A_L \begin{pmatrix} \tilde{\nu}_1 \\ \tilde{\nu}_2 \\ \cdots \\ \tilde{\nu}_n \end{pmatrix}.$$

Now we want to compute $\alpha \cup \alpha$ and for that we need to understand the self-intersection of Σ. First, since outside the branching locus Σ is orientable, we can push it off itself in a normal direction. So we only need to understand the self-intersections of Σ near the branching locus.

In order to compute the intersection of Σ with itself we consider a second copy of Σ slightly pushed off itself. Think of Σ near a branching circle m_k looking like a book with $a_k N$ leaves, as depicted in Figure 7.9. Arguing on Figure 7.9, assume we have a second copy of the "book" slightly pushed to the left. Then the first leaf of the second "book" does not intersect any leaf of the first, its second leaf intersects one leaf of the first, its third leaf intersects two leaves of the first and so on. These intersections happen along curves homologous to m_k.

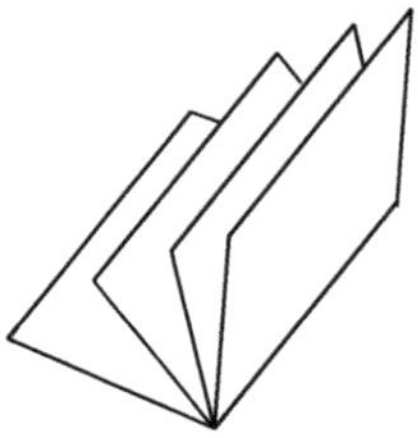

Fig. 7.9 Structure of the surface near the branching locus

Hence for the self-intersection of Σ we have

$$[\Sigma]\cdot[\Sigma] = \sum_{k=1}^{n}(1+2+\cdots+(a_i N-1))[m_k] = \sum_{k=1}^{n}\frac{a_k N(a_k N-1)}{2}[m_k]$$

$$= -\sum_{k=1}^{N}\frac{a_k N}{2}[m_k] \in H_1(M,\mathbb{Z}_N),$$

where the last equality holds because N is even and so $\frac{a_k N^2}{2}$ is a multiple of N. Since the algebraic intersection number of Σ and m_k is $-\tilde{\nu_k}$, we obtain

$$[\Sigma]\cdot([\Sigma]\cdot[\Sigma]) = \sum_{k=1}^{N} \frac{a_k N}{2}\tilde{\nu}_k = \frac{1}{2}\tilde{\boldsymbol{\nu}}^T A_L \tilde{\boldsymbol{\nu}} = \frac{1}{2}Q_0(\boldsymbol{\nu}).$$

We conclude that if $\alpha = \iota(\boldsymbol{\nu})$ then $\alpha\cup\alpha\cup\alpha = \frac{1}{2}Q_0(\boldsymbol{\nu})$, as desired. □

Theorem 7.3. *If there exists* $\alpha \in H^1(M, \mathbb{Z}_N)$ *with* $\alpha\cup\alpha\cup\alpha \neq 0$, *then* $Z_N(M) = 0$. *Otherwise*

$$|Z_N(M)| = |H^1(M, \mathbb{Z}_N)|^{\frac{1}{2}},$$

where the bars in the right-hand side denote the number of elements in the set.

Proof. Let L be a framed link such that $M = M_L$. From (7.4) we obtain

$$|Z_N(M)| = N^{-n/2}|\sum_{\boldsymbol{\nu}\in\mathbb{Z}_N^n} t^{\boldsymbol{\nu}^T A_L \boldsymbol{\nu}}|.$$

We compute

$$|\sum_{\boldsymbol{\nu}\in\mathbb{Z}_N^n} t^{\boldsymbol{\nu}^T A_L \boldsymbol{\nu}}|^2 = \sum_{\boldsymbol{\nu},\boldsymbol{\nu}'\in\mathbb{Z}} t^{\boldsymbol{\nu}^T A_L \boldsymbol{\nu} - \boldsymbol{\nu}'^T A_L \boldsymbol{\nu}'}.$$

Using the fact that A_L is symmetric, by substituting $\boldsymbol{\nu} = \boldsymbol{\nu}' + \boldsymbol{\nu}''$, we obtain that this is equal to

$$\sum_{\boldsymbol{\nu}''} t^{\boldsymbol{\nu}''^T A_L \boldsymbol{\nu}''} \sum_{\boldsymbol{\nu}'} t^{2\boldsymbol{\nu}'^T A_L \boldsymbol{\nu}''}. \tag{7.5}$$

The second sum can be written as an iteration of sums of the form $\sum_{\zeta:\,\zeta^m=1}\zeta$. Such a sum is either equal to N if $m = 0$ or to zero if $m > 0$. So the only way this second sum is not equal to zero is if $A_L\boldsymbol{\nu}'' = 0$, i.e. $\boldsymbol{\nu}'' \in \ker T_L$, in which case the whole sum is equal to N^n. We conclude that the expression in (7.5) is equal to

$$N^n \sum_{\boldsymbol{\nu}''\in\ker T_L} t^{\boldsymbol{\nu}''^T A_L \boldsymbol{\nu}''} = N^n \sum_{\boldsymbol{\nu}''\in\ker T_L} t^{Q_0(\boldsymbol{\nu}')}.$$

We have two situations. If $Q_0 \neq 0$, then the sum is equal to zero, because it consists of $+1$'s and -1's that cancel each other. If $Q_0 = 0$, then the sum consists of $+1$'s only. Hence

$$|Z_N(M)|^2 = \begin{cases} |\ker T_L|^{1/2} & \text{if } Q_0 \equiv 0 \\ 0 & \text{otherwise.} \end{cases}$$

The conclusion of the theorem follows now by applying Lemma 7.3 and Lemma 7.4. □

The phase of the invariant

Next, we examine the complex number of absolute value 1,

$$Z_N(M)/|Z_N(M)|,$$

which is the phase of the invariant. Clearly we have to assume that $Z_N(M) \neq 0$, which as we saw above happens precisely when $\alpha \cup \alpha \cup \alpha = 0$ for all $\alpha \in H^1(M, \mathbb{Z})$. We will analyze equation (7.4) from the number theoretical point of view, and we start by placing ourselves in a slightly more general setting.

Let N be an integer and let t be a primitive $2N$th root of unity if N is even, and a primitive Nth root of unity if N is odd. We introduce the **Gauss sum**

$$G_N(a;t) = \sum_{k \in \mathbb{Z}_N} t^{ak^2}.$$

For a symmetric $n \times n$ matrix A with integer entries denote

$$Z_N(A;t) = \left(\frac{G_N(1;t)}{|G_N(1,t)|} \right)^{-\sigma(A)} N^{-n/2} \sum_{\boldsymbol{\nu} \in \mathbb{Z}_N^n} t^{\boldsymbol{\nu}^T A \boldsymbol{\nu}}.$$

Of course, if $t = e^{-\frac{i\pi}{N}}$, $A = A_L$, the linking matrix of the framed link $L \subset S^3$, and $M = M_L$, then $Z_N(A;t) = Z_N(M)$.

Lemma 7.5. *Let A be a symmetric $n \times n$ matrix and t a $2N$th root of unity. If $N = N_1 N_2$ with N_1 and N_2 coprime integers, then*

$$Z_N(A;t) = Z_{N_1}(A; t^{N_2^2}) Z_{N_2}(A; t^{N_1^2}).$$

Proof. First note that, with the above convention for t, if N_1 and N_2 are odd then $t^{N_2^2}$ is a primitive N_1st root of unity while $t^{N_1^2}$ is a primitive N_2nd root of unity. And if N_1 is even and N_2 is odd, then $t^{N_2^2}$ is a primitive $2N_1$st root of unity while $t^{N_1^2}$ is a primitive N_2nd root of unity.

An element $\boldsymbol{\nu} \in \mathbb{Z}_N^n$ is uniquely expressed as $\boldsymbol{\nu} = N_2 \boldsymbol{\nu}_1 + N_1 \boldsymbol{\nu}_2$, for $\boldsymbol{\nu}_1 \in \mathbb{Z}_{N_1}^n$ and $\boldsymbol{\nu}_2 \in \mathbb{Z}_{N_2}^n$. We have

$$\sum_{\boldsymbol{\nu} \in \mathbb{Z}_N^n} t^{\boldsymbol{\nu}^T A \boldsymbol{\nu}} = \sum_{\boldsymbol{\nu}_1 \in \mathbb{Z}_{N_1}^n, \boldsymbol{\nu}_2 \in \mathbb{Z}_{N_2}^n} t^{N_2^2 \boldsymbol{\nu}_1^T A \boldsymbol{\nu}_1 + N_1^2 \boldsymbol{\nu}_2^T A \boldsymbol{\nu}_2 + 2N_1 N_2 \boldsymbol{\nu}_1^T A \boldsymbol{\nu}_2}.$$

Since $t^{2N_1N_2} = 1$, the third term in the exponent can be ignored, and so this is equal to

$$\sum_{\boldsymbol{\nu}_1 \in \mathbb{Z}_{N_1}^n} t^{N_2^2 \boldsymbol{\nu}_1^T A \boldsymbol{\nu}_1} \sum_{\boldsymbol{\nu}_2 \in \mathbb{Z}_{N_2}^n} t^{N_1^2 \boldsymbol{\nu}_2^T A \boldsymbol{\nu}_2}.$$

A similar computation yields

$$G_N(1;t) = G_{N_1}(1; t^{N_2^2}) G_{N_2}(1; t^{N_1^2}).$$

Also $N^{-1/2} = N_1^{-1/2} N_2^{-1/2}$ and the lemma is proved. □

Because of this we can concentrate only on computing $Z_N(A;t)$ when $N = p^m$, where p is a prime number and m is a positive integer.

If p is odd, we can diagonalize A as a matrix in $\mathbb{Z}_{p^m}$, so there is a matrix $M \in SL(n, \mathbb{Z})$ such that

$$M^T AM \equiv \begin{pmatrix} a_1 & 0 & \cdots & 0 \\ 0 & a_2 & \cdots & 0 \\ \cdots & \cdots & \cdots & \cdots \\ 0 & 0 & \cdots & a_n \end{pmatrix} \pmod{p^m}.$$

If $p = 2$ the situation is more complicated because we cannot divide by 2. While A is not always diagonalizable, for example

$$A = \begin{pmatrix} 0 & 1 \\ 1 & 0 \end{pmatrix}$$

is not diagonalizable, it is stably diagonalizable. This means that, if we introduce the matrix

$$B = \begin{pmatrix} 1 & 0 & \cdots & 0 \\ 0 & 2 & \cdots & 0 \\ \cdots & \cdots & \cdots & \cdots \\ 0 & 0 & \cdots & 2^{m-1} \end{pmatrix},$$

then there is $M \in SL(n+m, \mathbb{Z})$ such that

$$M^T(A \oplus B)M \equiv \begin{pmatrix} a_1 & 0 & \cdots & 0 \\ 0 & a_2 & \cdots & 0 \\ \cdots & \cdots & \cdots & \cdots \\ 0 & 0 & \cdots & a_{n+m} \end{pmatrix}$$

where the diagonal entries are considered modulo 2^{m+1} and the off-diagonal entries modulo 2^m.

Note that $\mathbb{Z}_{2^m}(B;t) \neq 0$ and $\mathbb{Z}_{2^m}(B;t) = \overline{\mathbb{Z}_{2^m}(-B;t)}$. For this reason we use instead the diagonalizable matrix $A \oplus B \oplus (-B)$, because the phase of $\mathbb{Z}_{2^m}(A)$ equals that of $\mathbb{Z}_{2^m}(A \oplus B \oplus (-B);t)$. We conclude that we can work with A as if it is always diagonalizable.

If we diagonalize A to have diagonal entries a_j, then the phase of $Z_{p^m}(A;t)$ is equal to that of

$$(G_N(1;t))^{-\sigma(A)} \prod_j \sum_{\nu \in \mathbb{Z}_N} t^{a_j \nu^2} = (G_N(1;t))^{-\sigma(A)} \prod_j G_N(a_j;t).$$

We now turn to the theory of Gauss sums, for which we refer the reader to [Serre (1973)] (see also [Murakami *et al.* (1992)]). Let p be prime, and let $t = \exp(\frac{i\pi d}{p^m})$ with $\gcd(d,p) = 1$ and $d+p$ odd (so that we can use Lemma 7.5). If p is odd, then we let $d = 2b$. Set $a = p^k c$ with $\gcd(p,c) = 1$. We have:

(i) if p is odd, then

$$G_{p^m}(a;t) = \begin{cases} p^m & \text{if } k-m \geq 0, \\ \sqrt{p^{m+k}} & \text{if } k-m < 0 \text{ and even}, \\ \left(\frac{c}{p}\right)\left(\frac{b}{p}\right)\sqrt{p^{m+k}} & \text{if } k-m<0 \text{ and odd, and } p \equiv 1 (\text{mod}\, 4), \\ i\left(\frac{c}{p}\right)\left(\frac{b}{p}\right)\sqrt{p^{m+k}} & \text{if } k-m<0 \text{ and odd, and } p \equiv 3 (\text{mod}\, 4); \end{cases}$$

(ii) if $p = 2$, then

$$G_{2^m}(a;t) = \begin{cases} 2^m & \text{if } k-m > 0, \\ 0 & \text{if } k-m = 0, \\ \exp(\frac{\pi i}{4} cd)\sqrt{2^{m_k}} & \text{if } k-m<0 \text{ and even}, \\ \exp(\frac{\pi i}{4} \cdot \frac{(c-1)(d-1)}{4})\sqrt{2^{m+k}} & \text{if } k-m<0 \text{ and odd}. \end{cases}$$

Here $\left(\frac{x}{p}\right)$ is the Legendre symbol, which is equal to 1 if x is a quadratic residue mod p and -1 otherwise.

It should be noted that in each situation the phase of the Gauss sum (i.e. the quotient of the Gauss sum by its absolute value) is an eight root of unity. Combining this observation with Lemma 7.5, we obtain the following result.

Theorem 7.4. *[Murakami* et al. *(1992)] The Murakami-Ohtsuki-Okada is given by the formula*

$$Z_N(M) = \begin{cases} \zeta_N(M)|H^1(M,\mathbb{Z}_N)|^{1/2} & \textit{if } \alpha \cup \alpha \cup \alpha = 0 \textit{ for all } \alpha \in H^1(M,\mathbb{Z}_N), \\ 0 & \textit{otherwise}, \end{cases}$$

where $\zeta_N(M)$ is an eight root of unity.

7.3 The reduced linking number skein module of a 3-dimensional manifold

7.3.1 *The Sikora isomorphism*

The second application of the exact Egorov identity is the construction of a **Sikora isomorphism**, which identifies the reduced linking number skein modules of two manifolds with homeomorphic boundaries. We point out that such an isomorphism was constructed for reduced Kauffman bracket skein modules in [Sikora (2000)].

Theorem 7.5. *Let M_1 and M_2 be two 3-dimensional manifold with homeomorphic boundaries. Then*

$$\mathcal{L}_N(M_1) \cong \mathcal{L}_N(M_2).$$

Proof. Because the manifolds M_1 and M_2 have homeomorphic boundaries, there is a framed link $L_1 \subset M_1$ such that M_2 is obtained by performing surgery on L_1 in M_1. Let N_1 be a regular neighborhood of L_1 in M, which is the union of several solid tori, and let N_2 be the union of the solid tori that are added in order to obtain M_2.

Identify the solid tori from N_1 and N_2 with $S^1 \times B^2$ so that we can talk about (p, q) curves. Then observe that in the surgery process the curves $(1, 0)$, $(0, 1)$ trade places, meaning that the curve on $\partial M_1 \backslash N_1$ that corresponds to $(1, 0)$ becomes the curve $(0, 1)$ on $\partial M_2 \backslash N_2$ and vice-versa. Hence the tori of N_2 become surgery tori in M_2, they define a framed link $L_2 \subset M_2$ such that M_1 is obtained from M_2 by surgery on L_2.

Every skein in M_1, respectively M_2, can be isotoped to one that misses N_1, respectively N_2. The homeomorphism $M_1 \backslash N_1 \cong M_2 \backslash N_2$ yields an isomorphism

$$\phi : \mathcal{L}_N(M_1 \backslash N_1) \to \mathcal{L}_N(M_2 \backslash N_2).$$

However, this does not induce a well defined map between $\mathcal{L}_N(M_1)$ and $\mathcal{L}_N(M_2)$ because a skein can be pushed through the N_i's, and skeins that were equal on one side become different on the other side. To make this map well defined, the skein should not change when pushed through the N_i's.

We use Theorem 7.1 and define $\mathcal{S}_1 : \mathcal{L}_N(M_1) \to \mathcal{L}_N(M_2)$ by

$$\mathcal{S}_1(\sigma) = \phi(\sigma) \cup \Omega(L_1)$$

and $\mathcal{S}_2 : \mathcal{L}_N(M_2) \to \mathcal{L}_N(M_1)$ by

$$\mathcal{S}_2(\sigma) = \phi^{-1}(\sigma) \cup \Omega(L_2).$$

By Proposition 7.1 b) we have

$$\Omega(L_1) \cup \Omega(\phi^{-1}(L_2)) = <\emptyset> \in \mathcal{L}_N(M_1),$$

since each of the components of $\phi^{-1}(L_2)$ is a meridian in the surgery torus, hence it surrounds exactly once the corresponding component in L_1. This implies that $\mathcal{S}_2 \circ \mathcal{S}_1 = Id$. A similar argument shows that $\mathcal{S}_1 \circ \mathcal{S}_2 = Id$. □

Lemma 7.6. *(i) Given two oriented 3-dimensional manifolds M_1 and M_2, let $M_1\#M_2$ be their connected sum. Then the homomorphism*

$$\mathcal{L}_N(M_1)\otimes\mathcal{L}_N(M_2)\to\mathcal{L}_N(M_1\#M_2)$$

defined by $<L_1>\otimes<L_2>\mapsto<L_1\cup L_2>$ is an isomorphism. Here L_1 and L_2 are pushed away from the balls that become identified.
(ii) Let M be a compact, connected, oriented, 3-dimensional manifold. Remove two balls from M and then glue the resulting manifold by an orientation reversing homeomorphism between the two spheres to obtain a manifold M'. Then the homomorphism $\mathcal{L}_N(M)\to\mathcal{L}_N(M')$ which maps each skein in M (pushed off the two balls) to a skein in $\mathcal{L}_N(M')$ is an isomorphism.
(iii) Let M_1 and M_2 be compact, oriented 3-dimensional manifolds with boundary, and let B_j^2 be a disk in the boundary of M_j, $j=1,2$. Glue M_1 and M_2 by an orientation preserving homeomorphism from B_1^2 to B_2^2 to obtain the manifold $M_1\cup M_2$. Then the homomorphism

$$\mathcal{L}_N(M_1)\otimes\mathcal{L}_N(M_2)\to\mathcal{L}_N(M_1\cup M_2)$$

defined by $<L_1>\otimes<L_2>\mapsto<L_1\cup L_2>$ is an isomorphism.

Proof. (i) In $M_1\#M_2$, the manifolds M_1 and M_2 are separated by a 2-dimensional sphere S^2_{sep}. If L is a link in $M_1\#M_2$ that crosses S^2_{sep}, then because the sphere is separating, there is a strand of L that is oriented from M_1 towards M_2, and a strand that is oriented from M_2 towards M_1 (the link must run back and forth between M_1 and M_2). We can include the two strands in a ball and then cross them (Figure 7.10). After resolving the crossings and deleting the trivial circle, the two strands no longer cross S_{sep}. Repeating we can turn L into a link disjoint from S^2_{sep}, hence into a union of a link in M_1 and a link in M_2. This shows that the map is onto.

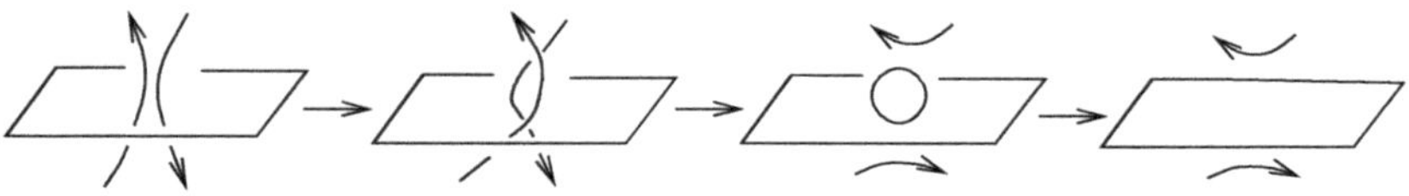

Fig. 7.10 How to separate the link from S^2_{sep}

Here is another way to see this, which we present because it can be generalized to other types of skein modules, such as the reduced Kauffman bracket skein modules. Every skein in $M_1\#M_2$ can be written as $\sum_{j=0}^{N-1}\sigma_j$, where each σ_j intersects S^2_{sep} in j strands pointing in the same direction. A trivial framed knot colored by Ω is equal to the empty link. But when we slide it over S^2_{sep} it turns $\sum_{j=0}^{N-1} \sigma_j$ into σ_0. Again we removed all stands that cross S^2_{sep}.

On the other hand, the reduced linking number skein module of a regular neighborhood of S^2_{sep} is $\mathbb{C}$ since every skein can be resolved to the empty link. This means that, in $M_1 \# M_2$, if a skein that lies entirely in M_1 can be isotoped to a skein that lies entirely in M_2, then this skein is a scalar multiple of the empty skein. This allows us to define an inverse of the map from the statement, hence the map is also one-to-one.

(ii) If M has no boundary, then neither has M'. Because the two relative skein modules are both isomorphic to $\mathbb{C}$, this map is trivially an isomorphism.

In general, if M does have a boundary, then we can include the two balls to be removed in a bigger ball B^3 inside M, so that $M = M \# S^3$ and the gluing phenomenon happens inside S^3. So $M' = M \# M_0$, where M_0 has no boundary. Then using (a) we conclude that

$$\mathcal{L}_N(M) = \mathcal{L}_N(M) \otimes \mathcal{L}_N(S^3) = \mathcal{L}_N(M) \otimes \mathcal{L}_N(M_0) = \mathcal{L}_N(M').$$

The proof of (iii) is identical to (i) with the disk replacing S^2_{sep}. □

7.3.2 *The computation of the reduced linking number skein module of a 3-dimensional manifold*

Now it is easy to describe the reduced linking number skein module of any compact, oriented, 3-dimensional manifold. The following result should be seen as an analogue, for the case of **reduced** modules, of Theorem 6.10.

Theorem 7.6. *Let M be a compact, oriented, 3-dimensional manifold.*
(i) If M has no boundary, then $\mathcal{L}_N(M) = \mathbb{C}$.
(ii) If M has the boundary components Σ_{g_i}, $i = 1, 2, \ldots, n$, then

$$\mathcal{L}_N(M) \cong \bigotimes_{i=1}^{n} \mathbb{C}^{Ng_i}.$$

Proof. It suffices to prove the theorem for the case where M is connected.

(i) If M has no boundary component then $\mathcal{L}_N(M) = \mathcal{L}_N(S^3) = \mathbb{C}$.

(ii) If M is bounded by a sphere, then $\mathcal{L}_N(M) = \mathcal{L}_N(B^3) = \mathbb{C}$, where B^3 denotes the 3-dimensional ball. If M has one genus g boundary component with $g \geq 1$, then $\mathcal{L}_N(M) \cong \mathcal{L}_N(H_g) = \mathbb{C}^{Ng}$ by Proposition 5.13 (b).

An oriented 3-dimensional manifold with n boundary components can be obtained as surgery on a connected sum of n handlebodies. The conclusion of the theorem follows by applying Lemma 7.6 (i). □

Pairings and the topological form of the inner product

Below we give a topological description of the inner product on the space of theta functions.

Theorem 7.7. *Let M_1 and M_2 be compact, oriented, 3-dimensional manifolds such that there is an orientation reversing homeomorphism $h : \partial M_1 \to \partial M_2$. Then there is a nondegenerate bilinear map*

$$\mathcal{L}_N(M_1) \times \mathcal{L}_N(M_2) \to \mathcal{L}_N(M_1 \cup_h M_2) = \mathbb{C},$$

defined by

$$(<L_1>, <L_2>) \mapsto <L_1 \cup L_2>$$

that identifies $\mathcal{L}_N(M_2)$ with the dual of $\mathcal{L}_N(M_1)$.

Notation: We denote the pairing defined in this theorem by $[\cdot, \cdot]_h$.

Note that this pairing also depends on how we identify $\mathcal{L}_N(M_1 \cup_h M_2)$ with $\mathbb{C}$. This is done via a Sikora homomorphism $\mathcal{L}_N(M_1 \cup_h M_2) \to \mathcal{L}_N(S^3) = \mathbb{C}$. The isomorphism is defined by a surgery diagram for $M_1 \cup_h M_2$), which is not unique, nor is there a canonical choice for it. Two surgery diagrams yield the same pairing if they can be changed into one another by (k2) moves. However, (k1) moves change the pairing. So the pairing is unique up to multiplication by a $2N$th root of unity.

Proof. The map is bilinear by construction, all we have to check is that it is nondegenerate.

Using Lemma 7.6 (ii) we deduce that it suffices to prove the theorem in the case where each boundary components of M_1 and M_2 lies in a different connected component of those manifolds. Removing the connected components of the M_1 and M_2 that have no boundary, and for which the reduced skein module is just $\mathbb{C}$, we deduce that it suffices to prove the property in the case where $\partial M_1 = \partial M_2 = \Sigma_g$.

Let

$$\mathcal{S}_1 : \mathcal{L}_N(M_1) \to \mathcal{L}_N(H_g), \quad \mathcal{S}_2 : \mathcal{L}_N(M_2) \to \mathcal{L}_N(H_g)$$

be Sikora isomorphisms that identify the reduced linking number skein modules of the two manifolds with the reduced linking number skein modules of genus g handlebodies. Under these isomorphisms, the pairing defined by $h : \partial M_1 \to \partial M_2$ becomes the pairing defined by the same homeomorphism but viewed now as

$$h : \sqcup_{k=1}^{n} \partial H_{g_k} \to \sqcup_{k=1}^{n} \partial H_{g_k}.$$

We reduced the problem to showing that if H_g^1 and H_g^2 are oriented genus g handlebodies, $g \geq 0$, and $h : \partial H_g^1 \to \partial H_g^2$ is an orientation reversing homeomorphism, then the map

$$\mathcal{L}_N(H_g^1) \times \mathcal{L}_N(H_g^2) \to \mathcal{L}_N(H_g^1 \cup_h H_g^2) = \mathbb{C},$$

defined by

$$(<L_1>, <L_2>) \mapsto <L_1 \cup L_2> \tag{7.6}$$

is a nondegenerate bilinear map.

Recall from §6.1 the instance when the gluing yields the 3-dimensional sphere S^3. In this case the handlebodies are regular neighborhoods of the two wedges of circles shown in Figure 7.11.

Fig. 7.11 The two wedges of circles that give rise to a Heegaard decomposition of S^3

The circles in each wedge define a canonical basis of the first homology group of the handlebody. Identify the two handlebodies so as to make the pairing a bilinear map on $\mathcal{L}(H_g)$. Then we have a basis $a_1, a_2, \ldots, a_g$ of $H_1(H_g, \mathbb{Z})$, which is shown in Figure 7.12 the way it sits in each handlebody. Frame the curves shown in this picture by the blackboard framing of the plane of the paper.

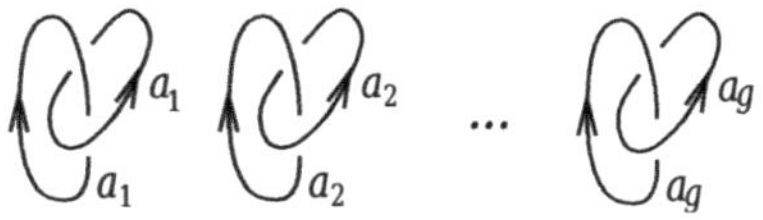

Fig. 7.12 The bases of the handlebody and its complement

By Proposition 5.13 (b),

$$a_1^{m_1} a_2^{m_2} \cdots a_g^{m_g}, \quad m_1, m_2, \ldots, m_g \in \mathbb{Z}_N$$

form a basis of $\mathcal{L}_N(H_g)$ and the pairing $[\cdot, \cdot]_h$ defined by (7.6) is

$$[a_1^{m_1} a_2^{m_2} \cdots a_g^{m_g}, a_1^{n_1} a_2^{n_2} \cdots a_g^{n_g}]_h = t^{2(m_1 n_1 + m_2 n_2 + \cdots + m_g n_g)}. \tag{7.7}$$

By Lemma 7.6 (iii), we have

$$\mathcal{L}_N(H_g) = \otimes_{j=1}^{g} \mathcal{L}_N(H_1),$$

where the basis of the reduced skein module of the jth solid torus is a_j^m, $m \in \mathbb{Z}_N$. Note that (7.7) implies that the pairing respects this tensor product decomposition, in the sense that

$$[\sigma_1 \otimes \sigma_2 \otimes \cdots \otimes \sigma_g, \sigma_1' \otimes \sigma_2' \otimes \cdots \otimes \sigma_g']_h = [\sigma_1, \sigma_1']_h [\sigma_2, \sigma_2']_h \cdots [\sigma_g, \sigma_g']_h.$$

So it suffices to check nondegeneracy for the solid torus. In this case the matrix with elements $[a_1^j, a_1^k] = t^{2jk}$ is symmetric, so it is diagonalizable[2]. Moreover, it is a Vandermonde matrix, hence invertible, so the diagonal entries are nonzero. This proves that the bilinear pairing is nondegenerate.

Let us now turn to the case of a general homeomorphism that glues the two handlebodies. In this case we can use another Sikora isomorphism of the first handlebody into itself which turns the homeomorphism h into the one that gives the Heegaard decomposition of S^3. This Sikora isomorphism is defined by the surgery diagram, in a neighborhood of ∂H_g, of a homeomorphism h_0 such that $h \circ h_0$ defines the Heegaard decomposition of S^3. Using once more Theorem 7.5 we reduce the problem to the case already discussed above. □

Remark 7.4. There are many decompositions of a handlebody into solid tori. In the proof of Theorem 7.7 we chose one that behaves nicely with respect to the given canonical basis.

We can now give a **topological interpretation of the inner product** on the space $\mathbf{\Theta}_N^{\Pi}(\Sigma_g)$ of theta functions. Endow Σ_g with the canonical basis $a_1, a_2, \ldots, a_g, b_1, b_2, \ldots, b_g$, and let H_g be the handlebody whose reduced skein module models $\mathbf{\Theta}_N^{\Pi}(\Sigma_g)$. Then $\Sigma_g = \partial H_g$ in such a way that the subspace $\mathbf{L} = \ker(H_1(\Sigma_g, \mathbb{R}) \to H_1(H_g, \mathbb{R}))$ is spanned by $b_1, b_2, \ldots, b_g$ and $a_1, a_2, \ldots, a_g$ is a canonical basis for $H_1(H_g, \mathbb{R})$.

Now consider two copies of the model, and let $h_{\text{inn}} : \partial H_g \to \partial H_g$ be an orientation reversing homeomorphism such that

$$h_{\text{inn}}(a_j) = -a_j, \quad h_{\text{inn}}(b_j) = b_j.$$

The result of the gluing, $H_g \cup_{h_{\text{inn}}} H_g$, is $\#_{j=1}^{g} S^1 \times S^2$, a connected sum of g copies of $S^1 \times S^2$. This is because for each j, b_j bounds a disk in each model, and the gluing of these disks by h_{inn} yields a sphere. Thus

[2]The matrix is symmetric not hermitian, so it is diagonalizable in the same way that symmetric matrices with real entries are.

each handle of H_g gives rise to a copy of $S^1 \times S^2$. To define the pairing, represent $H_g \cup_{h_{\mathrm{inn}}} H_g$ as surgery on the trivial link with g components.

Theorem 7.8. *The pairing*

$$(\sigma, \sigma') \mapsto N^{-g/2}[\sigma, \sigma']_{h_{\mathrm{inn}}}$$

is the inner product on $\mathcal{L}_N(H_g) = \boldsymbol{\Theta}_N^{\Pi}(\Sigma_g)$.

Proof. The pairing of two basis elements

$$[a_1^{m_1} a_2^{m_2} \cdots a_g^{m_g}, a_1^{k_1} a_2^{k_2} \cdots a_g^{k_g}]_{h_{\mathrm{inn}}}$$

is shown in Figure 7.13. We need the following result.

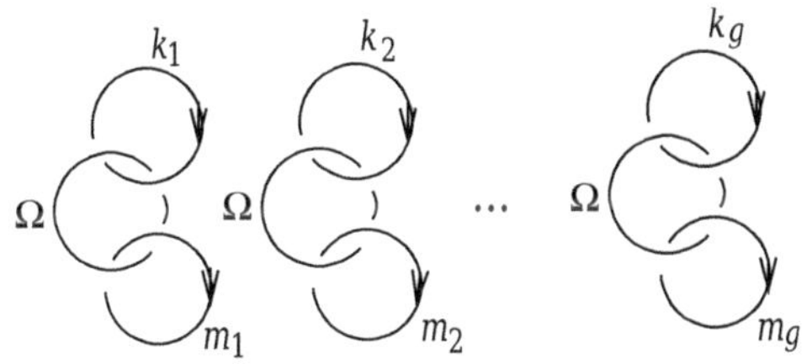

Fig. 7.13 Pairing that gives rise to the inner product

Lemma 7.7. *The identity from Figure 7.14 holds for every integer* j.

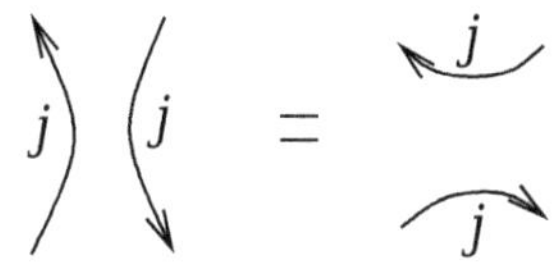

Fig. 7.14 Smoothing j strands with opposite orientation

Proof. Start with the innermost strands and proceed by using the smoothing from Figure 7.15. This proves the lemma. □

Returning to the proof of the theorem, by using Lemma 7.7 we can transform the skein from Figure 7.13 into a skein in which through the jth loop decorated by Ω pass $|k_j - m_j|$ strands. By Lemma 7.1 this is equal to 0, unless $k_j = m_j$ for all $j = 1, 2, \ldots, g$ in which case it is equal to $N^{g/2}$.

This shows that $a_1^{k_1} a_2^{k_2} \cdots a_g^{k_g}$ is an orthonormal basis with respect to the pairing $N^{-g/2}[\cdot, \cdot]_{h_{\mathrm{inn}}}$. As seen in the proof of Theorem 5.7,

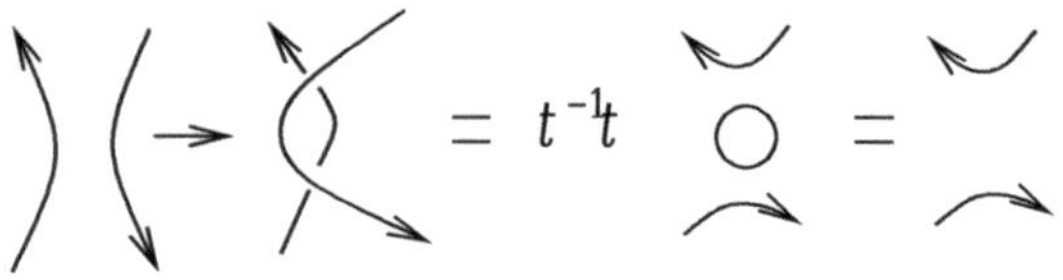

Fig. 7.15 Smoothing strands with opposite orientation

$a_1^{k_1} a_2^{k_2} \cdots a_g^{k_g}$ is the topological realization of the theta series $\theta^{\Pi}_{k_1,k_2,\ldots,k_g}(z)$. Since the theta series form an orthonormal basis with respect to the inner product (Theorem 4.13), we conclude that the pairing from the statement is the inner product. □

7.4 The 4-dimensional manifolds associated to discrete Fourier transforms

An interesting coincidence in mathematics is the fact that the cocycle that resolves the projective ambiguity of the representation of $Sp(2n, \mathbb{R})$ on $L^2(\mathbb{R}^n)$ defined by Fourier transforms (see §2.5.3) and the non-additivity of the signature of 4-dimensional manifolds (see §6.5.2) are both described in the same way in terms of the Maslov index.

It should be noted that the constructions from Chapters 4 and 5 hint that the same cocycle as the one in §2.5.3 should resolve the projective ambiguity of the representation of the mapping class group of the surface onto theta functions. And indeed, this is how it is usually done in standard texts. Our goal is to explain the above mentioned coincidence by showing how to resolve the projective ambiguity of the representation of the mapping class group of the Riemann surface on the space of theta functions in terms of 4-dimensional manifolds. We extracted the germ of the idea from Kevin Walker's unpublished notes on topological quantum field theory [Walker (1991)].

7.4.1 *Fourier transforms from general surgery diagrams*

Fix a Lagrangian subspace $\mathbf{L}$ of $H_1(\Sigma_g, \mathbb{R})$ spanned by the curves $b_1, b_2, \ldots, b_g$ in a canonical basis. This subspace defines a representation theoretical model for theta functions, the action of the finite Heisenberg group, as well as of the mapping class group. As explained in Chapter 5 we have a topological realization of the space of theta functions as $\mathcal{L}_N(H_g)$, and of the action of the finite Heisenberg group and of the mapping class group of Σ_g via skeins in $\mathcal{L}_N(\Sigma_g)$.

Let h be an element of the mapping class group $\mathrm{MCG}(\Sigma_g)$. We can represent h by surgery on a link $L_h \in \Sigma_g \times [0,1]$. The link L_h is not necessarily obtained from writing h as a composition of Dehn twists, as in Proposition 7.2. Nevertheless we can now prove the following result, which generalizes Proposition 7.2.

Theorem 7.9. *Let h be an element of the mapping class group of Σ_g obtained by performing surgery on the framed link L_h in $\Sigma_g \times [0,1]$. Then, up to multiplication by an eighth root of unity, the discrete Fourier transform $\rho(h) : \mathcal{L}_N(H_g) \to \mathcal{L}_N(H_g)$ is given by*

$$\rho(h)\beta = \Omega(L_h)\beta.$$

Proof. Write $h = T_1 T_2 \cdots T_n$, where T_j are Dehn twists obtained as surgeries along the nonseparating curves γ_j, $j = 1, 2, \ldots, n$. As explained in Example 6.17, h can be represented by a ribbon graph, namely its surgery diagram. This can be done for both L_h and $\gamma_1 \cup \gamma_2 \cup \cdots \cup \gamma_n$. Since the resulting mapping cylinder is the same, Theorem 6.7 implies that one diagram can be changed into the other by the moves (k1), (k2), (k3).

On the other hand, let us fill in the handlebodies at the two ends of the mapping cylinder in the standard way so as to turn the two surgery diagrams of h into framed links in S^3. The manifold $M = H_g \cup I_h \cup H_g$ has no boundary so its reduced skein module is $\mathbb{C}$. Using Theorem 7.7 we conclude that the gluing defines a nondegenerate pairing $[\cdot,\cdot] : \mathcal{L}_N(H_g) \times \mathcal{L}_N(H_g) \to \mathbb{C}$. If we consider basis elements e_j, e_k in $\mathcal{L}_N(H_g)$, then $[\Omega(L_h)e_j, e_k]$ completely determines the operator defined by $\Omega(L_h)$.

The skein $[\Omega(L_h)e_j, e_k] \in \mathcal{L}_N(M)$ is invariant under the Kirby moves (k2) and (k3) by Theorem 7.1. And by Lemma 7.2 it changes by a factor of $e^{\pm\frac{\pi i}{4}}$ under (k1). It follows that

$$[\Omega(L_h)e_j, e_k] = e^{\frac{\pi i}{4}m}[\Omega(\gamma_1)\Omega(\gamma_2)\cdots\Omega(\gamma_n)e_j, e_k],$$

where m is an integer that is the same for all j and k. Hence the operators defined by $\Omega(L_h)$ and $\Omega(\gamma_1 \cup \gamma_2 \cdots \cup \gamma_n)$ are equal up to a multiplication by an eighth root of unity. The conclusion now follows by applying Proposition 7.2. □

7.4.2 *A topological solution to the projectivity problem of the representation of the mapping class group on theta functions*

The discrete Fourier transform is only defined up to a multiplication by an eighth root of unity. We thus have a group homomorphism

$$\mathrm{MCG}(\Sigma_g) \to GL(\mathbf{\Theta}_N^{\Pi}(\Sigma_g))/\mathbb{Z}_8.$$

It would be desirable to turn this into a true representation

$$\mathrm{MCG}(\Sigma_g) \to GL(\mathbf{\Theta}_N^{\Pi}(\Sigma_g)),$$

but it is well known that this is impossible. Instead, to resolve this issue one passes to an extension of the mapping class group, as it was done in Chapter 2 for the symplectic group. There is a standard algebraic approach, which uses a $\mathbb{Z}_2$-extension that embeds inside the metaplectic group $Mp(2g, \mathbb{R})$.

However we follow a different approach, in order to maybe give a better understanding of this isssue. We resolve the projectivity of the representation of the mapping class group using topology. For this we define a **$\mathbb{Z}$-extension** of the mapping class group which has a true representation on the space of theta functions. While this approach uses a considerably larger extension of the mapping class group than necessary, it has the advantage that it generalizes to the Chern-Simons theories of gauge groups other than $U(1)$.

Four-dimensional manifolds associated to elements of the mapping class group

The idea is as follows. Let $h \in \mathrm{MCG}(\Sigma_g)$. Take the mapping cylinder I_h of h, glue H_g to its bottom, and $-H_g$ to its top. The result is a closed 3-dimensional manifold, which bounds a 4-dimensional manifold. The signature of this 4-dimensional manifold is the necessary additional piece of information to add to the mapping class group in order to resolve the projective ambiguity.

One way to do the construction is to identify I_h with $\Sigma_g \times [0, 1]$ with a surgery link inside, place $\Sigma_g \times [0, 1]$ in S^3, then glue the two handlebodies on one side and on the other as to obtain S^3, the way it was done in the proof of Theorem 7.9. Then, choose a 4-dimensional handlebody $\mathbf{M}$ defined by the surgery link. But this construction is ill behaved under composition!

We can improve this by gluing instead the two handlebodies to $\Sigma_g \times [0, 1]$ so as to obtain $\#_{k=1}^{g} S^1 \times S^2$, instead of S^3. This alternative construction has the advantage that if we glue two copies of $\#_{k=1}^{g} S^1 \times S^2$ by identifying the second handlebody, $-H_g$, of the first cylinder with the first handlebody, H_g, of the second cylinder by the identity map $Id : H_g \to H_g$, we obtain again $\#_{k=1}^{g} S^1 \times S^2$.

This construction can be realized using surgery diagrams for mapping cylinders. To do this rigorously, start with a surface Σ_g endowed with a canonical basis $a_1, a_2, \ldots, a_g, b_1, b_2, \ldots, b_g$ of $H_1(\Sigma_g, \mathbb{Z})$, with $b_1, b_2, \ldots, b_g$ spanning the Lagrangian subspace $\mathbf{L}$ of $H_1(\Sigma_g, \mathbb{R})$. Glue the handlebodies to $\Sigma_g \times [0,1]$ so that, for each boundary component, $b_1, b_2, \ldots, b_g$ become null-homologous and $a_1, a_2, \ldots, a_g$ become canonical bases of those handlebodies.

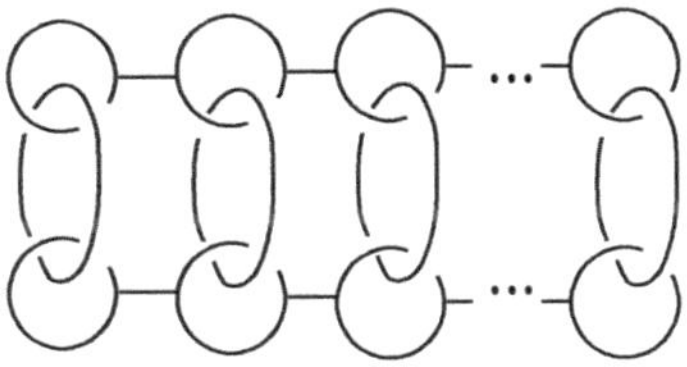

Fig. 7.16 Surgery diagram for the cylinder over a surface

As explained in Example 6.16, we can describe $\Sigma_g \times [0,1]$ by the diagram from Figure 7.16. If we glue H_g and $-H_g$ to the cylinder, then the two handlebody graphs disappear, and we are left with the trivial link with g components, which is the surgery diagram for $\#_{k=1}^{g} S^1 \times S^2$.

As for I_h, all we have to do is add to this diagram a surgery link L_h of h. We obtain a planar diagram that looks like the one on the left of Figure 7.17. If we fill in the handlebodies that correspond to the handlebody graphs in this diagram, we obtain a framed link in the plane. Denote this link by $L_{h,\mathbf{L}}$. It consists of L_h together with the surgery curves that yield $\#_{k=1}^{g} S^1 \times S^2$. Note that the while L_h depends on h only, $L_{h,\mathbf{L}}$ depends on the realization of h as a planar diagram, so it depends on the choice of the canonical basis. Since the gluing of the handlebodies really depends only on the Lagrangian subspace $\mathbf{L}$ (recall Lemma 5.5), by a slight abuse of notation, we encode the dependency of the link on the planar realization of the mapping cylinder by adding $\mathbf{L}$ as an index.

In this context, the surgery diagram of I_h turns into the surgery diagram of a closed 3-dimensional manifold M. The surgery link for this diagram is $L_{h,\mathbf{L}}$. This is shown schematically in Figure 7.17.

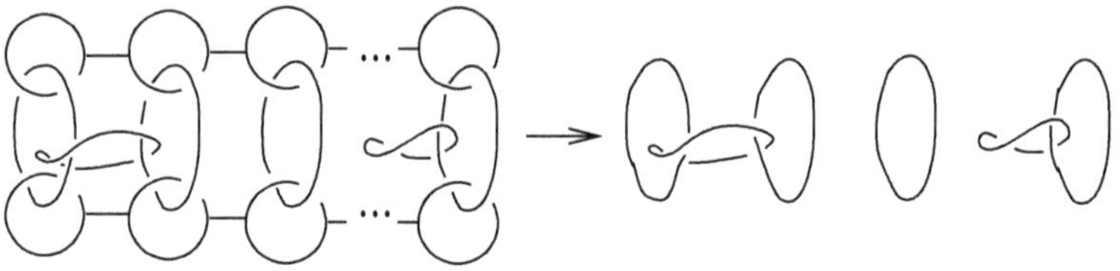

Fig. 7.17 Filling in the handlebodies to a mapping cylinder

Lemma 7.8. *The manifold M can be obtained by gluing to $\Sigma_g \times [0,1]$ the handlebodies H_g^0 and H_g^1 by homeomorphisms $f_j : \Sigma_g \times \{j\} \to \partial H_g^j$, $j = 0, 1$, such that $\mathbf{L}$ is the kernel of $(i_0 f_0)_* : H_1(\Sigma_g, \mathbb{R}) \to H_1(H_g^0, \mathbb{R})$ and $h_*^{-1}(\mathbf{L})$ is the kernel of the map $(i_1 f_1)_* : H_1(\Sigma_g, \mathbb{R}) \to H_1(H_g^1, \mathbb{R})$, where $i_j : \partial H_g^j \hookrightarrow H_g^j$ is the inclusion map.*

Proof. Note that the conditions that $\mathbf{L}$ respectively $h_*^{-1}(\mathbf{L})$ are the kernels of the specified maps determine[3] the homeomorphisms f_j, $j = 1, 2$, as proved in Lemma 5.5. It is straightforward to see that $\mathbf{L}$ is the kernel of the map that fills in the handlebody H_g^0, since this is how we produce the topological model for theta functions: by filling in the handlebody on the $\Sigma_g \times \{0\}$ side of the cylinder over the surface.

The fact that $h_*^{-1}(\mathbf{L})$ is the kernel of the map that fills in the handlebody H_g^1 follows from the fact that the mapping cylinder I_h is homeomorphic to $\Sigma_g \times [0,1]$ in such a way that the restriction of the homeomorphism to the top $\Sigma_g \times \{1\}$ is h itself. □

The manifold M depends only on h and $\mathbf{L}$. However, the surgery link of h tells us how to realize M as the boundary of a 4-dimensional manifold $\mathbf{M}$ obtained by adding 2-handles to the ball B^4. So the pair $(h, \mathbf{M})$ carries more information than just h.

If h and h' are elements of the mapping class group, to which 4-dimensional manifolds $\mathbf{M}$ and $\mathbf{M}'$ are associated via surgery diagrams, then to $h' \circ h$ we can associate the manifold $\mathbf{M}''$ defined by the surgery diagram of $h' \circ h$ obtained by the process described in Corollary 6.2. The manifold $\mathbf{M}''$ is obtained by gluing $\mathbf{M}$ and $\mathbf{M}'$ by identifying (in the standard way) $H_g^1 \subset \partial \mathbf{M}$ with $H_g^{0'} \subset \partial \mathbf{M}'$. The construction of the manifold $\mathbf{M}''$ is sketched in Figure 7.18.

[3]up to homeomorphisms of H_g^j

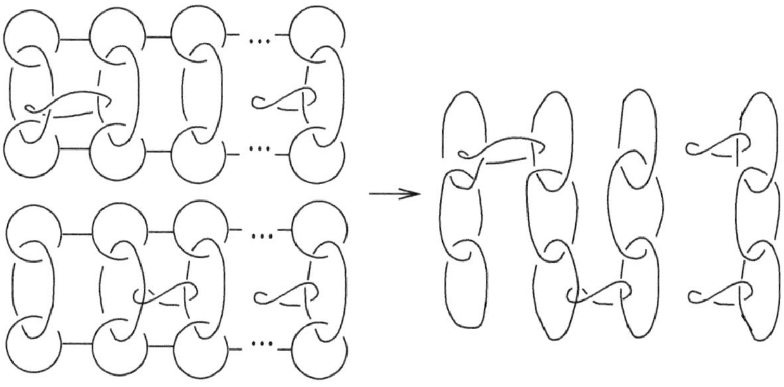

Fig. 7.18 Four-dimensional manifold associated to the composition of two homeomorphisms

The extended mapping class group

We should observe that the only ambiguity in defining $\mathcal{F}_{\mathbf{L}}(h)$ comes from adding trivial link components $U_{\pm}$ to the surgery diagram. So the only relevant information from the 4-dimensional manifold associated to h is its signature. This prompts us to make the following definition, in which we use the notations from above.

Definition 7.5. The **extended mapping class group** of the surface Σ_g is the $\mathbb{Z}$-extension of $\mathrm{MCG}(\Sigma_g)$ defined by the multiplication rule

$$(h', \sigma(\mathbf{M}'))(h, \sigma(\mathbf{M})) = (h'h, \sigma(\mathbf{M}' \cup \mathbf{M}))$$

where $\mathbf{M}'$ and $\mathbf{M}$ are glued in such a way that $H_g^1 \in \mathbf{M}$ is identified with $H_g^{0'}$ in $\mathbf{M}'$.

Note that by adding several $U_{\pm}$ components we can obtain every integer in the second entry. However, it is not entirely clear that what we defined is a group. For this we need to explicate the multiplication rule in terms of the homeomorphisms h, h', and the Lagrangian subspace $\mathbf{L}$ only. Wall's non-additivity formula (6.9) comes in handy.

In the notation of §6.5.2, we have

$$\mathbf{Y}_- = \mathbf{M}, \quad \mathbf{Y}_+ = \mathbf{M}'', \quad \mathbf{Y} = \mathbf{M}'',$$

$$X_0 = {H_g}^1 = H_g^{0'}, \quad X_- = I_h \cup H_g^0 \approx H_g^0, \quad X_+ = I_{h'} \cup H_g^{1'} \approx H_g^{1'},$$

$$Z = \Sigma_g.$$

As for the Lagrangian subspaces,

$$\mathbf{L}_- = \ker(H_1(Z, \mathbb{R}) \to H_1(X_-, \mathbb{R})) = \ker(H_1(\Sigma_g, \mathbb{R}) \to H_1(H_g^0, \mathbb{R})) = \mathbf{L}$$

and

$$\mathbf{L}_0 = \ker(H_1(Z,\mathbb{R}) \to H_1(X_-,\mathbb{R})) = \ker(H_1(\Sigma_g,\mathbb{R}) \to H_1(H_g^1,\mathbb{R}))$$
$$= h_*^{-1}(\mathbf{L}).$$

And, since $\ker(H_1(\Sigma_g,\mathbb{R}) \to H_1(H_g^{0'},\mathbb{R})) = h_*^{-1}(\mathbf{L})$ because of the way we identify H_g^1 with $(H_g^0)'$, we have

$$\mathbf{L}_+ = \ker(H_1(Z,\mathbb{R}) \to H_1(X_+,\mathbb{R})) = \ker(H_1(\Sigma_g,\mathbb{R}) \to H_1(H_g^{1'},\mathbb{R}))$$
$$= (h'^{-1}h^{-1})_*(\mathbf{L}).$$

Wall's formula for the non-additivity of the signature of 4-dimensional manifolds (Theorem 6.9) gives

$$\sigma(\mathbf{M}' \cup \mathbf{M}) = \sigma(\mathbf{M}') + \sigma(\mathbf{M}) - \tau(\mathbf{L}, h_*^{-1}(\mathbf{L}), (h'^{-1}h^{-1})_*(\mathbf{L})),$$

where τ is the Maslov index. But

$$\tau(\mathbf{L}, h_*^{-1}(\mathbf{L}), (h'^{-1}h^{-1})_*(\mathbf{L}))) = \tau(h_*(\mathbf{L}), \mathbf{L}, (h'^{-1})_*(\mathbf{L}))$$
$$= \tau((h'h)_*\mathbf{L}, h'_*(\mathbf{L}), \mathbf{L}).$$

We can now formulate a better definition.

Definition 7.6. The **extended mapping class group** of the surface Σ_g is the $\mathbb{Z}$-extension of $\mathrm{MCG}(\Sigma_g)$ defined by the multiplication rule

$$(h',n')(h,n) = (h'h, n+n' - \tau((h'h)_*(\mathbf{L}), h'_*(\mathbf{L}), \mathbf{L}))).$$

Notation: We denote the extended mapping class group by $\widetilde{\mathrm{MCG}(\Sigma_g)}$.

The cocycle condition for the Maslov index implies that the multiplication rule is associative. The identity element is $(Id, 0)$, because if two entries of the Maslov index are equal the index is zero, and for the same reason, the inverse of (h,n) is $(h^{-1}, -n)$. Hence we do indeed have a group, which is a central $\mathbb{Z}$-extension of the mapping class group.

Theorem 7.10. *There is a well-defined representation $\mathcal{F}_{\mathbf{L}}$ of from the extended mapping class group on $\mathcal{L}_N(H_g)$, given by*

$$\widetilde{MCG(\Sigma_g)} \to \mathcal{L}_N(\Sigma_g), \quad \mathcal{F}_{\mathbf{L}}(h,n) = e^{\frac{\pi i}{4}(n-\sigma(L_{h,\mathbf{L}}))}\Omega(L_h);$$

where L_h is the surgery link for h and $L_{h,\mathbf{L}}$ is the associated framed link in S^3 obtained from the surgery diagram of h by deleting the two handlebody graphs.

Proof. The skein

$$e^{-\frac{\pi i}{4}\sigma(\mathcal{L}_{h,\mathbf{L}})}\Omega(L_h)$$

does not depend on the chosen link L_h. Indeed, the matrix of $\Omega(L_h)$, as defined in the proof of Theorem 7.9, is invariant under the Kirby moves (k2) and (k3) of Theorem 6.7. On the other hand, taking the band sum of one component of $\mathcal{L}_{h,\mathbf{L}}$ that comes from L_h with another does not change the signature of the knot, since it does not change the signature of the 4-dimensional manifold that it bounds. Finally, performing (k1) moves in L_h produces the same (k1) moves in $L_{h,\mathbf{L}}$, and so the factor $e^{\pm\frac{\pi i}{4}}$ introduced in $\Omega(L_h)$ by this move is canceled with the factor $e^{\mp\frac{\pi i}{4}}$ induced by the change in the signature of $L_{h,\mathbf{L}}$. Therefore we have a well defined formula for $\mathcal{F}_{\mathbf{L}}$.

Also, if h, h' are described by the links L respectively L', then because Wall's non-additivity formula translates into

$$\sigma(L' \circ L) = \sigma(L) + \sigma(L') - \tau((h'h)_*(\mathbf{L}), h'(\mathbf{L}), \mathbf{L}),$$

we have

$$\begin{aligned}\mathcal{F}_{\mathbf{L}}(h', n')\mathcal{F}_{\mathbf{L}}(h, n) &= e^{\frac{\pi i}{4}(n+n'-\sigma(L_{h,\mathbf{L}})-\sigma(L'_{h,\mathbf{L}}))}\Omega(L'_h)\Omega(L_h) \\ &= e^{\frac{\pi i}{4}[n+n'+\tau((h'h)_*(\mathbf{L}),h'_*(\mathbf{L})\mathbf{L})-\sigma(L'\circ L)]}\Omega(L'_h \cup L_h) \\ &\mathcal{F}_{\mathbf{L}}(h'h, n+n'+\tau((h'h)_*(\mathbf{L}), h'_*(\mathbf{L})\mathbf{L})).\end{aligned}$$

Hence the map is multiplicative.

Also $\mathcal{F}_{\mathbf{L}}(Id, 0)$ is $\emptyset$, the identity element in $\mathcal{L}_N(\Sigma_g)$, so we obtain a group homomorphism onto the image. □

The map

$$(h, n) \mapsto \mathcal{F}_{\mathbf{L}}(h, n)$$

is now a true representation of the extended mapping class group on theta functions. We have

$$\begin{aligned}\mathcal{F}_{\mathbf{L}}((h', n')(h, n)) &= \mathcal{F}_{\mathbf{L}}(h'h, n+n'+\tau((h'h)_*(\mathbf{L}), h'_*(\mathbf{L}), (\mathbf{L})) \\ &= e^{\frac{i\pi}{4}\tau((h'h)_*\mathbf{L}, h'_*(\mathbf{L}), \mathbf{L})}\mathcal{F}_{\mathbf{L}}(h'h, n+n').\end{aligned}$$

We recognize the cocycle

$$c(h, h') = e^{\frac{i\pi}{4}\tau((h'h)_*(\mathbf{L}), h'_*(\mathbf{L}), \mathbf{L})}$$

that we used to resolve the projective ambiguity of the representation of $Sp(2n, \mathbb{R})$ in §2.5.3.

Let us examine the construction of $\mathcal{F}_{\mathbf{L}}(h, n)$ carefully, since we will repeat it in more generality later. Choose a surgery diagram such that

$$\sigma(L_{h,\mathbf{L}}) = n.$$

We can interpret the surgery diagram for I_h as defining a bilinear map

$$B_h : \mathcal{L}_N(H_g) \times \mathcal{L}_N(H_g) \to \mathcal{L}_N(S^3) = \mathbb{C},$$

given by filling in regular neighborhoods of the handlebody graphs with handlebodies with skeins inside, and multiplying the result by $N^{-g/2}$. Once we fix the canonical basis for $H_1(\Sigma_g, \mathbb{Z})$ we have a canonical basis for $\mathcal{L}_N(H_g)$, which identifies it with $\mathbb{C}^{Ng}$, and also a canonical way to insert H_g in the location of the handlebody graph. This allows us to compute the matrix of the pairing; call it $A_{h,n}$; so that the pairing reads

$$B_h(\mathbf{x}, \mathbf{y}) = N^{-g/2}\mathbf{y}^T A_{h,n}\mathbf{x}.$$

By Theorem 7.13,

$$B_h(\mathbf{x}, \mathbf{y}) = N^{-g/2}[\mathbf{x}, \mathbf{y}]_{h_{inn}} = \langle \mathcal{F}_{\mathbf{L}}(h,n)\mathbf{x}, \mathbf{y}\rangle,$$

so $A_{h,n}$ is the matrix of $\mathcal{F}_{\mathbf{L}}(h,n)$ in this basis.

Therefore if (h,n) is represented by a surgery diagram such as the one on the left of Figure 7.17, then the element

$$\left\langle \mathcal{F}_{\mathbf{L}}(h,n)a_1^{j_1}a_2^{j_2}\cdots a_g^{j_g}, a_1^{k_1}a_2^{k_2}\cdots a_g^{k_g}\right\rangle$$

is equal to the skein in S^3 described in Figure 7.19.

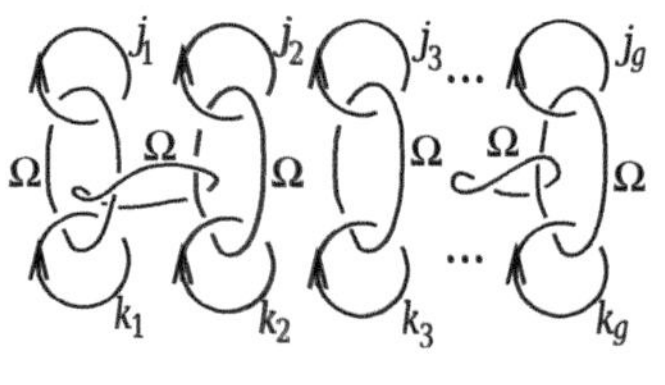

Fig. 7.19 Graphical computation of the matrix of the Fourier transform

We can compute $\mathcal{F}_{\mathbf{L}}(h',n')\mathcal{F}_{\mathbf{L}}(h,n)$ using Corollary 6.2. We have

$$\left\langle \mathcal{F}_{\mathbf{L}}(h',n')\mathcal{F}_{\mathbf{L}}(h,n)a_1^{k_1}a_2^{k_2}\cdots a_g^{k_g}, a_1^{l_1}a_2^{l_2}\cdots a_g^{l_g}\right\rangle$$
$$= \sum_{\mathbf{j}} \left\langle \mathcal{F}_{\mathbf{L}}(h',n')a_1^{j_1}\cdots a_g^{j_g}, a_1^{l_1}\cdots a_g^{l_g}\right\rangle \left\langle \mathcal{F}_{\mathbf{L}}(h,n)a_1^{k_1}\cdots a_g^{k_g}, a_1^{j_1}\cdots a_g^{j_g}\right\rangle.$$

This computation is shown in Figure 7.20. Here in the first step we performed Reidemeister II moves, in the second we used Lemma 7.7 and in the third step we used the definition of Ω. We recognize the surgery link $L' \circ L$, and this is the diagram which computes the value of

$$\left\langle \mathcal{F}_{\mathbf{L}}(h'h, \sigma(L' \circ L))a_1^{k_1}a_2^{k_2}\cdots a_g^{k_g}, a_1^{l_1}a_2^{l_2}\cdots a_g^{l_g}\right\rangle,$$

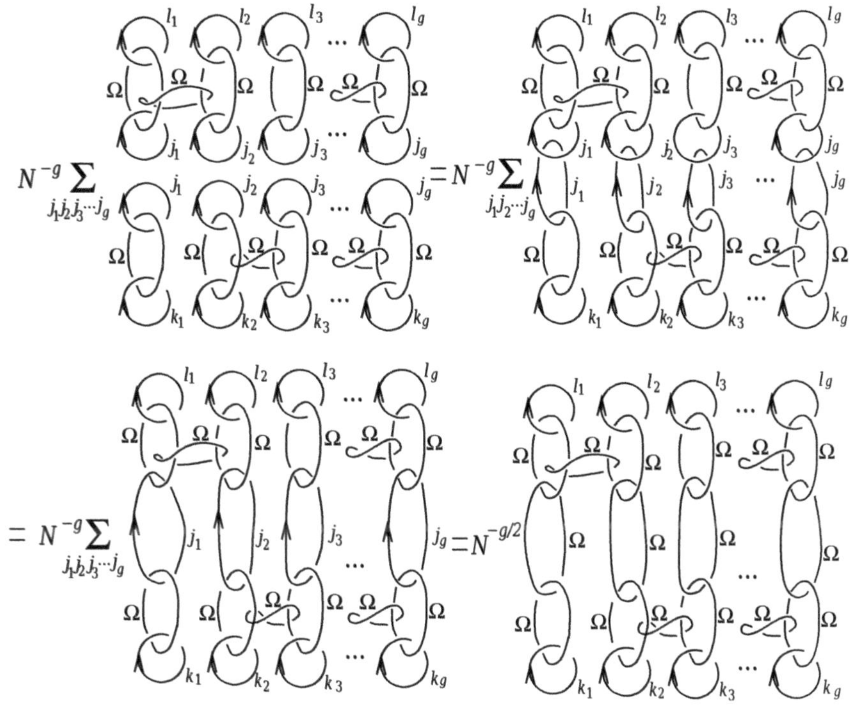

Fig. 7.20 The computation of the composition of two Fourier transforms

where

$$\begin{aligned}\sigma(L' \circ L) &= \sigma(L) + \sigma(L') - \tau((h'h)_*\mathbf{L}, h'_*(\mathbf{L}), \mathbf{L}) \\ &= n + n' - \tau((h'h)_*\mathbf{L}, h'_*(\mathbf{L}), \mathbf{L}).\end{aligned}$$

We should conclude with the remark that while the discrete Fourier transform $\mathcal{F}_{\mathbf{L}}(h, n)$ depends only on $\mathbf{L}$, the choice of a specific canonical basis for $H_1(\Sigma_g, \mathbb{Z})$ yields a choice of a basis that identifies $\mathcal{L}_N(H_g)$ with $\mathbb{C}^{Ng}$ and thus explicates the matrix matrix $A_{h,n}$ of this discrete Fourier transform.

Here are four examples on the torus Σ_1. In all figures, the top and bottom circles are handlebody graphs, while the circles in the middle are surgery link components.

Example 7.3. As before we denote by T the positive Dehn twist T_{b_1} about the curve $b_1 = (0, 1)$. Then the extended homeomorphism $(T, 0)$ has the surgery diagram from Figure 7.21.

Fig. 7.21 Surgery diagram for the Dehn twist

It follows that the j, k entry of the matrix of $N^{1/2}\mathcal{F}_{\mathbf{L}}(T, 0)$ is the skein in $\mathcal{L}_N(S^3) = \mathbb{C}$ given by the diagram from Figure 7.22. Here for the first equality we performed a Kirby (k0) move, while for the second we resolved the crossings in the j strands at the bottom and we canceled $\Omega(U_+)$ with $\Omega(U_-)$. The resulting diagram is equal to

$$t^{-j^2} N^{1/2} \left\langle \theta_j^{\tau}(z), \theta_k^{\tau}(z) \right\rangle = \delta_{jk} t^{-j^2},$$

where δ_{jk} is the Kronecker symbol. It follows that the j, k entry of the matrix of $\mathcal{F}_{\mathbf{L}}(T, 0)$ is $\delta_{jk} t^{-j^2}$. We recognize the formula from §4.4.3.

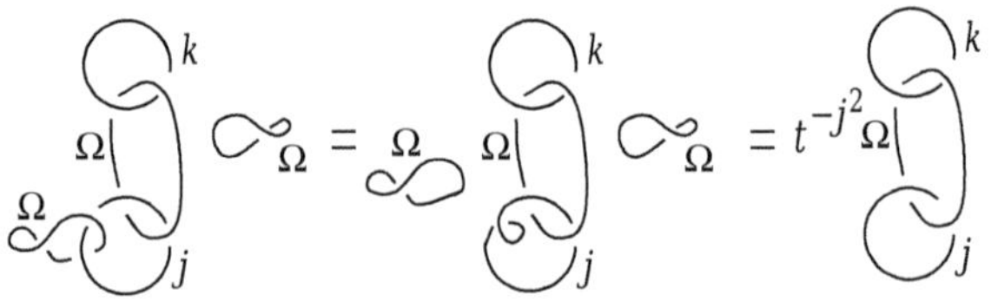

Fig. 7.22 The j, k entry of the matrix of $\mathcal{F}_{\mathbf{L}}(T, 0)$

Example 7.4. Let T_{a_1} be the positive Dehn twist about the curve $(1, 0)$. For $(T_{a_1}, 0)$, three surgery diagrams related by a (k0) move are shown in in Figure 7.23. Using the right-most diagram, one can compute that the j, k

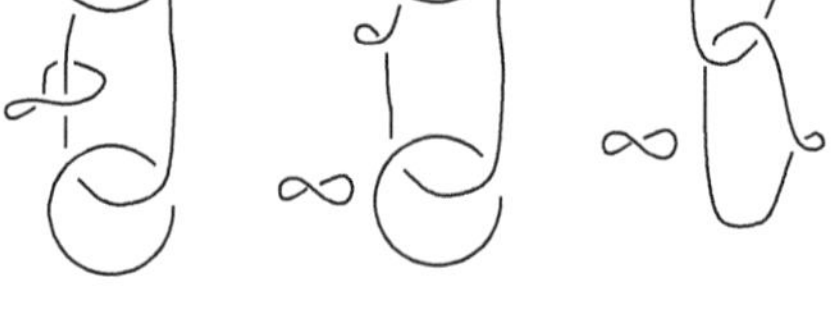

Fig. 7.23 Surgery diagram for the Dehn twist T_{a_1}

entry of the matrix of $\mathcal{F}_{\mathbf{L}}(T_{a_1}, 0)$ is

$$t^{-2jk-j^2-k^2} = t^{-(j+k)^2}.$$

One should also note, for the next example, that the surgery diagrams for $T_{a_1}^{-1}$ are mirror images of surgery diagrams for T_{a_1}.

Example 7.5. Again for the torus Σ_1, recall the S-map from §3.3.2. Then $(S, 0)$ has the surgery diagram from Figure 7.24, where the two circles are handlebody graphs.

Fig. 7.24 Surgery diagram for the S-map

This can be seen as follows. It is easy to check that

$$S = T_{b_1} T_{a_1}^{-1} T_{b_1}$$

for example by using the fact that $\mathrm{MCG}(\Sigma_1) = SL(2, \mathbb{Z})$ (see §3.3.2) and checking that

$$\begin{pmatrix} 1 & 0 \\ 1 & 1 \end{pmatrix} \begin{pmatrix} 1 & -1 \\ 0 & 1 \end{pmatrix} \begin{pmatrix} 1 & 0 \\ 1 & 1 \end{pmatrix} = \begin{pmatrix} 0 & -1 \\ 1 & 0 \end{pmatrix}.$$

Composing surgery diagrams for T_{b_1}, $T_{a_1}^{-1}$ and T_{b_1}, we obtain the surgery diagram for $(S, 0)$ from Figure 7.25. Now apply a series of (k0) and trans-

Fig. 7.25 Surgery diagram for the S-map

form it as shown in Figure 7.26.

Let $L_1 \cup L_2$ be the Hopf link depicted in Figure 7.24, and orient L_1 and L_2 so that in the planar projection that yields the link diagram they are both oriented counterclockwise. Then

$$< L_1^{\|j} \cup L_2^{\|k} >= t^{-2jk}.$$

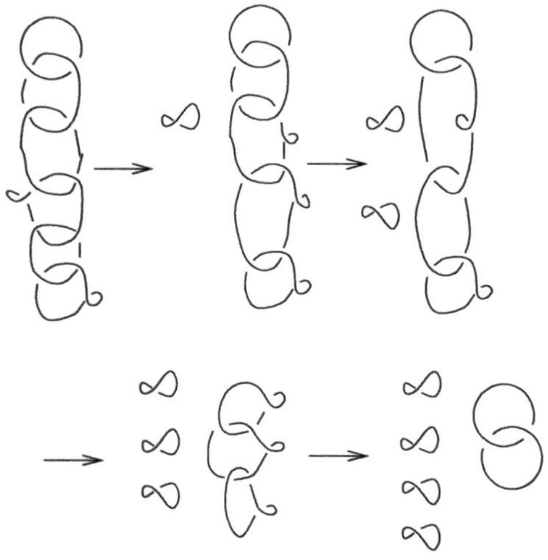

Fig. 7.26 Transformation of the surgery diagram for the S-map

It follows that the matrix of $\mathcal{F}_{\mathbf{L}}(S,0)$ has entries equal to $N^{-1/2}t^{-2jk}$, which is the standard discrete Fourier transform from §4.4.2.

Example 7.6. Consider a general $h \in \mathrm{MCG}(\Sigma_1)$. Write the map h_* : $H_1(\Sigma_1,\mathbb{Z}) \to H_1(\Sigma_1,\mathbb{Z})$ as the matrix

$$\begin{pmatrix} a & b \\ c & d \end{pmatrix}.$$

The theory of linear Diophantine equations (see [Olds (1963)]) shows that

$$\begin{pmatrix} a & b \\ c & d \end{pmatrix} = ST^{a_1}ST^{a_2}\cdots ST^{a_n}S.$$

where we have the continued fraction expansion

$$\frac{b}{a} = -\cfrac{1}{a_1 - \cfrac{1}{a_2 - \cdots \cfrac{1}{a_n}}}.$$

This means that (h,m) is given by the surgery diagram from Figure 7.27, where $m = a_1 + a_2 + \cdots + a_n$.[4]

The j,k entry of $\mathcal{F}_{\mathbf{L}}(h,m)$ is obtained from this diagram by

[4]This diagram is drawn rotated by 90° in order to save space; it should be understood that the genus 1 handlebody graphs are the left-most and the right-most circles.

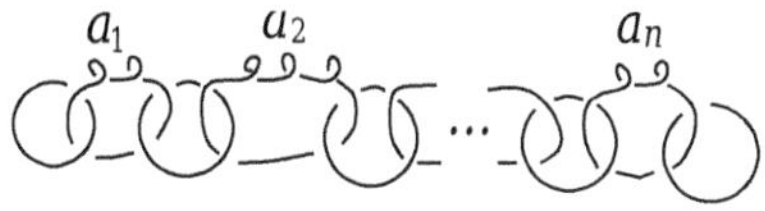

Fig. 7.27 Surgery diagram for h

- orienting the left-most and right-most link components counterclockwise;
- taking jth parallel copies of the left-most link component, k parallel copies of the right-most link component;
- decorating the other link components by Ω;
- evaluating the resulting skein in $\mathcal{L}_N(S^3) = \mathbb{C}$, and multiplying the result by $N^{-1/2}$.

7.5 Theta functions and topological quantum field theory

In this section we explain how the above considerations lead to a construction of what is known as a **topological quantum field theory**. We start with a heuristical discussion.

7.5.1 *Empty skeins and the emergence of topological quantum field theory*

If M is an oriented, compact, 3-dimensional manifold, then Theorem 7.6 allows us to associate to M a finite dimensional vector space that depends only on ∂M. This vector space is the reduced linking number skein module of M.

Let L be a framed link in S^3 and let M_L be the manifold without boundary obtained by performing surgery on L. The skein $\Omega(L)$ used for defining the Murakami-Ohtsuki-Okada invariant of M_L is the image through a Sikora isomorphism

$$\mathcal{S} : \mathcal{L}_N(M_L) \to \mathcal{L}_N(S^3) = \mathbb{C}$$

of the empty skein in M_L. We can therefore think of the **empty skein**

$$<\emptyset> \in \mathcal{L}_N(M)$$

as being the Murakami-Ohtsuki-Okada invariant of M_L. This is not quite correct, because if we want the numerical realization of this invariant in $\mathcal{L}_N(S^3) = \mathbb{C}$, then there is an ambiguity consisting of a multiplication by an eight root of unity. This is because the Sikora isomorphism depends on the surgery link L, and the skeins that come from two links L_1 and L_2 differ by a factor of $e^{-\frac{\pi i}{4}(\sigma(L_1)-\sigma(L_2))}$. Nevertheless the empty skein gives us a good intuition about the Murakami-Ohtsuki-Okada invariant and in particular tells us that the vector space that is associated to a manifold as its reduced linking number skein module has a special vector in it.

If M is a connected manifold with boundary, then, as explained in §6.4, it has a surgery diagram consisting of the embedding of a handlebody graph Γ and a framed link L in S^3. The reduced linking number skein module of M is isomorphic to the reduced linking number skein module of the complement of a regular neighborhood of Γ in S^3. Abusing slightly the notation let us denote this reduced skein module by $\mathcal{L}_N(S^3\backslash\Gamma)$. We let $\mathcal{L}_N(S^3\backslash\Gamma)$ be the vector space associated to M, and observe that the vector spaces of manifolds with homeomorphic boundaries are the same. It is important to keep in mind that the association of the vector space $\mathcal{L}_N(S^3\backslash\Gamma)$ to M, obtained by identifying $\mathcal{L}_N(M)$ with $\mathcal{L}_N(S^3\backslash\Gamma)$ via a Sikora isomorphism, is not canonical. Nevertheless we can think that $\mathcal{L}_N(S^3\backslash\Gamma)$ is the vector space associated to the surface ∂M.

We can let the Murakami-Ohtsuki-Okada invariant of the manifold with boundary M be the empty skein in M, and realize this concretely as the vector $\Omega(L) \in \mathcal{L}_N(S^3\backslash\Gamma)$. We can moreover identify $\mathcal{L}_N(S^3\backslash\Gamma)$ with some $\mathbb{C}^m$ for some m using Theorem 7.6 and then obtain the manifold invariant as a vector with numerical entries. Again there is a problem with the non-uniqueness of the surgery diagram, which makes the invariant ambiguous. We will resolve this ambiguity, but before doing that we want to argue that it is much better to think in terms of cobordisms.

The key idea is that if we compose two cobordisms then the empty links in the cobordisms are mapped to the empty link inside the resulting cobordism (see Figure 7.28). So we have a "composition" of empty skeins that parallels that of cobordisms.

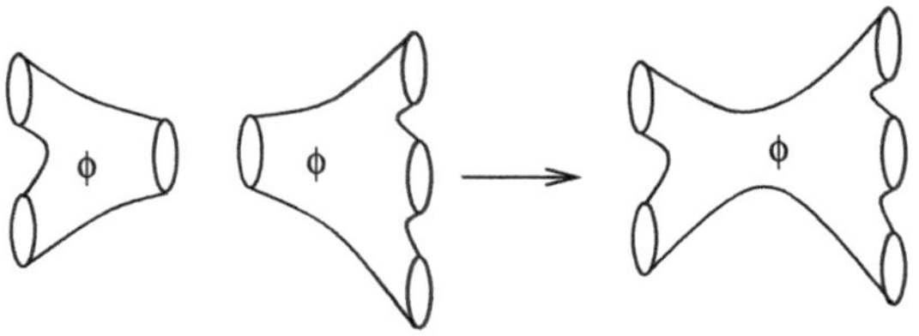

Fig. 7.28 "Composition" of empty skeins

On the other hand, if $(M, \partial_- M, \partial_+ M)$ is a cobordisms, then the parametrizations of $\partial_\pm M$ determine a unique gluing, up to isotopy, of a disjoint union $H^\pm = \sqcup_j H^\pm_{g_j}$ of handlebodies to M along $\partial_\pm M$. By Theorem 7.7 the map

$$\mathcal{L}_N(H^-) \times \mathcal{L}(M) \times \mathcal{L}_N(H^+) \mapsto \mathcal{L}_N(H^- \cup M \cup H^+) = \mathbb{C},$$
$$(< L_- >, \emptyset, < L_+ >) \mapsto < L_- \cup \emptyset \cup L_+ >$$

identifies $\emptyset$ with an element of $\mathcal{L}_N(H^-)^* \otimes \mathcal{L}_N(H^+)$, namely with a linear map

$$\mathcal{L}_N(H^-) \to \mathcal{L}_N(H^+).$$

And $\mathcal{L}_N(H^-)$ and $\mathcal{L}_N(H^+)$ are (isomorphic to) the vector spaces associated to the surfaces $\partial_- M$ and $\partial_+ M$.

In short, the above heuristics leads us to the following:

- to each surface we associate a finite dimensional vector space,
- to each cobordism we associate a linear map between the vector space of the initial boundary and that of the terminal boundary such that the composition of the cobordisms gives rise to the composition of the linear maps.

With this in mind we will now create a universe that incorporates the theories of theta functions for all surfaces.

7.5.2 *Atiyah's axioms for a topological quantum field theory*

We now state Atiyah's axioms for a topological quantum field theory, which give the formal framework for the above heuristical discussion. Sir Michael Atiyah derived these axioms in [Atiyah (1988)] and [Atiyah (1990)] to give an algebraic expression to Witten's physical constructs [Witten (1989)]. It is common to use the abbreviation **TQFT** for topological quantum field theory.

In Atiyah's formalism, a TQFT consists of:

- A functor V from the category whose objects are closed oriented surfaces and whose morphisms are orientation preserving homeomorphisms to the category whose objects are finite-dimensional vector spaces and whose morphisms are isomorphisms of vector spaces.
- A way to associate to each oriented, compact, 3-dimensional cobordisms $(M, \partial_- M, \partial_+ M)$ a morphism $Z(M) : V(\partial_- M) \to V(\partial_+ M')$.

These should satisfy the following axioms:

A1 (Multiplicativity) $V(\Sigma \sqcup \Sigma') = V(\Sigma) \otimes V(\Sigma')$ and $Z(M \sqcup M') = Z(M) \otimes Z(M')$, where we recall that $\sqcup$ denotes the disjoint union.

A2 (Involutory) $V(-\Sigma) = V(\Sigma)^*$ where $-\Sigma$ is the surface with opposite orientation and the $*$ denotes the dual of the vector space.
Also, if $(M, \partial_- M, \partial_+ M)$ is a cobordism and if Σ is a boundary component we can change the orientation of Σ thus moving it from $\partial_- M$ to $\partial_+ M$ or vice-versa. The homomorphism $Z(M, \partial_- M, \partial_+ M)$ can be viewed as an element of $V(\partial_- M)^* \otimes V(\partial_+ M)$. When changing the orientation of Σ, $Z(M, \partial_- M, \partial_+ M)$ is mapped by the natural transformation that takes $V(\Sigma)$ to $V(\Sigma)^*$ in the above tensor product.

A3 (Associativity) If $(M, \partial_- M, \partial_+ M)$ and $(M', \partial_- M', \partial_+ M')$ are cobordisms that can be composed, and if

$$(M'', \partial_- M'', \partial_+ M'') = (M', \partial_- M', \partial_+ M')(M, \partial_- M, \partial_+ M)$$

then

$$Z(M'', \partial_- M'', \partial_+ M'') = Z(M', \partial_- M', \partial_+ M')Z(M, \partial_- M, \partial_+ M),$$

meaning that $Z(M'', \partial_- M'', \partial_+ M'')$ is the composition of the morphisms $Z(M', \partial_- M', \partial_+ M')$ with the morphism $Z(M, \partial_- M, \partial_+ M)$.

A4 $V(\emptyset) = \mathbb{C}$, where $\emptyset$ denotes the empty surface.

A5 $Z(\Sigma \times [0,1], \Sigma \times \{0\}, \Sigma \times \{1\})$ is the identity morphism of $V(\Sigma)$.

The associativity axiom A3 is also referred to as **functoriality**, because, along with A5, it requires the TQFT to be a functor from the category of cobordisms to the category of finite-dimensional vector spaces.

The "trivial" theta functions

A fact that can be noticed immediately by examining Atiyah's axioms, is that one has to make sense of the "theta functions" associated to the 2-dimensional sphere and to the empty surface.

In the first case we can use skein modules, and call $\mathcal{L}_N(B^3) = \mathbb{C}$ the space of theta functions associated to S^2. This is because the 3-dimensional ball is the handlebody bounded by S^2. The role of the group algebra of the finite Heisenberg group is played by $\mathcal{L}_N(S^2)$ which is $\mathbb{C}$. Finally, there is only one element in $\mathrm{MCG}(S^2)$ (Theorem 3.6); its corresponding discrete Fourier transform is the identity map.

We stress out that the identification of $\mathcal{L}_N(B^3)$ with $\mathbb{C}$ is canonical. Indeed the Alexander Lemma (see §3.3.2) can be extended to 3 dimensions to show that every orientation preserving automorphism of B^3 is isotopic to the identity. So we can identify B^3 in a standard way with the unit ball in $\mathbb{R}^3$ and then the map which associates to a link the coefficient of $\emptyset$ obtained after applying skein relations defines uniquely the isomorphism between $\mathcal{L}_N(B^3)$ and $\mathbb{C}$.

In the second case, axiom A4 implies that the vector space associated to the trivial surface should be $\mathbb{C}$ as well. Again we let its associated finite Heisenberg group be trivial, so that the group algebra is $\mathbb{C}$.

In either case, there is no such thing as a canonical basis, so we agree that both S^2 and $\emptyset$ are endowed always with the Lagrangian subspace $\mathbf{L} = \{0\}$.

7.5.3 *The functor from the category of extended surfaces to the category of finite-dimensional vector spaces*

The topological quantum field theory of theta functions requires a slightly finer structure than that specified by Atiyah's axioms. Its 2-dimensional aspect consists of a functor from the category of what we will call, following Kevin Walker [Walker (1991)], extended surfaces to the category of finite-dimensional vector spaces.

The category of extended surfaces

Intuitively, the category of extended surfaces is a larger universe that contains all extended mapping class groups.

Definition 7.7. The **category of extended surfaces** has

(a) as objects the **extended surfaces** which consist of pairs $(\Sigma, \mathbf{L})$ where Σ is a compact oriented surface without boundary and $\mathbf{L}$ is a Lagrangian subspace of $H_1(\Sigma, \mathbb{R})$ which in each component of Σ is spanned by half of the elements of a canonical basis of $H_1(\Sigma, \mathbb{Z})$;
(b) as morphisms the **extended homeomorphisms** which consist of pairs (h, n) where h is an element of the mapping class group of a surface and n is an integer.

The **multiplication rule** for extended homeomorphisms is defined as follows. If

$$(h, n) : (\Sigma_g, \mathbf{L}) \to (\Sigma_g, (\mathbf{L}') \text{ and } (h', n') : (\Sigma_g, \mathbf{L}') \to (\Sigma_g, \mathbf{L}''),$$

then

$$(h', n')(h, n) = (h'h, n + n' - \tau((h'h)_*(\mathbf{L}), h'_*(\mathbf{L}'), \mathbf{L}'').$$

Here, if $\Sigma = \sqcup_{k=1}^{n} \Sigma_{g_k}$, then the Maslov index τ is computed in

$$\bigoplus_{k=1}^{n} H_1(\Sigma_{g_k}, \mathbb{R})$$

endowed with the symplectic form which is the sum of the intersection forms in each component.

The 2-dimensional functor

It is time to describe the functor V. First, if Σ is connected and of genus at least 1, that is $\Sigma = \Sigma_g$, and if $\mathbf{L}$ is a Lagrangian subspace of $H_1(\Sigma_g, \mathbb{R})$, then the functor associates to the pair $(\Sigma_g, \mathbf{L})$ the vector space

$$V(\Sigma_g, \mathbf{L}) = \mathcal{H}_{N,g}(\mathbf{L})$$

of the representation induced by

$$\chi_{\mathbf{L}} : \exp(\mathbf{L} + \mathbb{R}E) \to \mathbb{C}, \quad \chi_{\mathbf{L}}(l + kE) = e^{\frac{\pi i}{N} k}.$$

In view of Theorem 5.7 and Lemma 5.5, $\mathcal{H}_{N,g}(\mathbf{L})$ has a topological incarnation obtained by attaching Σ_g to a handlebody H_g in such a way that the curves of the canonical basis spanning $\mathbf{L}$ bound embedded disks; then $\mathcal{H}_{N,g}(\mathbf{L}) = \mathcal{L}_N(H_g)$.

By our convention $V(\emptyset, \{0\}) = \mathbb{C}$ and $V(S^2, \{0\}) = \mathbb{C}$.

The functor extends to disconnected surfaces by the rule

$$V(\Sigma \sqcup \Sigma', \mathbf{L} \oplus \mathbf{L}') = V(\Sigma, \mathbf{L}) \otimes V(\Sigma', \mathbf{L}').$$

Thus if

$$\Sigma = (\sqcup_{j=1}^{m} S^2) \sqcup (\sqcup_{k=1}^{n} \Sigma_{g_k}) \text{ and } \mathbf{L} = \oplus_{k=1}^{n} \mathbf{L}_k,$$

where $g_k \geq 1$, then

$$V(\Sigma, \mathbf{L}) = \mathcal{H}_{N,g_1}(\mathbf{L}_1) \otimes \mathcal{H}_{N,g_2}(\mathbf{L}_2) \otimes \cdots \otimes \mathcal{H}_{N,g_n}(\mathbf{L}_n).$$

Or, in a more topological setting,

$$V(\Sigma, \mathbf{L}) = (\otimes_{j=1}^{m} \mathcal{L}_N(B^3)) \otimes \mathcal{L}_N(H_{g_1}) \otimes \mathcal{L}_N(H_{g_2}) \otimes \cdots \otimes \mathcal{L}_N(H_{g_n}),$$

with the *caveat* that the boundaries of the handlebodies H_{g_j}, $j = 1, 2, \ldots, n$, are identified with the surfaces Σ_{g_j}, $j = 1, 2, \ldots, n$, so as to obtain the models for theta functions defined by the Lagrangian subspaces $\mathbf{L}_j$, $j = 1, 2, \ldots, n$.

Regarding morphisms, if

$$(h, n) : (\Sigma_g, \mathbf{L}) \to (\Sigma_g, \mathbf{L}')$$

is an extended homeomorphism between extended surfaces, then the functor associates to it a unitary map that combines $\mathcal{F}_h$ and $\mathcal{F}_{\mathbf{L},\mathbf{L}'}$. In the context of §5.4.3 it acts on a skein in $\mathcal{L}_N(H_g)$ as follows:

- lift the skein to $\mathcal{L}_N(\Sigma_g)$ using the section $s_{\mathbf{L}}$ of the projection $\pi_{\mathbf{L}}$;
- map the result by h;
- project the result to $\mathcal{L}_N(H_g)$ by $\pi_{\mathbf{L}'}$.

In other words, we lift the skein to $\mathcal{L}_N(\Sigma_g)$ in all possible nonequivalent ways using the identification of ∂H_g with Σ_g so that $\mathbf{L}$ is null-homologous, map these by h, take the average, then map the result to $\mathcal{L}_N(H_g)$ but this time by the map that identifies ∂H_g with Σ_g so that $\mathbf{L}'$ is null-homologous.

The only problem is that we need to resolve the projective ambiguity in the definition of the map in order to have a true functor. For this we need to revert to the situation of surgery diagrams and look for a topological solution to this problem.

Inspired by 7.4.2, we associate to the extended homeomorphism (Id, n) a surgery diagram such that the deletion of the handlebody graphs corresponds to gluing to the cylinder $\Sigma_g \times [0,1]$ the handlebodies H_g and $-H_g$ such that

$$\mathbf{L} = \ker(H_1(\Sigma_g \times \{0\}, \mathbb{R}) \to H_1(H_g, \mathbb{R})$$
$$\mathbf{L}' = \ker(H_1(\Sigma_g \times \{1\}, \mathbb{R}) \to H_1(-H_g, \mathbb{R}).$$

Example 7.7. If Σ_g is endowed with the canonical basis $a_1,a_2,\ldots,a_g,b_1,$ $b_2,\ldots,b_g$ of $H_1(\Sigma_g, \mathbb{Z})$, and if we let

$$\mathbf{L} = \text{Span}(b_1, b_2, \cdots, b_g) \text{ and } \mathbf{L}' = \text{Span}(a_1, a_2, \ldots, a_g),$$

then $\Sigma_g \times [0,1]$ is represented by the diagram from Figure 7.29.

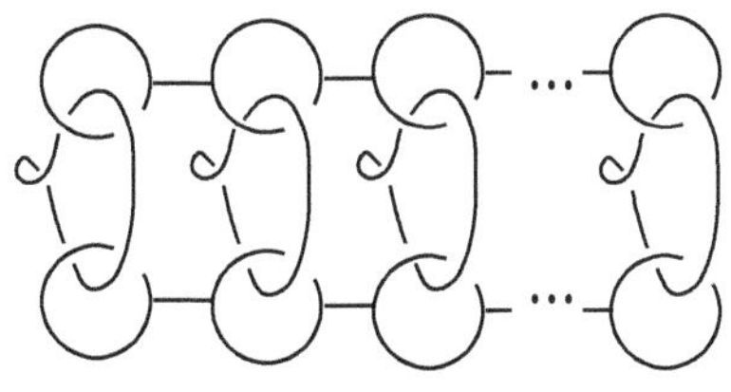

Fig. 7.29 Surgery diagram for $\Sigma_g \times [0, 1]$

If (h,n) is an extended homeomorphism and L_h is a surgery link for h, let $L_{h,\mathbf{L},\mathbf{L}'}$ be the link obtained after placing L in the surgery diagram of $\Sigma_g \times [0,1]$ described above and deleting the two handlebody graphs. We define

$$V(h,n) = e^{\frac{\pi i}{4}(n-\sigma(L_{h,\mathbf{L},\mathbf{L}'}))}\Omega(L_h) \in \mathcal{L}_N(\Sigma_g \times [0,1]) = L(\mathcal{L}_N(H_g)).$$

In the case where the surface is disconnected, say

$$\Sigma = S^2 \sqcup S^2 \sqcup \cdots \sqcup S^2 \sqcup \Sigma_{g_1} \sqcup \Sigma_{g_2} \sqcup \cdots \sqcup \Sigma_{g_n},$$
$$\mathbf{L} = \mathbf{L}_1 \oplus \mathbf{L}_2 \oplus \cdots \oplus \mathbf{L}_n, \quad \mathbf{L}' = \mathbf{L}'_1 \oplus \mathbf{L}'_2 \oplus \cdots \oplus \mathbf{L}'_n$$

and if

$$(h,n) : (\Sigma, \mathbf{L}) \to (\Sigma, \mathbf{L}')$$

is an extended homeomorphism, then

$$V(\Sigma, \mathbf{L}) = \otimes_{j=1}^{n} V(\Sigma_{g_j}, \mathbf{L}_j) \text{ and } V(\Sigma, \mathbf{L}') = \otimes_{j=1}^{n} V(\Sigma_{g_j}, \mathbf{L}_j),$$

and we define

$$V(h,n) : \otimes_{j=1}^{n} V(\Sigma_{g_j}, \mathbf{L}_j) \to \otimes_{j=1}^{n} V(\Sigma_{g_j}, \mathbf{L}_j),\ V(h,n) = \otimes_{j=1}^{n} V(h|\Sigma_{g_j}, n_j),$$

where $n_1, n_2, \ldots, n_g$ are any integers such that $n_1 + n_2 + \cdots + n_j = n$.

Proposition 7.3. *The association*

$$(h,n) \mapsto V(h,n)$$

is well defined and functorial.

Rather than proving this proposition here, we will make a far more general construction in §7.5.4 and check functoriality in its full generality.

Example 7.8. If we consider extended homeomorphisms of the form

$$(h,n) : (\Sigma_g, \mathbf{L}) \to (\Sigma_g, \mathbf{L})$$

then $V(h,n) = \mathcal{F}_{\mathbf{L}}(h,n)$.

Example 7.9. If we consider extended homeomorphisms of the form

$$(Id,n) : (\Sigma_g, \mathbf{L}) \to (\Sigma_g, \mathbf{L}')$$

then up to multiplication by a constant $V(Id,n) = \mathcal{F}_{\mathbf{L},\mathbf{L}'}$. We can make this equality unambiguous, by introducing the integer parameter n in the definition of the Fourier transform, thus defining

$$\mathcal{F}_{\mathbf{L},\mathbf{L}'}(n) = V(Id,n).$$

7.5.4 *The topological quantum field theory underlying the theory of theta functions*

The category of extended cobordisms with embedded oriented framed links

The topological quantum field theory associated to theta functions is a general object that incorporates theta functions, the action of the finite Heisenberg group, the maps $\mathcal{F}_{\mathbf{L}}(h)$, and the maps $\mathcal{F}_{\mathbf{L},\mathbf{L}'}$, and brings together surfaces of all genera.

We need to create a category that is finer than the category of cobordisms itself, which was introduced in §6.4.2. First, because of our solution to the projectivity problem of the representation of the mapping class group, the objects of the category should be extended surfaces. Second, because theta functions and the finite Heisenberg group are modeled by framed links, the morphisms of the category should be cobordisms with embedded oriented framed links in their interiors. Among these cobordisms should be the ones arising from mapping cylinders, which give rise to discrete Fourier transforms, so cobordisms should be framed by integers.

Thus we have a category of **extended cobordisms with embedded oriented framed links**, which has

- as objects **extended surfaces**
$$(\Sigma, \mathbf{L}),$$
where
  - Σ is a surface, not necessarily connected and maybe empty, and
  - $\mathbf{L}$ is a Lagrangian subspace of $H_1(\Sigma, \mathbf{L})$ generated by elements of a canonical basis in each connected component of Σ that has genus at least 1;
- as morphisms **extended cobordisms with framed links**:
$$(M, \partial_- M, \partial_+ M, \mathbf{L}_-, \mathbf{L}_+, L, n),$$
where
  - M is an oriented, compact, 3-dimensional manifold;
  - $\partial_- M$ and $\partial_+ M$ are parametrized oriented closed subsurfaces which partition ∂M, and the orientation of $\partial_+ M$ agrees with that of ∂M while that of $\partial_- M$ disagrees;
  - $\mathbf{L}_\pm$ is a Lagrangian subspace of $H_1(\partial_\pm M, \mathbb{R})$ generated by elements of a canonical basis in each connected component of genus at least 1;

– L is a framed link embedded oriented in the interior of M;
– n is an integer.

Two extended cobordisms with embedded oriented framed links

$$(M', \partial_{\pm} M', \mathbf{L}'_{\pm}, L', n') \text{ and } (M, \partial_{\pm} M, \mathbf{L}_{\pm}, L, n)$$

can be composed if and only if $-\partial_- M'$ and $\partial_+ M$ are homeomorphic under a homeomorphism that preserves orientations and parametrizations, and moreover if this homeomorphism maps $\mathbf{L}'_-$ to $\mathbf{L}_+$. Let $f : \Sigma_{fixed} \to \partial_+ M$ and $f' : \Sigma_{fixed} \to -\partial_- M'$ be the two parametrizations and let $\mathbf{L} = f^{-1}(\mathbf{L}_+) = f'^{-1}(\mathbf{L}'_-)$.

The composition

$$(M'', \partial_{\pm} M'', \mathbf{L}''_{\pm}, L'', n'') := (M', \partial_{\pm} M', \mathbf{L}'_{\pm}, L', n')(M, \partial_{\pm} M, \mathbf{L}_{\pm}, L, n)$$

is defined by

$$M'' = M' \cup M, \quad \partial_- M'' = \partial_- M, \quad \partial_+ M'' = \partial_+ M',$$
$$\mathbf{L}''_- = \mathbf{L}_-, \quad \mathbf{L}''_+ = \mathbf{L}'_+, \quad L'' = L' \cup L,$$
$$n'' = n + n' - \tau(\mathbf{L}_1, \mathbf{L}, \mathbf{L}_2),$$

where $\mathbf{L}_1$ consists of those elements $x \in H_1(\Sigma_{fixed}, \mathbb{R})$ for which there is $y \in \mathbf{L}_-$ such that $f(x)$ is homologous in $H_1(M, \mathbb{R})$ to y and $\mathbf{L}_2$ consists of those elements $x' \in H_1(\Sigma_{fixed}, \mathbb{R})$ for which there is $y' \in \mathbf{L}'_+$ such that $f'(x')$ is homologous in $H_1(M', \mathbb{R})$ to y'.

The reader who followed our previous explanations can unravel the composition rule for framings that we gave above. There are again 4-dimensional manifolds governing this rule. Let us consider only the case where the cobordisms are connected, since disconnected cobordisms can be composed component by component.

Like in §6.4.2, represent the two cobordisms by surgery diagrams. Do this in such a way that the deletion of the handlebody graphs obtained by gluing handlebodies to the boundary components induces maps in the first homology with kernels equal to $\mathbf{L}_{\pm}$ respectively $\mathbf{L}'_{\pm}$. After the deletion of the handlebodies, the surgery diagrams define 4-dimensional handlebodies $\mathbf{M}$ respectively $\mathbf{M}'$. Add trivial link components with framing ± 1 until the signature of $\mathbf{M}$ becomes n, and of $\mathbf{M}'$ becomes n'. Then, by Wall's non-additivity formula,

$$n + n' - \tau(\mathbf{L}_1, \mathbf{L}, \mathbf{L}_2)$$

is the signature of the 4-dimensional manifold obtained by gluing $\mathbf{M}'$ to $\mathbf{M}$ along the 3-dimensional handlebodies attached to $\partial_+ M$ respectively

$-\partial_- M'$. With this description we see that $\mathbf{L}_1$ and $\mathbf{L}_2$ are indeed Lagrangian subspaces of $H_1(\Sigma_{fixed}, \mathbb{R})$, because they are respectively the kernels of the maps from $H_1(\Sigma_{fixed}, \mathbb{R})$ to the first homology group of the manifold obtained from M by filling in $-\partial_- M$ respectively the first homology group of the manifold obtained from M' by filling in $\partial_+ M'$.

This 4-dimensional picture demonstrates that the composition of such extended cobordisms is associative.

The topological quantum field theory

It is now time to bring the theory of theta functions under the unified framework of topological quantum field theory. The main result of this section proves not just the existence of a TQFT associated to theta functions, but also that such a theory is **unique**. The construction of the TQFT is done in the spirit of [Blanchet *et al.* (1995)].

The TQFT is a functor from the category of extended cobordisms with embedded oriented framed links to the category of finite dimensional vector spaces. This functor should satisfy Atiyah's conditions reformulated for this setting:

A1 (Multiplicativity) $V(\Sigma \sqcup \Sigma', \mathbf{L} \oplus \mathbf{L}') = V(\Sigma, \mathbf{L}) \otimes V(\Sigma', \mathbf{L}')$ and

$$Z(M \sqcup M', \partial_\pm M \sqcup \partial_\pm M', \mathbf{L}_\pm \oplus \mathbf{L}'_\pm, L \sqcup L', n + n')$$
$$= Z(M, \partial_\pm M, \mathbf{L}_\pm, L, n) \otimes Z(M', \partial_\pm M', \mathbf{L}'_\pm, L', n').$$

A2 (Involutory) $V(-\Sigma, \mathbf{L}) = V(\Sigma, \mathbf{L})^*$ where $-\Sigma$ is the surface with opposite orientation and the $*$ denotes the dual of the vector space. Also, if $(M, \partial_\pm M, \mathbf{L}_\pm, L, n)$ is an extended cobordism with embedded framed link and if $(\Sigma, \mathbf{L}_\Sigma)$ is a boundary component, we can change the orientation of Σ thus moving it from $\partial_- M$ to $\partial_+ M$ or vice-versa. The homomorphism

$$Z(M, \partial_\pm M, \mathbf{L}_\pm, L, n)$$

can be viewed as an element of $V(\partial_- M, \mathbf{L}_-)^* \otimes V(\partial_+ M, \mathbf{L}_+)$. When changing the orientation of Σ, $Z(M, \partial_\pm M, \mathbf{L}, L, n)$ is mapped by the natural transformation that takes the Hilbert space $V(\Sigma, \mathbf{L}_\Sigma)$ to $V(\Sigma, \mathbf{L}_\Sigma)^*$ in the above tensor product.

A3 (Associativity) Given the cobordisms $(M, \partial_- M, \partial_+ M, \mathbf{L}_-, \mathbf{L}_+, L, n)$ and $(M', \partial_- M', \partial_+ M', \mathbf{L}'_-, \mathbf{L}'_+, L', n')$ that can be composed, if their composition is $(M'', \partial_- M'', \partial_+ M'', \mathbf{L}''_-, \mathbf{L}''_+, L'', n'')$, then

$$Z(M'', \partial_\pm M'', \mathbf{L}''_\pm, L'', n'')$$
$$= Z(M', \partial_\pm M', \mathbf{L}'_\pm, L', n') Z(M, \partial_\pm M, \mathbf{L}_\pm, L, n).$$

A4 $V(\emptyset) = \mathbb{C}$, where $\emptyset$ denotes the empty surface.

A5 $Z(\Sigma \times [0,1], \Sigma \times \{0\}, \Sigma \times \{1\}, \mathbf{L}, \mathbf{L}, \emptyset, 0)$ is the identity morphism of $V(\Sigma, \mathbf{L})$.

Now we can formulate the main result of this chapter.

Theorem 7.11. *[Gelca and Hamilton (2014)] There exists a unique TQFT such that for all $g \geq 0$, the following conditions hold:*

(1) $V(\Sigma_g, \mathbf{L}) = \mathcal{L}_N(H_g) = \mathcal{H}_{N,g}(\mathbf{L}) = \mathbf{\Theta}_N^{\Pi}(\Sigma_g)$.

(2) Let H_g be the genus g handlebody, $\Sigma_g = \partial H_g$, and $\mathbf{L} = ker(H_1(\Sigma_g, \mathbb{R}) \to H_1(H_g, \mathbb{R}))$. For every framed link L in H_g,

$$Z(H_g, \emptyset, \Sigma_g, \{0\}, \mathbf{L}, L, 0) : V(\emptyset) \to \mathcal{L}_N(H_g),$$
$$Z(H_g, \emptyset, \Sigma_g, \{0\}, \mathbf{L}, L, 0)1 = <L> .$$

(3) For $(\Sigma_g, \mathbf{L})$ and $(h, n) \in \widetilde{MCG}(\Sigma_g)$,

$$Z(I_h, \partial_- I_h, \partial_+ I_h, \mathbf{L}, \mathbf{L}, \emptyset, n) = \mathcal{F}_{\mathbf{L}}(h, n),$$

where $\partial_- I_h$ and $\partial_+ I_h$ are the bottom, respectively top of the mapping cylinder, identified with $\Sigma_g \times \{0\}$ respectively $\Sigma_g \times \{1\}$ in $\Sigma_g \times [0,1]$.

(4) Given the extended homeomorphism $(Id, n) : (\Sigma_g, \mathbf{L}) \to (\Sigma_g, \mathbf{L}')$, then

$$(\Sigma_g \times [0,1], \Sigma_g \times \{0\}, \Sigma_g \times \{1\}, \mathbf{L}, \mathbf{L}', \emptyset, n) = \mathcal{F}_{\mathbf{L},\mathbf{L}'}(n).$$

(5) There is a constant $\kappa \in \mathbb{C}$ such that for every framed link L in S^3,

$$Z(S^3, \emptyset, \emptyset, \{0\}, \{0\}, L, 0) = \kappa <L> \in \mathcal{L}_N(S^3) = \mathbb{C}.$$

The value of κ is uniquely determined by the above and is $\kappa = \frac{1}{\sqrt{N}}$.

Remark 7.5. Conditions (1) and (2) express the fact that the TQFT incorporates the spaces of theta functions as spaces of links. Condition (2) also contains all the information about the action of the finite Heisenberg group, by splitting $H_g = H_g \cup \Sigma_g \times [0,1]$ and considering part of the link L inside H_g (as a theta function) and part inside $\Sigma_g \times [0,1]$ (as an element of the Heisenberg group).

Another way to look at (2) is to think that the TQFT associates to each manifold with an embedded framed link a vector in the vector space of its boundary. Then (2) states that if the manifold is the handlebody H_g, then to the link L inside H_g one associates the skein $<L>$ in $\mathcal{L}_N(H_g)$.

Conditions (3) and (4) express the fact that the TQFT incorporates the discrete Fourier transforms defined by homeomorphisms respectively by pairs of Lagrangian subspaces.

Condition (5) tells us how normalize the pairing on $\mathcal{L}_N(H_g) \times \mathcal{L}_N(H_g)$ defined by the Heegaard decomposition of S^3. By performing surgery, which can be seen as a composition of two cobordisms (see proof below), we can turn S^3 into $\#_{k=1}^g S^1 \times S^2$. Then this condition tells us how to normalize the pairing

$$\mathcal{L}_N(H_g) \times \mathcal{L}_N(H_g) \to \mathcal{L}_N(\#_{k=1}^g S^1 \times S^2) = \mathbb{C},$$

thus it contains information about how to define the inner product for theta functions.

In short (1) and (2) are about theta functions and the action of the finite Heisenberg group, (3) and (4) are about the action of the mapping class group, and (5) is about the inner product. Hence the theorem proves the existence of a unique TQFT that incorporates the constructs of the theory of theta functions emphasized in this book.

Proof. We divide the proof of the theorem into two parts.

Proof of the existence of the TQFT

Condition (1) together with axiom A2 completely determine the vector space associated to an extended surface. Hence we are left with defining the vector space homomorphism associated by the TQFT to an extended cobordism with an embedded oriented framed link. The construction of the discrete Fourier transform $\mathcal{F}_{\mathbf{L}}(h, n)$ discloses how this should be done.

We discuss first the case of a connected cobordism. Let therefore $(M, \partial_\pm M, \mathbf{L}_\pm, L, n)$ be an extended cobordism with an embedded oriented framed link inside so that the 3-dimensional manifold M is connected. Represent $(M, \partial_\pm M)$ by a surgery diagram as in §6.4.2. Choose the surgery diagram so that the deletion of the handlebody graphs obtained by gluing handlebodies to the boundary components induces maps in the first homology with kernels equal to $\mathbf{L}_\pm$. Choose the surgery link L_{surg} so that after the deletion of the handlebody graphs, the resulting link in S^3 has signature n (and hence the associated 4-dimensional handlebody has signature n). Next add the link L to the surgery diagram, remembering that L is not part of the surgery link L_{surg}. We turn this diagram into a skein in the complement of the handlebody graph by decorating each surgery link component by Ω, and let the functor associate this skein to the extended cobordism with embedded oriented framed link:

$$Z(M, \partial_\pm M, \mathbf{L}_\pm, L, n) := < L \cup \Omega(L_{surg}) > .$$

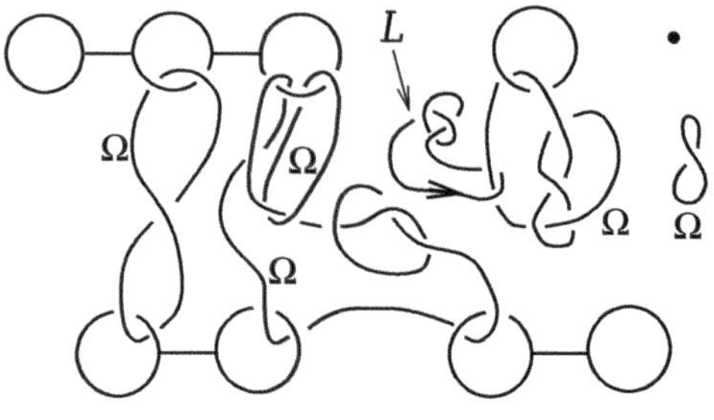

Fig. 7.30 The skein associated to an extended cobordism with an embedded oriented framed link

An example is shown in Figure 7.30.

Let us see how this defines a linear map

$$V(\partial_- M, \mathbf{L}_-) \to V(\partial_+ M, \mathbf{L}_+).$$

Again, as in the case of the Fourier transform, the skein $<L \cup \Omega(L_{surg})>$ defines a bilinear map

$$B : V(\partial_- M, \mathbf{L}_-) \times V(\partial_+ M, \mathbf{L}_+) \to \mathbb{C}.$$

This bilinear map arises by gluing to M along $\partial_- M$ and $\partial_+ M$ disjoint unions of handlebodies H_- respectively H_+ so that

$$\mathbf{L}_- = \ker(H_1(\partial M_-, \mathbb{R}) \to H_1(H_-, \mathbb{R})),$$
$$\mathbf{L}_+ = \ker(H_1(\partial M_+, \mathbb{R}) \to H_1(H_+, \mathbb{R})),$$

and is defined by

$$B : \mathcal{L}_N(H_-) \times \mathcal{L}_N(H_+) \to \mathcal{L}_N(S^3) = \mathbb{C},$$
$$B(<L_1>, <L_2>) = <L_1 \cup L \cup \Omega(L_{surg}) \cup L_2> .$$

Then the linear transformation associated to the extended cobordism with embedded oriented framed link is defined by the identity

$$\langle Z(M, \partial_\pm M, \mathbf{L}_\pm, L, n)\mathbf{x}, \mathbf{y}\rangle = N^{-g_+/2} B(\mathbf{x}, \mathbf{y}),$$

where g_+ is the sum of the genera of the handlebodies that comprise H_+. The bilinear map is defined by the identification of

$$\mathcal{L}_N(H_-) \otimes_{\mathcal{L}_N(\partial_- M)} \mathcal{L}_N(M) \otimes_{\mathcal{L}_N(\partial_+ M)} \mathcal{L}_N(H_+)$$

with $\mathbb{C}$, and this really depends on how $\mathcal{L}_N(\partial_- M)$ acts on $\mathcal{L}_N(H_-)$ and $\mathcal{L}_N(\partial_- M)$ acts on $\mathcal{L}_N(H_-)$. In the representation theoretical model these two actions depend only on the choices of $\mathbf{L}_-$ respectively $\mathbf{L}_+$, which implies that the bilinear map B and hence linear transformation

$$Z(M, \partial_\pm M, \mathbf{L}_\pm, L, n)$$

are well defined.

For disconnected cobordisms we just draw separate diagrams for each connected component.

Axiom A5 is straightforward since with the above definition,

$$Z(\Sigma \times [0,1], \Sigma \times \{0\}, \Sigma \times \{1\}, \mathbf{L}, \mathbf{L}, \emptyset, 0) = \mathcal{F}_{\mathbf{L}}(Id, 0) = Id.$$

The only thing left to check is the associativity axiom A3. The proof becomes more intuitive if we choose a basis for the vector spaces and check the multiplication rule for matrices, the way we did it for the representation of the extended mapping class group. So let

$$(M', \partial_\pm M', \mathbf{L}'_\pm, L', n') \text{ and } (M, \partial_\pm M, \mathbf{L}_\pm, L, n)$$

be extended cobordisms with embedded oriented framed links that can be composed, and let $(M'', \partial_\pm M'', \mathbf{L}''_\pm, L'', n'')$ be their composition.

It suffices to prove the case of connected cobordisms, since the disconnected ones can be composed component by component.

Choose a canonical basis for each of the connected components with genus at least 1 of $\partial_- M$, $\partial_+ M$, $\partial_- M'$, $\partial_+ M'$, associated to the Lagrangian subspaces $\mathbf{L}_-$, $\mathbf{L}_+$, $\mathbf{L}'_-$, $\mathbf{L}'_+$. Moreover, do this so that under the identification of $\partial_+ M$ with $\partial_- M'$ the corresponding canonical bases are also identified. Represent the cobordisms by surgery diagrams $(L_{surg}, \Gamma_-, \Gamma_+)$ $(L'_{surg}, \Gamma'_-, \Gamma'_+)$ in which the circles of the handlebody graphs are the a_j's of the canonical bases. Now the gluing of surgery diagrams for the cobordisms $(M', \partial_- M', \partial_+ M')$ and $(M, \partial_- M, \partial_+ M)$ obeys the rule from Theorem 6.8.

So if L'_{surg} and L_{surg} are the surgery links of the two extended cobordisms, then $L'_{surg} \circ L_{surg}$ (defined in §6.4.2) is the surgery link of their composition.

We have a basis for $V(\partial_- M, \mathbf{L}_-)$, one for $V(\partial_+ M, \mathbf{L}_+) = V(\partial_- M', \mathbf{L}'_-)$, and one for $V(\partial_+ M', \mathbf{L}_+)$ obtained by listing the "circles" of the handlebody graphs Γ_-, Γ_+, Γ'_+ in the order they appear from left to right. Let these bases be

$$\begin{aligned} &\underline{a}_1^{k_1} \underline{a}_2^{k_2} \cdots \underline{a}_m^{k_m}, \quad k_1, k_2, \ldots, k_m \in \mathbb{Z}_N \\ &a_1^{j_1} a_2^{j_2} \cdots a_n^{j_n}, \quad j_1, j_2, \ldots, j_m \in \mathbb{Z}_N \\ &\bar{a}_1^{l_1} \bar{a}_2^{l_2} \cdots \bar{a}_p^{l_p}, \quad l_1, l_2, \ldots, l_m \in \mathbb{Z}_N, \end{aligned}$$

where m, n, p are the sum of the genera of the connected components of $\partial_- M$, $\partial_+ M$, respectively $\partial_+ M'$.

The matrix element

$$\left\langle Z(M, \partial_\pm M, \mathbf{L}_\pm, L, n)\underline{a}_1^{k_1}\underline{a}_2^{k_2}\cdots\underline{a}_m^{k_m}, a_1^{j_1}a_2^{j_2}\cdots a_n^{j_n}\right\rangle \tag{7.8}$$

is obtained by inserting $k_1, k_2, \ldots k_m$ parallel circles respectively in place of the "circles" of Γ_- and $j_1, j_2, \ldots, j_m$ parallel circles respectively in place of the circles of Γ_+, then deleting the edges of $\Gamma_\pm$ as well as the isolated vertices, and finally, multiplying the resulting skein in $\mathcal{L}_N(S^3)$ by $N^{-g_+/2}$, where g_+ is the sum of the genera of the connected components of $\partial_+ M$. Similarly, the matrix element

$$\left\langle Z(M', \partial_\pm M', \mathbf{L}'_\pm, L', n')a_1^{j_1}a_2^{j_2}\cdots a_n^{j_n}, \bar{a}_1^{l_1}\bar{a}_2^{l_2}\cdots\bar{a}_p^{l_p}\right\rangle \tag{7.9}$$

is obtained by inserting $j_1, j_2, \ldots j_n$ parallel circles respectively in place of the "circles" of Γ'_- and $l_1, l_2, \ldots, l_p$ parallel circles respectively in place of the circles of Γ'_+, then deleting the edges of $\Gamma'_\pm$ as well as the isolated vertices, and finally, multiplying the resulting skein in $\mathcal{L}_N(S^3)$ by $N^{-g'_+/2}$, where g'_+ is the sum of the genera of the connected components of $\partial_+ M'$. The matrix element

$$\left\langle Z(M'', \partial_\pm M'', \mathbf{L}''_\pm, L'', n'')\underline{a}_1^{k_1}\underline{a}_2^{k_2}\cdots\underline{a}_m^{k_m}, \bar{a}_1^{l_1}\bar{a}_2^{l_2}\cdots\bar{a}_p^{l_p}\right\rangle \tag{7.10}$$

is obtained by summing the products of (7.8) and (7.9) over all $j_1, j_2, \ldots, j_n \in \mathbb{Z}_N$. This computation is similar to that from Figure 7.20 and, with the notation $u = (-g_+ - g'_+)/2$ and $w = u + n/2$, is outlined in Figure 7.31.

We are not yet done since the last diagram does not equal (7.10), for example because the exponent of N in the last term is not equal to $-g'_+/2$. It differs from this number by $(n - g_+)/2$. But $n - g_+$ is precisely $n(\Gamma_+) - 1$, where $n(\Gamma_+)$ is the number of connected components of Γ_+. Also, the surgery link in the last term is not quite L''_{surg}, the surgery link of the composition cobordism. According to Theorem 6.8 this link is missing a number of $n(\Gamma_+) - 1$ circles decorated by Ω, which surround groups of vertical circles in the middle of the diagram. We show now that adding those circles amounts to precisely introducing a term of $N^{-(n-g_+)/2}$, thus producing the desired result. Indeed, examining the picture locally, and using Lemma 7.7 we can perform locally the computation from Figure 7.32, which proves the claim.

This completes the proof of associativity and hence we are in the presence of a TQFT.

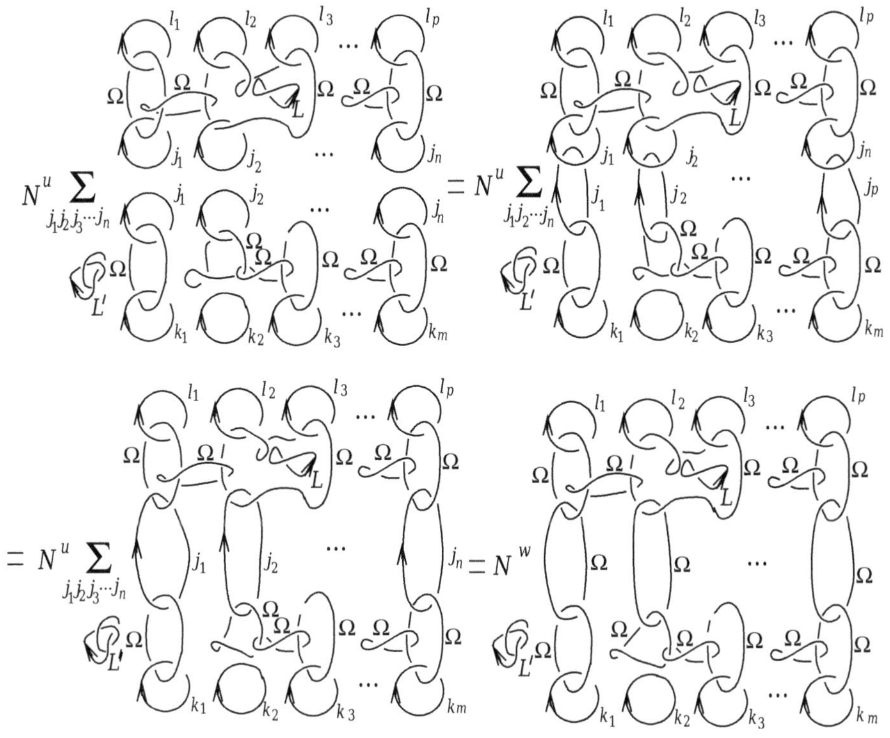

Fig. 7.31 The proof of associativity

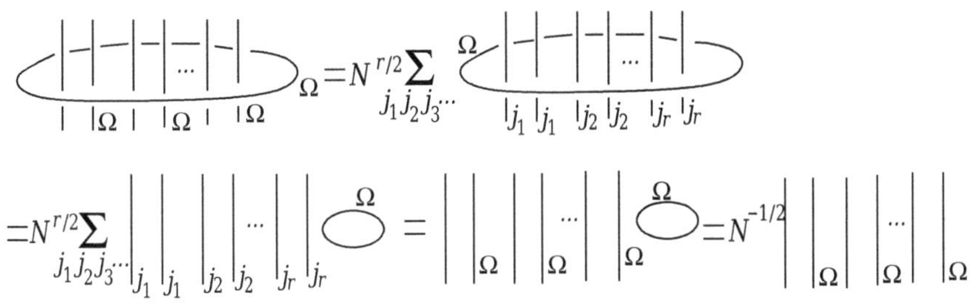

Fig. 7.32 Local computation in checking associativity

Proof of the uniqueness of the TQFT

Consider a second TQFT satisfying the conditions from the statement. Condition (1) implies that the two associate the same vector spaces to extended surfaces. Let Z' be the new map that associates vector space homomorphisms to extended cobordisms with embedded oriented framed links. We want to show that

$$Z' = Z.$$

It suffices to check equality for connected cobordisms because of axiom A1.

Let us first consider the case where M is closed, namely let us check first that

$$Z'(M, \emptyset, \emptyset, \{0\}, \{0\}, L, n) = Z(M, \emptyset, \emptyset, \{0\}, \{0\}, L, n).$$

Describe M as surgery on a framed link

$$L_{surg} = L_1 \cup L_2 \cup \cdots \cup L_m \in S^3$$

with $\sigma(L_{surg}) = n$. As such, M is obtained by gluing solid tori to the boundary of $S^3 \backslash \sqcup_k int(N_k)$, where N_k is a compact regular neighborhood of the link component L_k, $k = 1, 2, \ldots, n$, and the circle denotes interior. Let us consider the extended cobordisms with embedded oriented framed links

$$(S^3 \backslash \sqcup_k int(N_k), \sqcup_k \partial N_k, \emptyset, \oplus_k \mathbf{L}_k, \{0\}, L, 0) \text{ and } (\sqcup_k H_1, \emptyset, \sqcup_k \partial H_1, \{0\}, \mathbf{L}'_k, \emptyset, 0)$$

where for each k,

$$\mathbf{L}_k = \ker(H_1(\partial N_k, \mathbb{R}) \to H_1(N_k, \mathbb{R}))$$
$$\mathbf{L}'_k = \ker(H_1(\partial H_k, \mathbb{R}) \to H_1(H_k, \mathbb{R})).$$

The gluing map does not map $\mathbf{L}'_k$ to $\mathbf{L}_k$, so we cannot glue these extended cobordisms to each other. Specifically, $\mathbf{L}_k$ are both spanned by b_1, but the gluing map sends b_1 to the framing curve of the kth link component, which in view of Lemma 5.6 can be thought of as an a_1, with (a_1, b_1) a canonical basis. Thus, for the gluing to be possible, we need to insert in between the cobordism

$$(\sqcup_k \Sigma_1 \times [0,1], \sqcup_k \Sigma_1 \times \{0\}, \oplus_k \mathrm{Span}(a_1), \oplus_k \mathrm{Span}(b_1), \emptyset, 0).$$

Using (2) we obtain that

$$Z'(\sqcup_k H_1, \emptyset, \sqcup_k \partial H_1, \{0\}, \oplus_k \mathbf{L}'_k, \emptyset, 0)$$

is the map $\mathbb{C} \to \oplus_k \mathcal{L}_N(H_1)$, $1 \mapsto \emptyset$, and using (4) we obtain that

$$Z'(\sqcup_k \Sigma_1 \times [0,1], \sqcup_k \Sigma_1 \times \{0\}, \oplus_k \mathrm{Span}(a_1), \oplus_k \mathrm{Span}(b_1), \emptyset, 0)$$

is the standard Fourier transform. Hence the composition of

$$Z'(\sqcup_k \Sigma_1 \times [0,1], \sqcup_k \Sigma_1 \times \{0\}, \oplus_k \mathrm{Span}(a_1), \oplus_k \mathrm{Span}(b_1), \emptyset, 0)$$

with

$$Z'(\sqcup_k H_1, \emptyset, \sqcup_k \partial H_1, \{0\}, \oplus_k \mathbf{L}'_k, \emptyset, 0)$$

is the map that sends $1 \in \mathbb{C}$ to

$$\oplus_k \mathcal{F}_S \theta_0^\tau(z) = N^{-m/2} \sum_{k=1}^{m} \sum_{j=0}^{N-1} \theta_j^\tau(z) = N^{-m/2} \sum_{k=1}^{m} \sum_{j=0}^{N-1} < a_1^{\|j} > .$$

When composing $Z'(S^3 \setminus \sqcup int(N_k), \sqcup_k \partial N_k, \emptyset, \oplus_k \mathbf{L}_k, \{0\}, L, 0)$ with this we obtain, using the axioms of the TQFT,

$$N^{-m/2} \sum_{j_1, j_2, \ldots, j_k \in \mathbb{Z}_N} Z'(S^3, \emptyset, \emptyset, \{0\}, \{0\}, L \cup \cup_{k=1}^m L_k^{\|j_k}, n')$$
$$= Z(S^3, \emptyset, \emptyset, \{0\}, \{0\}, L \cup \Omega(L_{surg}), n'),$$

for some framing n'. Let us compute this framing.

The framing remains 0 when gluing

$$(\sqcup_k \Sigma_1 \times [0,1], \sqcup_k \Sigma_1 \times \{0\}, \oplus_k \mathrm{Span}(a_1), \oplus_k \mathrm{Span}(b_1), \emptyset, 0)$$

to

$$(\sqcup_k H_1, \emptyset, \sqcup_k \partial H_1, \{0\}, \oplus_k \mathbf{L}'_k, \emptyset, 0)$$

because in the gluing formula, the non-additivity term $\tau(\mathbf{L}_1, \mathbf{L}, \mathbf{L}_2)$ is equal to zero because

$$\mathbf{L}_1 = \mathbf{L} = \oplus_k \mathrm{Span}(b_1).$$

On the other hand, when gluing the result to the link complement, we create the surgery manifold $M_{L_{surg}}$. This manifold is the boundary of a 4-dimensional handlebody $\mathbf{M}_{L_{surg}}$ and the signature of the latter determines the framing of $M_{L_{surg}}$. Consequently $n' = n$.

It remains to determine the value of the constant κ. Combining the axioms A2 and A5, we conclude that

$$Z'(S^2 \times [0,1], \emptyset, S^2 \times \{0\} \sqcup -S^2 \times \{1\}, \{0\}, \{0\}, \emptyset, 0)1 = 1$$

and

$$Z'(S^2 \times [0,1], -S^2 \times \{0\} \sqcup S^2 \times \{1\}, \emptyset, \{0\}, \{0\}, \emptyset, 0)1 = 1.$$

Gluing these two extended cobordisms we obtain

$$Z'(S^2 \times S^1, \emptyset, \emptyset, \{0\}, \{0\}, \emptyset, 0) = 1.$$

On the other hand, $\Sigma_1 \times S^1$ is surgery on the trivial knot, and so by the above computation

$$Z'(S^2 \times S^1, \emptyset, \emptyset, \{0\}, \{0\}, \emptyset, 0) = \kappa \Omega(O) = \kappa \sqrt{N},$$

where O is the trivial framed knot. Hence

$$\kappa = N^{-1/2}.$$

This proves that $Z' = Z$ on manifolds without boundary.

The case of manifolds with boundary is not much more complicated and the above discussion gives away the idea. We can turn a cobordism into a 3-dimensional manifold without boundary by filling in handlebodies along the boundary components. We can place inside these handlebodies oriented framed links that represent basis elements of the reduced linking number skein modules of these handlebodies, and then the invariant of the closed manifold obtained this way is the corresponding entry of the matrix of the linear homomorphism associated to the cobordism. Since Z' coincides with Z on closed manifolds, these entries are the same as those calculated using Z. We conclude that $Z = Z'$ for all cobordisms. The theorem is proved. $\square$

Remark 7.6. Note that for an extended homomorphism

$$(h, n) : (\Sigma, \mathbf{L}) \to (\Sigma, \mathbf{L}'),$$

the TQFT requires that

$$V(h, n) = Z(I_h, -\Sigma, \Sigma, \mathbf{L}, \mathbf{L}', \emptyset, n).$$

So as a corollary of this theorem we obtain the proof of Proposition 7.3.

Example 7.10. Let K be an oriented framed knot in S^3. Consider a compact regular neighborhood, $N(K)$, of K. This neighborhood can be identified with the standard solid torus H_1 so that

- the framing curve of K on ∂K (oriented as to agree with K) is identified with $a_1 \in \Sigma_1 = \partial H_1$, and
- the curve that spans

$$\mathbf{L} = \ker(H_1(\partial N(K), \mathbb{R}) \to H_1(N(K), \mathbb{R})),$$

is identified with $b_1 \in \Sigma_1 = \partial H_1$.

We turn the knot complement into the extended cobordism

$$(S^3 \backslash int(N(K)), \partial N(K), \emptyset, \mathbf{L}, \{0\}, \emptyset, 0).$$

Next we interpret

$$Z(S^3 \backslash int(N(K)), \partial N(K), \emptyset, \mathbf{L}, \{0\}, \emptyset, 0)$$

as the linear functional $\mathcal{L}_N(H_1) \to \mathbb{C}$ defined by gluing back $N(K)$ as to obtain S^3. Evaluating this functional on the basis elements

$$< a_1^{\| j} > = Z(H_1, \emptyset, \partial H_1, \{0\}, \mathbf{L}, a_1^{\| j}, 0)$$

we obtain

$$Z(S^3\backslash int(N(K)), \partial N(K), \emptyset, \mathbf{L}, \{0\}, \emptyset, 0) < a_1^{\|j} > = < K^{\|j} > .$$

Given the identification

$$\mathcal{L}_N(H_1)^* = \mathcal{L}_N(H_1)$$

defined by the inner product, the linear functional

$$Z(S^3\backslash int(N(K)), \partial N(K), \emptyset, \mathbf{L}, \{0\}, \emptyset, 0)$$

can be turned into a vector, hence into a theta function. This theta function is

$$\theta_K^{\mathcal{T}}(z) = \sum_{j=0}^{N-1} < K^{\|j} > \theta_j^{\mathcal{T}}(z) = \sum_{j=0}^{N-1} t^{w(K)j^2} \theta_j^{\mathcal{T}}(z),$$

where $w(K)$ is the writhe of K (see Chapter 1). This theta function does not depend on the orientation of K, since $< K^{\|j} >$, $j = 0, 1, \ldots, N-1$, do not.

For example, the theta function associated to the complement of the right-handed trefoil knot (Figure 7.33) is

$$\theta_{trefoil}^{\mathcal{T}}(z) = \sum_{j=0}^{N-1} t^{3j^2} \theta_j^{\mathcal{T}}(z).$$

Fig. 7.33 The right-handed trefoil knot

Conclusions to the seventh chapter

In this chapter we examined the **exact Egorov condition** satisfied by the discrete Fourier transform, and using it we were able to derive a **surgery formula** for the discrete Fourier transform. Then, by interpreting the exact Egorov condition in terms of **handle slides**, we were led to the construction of an **invariant** of 3-dimensional manifolds without boundary (the Murakami-Ohtsuki-Okada invariant).

The topological insights derived from the exact Egorov identity allowed us to give a **topological solution to the projective ambiguity** of the discrete Fourier transform, based on the signature of 4-dimensional manifolds.

Finally, using **3-dimensional cobordisms**, **framed links**, and **4-dimensional handlebodies** we were able to place within the same **unified framework** the theory of theta functions, namely the **space of theta functions**, the action of the **finite Heisenberg group**, and the action of the **mapping class group**. This unified framework is a topological quantum field theory. It is important to notice that the TQFT is not imposed artificially upon theta functions, but that it arises naturally in their study.

Chapter 8

Theta functions in the quantum group perspective

Chapter 5 turned theta functions from analytical into combinatorial objects, rephrasing constructs from the theory of theta functions in the language of skein modules. By their own nature, skeins are endowed with symmetries that arise from ambient isotopies.[1] Of these symmetries, we focus on the third Reidemeister move, an instance of which is depicted in Figure 8.1.

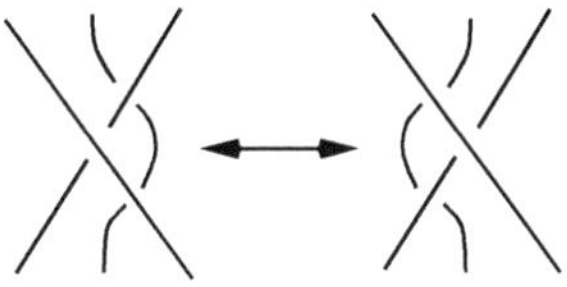

Fig. 8.1 Third Reidemeister Move

The third Reidemeister move showed up elsewhere in mathematics, namely in models from 2-dimensional physics. It is this occurrence that allows us to turn skein modules into representation theoretical objects using quantum groups.

We give below a short description of those models in mathematical physics from which Vladimir Drinfeld distilled the concept of a **quantum group**, and then present the general ideas and concepts from the theory of quantum groups that are relevant for us. Then we introduce the quantum group associated to theta functions and use it to model theta functions and the actions of the Heisenberg and of the mapping class group.

Of the many books on the subject of quantum groups we recommend

[1]Here the word symmetry is used as in physics, meaning the presence of a group action, in this case the group of isotopies of the ambient space.

to the reader [Gomez *et al.* (1996)], which presents the physical point of view, and [Kassel (1995)], which gives the mathematical details. We also recommend the original article of [Drinfeld (1986a)], as well as the better known [Drinfeld (1986b)].

8.1 Quantum groups

8.1.1 *The origins of quantum groups*

Let us begin with two examples in 2-dimensional physics that motivate the introduction of quantum groups.

The scattering of relativistic massive particles in a $1 + 1$-*dimensional spacetime*

The spacetime of the system is $\mathbb{R}^2$, so the spacial dimension is 1. We consider an ideal situation of a multiscattering with no particle production or annihilation. This means that several particles collide and then separate. Let n be the number of such particles. We are interested in the S-**matrix** (or **scattering matrix**) of the scattering process.[2] This matrix relates the initial and the final state of the physical system and is a (unitary) matrix of amplitudes which compute the probability for the system to go from one state to another. In short the system starts with a Hilbert space of states and ends with the same Hilbert space of states and the scattering matrix is a unitary operator on this Hilbert space.

For the jth particle we let $\theta_j = \arctan(v_j/c)$ be its **rapidity**, where v_j is its velocity and c is the speed of light. The condition that the scattering of two particles is relativistically invariant translates to the fact that the entries of the S-matrix of their scattering depend only on the difference between their rapidities. We assume that the particles have distinct rapidities.

The computation of the scattering matrix can be complicated but there is one instance where this computation is greatly simplified, and that is when the matrix is **factorizable**. This means that:

The n-particle S-matrix can be written as the product of $\binom{n}{2}$ 2-particle S-matrices.

We can represent the multiparticle factorized scattering like in Fig-

[2]Not to be confused with the matrix of the S-map on the torus.

ure 8.2. In this figure each line is the world-line of the particle (its path through the spacetime), and the slope of each line is the rapidity of the particle. The system evolves in the positive direction of time, from the bottom to the top of the figure and each time two lines cross, the corresponding particles collide.

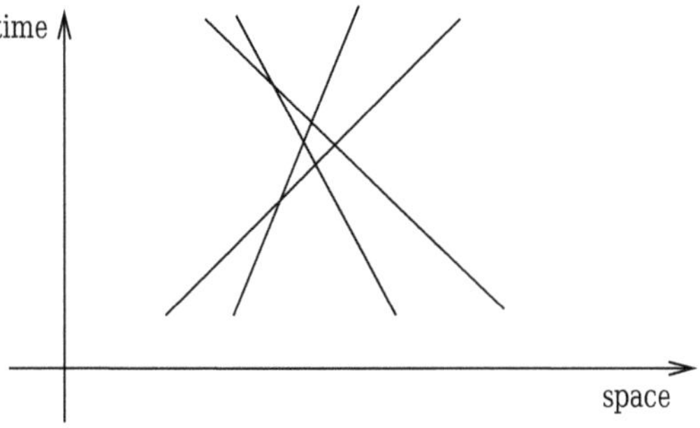

Fig. 8.2 Scattering of n particles

In the assumption of factorizability hides a symmetry. This is because the scattering of the n particles can be decomposed into 2-particle scatterings in various ways. And this translates into the invariance of the model under time translations, meaning that in this model each line can be translated up or down. The time symmetry is represented in Figure 8.3, and we recognize the third Reidemeister move.

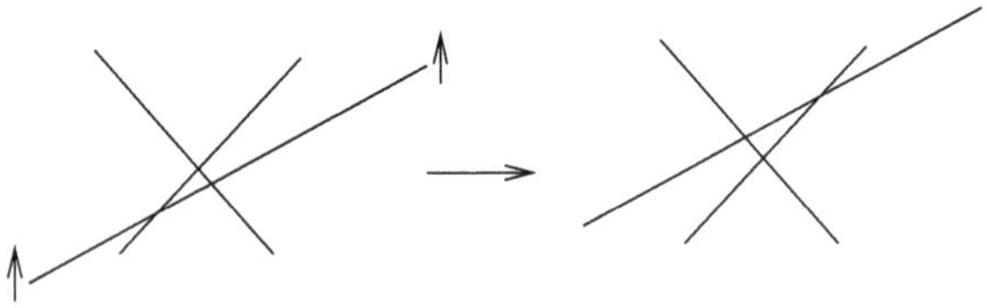

Fig. 8.3 Symmetry with respect to time translation

Phrasing it algebraically, the 2-particle scattering matrix S should satisfy

$$(S \otimes I)(I \otimes S)(S \otimes I) = (I \otimes S)(S \otimes I)(I \otimes S).$$

This is the **Yang-Baxter equation.**

Integrable vertex models in classical statistical mechanics

Now we concentrate on a classical statistical system in two dimensions. A **vertex model** is a statistical model defined on a lattice given by a set of straight lines in the plane. Viewing this as a graph, each node represents a particle, and the edges represent bonds. To each bond one associates a state and the energy of a vertex depends on the states of the adjacent edges. The simplest situation (occurring for example for crystal lattices with hydrogen bonds) is where there are only two states, which we represent by orienting the edges. Of interest to us is the six-vertex model, where at each vertex the number of incoming edges equal the number of outgoing edges.[3] These constraints have physical motivation, they arise from conservation laws.

It is customary to consider crystals with rectangular lattice, and an example of a six-vertex model is given in Figure 8.4. An exercise in intuition makes us see "smoothings" and "crossings" in Figure 8.4, and this is exactly what happened to V.F.R. Jones and Nikolai Reshetikhin when they discovered the relationship between statistical physics and knot theory.

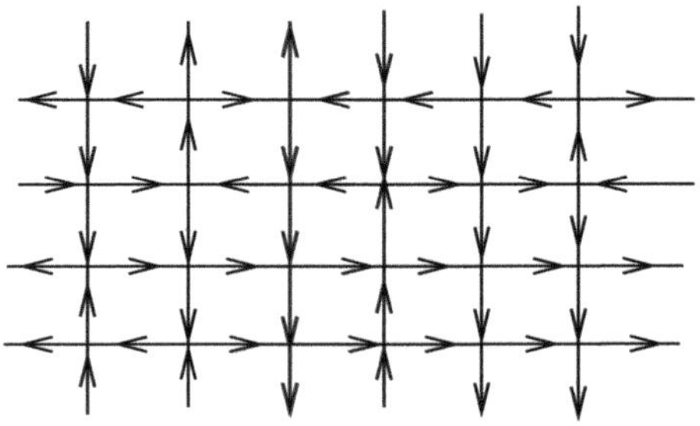

Fig. 8.4 Six-vertex model

To each vertex V and each possible choice of states for neighboring edges one associates a Boltzmann weight W_V. These weights appear in the computation of the partition function, which is used for finding the probability that the system lies in a certain state. The partition function is given by

$$Z = \sum_{\mathcal{S}} \prod_{V} W_V,$$

where the sum is taken over all possible states $\mathcal{S}$. It was E.H. Lieb's idea to turn this into a $1+1$-dimensional problem, by viewing the lattice as

[3]The name is motivated by the fact that there are six possible configurations at each vertex.

the time evolution of a 1-dimensional chain. This is done by rewriting the partition function as

$$Z = \sum_{\text{vertical states}} \prod_{\text{rows}} \left(\sum_{\text{horizontal states}} \prod_{V \in \text{row}} W_V \right).$$

The quantity in the parentheses depends on the two sets of vertical states above and below the row of horizontal variables, and can be interpreted as the row-to-row transfer matrix of the system. The operations outside of the parentheses define the multiplication of these matrices. So we changed the system into one similar to the evolution of relativistic particles discussed above. We can imagine each row as a "scattering" of n particles, where n is the number of vertical lines. Of course the assumption that the transfer matrix is factorizable simplifies considerably the study of the model allowing us to find explicit solutions. Again this leads to the Yang-Baxter equation, albeit in a less straightforward manner than in the n-particle model described before.

There is a difference with the first example, here the "particles" do not cross paths. So the scattering matrices that solve the Yang-Baxter equation will look slightly different. In what follows the scattering matrix of this situation will be the R-matrix, while the scattering matrix of the situation where particles cross paths will be $\check{R}$ (flip R).

It is in this setting that quantum groups arose. They first showed up as particular examples in the works of the Russian school of mathematical physics, and then Drinfeld produced the general framework. One should also mention the work of Michio Jimbo [Jimbo (1985)], who discovered independently some of Drinfeld's constructs. Quantum groups were obtained as Hopf algebras build from Lie algebras, with the states being defined using representations of the quantum group and the Yang-Baxter equation being solved by certain homomorphisms between such representations.

8.1.2 *Quantum groups as Hopf algebras*

It is time to introduce the concept of a quantum group. The name is somewhat of a misnomer, since a quantum group is associated to a (classical) Lie group as a deformation of the universal enveloping algebra of its Lie algebra. Drinfeld created a larger realm for quantum groups, but it is still this original setting that is of interest to us.

For us a **quantum group** is a Hopf algebra that is a "deformation" of the universal enveloping algebra of a Lie algebra. We let the word "deformation" be imprecise, it means a modification of the universal enveloping algebra that depends on a parameter (the Planck's constant), such that the original algebra is recovered when the parameter is set equal to zero.

Hopf algebras

At the heart of the theory of quantum groups lies the notion of a Hopf algebra, named after Heinz Hopf. A Hopf algebra combines an algebra and a coalgebra structure. In a more abstract phrasing, an algebra over $\mathbb{C}$ is a triple (A, m, η) where A is a vector space and $m : A \otimes A \to A$ and $\eta : \mathbb{C} \to A$ are linear maps for which the following diagrams commute:

$$\begin{array}{ccc} A\otimes A\otimes A & \xrightarrow{m\otimes Id} & A\otimes A \\ {\scriptstyle Id\otimes m}\downarrow & & \downarrow{\scriptstyle m} \\ A\otimes A & \xrightarrow{m} & A \end{array} \qquad \begin{array}{ccccc} \mathbb{C}\otimes A & \xrightarrow{\eta\otimes Id} & A\otimes A & \xleftarrow{Id\otimes\eta} & A\otimes\mathbb{C} \\ & \searrow{\scriptstyle =} & \downarrow{\scriptstyle m} & {\scriptstyle =}\swarrow & \\ & & A & & \end{array}$$

We recognize m to be the multiplication and η to be the inclusion of the scalars $c \mapsto c{\cdot}1$, and the diagrams express the associativity of multiplication and the properties of the unit element.

It becomes natural to define a **coalgebra** as a triple (A, Δ, ϵ) where A is a $\mathbb{C}$-vector space and $\Delta : A \to A \otimes A$ and $\epsilon : A \to \mathbb{C}$ are linear maps such that the following diagrams commute:

$$\begin{array}{ccc} A\otimes A\otimes A & \xleftarrow{\Delta\otimes Id} & A\otimes A \\ {\scriptstyle Id\otimes\Delta}\uparrow & & \uparrow{\scriptstyle \Delta} \\ A\otimes A & \xleftarrow{\Delta} & A \end{array} \qquad \begin{array}{ccccc} \mathbb{C}\otimes A & \xleftarrow{\epsilon\otimes Id} & A\otimes A & \xrightarrow{Id\otimes\epsilon} & A\otimes\mathbb{C} \\ & \nwarrow{\scriptstyle =} & \uparrow{\scriptstyle \Delta} & {\scriptstyle =}\nearrow & \\ & & A & & \end{array}$$

The maps Δ and ϵ are called **comultiplication** and **counit**. The first diagram expresses the **coassociativity** of Δ.

Definition 8.1. A **bialgebra** is a vector space A endowed with a multiplication m, unit η, comultiplication Δ and counit ϵ such that

- (A, m, η) is an algebra;
- (A, Δ, ϵ) is a coalgebra;
- the multiplication and the comultiplication are such that the following diagram is commutative

$$\begin{array}{ccccc} A\otimes A & \xrightarrow{m} & A & \xrightarrow{\Delta} & A\otimes A \\ {\scriptstyle \Delta\otimes\Delta}\downarrow & & & & \uparrow{\scriptstyle m\otimes m} \\ A\otimes A\otimes A\otimes A & & \xrightarrow{I\otimes\sigma\otimes I} & & A\otimes A\otimes A\otimes A \end{array}$$

where $\sigma : A \otimes A \to A \otimes A$, $\sigma(x \otimes y) = y \otimes x$.

- the multiplication and the counit are such that the following diagram is commutative

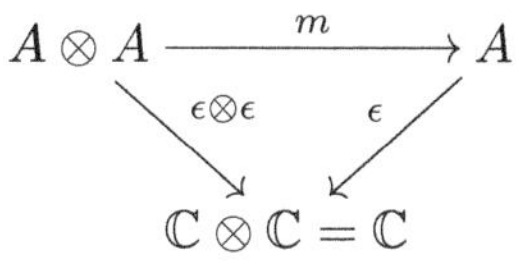

- the comultiplication and the unit are such that the following diagram is commutative

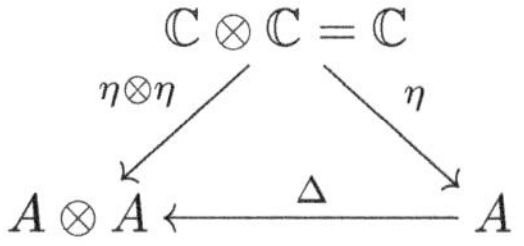

- the unit and counit are such that the following diagram is commutative

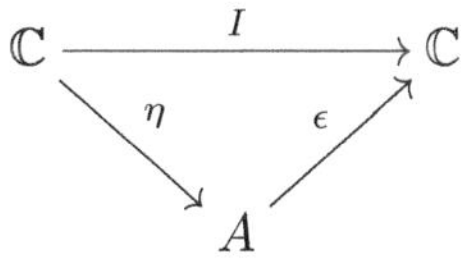

Remark 8.1. These diagrams express the fact that the comultiplication and counit are homomorphisms of algebras, or equivalently that the multiplication and unit are homomorphisms of coalgebras.

Finally we can define the main object of this chapter.

Definition 8.2. A **Hopf algebra** is a bialgebra $A, m, \eta, \Delta, \epsilon$ endowed with antihomomorphism $S : A \to A$ called **antipode** such that the following diagrams are commutative:

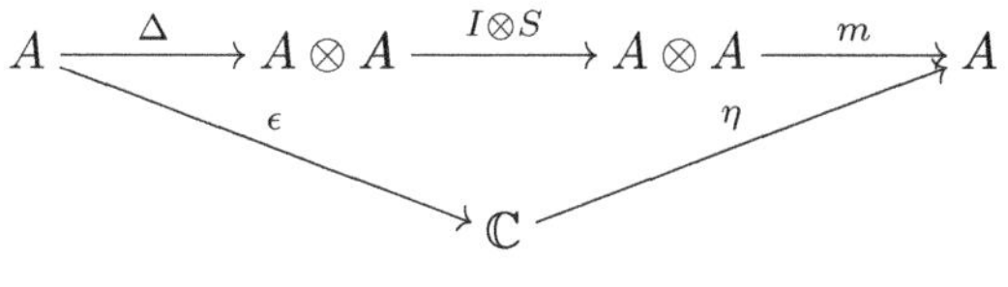

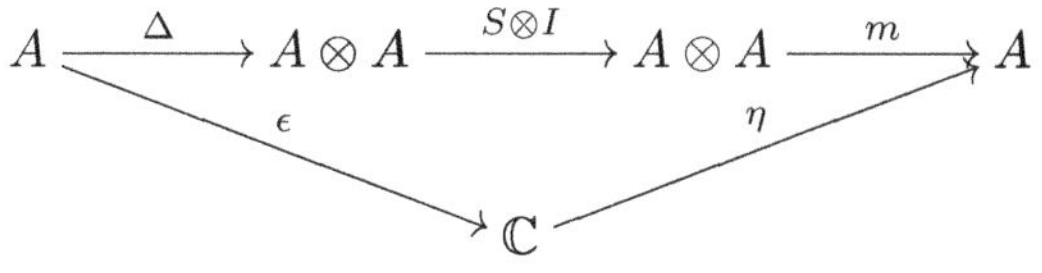

A Hopf algebra is commutative if the underlying algebra is commutative. By analogy, a Hopf algebra is **cocomutative** if $\sigma \circ \Delta = \Delta$, where $\sigma(a \otimes b) = b \otimes a$.

Example 8.1. Every group algebra is a Hopf algebra. If G is a group, then $\mathbb{C}[G]$ is endowed with the comultiplication

$$\Delta : \mathbb{C}[G] \to \mathbb{C}[G] \otimes \mathbb{C}[G], \quad \Delta(g) = g \otimes g, \quad \text{for all } g \in G,$$

counit

$$\epsilon : \mathbb{C}[G] \to \mathbb{C}, \quad \epsilon(g) = 1, \quad \text{for all } g \in G,$$

and antipode

$$S : \mathbb{C}[G] \to \mathbb{C}[G], \quad S(g) = g^{-1}, \text{ for all } g \in G,$$

where it is understood that these maps are extended linearly. It is an easy exercise to check that these satisfy the defining properties of Hopf algebras.

In particular, the linking numbers skein algebra

$$\mathcal{L}(\Sigma_g) = \mathbb{C}[\mathbf{H}(\mathbb{Z}^{2g})]$$

is a Hopf algebra. In topological language, for every framed oriented link L in $\Sigma_g \times [0,1]$, we have

$$\begin{aligned} &\Delta : \mathcal{L}(\Sigma_g) \to \mathcal{L}(\Sigma_g) \otimes \mathcal{L}(\Sigma_g), \quad \Delta(<L>) = <L> \otimes <L> \\ &\epsilon : \mathcal{L}(\Sigma_g) \to \mathbb{C}, \quad \epsilon(<L>) = 1 \\ &S : \mathcal{L}(\Sigma_g) \to \mathcal{L}(\Sigma_g), \quad S(<L>) = <L^{-1}> \end{aligned}$$

where L^{-1} is the link L with orientation changed on each component.

This Hopf algebra structure descends to $\mathbb{C}[\mathbf{H}(\mathbb{Z}_N^{2g})]$, the latter being also a group algebra. However the Hopf algebra structure does not descend to

$$\mathcal{L}_N(\Sigma_g) = L(\mathbf{\Theta}_N^{\Pi}(\Sigma_g)).$$

We can see for example that the counit is ill-behaved under the factorization, since on the one hand $\epsilon((0,0,1)) = 1$, and on the other hand $\epsilon(e^{\frac{\pi i}{N}}(0,0,0)) = e^{\frac{\pi i}{N}}$. The algebra $\mathcal{L}_N(\Sigma_g)$ is the quantum torus at a root of unity, and it is known that the quantum torus does not admit a Hopf algebra structure (see for example [Tang *et al.* (2007)]).

Example 8.2. The universal enveloping algebra of a Lie algebra is a Hopf algebra. Let $\mathfrak{G}$ be a Lie algebra and let $U(\mathfrak{G})$ be its universal enveloping algebra. For readers not familiar with the definition of the latter, let us say that if $\mathfrak{G}$ is a Lie algebra of matrices, then $U(\mathfrak{G})$ is the algebra generated by those matrices, so that the Lie bracket is the commutator.

$U(\mathfrak{G})$ has the Hopf algebra structure with comultiplication

$$\Delta(X) = X \otimes 1 + 1 \otimes X, \quad X \in \mathfrak{G},$$

counit

$$\epsilon(X) = 0, \quad X \in \mathfrak{G}$$

and antipode

$$S(X) = -X, \quad X \in \mathfrak{G},$$

where this maps are extended linearly to the entire $U(\mathfrak{G})$.

If G is a Lie group such that $\mathfrak{G}$ is its Lie algebra, then $\mathfrak{G}$ is in fact the infinitesimal version of G, with the elements of the Lie group and the Lie algebra related by

$$g = \exp(X).$$

We can think of the Hopf algebra $U(\mathfrak{G})$ as the infinitesimal version of the Hopf algebra $\mathbb{C}[G]$ and the comultiplication, counit, and antipode of the first are obtained by "differentiating" the comultiplication, counit, and antipode of the second.

Quantum groups

Drinfeld's philosophy is that a quantum group is a Hopf algebra that is a quantum deformation of a classical Hopf algebra such as the group algebra of a Lie group or the universal enveloping algebra of a Lie algebra (examples 8.1, 8.2). This is a surprising idea because the Lie algebra structure itself is rigid; it cannot be deformed! But its universal enveloping algebra can. Usually, quantum groups are exemplified by the Drinfeld-Jimbo construction of the quantum group of the universal enveloping algebra of a semisimple Lie algebra (see Chapter XVII in [Kassel (1995)]). Here is a different example.

Example 8.3. Let Σ_g be a closed genus g Riemann surface with Jacobian variety $\mathcal{J}(\Sigma_g)$. In Chapter 4 we saw that $H_1(\Sigma_g, \mathbb{Z})$ can be identified with the group of exponential functions on the Jacobian variety. Then the group algebra

$$A = \mathbb{C}[H_1(\Sigma_g, \mathbb{Z})]$$

is the algebra of trigonometric polynomials on the $2g$-dimensional torus that is $\mathcal{J}(\Sigma_g)$. Being a group algebra, A has a Hopf algebra structure. This is our "classical" Hopf algebra.

Now we introduce a formal variable t, which we can think of as being $e^{i\pi h}$ where h is a formal Planck's constant. Consider the free $\mathbb{C}[t, t^{-1}]$-module

$$A_t = \mathbb{C}[t, t^{-1}]H_1(\Sigma_g, \mathbb{Z})$$

with multiplication

$$(\mathbf{p},\mathbf{q}) *_t (\mathbf{p}',\mathbf{q}') = t^{\pi i(\mathbf{p}^T\mathbf{q}'-\mathbf{q}^T\mathbf{p}')}(\mathbf{p}+\mathbf{p}',\mathbf{q}+\mathbf{q}'),$$

or in exponential notation,

$$e^{2\pi i\mathbf{p}^T\mathbf{x}+2\pi i\mathbf{q}^T\mathbf{y}} *_t e^{2\pi i\mathbf{p}^T\mathbf{x}+2\pi i\mathbf{q}^T\mathbf{y}}$$
$$= e^{i\pi((\mathbf{p}^T\mathbf{q}'-\mathbf{q}^T\mathbf{p}')h}e^{2\pi i(\mathbf{p}+\mathbf{p}')^T\mathbf{x}+2\pi i(\mathbf{q}+\mathbf{q}')^T\mathbf{y}}.$$

This is a deformation of A with respect to the formal parameter h. It is not hard to see that A_h is the group algebra of the Heisenberg group, $\mathbb{C}[\mathbf{H}(\mathbb{Z}^g)]$, and so it is a Hopf algebra, and now in Drinfeld's perspective it is a quantum group.

Setting $t^{2N} = 1$ (but only formally, no numerical values), we obtain the ring of coefficients

$$\mathbb{C}[t,t^{-1}]/(t^{2N}-1) = \mathbb{C}[t]/(t^{2N}-1) = \mathbb{C}[\mathbf{Z}_{2N}].$$

Working with this ring of coefficients and using the same multiplication rule we obtain a new algebra. In this algebra we further impose the factorization relations

$$(\mathbf{p},\mathbf{q})^N = (\mathbf{0},\mathbf{0}) \text{ for all } (\mathbf{p},\mathbf{q}) \in H_1(\Sigma_g,\mathbb{Z}).$$

The result is the group algebra of the finite Heisenberg group, $\mathbb{C}[\mathbf{H}(\mathbb{Z}_N^g)]$. We think of it as the quantum group at a root of unity of the algebra of trigonometric polynomials on the Jacobian variety.

Those familiar with the quantum group associated to $sl(2,\mathbb{C})$ constructed by Kirillov and Reshetikhin will recognize the pattern for passing from the generic case of the deformation parameter to the root of unity: besides setting the parameter equal to a root of unity, additional factorization relations have to be introduced. This is what makes quantum groups at roots of unity considerably harder to study.

We conclude that we can view the Schrödinger representation of $\mathbb{C}[\mathbf{H}(\mathbb{Z}_N^g)]$ as the representation of a quantum group. However it is not this quantum group that prompted us to write this chapter of the book, but the one that allows us to turn link diagrams into vertex models. We will encounter it later.

The representation theory of a Hopf algebra

Hopf algebras are particularly handy from the representation theoretical point of view. In what follows, all representations are finite dimensional.

So let A be a Hopf algebra.

- If V and W are representations, then $V \oplus W$ is a representation. This is true in general for algebras, no need for the Hopf algebra structure.
- If V and W are two of its representations, then $V \otimes W$ is obviously a representation of $A \otimes A$. But by using the Hopf algebra structure, $V \otimes W$ can be turned into a representation of A itself by setting
$$a(v \otimes w) := \Delta(a) v \otimes w.$$
- The one-dimensional vector space $\mathbb{C}$ can be turned into a representation of A by setting
$$a \cdot 1 := \epsilon(a)1.$$
- If V is a representation of A, then V^*, the algebraic dual of V, is made into a representation of A by
$$a \cdot v^*(v) := v^*(Sv).$$

We conclude that the representations of a Hopf algebra form a ring with unit, endowed with an involution defined by taking the dual.

8.1.3 *The Yang-Baxter equation and the universal R-matrix*

Let us show how quantum groups can be used to construct vertex models. We do this in the language of particle scattering from the beginning of the chapter, though the models constructed this way address problems in statistical physics. We prefer this scattering model because it brings us closer to knots. What we need is to find an S-matrix for 2-particle collision that satisfies the Yang-Baxter equation. It turns out that the quantum group examples constructed by Drinfeld and his predecessors are particularly good for solving this problem.

The idea is to encode the states of particles by representations of a quantum group A. The state of an n-particle system would be the tensor product of the representations defining individual states.

If V and W are the states of two colliding particles, then the scattering matrix should be an A-module homomorphism

$$R_{V,W} : V \otimes W \to W \otimes V.$$

This is represented graphically in Figure 8.5.

The **Yang-Baxter equation** for the scattering of particles then reads

$$\begin{aligned}&(R_{U,V} \otimes I)(I \otimes R_{U,W})(R_{V,W} \otimes I)\\ &\quad = (I \otimes R_{V,W})(R_{U,W} \otimes I)(I \otimes R_{U,V}),\end{aligned} \tag{8.1}$$

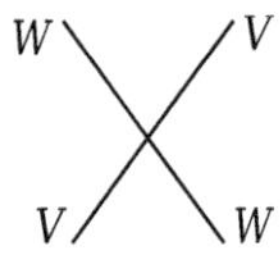

Fig. 8.5 Scattering of 2 particles

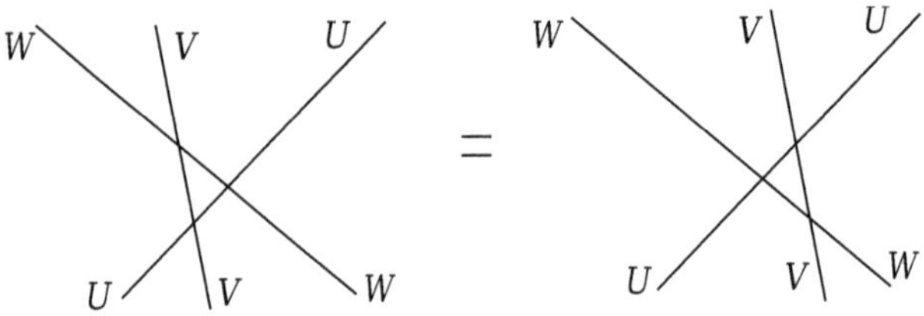

Fig. 8.6 The Yang-Baxter equation

shown graphically in Figure 8.6.

It turns out that many examples, including the one we have in mind, have a universal element $R \in A \otimes A$ that induces all these homomorphisms $R_{V,W}$ between representations. In order to explain what such an element should be, we need some notation.

Notation: Let A be a Hopf algebra. Recall the map

$$\sigma : A \otimes A \to A \otimes A, \quad \sigma(x \otimes y) = y \otimes x.$$

We let

$$\iota_{12}, \iota_{13}, \iota_{23} : A \otimes A \to A \otimes A \otimes A$$

be defined by

$$\iota_{12}(a \otimes b) = a \otimes b \otimes 1, \quad \iota_{13}(a \otimes b) = a \otimes 1 \otimes b, \quad \iota_{23}(a \otimes b) = 1 \otimes a \otimes b.$$

For the other situations, we let $\iota_{jk} = \iota_{kj}\sigma$.

Definition 8.3. A Hopf algebra $(A, m, \eta, \Delta, \epsilon, S)$ is called **quasitriangular** if there exists an invertible element $R \in A \otimes A$ such that the following conditions are satisfied

(qtha1) $\sigma \circ \Delta(a) = R\Delta(a)R^{-1}$ for all $a \in A$;
(qtha2) $(\Delta \otimes I)(R) = R_{13}R_{23}$;
(qtha3) $(I \otimes \Delta)(R) = R_{13}R_{12}$;

where $R_{jk} = \iota_{jk}(R)$. The element R is called a **universal R-matrix**.

For two representations V, W of A, we define the permutation operator

$$P : V \otimes W \to W \otimes V, \quad P(v \otimes w) = w \otimes v.$$

Theorem 8.1. *Let R be the universal R-matrix of a quasitriangular Hopf algebra. Then the linear operators*

$$\check{R} : U \otimes V \to V \otimes U, \quad \check{R} = PR$$

yield a solution to the Yang-Baxter equation (8.1).

Proof. First let us check that

$$R_{12}R_{13}R_{23} = R_{23}R_{13}R_{12}, \tag{8.2}$$

which is the Yang-Baxter equation in its statistical physics form.

Notation: Here and elsewhere we let

$$R = \sum_j \alpha_j \otimes \beta_j.$$

We have

$$[(\sigma \circ \Delta) \otimes I](R) = \sum_j (\sigma \circ \Delta)(\alpha_j) \otimes \beta_j = \sum_j R_{12}\Delta(\alpha_j)R_{12}^{-1} \otimes \beta_j$$

$$= R_{12}(\sum_j \Delta(\alpha_j) \otimes \beta_j)R_{12}^{-1} = R_{12}[(\Delta \otimes I)(R)]R_{12}^{-1} = R_{12}R_{13}R_{23}R_{12}^{-1}.$$

In this computation, for the second equality we used (qtha1) and for the last we used (qtha2).

Also

$$[(\sigma \circ \Delta) \otimes I](R) = (\iota_{12}\sigma)[(\Delta \otimes I)(R)] = (\iota_{12}\sigma)(R_{13}R_{23}) = R_{23}R_{13},$$

where for the second equality we used (qtha3). Comparing the two equalities we obtain (8.2).

Now let us write (8.1) as

$$(P \otimes I)(R \otimes I)(I \otimes P)(I \otimes R)(P \otimes I)(R \otimes I)$$
$$= (I \otimes P)(I \otimes R)(P \otimes I)(R \otimes I)(I \otimes P)(I \otimes R).$$

We have

$$(P \otimes I)(R \otimes I)(I \otimes P)(I \otimes R)(P \otimes I)(R \otimes I)$$
$$= (P \otimes I)[(R \otimes I)(I \otimes P)][(I \otimes R)(P \otimes I)](R \otimes I)$$
$$= [(P \otimes I)R_{13}]R_{13}R_{12} = R_{23}R_{13}R_{12}.$$

Similarly

$$(I \otimes P)(I \otimes R)(P \otimes I)(R \otimes I)(I \otimes P)(I \otimes R)$$
$$= (I \otimes P)[(I \otimes R)(P \otimes I)][(R \otimes I)(I \otimes P)]R_{23}$$
$$= [(I \otimes P)R_{13}]R_{13}R_{23} = R_{12}R_{13}R_{23}.$$

Equality now follows from (8.2). □

We will check below (Lemma 8.4) that the operators $\check{R}$ are A-linear, namely that they are homomorphisms between representations.

Because the Yang-Baxter equation and the third Reidemeister move are related to braids, some mathematicians (e.g. [Kassel (1995)]) replace the word quasitriangular with **braided**, probably because it can be used to construct representations of the braid group.

The following result has as a consequence the fact that R^{-1} can also be used to solve the Yang-Baxter equation.

Lemma 8.1. *The following identities hold*

$$(\epsilon \otimes I)(R) = (I \otimes \epsilon)(R) = I$$
$$R^{-1} = (S \otimes I)(R) = (I \otimes S^{-1})(R)$$
$$(S \otimes S)(R) = R.$$

Proof. For the first identity we argue as follows. From (qtha1) and the identity $(\epsilon \otimes I)\Delta = I$ we obtain

$$R = (\epsilon \otimes I \otimes I)(\Delta \otimes I)(R) = (\epsilon \otimes I \otimes I)(R_{13}R_{23} = [(\epsilon \otimes I)(R)]\epsilon(1)R,$$

where for the last equality we used (qtha2). Because $\epsilon(1) = 1$ and R is invertible, we deduce from the above equality that

$$(\epsilon \otimes I)(R) = 1. \tag{8.3}$$

Next, we know that the antipode satisfies $m(S \otimes I)\Delta(a) = \epsilon(a)1$ for all $a \in A$. This implies

$$(m \otimes I)(S \otimes I \otimes I)(\Delta \otimes I)(R) = (\epsilon \otimes I)(R).$$

The latter is equal to 1 by (8.3). Consequently

$$1 = (m \otimes I)(S \otimes I \otimes I)(R_{13}R_{23}) = (S \otimes I)(R)S(1)R.$$

Since $S(1) = 1$, we obtain

$$(S \otimes I)(R) = R^{-1}.$$

The second equality follows from the fact that if we replace Δ by $\sigma \circ \Delta$ and S by S^{-1} and R by $\sigma(R)$ we still obtain a quasitriangular Hopf algebra.

Finally,

$$(S \otimes S)(R) = (I \otimes S)(S \otimes I)(R) = (I \otimes S)(R^{-1})$$
$$= (I \otimes S)(I \otimes S^{-1})(R) = (1 \otimes 1)(R) = R.$$

□

We introduce now a special element of a quasitriangular Hopf algebra, which will be essential for the construction of the Reshetikhin-Turaev invariant of a link.

Definition 8.4. Let the universal R-matrix be as before $R = \sum_j \alpha_j \otimes \beta_j$. We define

$$u = m((S \otimes I)(\sigma(R))) = \sum_j S(\beta_j)\alpha_j.$$

Here are two properties of this element.

Lemma 8.2. *(i) The element u is invertible and*

$$u^{-1} = \sum_j S^{-1}(\beta_j)S(\alpha_j) = \sum_j \beta_j S^2(\alpha_j).$$

(ii) For every $a \in A$,

$$S^2(a) = uau^{-1}.$$

Before we proceed with the proof, we establish some notation, which is standard in the theory of Hopf algebras and which simplifies considerably formulas. This is called **Sweedler's notation**.

Notation: If a is an element of the Hopf algebra A then $\Delta(a)$ is of the form $\sum_j a'_j \otimes a''_j$. We write simply

$$\Delta(a) = \sum_{(a)} a' \otimes a''.$$

The coassociativity of Δ is expressed by the equality

$$\sum_{(a)} \left(\sum_{(a')} (a')' \otimes (a')'' \right) \otimes a'' = \sum_{(a)} a' \otimes \left(\sum_{(a'')} (a'')' \otimes (a'')'' \right).$$

In Sweedler's notation we write both sides simply as

$$\sum_{(a)} a' \otimes a'' \otimes a'''.$$

Proof. We first show that $S^2(a)u = ua$ for all $a \in A$. Equation (qtha1) implies

$$[((\sigma \circ \Delta) \otimes I)(a \otimes b)](R \otimes I) = (R \otimes I)(\Delta \otimes I)(a \otimes b), \text{ for all } a, b \in A,$$

and since Δ is an algebra homomorphism (Remark 8.1),

$$[((\sigma \circ \Delta) \otimes I)(x)](R \otimes I) = (R \otimes I)(\Delta \otimes I)(x), \text{ for all } x \in A \otimes A.$$

Writing this relation for $x = \Delta(a)$ and using Sweedler's notation, we have

$$\sum_{j,(a)} a'' \alpha_j \otimes a' \beta_j \otimes a''' = \sum_{j,(a)} \alpha_j a' \otimes \beta_j a'' \otimes a'''.$$

Apply to this the linear map

$$[m\sigma(I \otimes (m\sigma))] \circ (I \otimes S \otimes S^2) : A \otimes A \otimes A \otimes A \to A,$$

to obtain

$$\sum_{j,(a)} S^2(a''')S(a'\beta_j)a''\alpha_j = \sum_{j,(a)} S^2(a''')S(\beta_j a'')\alpha_j a'.$$

Using the fact that S is an algebra antihomomorphism we can rewrite this as

$$\sum_{j,(a)} S^2(a''')S(\beta_j)S(a')a''\alpha_j = \sum_{j,(a)} S^2(a''')S(a'')S(\beta_j)\alpha_j a'. \qquad (8.4)$$

Let us find a simpler formula for the left-hand side of (8.4). Using the relationship between the antipode and the counit we have

$$\sum_{(a)} S(a')a'' \otimes a''' = \sum_{(a)} \epsilon(a')1 \otimes a'' = 1 \otimes a.$$

Hence

$$\sum_{(a)} S(a')a'' \otimes S^2(a''') = 1 \otimes S^2(a).$$

Multiplying both sides on the right by $\sum_j \alpha_j \otimes S(\beta_j)$ we obtain

$$\sum_{j,(a)} S(a')a''\alpha_j \otimes S^2(a''')S(\beta_j) = \sum_j \alpha_j \otimes S^2(a)S(\beta_j).$$

We deduce that the left-hand side of (8.4) is equal to

$$\sum_j S^2(a)S(\beta_j)\alpha_j = S^2(a)u.$$

For the right-hand side we use again the relationship between the antipode and the counit to write

$$\sum_{(a)} a' \otimes S(a''S(a''')) = \sum_{(a)} a' \otimes S(\epsilon(a'')1) = \sum_{(a)} a'\epsilon(a'') \otimes S(1) = a \otimes 1.$$

Multiplying by $u \otimes 1$ on the left we obtain

$$\sum_{j,(a)} S(\beta_j)\alpha_j a' \otimes S^2(a''')S(a'') = ua \otimes 1.$$

Apply $m\sigma$ to both sides to obtain

$$\sum_{j,(a)} S^2(a''')S(a'')S(\beta_j)\alpha_j a' = ua.$$

If we combine the expression we obtained for the left-hand side of (8.4) with the expression we obtained for the right-hand side we obtain

$$S^2(a)u = ua,$$

as desired.

Now let us show that u is invertible and that its inverse satisfies the formula from the statement. By the first half of the proof, $S(\beta_j)u = S^2(S^{-1}(\beta_j))u = uS^{-1}(\beta_j)$ hence

$$\begin{aligned} u\textstyle\sum_j S^{-1}(\beta_j)S(\alpha_j) &= \textstyle\sum_j uS^{-1}(\beta_j)S(\alpha_j) = \sum_j S(\beta_j)uS(\alpha_j) \\ &= \textstyle\sum_{j,k} S(\beta_j)S(\beta_k)\alpha_k S(\alpha_j) = \sum_{j,k} S(\beta_k\beta_j)\alpha_k S(\alpha_j). \end{aligned} \tag{8.5}$$

But by Lemma 8.1,

$$\sum_{j,k} \alpha_j S(\alpha_k) \otimes \beta_j\beta_k = RR^{-1} = 1 \otimes 1.$$

So (8.5) is equal to $m(1 \otimes 1) = 1$. We conclude that u^{-1} defined as in the statement is a right inverse of u. But by the first part $S^2(u^{-1})u = uu^{-1} = 1$, showing that u is left invertible as well. Then necessarily the expression from (i) defines the inverse of u. The fact that $u^{-1} = \sum_j \beta_j S^2(\alpha_j)$ follows from Lemma 8.1. □

Corollary 8.1. *We have*

$$S^2(u) = u \text{ and } S^2(u^{-1}) = u^{-1}.$$

Proof. Set $a = u$ in (ii). □

Lemma 8.3. *The element u satisfies*

$$\Delta(u) = (\sigma(R)R)^{-1}(u \otimes u) = (u \otimes u)(\sigma(R)R)^{-1} \text{ and } \epsilon(u) = 1.$$

Proof. The second relation follows from Lemma 8.1. Let us focus on proving the first relation. The proof is from [Kassel (1995)].

Using (qtha1) we have

$$\Delta(a) = \sigma(R)(\sigma \circ \Delta(a))\sigma(R)^{-1} \text{ for all } a \in A.$$

Hence

$$\Delta(a)\sigma(R)R = \sigma(R)R\Delta(a).$$

So it suffices to show $\Delta(u)\sigma(R)R = u \otimes u$. Using the properties of the comultiplication and antipode, we can write

$$\begin{aligned}\Delta(u)\sigma(R)R &= \sum_j \Delta(S(\beta_j))\Delta(\alpha_j)\sigma(R)R \\ &= \sum_j (S\otimes S)(\sigma\circ\Delta(\beta_j))\Delta(\alpha_j)\sigma(R)R \\ &= \sum_j (S\otimes S)(\sigma\circ\Delta(\beta_j))\sigma(R)R\Delta(\alpha_j).\end{aligned}$$

Define a right action $\cdot$ of $A\otimes A\otimes A\otimes A$ on $A\otimes A$ by

$$(a\otimes b)\cdot(\mathbf{a}\otimes\mathbf{b}) = (S\otimes S)(\mathbf{b})(a\otimes b)\mathbf{a}, \text{ for } a,b\in A, \mathbf{a},\mathbf{b}\in A\otimes A.$$

We introduce the elements R_{jk} in $A\otimes A$ and $A\otimes A\otimes A\otimes A$ in the similar manner that we introduced such elements in $A\otimes A\otimes A$ at the beginning of this unit (for example $\sigma(R) = R_{21}$). The formula that we derived above can be written in the new language as

$$\Delta(u)R_{21}R = R_{21}\cdot(R_{12}R_{13}R_{23}R_{14}R_{24}).$$

By the Yang-Baxter equation (8.2) the right-hand side equals

$$R_{21}\cdot(R_{23}R_{13}R_{12}R_{14}R_{24}).$$

Let us compute this expression. Using the formula for R^{-1} (Lemma 8.1), we have

$$\begin{aligned}R_{21}\cdot R_{23} &= \sum_{j,k} S(\beta_k)\beta_j\otimes\alpha_j\alpha_k = (S\otimes I)\left(\sum_{j,k} S^{-1}(\beta_j)\beta_k\otimes\alpha_j\alpha_k\right) \\ &= (S\otimes I)(R_{21}^{-1}R_{21}) = (S\otimes I)(1\otimes 1) = 1\otimes 1.\end{aligned}$$

Further

$$R_{21}\cdot(R_{23}R_{13}) = (1\otimes 1)\cdot R_{13} = \sum_j S(\beta_j)\alpha_j\otimes 1 = u\otimes 1.$$

Next

$$R_{21}\cdot(R_{23}R_{13}R_{12}) = (u\otimes 1)\cdot R_{12} = (u\otimes 1)R$$

and

$$R_{21} \cdot (R_{23}R_{13}R_{12}R_{14}) = (u \otimes 1) \left(\sum_{j,k} \alpha_j \alpha_k \otimes S(\beta_k)\beta_j \right)$$

$$= (u \otimes 1)(I \otimes S) \left(\sum_{j,k} \alpha_j \alpha_k \otimes S^{-1}(\beta_k)\beta_j \right) = (u \otimes 1)(I \otimes S)(R^{-1}R)$$

$$= (u \otimes 1)(I \otimes S)(1 \otimes 1) = u \otimes 1.$$

Finally,

$$R_{21} \cdot (R_{23}R_{13}R_{12}R_{14}R_{24}) = (u \otimes 1) \cdot R_{24} = (u \otimes 1)(1 \otimes u) = u \otimes u.$$

The lemma is proved. □

Corollary 8.2. *We have*

$$\Delta(S(u)) = (\sigma(R)R)^{-1}(S(u) \otimes S(u)) = (S(u) \otimes S(u))(\sigma(R)R)^{-1}.$$

Hence the central element $uS(u)$ satisfies

$$\Delta(uS(u)) = (\sigma(R)R)^{-1}(uS(u) \otimes uS(u)) = (uS(u) \otimes uS(u))(\sigma(R)R)^{-1}.$$

8.1.4 *Link invariants and ribbon Hopf algebras*

The idea behind the construction of link invariants from quantum groups

Quantum groups appeared around the same time as the Jones polynomial, and it did not take long until they were related to knots and links. To distinguish overcrossings from undercrossings one can use $\check{R} = PR$ for the crossing on the left of Figure 8.7 and $\check{R}^{-1} = R^{-1}P$ for the crossing on the right of Figure 8.7.

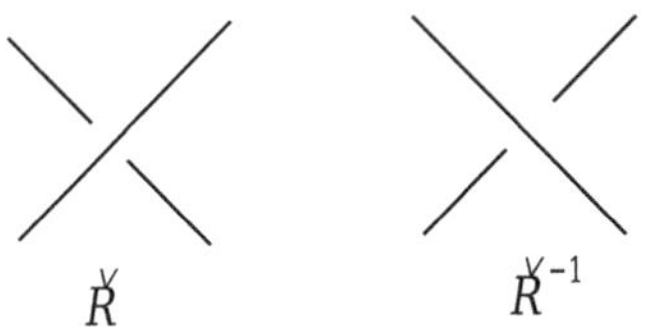

Fig. 8.7 $\check{R}$ and $\check{R}^{-1}$

An easy computation can transform 8.2 into

$$R_{12}^{-1}R_{13}^{-1}R_{23}^{-1} = R_{23}^{-1}R_{13}^{-1}R_{12}^{-1},$$

and the computation from the proof of Theorem 8.1 can be repeated to show that $\check{R}^{-1}$ satisfies the Yang-Baxter equation 8.1. So now in the diagrams depicted in Figure 8.8 the morphisms

$$U \otimes V \otimes W \to W \otimes V \otimes U$$

do not change when performing Reidemeister III moves. It is not hard to check the invariance under the third Reidemeister move for any choice of under/overcrossings in these diagrams.

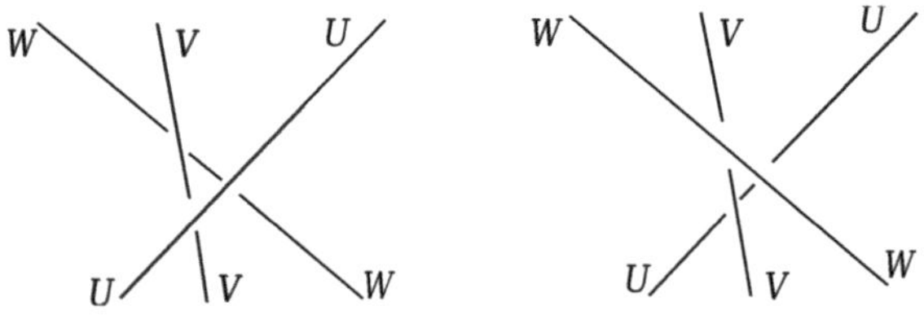

Fig. 8.8 Invariance under Reidemeister III moves

We also have invariance under the Reidemeister II move because

$$\check{R}\check{R}^{-1} = \check{R}^{-1}\check{R} = I \otimes I$$

as can be seen in Figure 8.9.

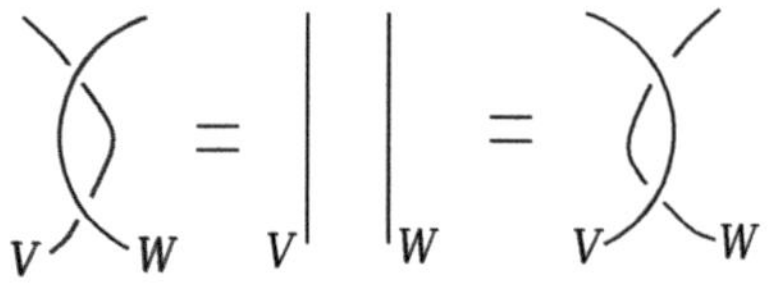

Fig. 8.9 Invariance under Reidemeister II moves

We can avoid worrying about Reidemeister I moves by producing just invariants for framed knots and links. In fact in the previous chapters we only encountered invariants of framed knots and links, so this is precisely what we want. But we have to check that a positive twist cancels a negative twist. And this is where the difficulty lies.

To obtain true knot and link invariants we have to make sense of the phenomena in Figure 8.10. In our interpretation of strands as world-lines of particles, it is natural to interpret these phenomena as particle/antiparticle creation and annihilation. We can think of an upwards arrow as being the world-line of a particle and an downwards arrow as the world-line of an antiparticle. This is consistent with the observation from mathematical physics that for particles that are not their own antiparticles the mathematical model of the particle moving forwards in time is the same as the mathematical model of the antiparticle moving backwards in time. Since it is easy to make use of duality between representations to produce homomorphisms associated to these pictures, we use for the particle-antiparticle picture the duality between V and V^*. It is somewhat counter-intuitive but the standard convention is to associate

- to down-going strands decorated by V the representation V itself;
- to up-going strands decorated by V the dual representation V^*.

Then the diagrams from Figure 8.10 represent homomorphisms of the form

$$\mathbb{C} \to V^* \otimes V, \quad \mathbb{C} \to V \otimes V^*, \quad V \otimes V^* \to \mathbb{C}, \quad V^* \otimes V \to \mathbb{C}.$$

Fig. 8.10 Interpreting maxima and minima as particle-antiparticle creation and annihilation

A first choice would be to choose a basis (e_k) for V and the dual basis (e^k) for V^* and simply let these maps be defined by $1 \mapsto \sum_k e^k \otimes e_k$, $1 \mapsto \sum_k e_k \otimes e^k$, $x \otimes f \mapsto f(x)$, $f \otimes x \mapsto f(x)$. But the first and the third of these maps are not A-linear. Below we explain how to alter these maps to produce A-linear maps in all cases, and then how this yields link invariants.

We point out that in order for our construction to yield link invariants, besides invariance under Reidemeister II and III moves we should also have invariance under the moves described in Figure 8.11. One can show that checking invariance under Reidemeister moves II and III, as well as the moves IV, V, VI from this diagram suffices (see [Reshetikhin and Turaev (1990)]), but our construction from §8.3 will produce the linking number, so we don't need this result to prove topological invariance.

Ribbon Hopf algebras

It is now time to turn to rigorous constructions. In order to guarantee that positive and negative twists cancel each other, we introduce a more restrictive notion of Hopf algebra.

IV
V
VI

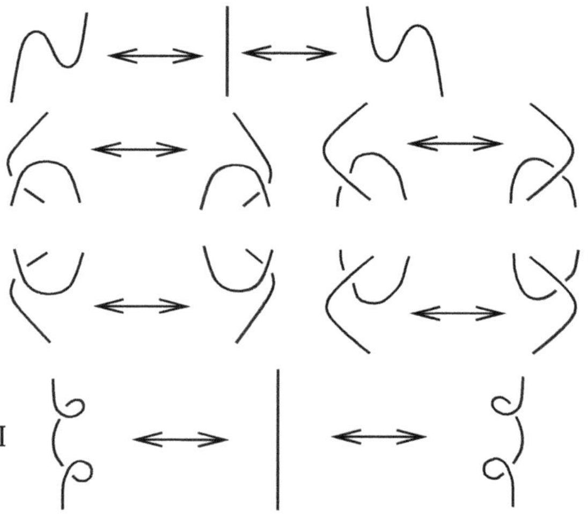

Fig. 8.11 Moves between link diagrams

Definition 8.5. A **ribbon Hopf algebra** is a quasitriangular Hopf algebra $(A, m, \eta, \Delta, \epsilon, S, R)$ that possesses an invertible element v in its center such that the following conditions hold

(rha1) $v^2 = uS(u)$;
(rha2) $\Delta(v) = (\sigma(R)R)^{-1}(v \otimes v)$;
(rha3) $S(v) = v$;
(rha4) $\epsilon(v) = 1$.

Remark 8.2. The element $uS(u)$ is central and satisfies (rha2), (rha3), (rha4), but (rha1) means that $uS(u)$ has a central square root.

Following this discussion we make the following construction, which will yield framed link invariants:

We start with a **ribbon Hopf algebra**

$$(A, m, \eta, \Delta, \epsilon, S),$$

whose antipode S is **invertible.** Let the universal R-matrix be $R = \sum_j \alpha_j \otimes \beta_j$, let $u = \sum_j S(\beta_j)\alpha_j$, and let v be the central element which is the square root of $uS(u)$, provided by the definition.

Given an oriented framed link L with link components $L_1, L_2, \ldots, L_m$ assign to each link component a representation V_j, $j = 1, 2, \ldots, m$, of A.

Definition 8.6. This assignment

$$\mathbf{V} : \{L_1, L_2, \ldots, L_m\} \to \{V_1, V_2, \ldots, V_m\}$$

of representations to the link components is called a **coloring** of the link by representations.

Notation: We denote the data consisting of L and its coloring $\mathbf{V}$ by $\mathbf{V}(L)$.

View S^3 as $\mathbb{R}^3$ compactified with the point at infinity, and in it fix a plane and a direction in the plane called the vertical direction. Given an oriented framed link L in S^3, deform it through an isotopy to a link whose framing is parallel to the plane, and whose projection onto the plane is a link diagram that can be sliced by finitely many horizontal lines into pieces, each of which consists of several vertical strands and exactly one of the events from Figure 8.12. An example is shown in Figure 8.13. The lines in these figures will be oriented, inheriting their orientations from the original link.

Fig. 8.12 Simple events

Now assume that the link is colored by representations. To every horizontal line that separates two events (the dotted lines in Figure 8.13) associate a representation of A as follows. For each intersection of the horizontal separation line with a strand colored by V, we associate the representation V if the arrow points downwards and V^* if the arrow points upwards. The representation associated to the horizontal line is the tensor product of the representations associated to the intersection points, taken from left to right. This defines a representation of the Hopf algebra. If there is no strand that intersects that horizontal line, then the representation is by default the trivial representation, $\mathbb{C}$, defined by the counit.

To the events in Figure 8.12 we associate homomorphisms between representations as follows:

- To a crossing like the first event in Figure 8.12 we associate $\check{R}$.
- To a crossing like the second event in Figure 8.12 we associate $\check{R}^{-1}$.
- If the event is like the third in Figure 8.12, then there are two possibilities depending on the orientation of the strand:
  - if the strand is oriented left to right, so that the homomorphism should be $V^* \otimes V \to \mathbb{C}$, then the homomorphism is $E(f \otimes x) = f(x)$;

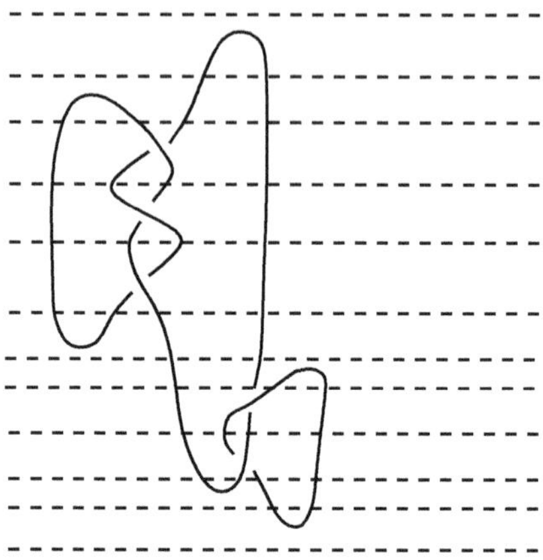

Fig. 8.13 The decomposition of a link into simple events

- if the strand is oriented right to left, so that the homomorphism should be $V \otimes V^* \to \mathbb{C}$, then the homomorphism is $E_{op}(x \otimes f) = f(v^{-1}ux)$.

- If the event is like the fourth in Figure 8.12, then there are two possibilities depending on the orientation of the strand:
  - if the strand is oriented right to left, so that the homomorphism should be $\mathbb{C} \to V^* \otimes V$, then the homomorphism is defined by

$$N_{op}(1) = \sum_k e^k \otimes u^{-1} v e_k;$$

  - if the strand is oriented left to right, so that the homomorphism should be $\mathbb{C} \to V \otimes V^*$, then the homomorphism is defined by

$$N(1) = \sum_k e_k \otimes e^k.$$

Here (e_k) denotes a basis of the vector space V and (e^k) is the dual basis in V^*, meaning that $e^k(e_m) = \delta_{km}$.

Lemma 8.4. *The maps defined above are A-linear.*

Proof. The fact that $\check{R}$ and $\check{R}^{-1}$ are A-linear is proved as follows:

$$\check{R}(a \cdot (x \otimes y)) = \check{R}(\Delta(a)(x \otimes y)) = P(R\Delta(a)(x \otimes y)) = P(\sigma \circ \Delta(a) R(x \otimes y))$$

where for the last equality we used the relation (qtha1) from the definition of a quasitriangular Hopf algebra. In Sweedler's notation, this is further equal to

$$P(\sum_{(a)}(a''\otimes a')R(x\otimes y)) = \sum_{(a)}(a'\otimes a'')P(R(x\otimes y)) = a\cdot \check{R}(x\otimes y).$$

This proves that $\check{R}$ is A-linear. Its inverse $\check{R}^{-1}$ must therefore be A-linear as well.

For the map $E(f\otimes x) = f(x)$ with Sweedler's notation we can write

$$\begin{aligned} E(a(f\otimes x)) &= E(\Delta(a)(f\otimes x)) = E(\sum_{(a)} a'f\otimes a''x) = f(S(a')a''x) \\ &= f(m(S\otimes I)\Delta(a)x) = f(\epsilon(a)x) = \epsilon(a)f(x) = aE(f\otimes x), \end{aligned}$$

where we used the defining property of the antipode.

For the map $E_{op}(x\otimes f) = f(v^{-1}ux)$ we have

$$\begin{aligned} E_{op}(a(x\otimes f)) &= E_{op}(\Delta(a)(x\otimes f)) = E_{op}(\sum_{(a)} a'x\otimes a''f) \\ &= f(\sum_{(a)} S(a'')v^{-1}ua'x) = f(\sum_{(a)} S(a'')v^{-1}S^2(a')ux) \\ &= f(\sum_{(a)} S(a'')S^2(a')v^{-1}ux) \end{aligned}$$

where we used the fact that $ua' = S^2(a')u$ (Lemma 8.2) and that v^{-1} is central. We compute this using the defining property of S as follows

$$\begin{aligned} f(\sum_{(a)} S(a'')S^2(a')v^{-1}ux) &= f(S(m(S\otimes I))\Delta(a)v^{-1}ux) \\ &= f(S(\epsilon(a)1)v^{-1}ux) = f(\epsilon(a)S(1)v^{-1}ux) = \epsilon(a)f(v^{-1}ux) \\ &= aE_{op}(x\otimes f). \end{aligned}$$

One should stress out that without inserting $v^{-1}u$ in the definition of the map E_{op}, this map would not be A-linear.

Let us now check the A-linearity of

$$N : \mathbb{C}\to V\otimes V^*, \quad N(z) = z\sum e_k\otimes e^k.$$

On the one hand, we have

$$N(a\cdot z) = N(\epsilon(a)z) = z\epsilon(a)\sum e_k\otimes e^k.$$

On the other hand, we have

$$a\cdot N(z) = z\Delta(a)\sum_k e_k\otimes e^k.$$

We need to check that

$$\Delta(a)\sum_k e_k \otimes e^k = \epsilon(a)\sum_k e_k \otimes e^k.$$

View both sides as elements in End(V), and apply them to an element $x \in V$. With Sweedler's notation, we can write

$$[\Delta(a)\sum_k e_k \otimes e_k](x) = [\sum_k\sum_{(a)} a' e_k \otimes a'' e^k](x) = \sum_{(a)}\sum_k a'[e^k(S(a'')x)]e_k.$$

Note that if T is a linear transformation, then $\sum_k e^k(Tx)e_k = Tx$, so the above expression is equal to

$$\sum_{(a)} a' S(a'')x.$$

But

$$\sum_{(a)} a' S(a'') = [m(I \otimes S)\Delta](a) = \epsilon(a)1$$

by the axiom of the antipode. This proves that N is A-linear. The proof that N_{op} is linear is similar. The lemma is proved. □

The simple events define A-linear transformations between the representations associated to consecutive horizontal lines. Such transformations go from the representation associated to the line below the event to the representation of the line above the event. It is obtained by tensoring the identity operator for the vertical stands that do not participate to the event with the operator of the event. For example in Figure 8.13, the third event from the bottom is a crossing, and the operator corresponding to it is $I \otimes \check{R} \otimes I$.

Now we trace the diagram of the oriented framed link from bottom to top (in the "time" direction), and we compose the consecutive homomorphisms. The end result is a homomorphism between trivial representations

$$\mathbb{C} \to \mathbb{C},$$

which is necessarily of the form $z \mapsto \lambda z$ for some $\lambda \in \mathbb{C}$.

Notation: We denote this number λ by $<\mathbf{V}(L)>$.

Definition 8.7. The number $<\mathbf{V}(L)>$ associated to the oriented framed link L and coloring $\mathbf{V}$ is called the **Reshetikhin-Turaev link invariant.**

The following result shows that the Reshetikhin-Turaev invariant is indeed a link invariant. The result is based on an analogue of Reidemeister's theorem for link diagrams of the type describe above, which we do not prove in this book, because the Reshetikhin-Turaev link invariant associated to theta functions is computable from the linking number and hence is trivially a topological invariant. The reader might want to prove the result in the piece-wise linear setting, where it is an easy exercise in combinatorial geometry. But we point out that this book requires curves to be smooth, so that everything can be related to Riemann surfaces and theta functions. And in this setting the proof is harder.

Theorem 8.2. *[Reshetikhin and Turaev (1990)] Let* $\mathbf{V}$ *be a coloring of a link by representations of a quasitriangular Hopf algebra* A*. Then* $<\mathbf{V}(L)>$ *is invariant under the Reidemeister moves II and III as well as the moves IV, V, VI from Figure 8.11.*

Proof. The proof blends the one from [Reshetikhin and Turaev (1990)] and the one from [Kirby and Melvin (1991)]. We have already checked invariance under the Reidemeister moves II and III.

For IV we only check invariance under the move on the left, the other being similar. Orient the strand downwards, or else work with the dual representation. The morphism defined by the strand on the left is

$$(E_{op} \otimes I)(I \otimes N_{op}) : V \to V.$$

Evaluating this on a vector we have

$$\begin{aligned}(E_{op} \otimes I)(I \otimes N_{op})x &= (E_{op} \otimes I)\sum_k x \otimes e^k \otimes u^{-1}ve_k \\ &= \sum_k [e^k(v^{-1}ux)]u^{-1}ve_k = u^{-1}v\sum_k [e^k(v^{-1}ux)]e_k,\end{aligned}$$

where the last equality follows from linearity, since $e^k(v^{-1}ux)$ is a scalar. Applying once more the observation that if T is a linear transformation then $\sum_k e^k(Tx)e_k = Tx$, we can write this as

$$u^{-1}v(v^{-1}ux) = x.$$

This shows that the homomorphism is the identity, which proves IV.

For move V we check again only the first situation. Orient the two strands so that the vertical is oriented downwards and the horizontal points to the left. Then the identity to be proved is

$$(I \otimes E_{op})(\check{R} \otimes I) = (E_{op} \otimes I)(I \otimes \check{R}^{-1}).$$

Let us evaluate the left-hand side on an element of the form $x \otimes y \otimes f \in V \otimes V \otimes V^*$. We have

$$(I \otimes E_{op})(\check{R} \otimes I)(x \otimes y \otimes f) = (I \otimes E_{op})(\sum_j \beta_j y \otimes \alpha_j x \otimes f)$$
$$= \sum_j \beta_j y f(v^{-1}ua_j x).$$

Using the fact that $R^{-1} = (S \otimes I)(R) = \sum_j S(\alpha_j) \otimes \beta_j$, the right-hand side evaluated on the same element gives

$$(E_{op} \otimes I)(I \otimes \check{R}^{-1})(x \otimes y \otimes f) = (E_{op} \otimes I)(\sum_x \otimes S(\alpha_j) f \otimes \beta_j y)$$
$$= \sum_j \beta_j y f(S^2(\alpha_j) v^{-1} ux).$$

This equals the expression we got for the left-hand side because by Lemma 8.2, $ua_j = S^2(a_j)u$. The identity is proved.

Finally, for the left-hand side of VI we discuss again the case where the strand is oriented downwards. The lower twist defines the homomorphism

$$\phi^+ : V \to V, \quad \phi^+ = (I \otimes E_{op})(\check{R} \otimes I)(I \otimes N).$$

Computing we obtain

$$\begin{aligned}
\phi^+(x) &= (I \otimes E_{op})(PR \otimes I)(I \otimes N)x \\
&= \sum_k (I \otimes E_{op})(PR \otimes I)x \otimes e_k \otimes e^k \\
&= \sum_{k,j} (I \otimes E)\beta_j e_k \otimes \alpha_j x \otimes e^k = \sum_{l,j} \beta_j (e^k(v^{-1}u\alpha_j x)e_k).
\end{aligned}$$

Again, since for a linear transformation T, $\sum_k e^k(Tx)e_k = Tx$, we can write this as

$$\sum_j \beta_j v^{-1} u \alpha_j x.$$

Hence

$$\begin{aligned}
\phi^+ &= \sum_j \beta_j v^{-1} u \alpha_j = \sum_j \beta_j u \alpha_j v^{-1} = \sum_j \beta_j u \alpha_j u^{-1} u v^{-1} \\
&= (\sum_j \beta_j S^2(\alpha_j)) u v^{-1} = u^{-1} u v^{-1} = v^{-1},
\end{aligned}$$

where we used the fact that v is central and Lemma 8.2.

The upper twist defines the homomorphism

$$\phi^- : V \to V, \quad \phi^- = (I \otimes E_{op})(\check{R}^{-1} \otimes I)(I \otimes N).$$

The computation of $\phi^-(x)$ is similar except that instead of R we have R^{-1}, which by Lemma 8.1 is equal to $(S \otimes I)(R)$ and then $\check{R}^{-1} = R^{-1}P$. The computation yields

$$\phi^-(x) = (I \otimes E_{op})(R^{-1}P \otimes I)(I \otimes N)x = \sum_j S(\alpha_j)v^{-1}u\beta_j x.$$

Again by using Lemma 8.2 and Corollary 8.1 we can write

$$\phi^- = \sum_j S(\alpha_j)v^{-1}u\beta_j = v^{-1}\sum_j S(\alpha_j)S^2(\beta_j)u = v^{-1}S(u)u.$$

But $S(u)u = v^2$ by the relation (rha1) in the definition of a ribbon Hopf algebra, so $\phi^- = v$. This shows that ϕ^+ and ϕ^- are one the inverse of the other, proving VI. □

If we accept that any two diagrams of an oriented framed link can be transformed into one another by isotopies that do not change the events in the diagram as well as the Reidemeister moves II and III and the moves IV, V, VI, then we have the following consequence of Proposition 8.2.

Corollary 8.3. *Let* $\mathbf{V}$ *be a coloring of the framed oriented link* L *by representations of the quasitriangular Hopf algebra* A. *Then* $<\mathbf{V}(L)>$ *is a link invariant.*

Theorem 8.3. *Let* L *be an oriented framed link with link components* $L_1, L_2, \ldots, L_n$ *and let*

$$\mathbf{V} : \{L_1, L_2, \ldots, L_n\} \to Rep(A), \quad \mathbf{V}(L_j) = V_j,$$

be a coloring of L *by representations of the ribbon Hopf algebra* A.
(i) Assume that for a link component L_j, *the representation* V_j *can be written as the direct sum* $W_1 \oplus W_2$ *of the representations* W_1 *and* W_2 *of* A. *Let* $\mathbf{V}_1$ *and* $\mathbf{V}_2$ *be the colorings of* L *obtained from* $\mathbf{V}$ *by replacing* V_j *by* W_1 *respectively* W_2. *Then*

$$<\mathbf{V}(L)>=<\mathbf{V}_1(L)> + <\mathbf{V}_2(L)> .$$

(ii) (cabling principle) Assume that for a link component L_j, *the representation* V_j *can be written as the tensor product* $W_1 \otimes W_2$ *of the representations* W_1 *and* W_2 *of* A. *Let* L' *be the link obtained by replacing* L_j *by* $L_j^{\|2}$, *and let* $\mathbf{V}'$ *be the coloring of* L' *obtained from the coloring of* L *by keeping the*

same colors for all components L_k, $k \neq j$, *coloring the first component of* $L_j^{\|2}$ *by* W_1 *and the second by* W_2. *Then*

$$<\mathbf{V}(L)>=<\mathbf{V}'(L')>.$$

(iii) If L' *is obtained from* L *by changing the orientation of one link component and* $\mathbf{V}'$ *is obtained from* $\mathbf{V}$ *by changing the color of that respective link component to its dual, then*

$$<\mathbf{V}(L)>=<\mathbf{V}'(L')>.$$

(iv) If V_j *is the trivial representation let* L' *be the link obtained from* L *by deleting* L_j *and let* $\mathbf{V}'$ *be the restriction of* $\mathbf{V}$ *to* $\mathbf{L}'$. *Then*

$$<\mathbf{V}(L)>=<\mathbf{V}'(L')>.$$

Proof. The proof is inspired by [Kirby and Melvin (1991)]. Part (i) follows from the fact that an element $a \in A$ acts on a direct sum $V \oplus W$ by

$$a \cdot (x \oplus y) = (ax) \oplus (ay).$$

Then the homomorphisms defined by elementary events (Figure 8.12) split with respect to the direct sums. For a crossing with strands colored by V and $W = W_1 \oplus W_2$, using the canonical identification given by the distributivity of the tensor product $V \otimes (W_1 \oplus W_2) = V \otimes W_1 \oplus V \otimes W_2$, we have a natural splitting of the endomorphism of $V \oplus W$ defined by R and the endmorphisms of $V \oplus W_1$ and $V \oplus W_2$ of R. Hence for $\check{R}$ we have the equality from Figure 8.14. The case of $\check{R}^{-1}$ is identical.

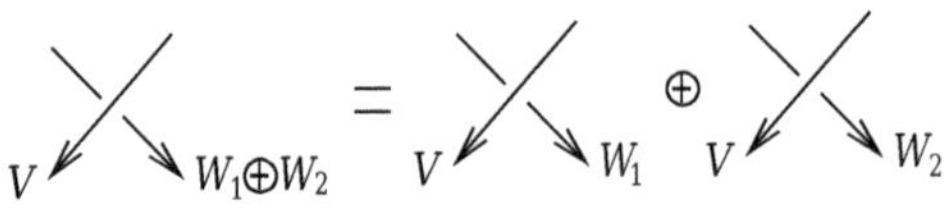

Fig. 8.14 Splitting of the homomorphism $\check{R}$

For the homomorphisms E, E_{op}, N, N_{op} we use the basis of W obtained by combining a basis for W_1 and a basis for W_2. Then $E = E|W_1 \otimes W_1^* \oplus W_2 \otimes W_2^*$, and same for the other homomorphisms. Hence the elementary homomorphisms split as direct sums. Using the distributivity of the tensor product with respect to the direct sum we obtain distributivity for general homomorphisms defined by horizontal "slices". We conclude that the homomorphism defined by the entire diagram for the coloring of the L_j component by V_j is the sum of the homomorphisms obtained by coloring L_j by W_1 and W_2. This proves (i).

Properties (ii), (iii), and (iv) are consequences of the "finer" properties of the universal R-matrix, and of the elements u and v. Note that in the proof of Theorem 8.2 we only used property (rha1) of the twist element v; it is here were the other three properties are required.

Again it suffices to check the properties just for the simple events in Figure 8.12. Let us look first at (ii). For the crossing corresponding to $\check{R}$ we have

$$\begin{aligned}\check{R}_{W_1\otimes W_2,V} &= P_{W_1\otimes W_2,V}R_{W_1\oplus W_2,V} = P_{W_1\otimes W_2,V}(\Delta\otimes I)(R)_{(W_1\otimes W_2)\otimes V}\\ &= P_{W_1\otimes W_2,V}(R_{13}R_{23})_{W_1\otimes W_2\otimes V} = \sum_{j,k}\beta_j\beta_k\otimes\alpha_j\otimes\alpha_k\\ &= (\check{R}_{W_1,V}\otimes I)(I\otimes\check{R}_{W_2,V}),\end{aligned}$$

where for the third equality we used the relation (qtha2) from the definition of a quasitriangular Hopf algebra. We obtain the equality on the left in Figure 8.15. The equality on the right of this figure stands for

$$E_{W_1\otimes W_2} = E_{W_1}(I\otimes E_{W_2}\otimes I)$$

which is an easy check.

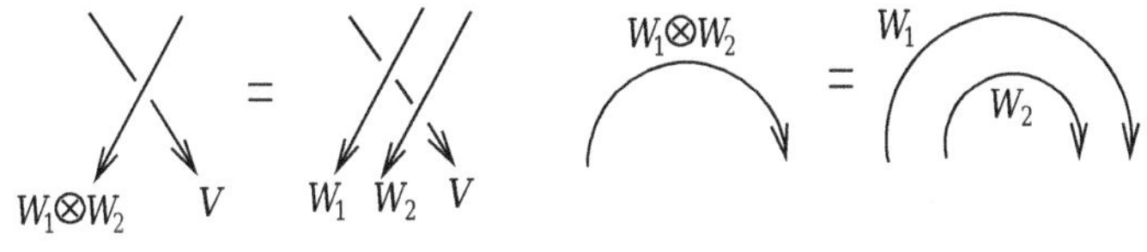

Fig. 8.15 Cabling identities for $\check{R}$ and E

The identity

$$(E_{op})_{W_1\otimes W_2} = (E_{op})_{W_1}(I\otimes(E_{op})_{W_2}\otimes I)$$

which arises when reversing the arrow in the identity on the right of Figure 8.15 requires relation (rha2). We have

$$\begin{aligned}&(E_{op})_{W_1\otimes W_2}(x_1\otimes x_2\otimes f_2\otimes f_1) = (f_1\otimes f_2)[(v^{-1}u)(x_1\otimes x_2)]\\ &\quad = (f_1\otimes f_2)[\Delta(v^{-1}u)(x_1\otimes x_2)] = (f_1\otimes f_2)[\Delta(v^{-1})\Delta(u)(x_1\otimes x_2)].\end{aligned}$$

Now $\Delta(v) = (\sigma(R)R)^{-1}(v\otimes v)$ implies that $\Delta(v^{-1}) = (v^{-1}\otimes v^{-1})(\sigma(R)R)$. Also Lemma 8.3 gives $\Delta(u) = (\sigma(R)R)^{-1}(u\otimes u)$. So the last expression is equal to

$$\begin{aligned}&(f_1\otimes f_2)[(v^{-1}\otimes v^{-1})(\sigma(R)R)(\sigma(R)R)^{-1}(u\otimes u)(x_1\otimes x_2)]\\ &\quad = (f_1\otimes f_2)[(v^{-1}\otimes v^{-1})(u\otimes u)(x_1\otimes x_2)] = f_1(v^{-1}ux_1)f_2(v^{-1}ux_2).\end{aligned}$$

And this is equal to

$$(E_{op})_{W_1}(I \otimes (E_{op})_{W_2} \otimes I)(x_1 \otimes x_2 \otimes f_2 \otimes f_1) = f_1(v^{-1}ux_1)f_2(v^{-1}ux_2).$$

proving the identity. The proof of the corresponding identities for N and N_{op} is similar. This proves (ii).

In the case of (iii), crossings are taken care by our convention on the relationship between the orientation of the strand and the representation that colors the strand; if we reverse the orientation and change the coloring representation to its dual, we still perform the same computation. There is an issue at "maxima" and "minima" of the link diagram but those are taken care of by the identity (rha3).

Finally, for (iv), let us identify the trivial representation V^0 with $\mathbb{C}$. Then

$$R : V \otimes \mathbb{C} \to V \otimes \mathbb{C}$$

is in fact the map

$$V \to V, \quad x \mapsto \sum_j \epsilon(\beta_j)\alpha_j x.$$

But $\sum_j \epsilon(\beta_j)\alpha_j = m((\epsilon \otimes I)(R)) = 1$, by Lemma 8.1. So $\check{R} : V \otimes V^0 \to V^0 \otimes V$ is just the identity operator on V. The case $\check{R} : V^0 \otimes V \to V \otimes V^0$ is similar.

On the other hand, the equality $\epsilon(v) = \epsilon(u) = 1$ implies that the maps E, E_{op}, N, N_{op} are the identity operator under the standard identification of $V^0 \otimes V^0 = \mathbb{C} \otimes \mathbb{C}$ with $\mathbb{C}$.

Hence any component colored by an irreducible representation is "invisible"; that component can be deleted. The theorem is proved. □

It would be natural to continue with the definition of **modular Hopf algebras**. But it was shown by Spencer Sterling [Sterling (2011)] that there are **no** modular Hopf algebras that model the theory of classical theta functions, so our discussion stops here. We also avoid the discussion of braided categories and focus just on the application of the Reshetikhin-Turaev invariant to the theory of theta functions. For a detailed discussion on how to construct an abstract modular category, not based on representations of modular Hopf algebras, that is associated to abelian Chern-Simons theory, the reader can consult [Sterling (2011)].

8.2 The quantum group associated to classical theta functions

It is now time to return to theta functions. We will model the space of theta functions, the action of the Heisenberg group, and the action of the mapping class group using representations of a quantum group.

8.2.1 *The quantum group and its representation theory*

The construction of the quantum group

The quantum group used in our model was introduced briefly in [Murakami *et al.* (1992)]. A detailed description of this group and its application to theta functions was given in [Gelca and Hamilton (2012)], which we now outline.

In our discussion from §1.3, we pointed out that Edward Witten related the linking number to the Chern-Simons theory with gauge group $U(1)$, so the quantum group should somehow be related to this Lie group. There is a general approach of Drinfeld and Jimbo, in which the quantum group is obtained by deforming the universal enveloping algebra of the complexification of the Lie algebra of the given Lie group. Their construction works only when the Lie algebra is semisimple, and this is not the case with $\mathfrak{u}(1)$. The construction we give below is nevertheless inspired by Drinfeld and Jimbo, and mostly by [Kirillov and Reshetikhin (1989)] and [Reshetikhin and Turaev (1991)]. A detailed account of how their work inspired our construction can be found in [Gelca and Hamilton (2012)].

Here is how the quantum group is constructed. We start with the Lie algebra $\mathfrak{u}(1)$, then complexify it:

$$\mathfrak{u}_{\mathbb{C}}(1) = \mathfrak{u}(1) \otimes_{\mathbb{R}} \mathbb{C} = \mathbb{C}.$$

The universal enveloping algebra of this is

$$\mathcal{U}(\mathfrak{u}_{\mathbb{C}}(1)) = \mathbb{C}[H]$$

where H corresponds to $1 \in \mathbb{C} = \mathfrak{u}_{\mathbb{C}}(1)$. Here $\mathbb{C}[H]$ denotes the polynomial algebra in the variable H. The quantum enveloping algebra of this algebra is

$$\mathcal{U}_h(\mathfrak{u}_{\mathbb{C}}(1)) := \mathbb{C}[H][[h]]$$

the algebra of formal power series in the abstract variable h (Planck's constant) with coefficients in H. It is not hard to check that the multiplication and unit of this algebra together with the comultiplication defined by

$\Delta(H) = H \otimes 1 + 1 \otimes H$, counit defined by $\epsilon(H) = 0$, and antipode defined by $S(H) = -H$ yield a Hopf algebra structure on $\mathcal{U}_h(\mathfrak{u}_{\mathbb{C}}(1))$.

Next, let $K = e^{\frac{hH}{4}}$ and $q = e^{\frac{h}{2}}$ and define $\mathcal{U}_q(\mathfrak{u}_{\mathbb{C}}(1))$ as the quotient

$$\mathcal{U}_q(\mathfrak{u}_{\mathbb{C}}(1)) := \mathbb{C}[K, K^{-1}]\rangle[q, q^{-1}]/(KK^{-1} = 1 = K^{-1}K).$$

This is a Hopf subalgebra of $\mathcal{U}_h(\mathfrak{u}_{\mathbb{C}}(1))$, which would be the quantum group associated to $U(1)$ when the Planck's constant is an abstract variable. To obtain the quantum group when $h = \frac{1}{N}$ with N an even integer, which is the case throughout this book, we make the following definition.

Definition 8.8. The quantum group $\mathcal{U}_t(\mathfrak{u}_{\mathbb{C}}(1))$ is defined as the quotient of the quantum group $\mathcal{U}_q(\mathfrak{u}_{\mathbb{C}}(1))$ by the relations

$$K^{2N} = 1, \quad q = t^2 = e^{\frac{2\pi i}{N}}.$$

Remark 8.3. The quantum group $\mathcal{U}_t(\mathfrak{u}_{\mathbb{C}}(1))$ may be identified with $\mathbb{C}[\mathbb{Z}_{2N}]$, the group algebra of $\mathbb{Z}_{2N}$, by identifying K with the generator of $\mathbb{Z}_{2N}$. In other words

$$\mathcal{U}_t(\mathfrak{u}_{\mathbb{C}}(1)) = \mathbb{C}[K]/(K^{2N} = 1).$$

Theorem 8.4. *$\mathcal{U}_t(\mathfrak{u}_{\mathbb{C}}(1)) = \mathbb{C}[\mathbb{Z}_{2N}]$ is a Hopf algebra with the multiplication and unit of the group algebra $\mathbb{C}[\mathbb{Z}_{2N}]$, and with comultiplication, counit, and antipode defined by*

$$\Delta(K^j) = K^j \otimes K^j, \quad \epsilon(K^j) = 1, \quad S(K^j) = K^{-j}, \quad j = 0, 1, \ldots, N-1.$$

Proof. Coassociativity follows by noticing that

$$(\Delta \otimes I)\Delta(K^j) = K^j \otimes K^j \otimes K^j = (I \otimes \Delta)\Delta(K^j), \text{ for all } j.$$

Also

$$(\epsilon \otimes I)\Delta(K^j) = K^j = (I \otimes \epsilon)\Delta(K^j).$$

This shows that it is a coalgebra. To check that it is a bialgebra, we have

$$\Delta m(K^j \times K^k) = \Delta(K^{j+k}) = K^{j+k} \otimes K^{j+k}$$

and

$$(m \otimes m)(I \otimes \sigma \otimes I)(\Delta \otimes \Delta)(K^j \times K^k)$$
$$= (m \otimes m)(K^j \otimes K^k \otimes K^j \otimes K^k) = K^{j+k} \otimes K^{j+k}.$$

Then $(\epsilon \otimes \epsilon)(K^j \otimes K^k) = 1 = \epsilon(K^{j+k})$, $\Delta(1) = \Delta(K^0) = K^0 \otimes K^0 = 1 \otimes 1$, and $\epsilon(1) = 1$.

The requirement that this bialgebra is a Hopf algebra is checked as follows

$$m(I \otimes S)(\Delta(K^j) = m(I \otimes S)(K^j \otimes K^j) = m(K^j \otimes K^{-j}) = 1$$

and

$$m(S \otimes I)(\Delta(K^j) = m(S \otimes I)(K^j \otimes K^j) = m(K^{-j} \otimes K^j) = 1. \qquad \square$$

The representation theory

The irreducible representations of $\mathcal{U}_t(\mathfrak{u}_{\mathbb{C}}(1)) = \mathbb{C}[\mathbb{Z}_{2N}]$ are

$$V^j; \quad j = 0, 1, \ldots, 2N-1,$$

where $V^j \cong \mathbb{C}$ and K acts by

$$K \cdot v = t^j v = e^{\frac{\pi i}{N}} v.$$

We denote by e_j the canonical basis element of V^j.

Because of the Hopf algebra structure, finite-dimensional representations form a ring (with the underlying abelian group defined á la Grothendieck from the monoid whose addition is the direct sum of representations) in which the product is the tensor product of representations. Since $(V^1)^{\otimes j} \cong V^j$, the representation ring is

$$\mathrm{Rep}(\mathcal{U}_t(\mathfrak{u}_{\mathbb{C}}(1)) = \mathbb{C}[V^1] / \left((V^1)^{2N} - 1\right) = \mathbb{C}[\mathbb{Z}_{2N}].$$

The fact that, in this case, the representation ring coincides with the quantum group, is purely coincidental. Since $\mathcal{U}_t(\mathfrak{u}_{\mathbb{C}}(1))$ is a Hopf algebra, this implies that the dual space of each representation is itself a representation. It is easy to see that there are natural isomorphisms

$$D : (V^j)^* \to V^{2N-j}, \quad De^j = e_j, \quad j = 1, \ldots, 2N-1,$$

where the functional e^j is defined by $e^j(e_j) = 1$.

In what follows, we will explain how $\mathcal{U}_t(\mathfrak{u}_{\mathbb{C}}(1))$ may be given the structure of a ribbon Hopf algebra. Most of our computations are based on the simple fact that if ζ is a kth root of unity, then

$$1 + \zeta + \zeta^2 + \cdots + \zeta^{k-1} = \begin{cases} k \text{ if } \zeta = 1 \\ 0 \text{ if } \zeta \neq 1. \end{cases} \tag{8.6}$$

8.2.2 *The quantum group of theta functions is a quasitriangular Hopf algebra*

The representations of the quantum group will be used for coloring oriented framed links in a 3-dimensional manifold. The manifold will be S^3, H_g, or $\Sigma_g \times [0,1]$.

We normalize the entire theory so that an oriented framed knot in the skein theory should correspond to the knot colored by the irreducible representation V^1. Then $K^{\| n}$ corresponds to K colored by $(V^1)^n = V^n$, where the exponent is taken modulo $2N$.

We want to model the crossing of n parallel strands by m parallel strands. In view of the discussion from §8.1.4 and especially the cabling principle of Theorem 8.3, this should be modeled by

$$\check{R} : V^m \otimes V^n \to V^n \otimes V^m.$$

Examining Figure 8.16, we deduce that this homomorphism between representations should be multiplication by t^{mn} for all $m, n \in \{0, 1, \ldots, 2N-1\}$.

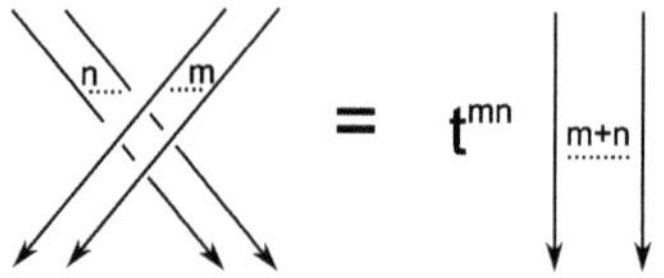

Fig. 8.16 The skein relation for the crossing of n strands by m strands

The derivation of the formula for R

We claim that there is a universal R-matrix that gives rise to this family of operators $\check{R}$. The universal R-matrix should be of the form

$$R = \sum_{j,k=0}^{2N-1} c_{jk} K^j \otimes K^k.$$

Let us compute the coefficients c_{jk}. Because K^j acts as $t^{jn} I$ on V^n for all j and n, we obtain the system of equations

$$\sum_{j,k=0}^{2N-1} c_{jk} t^{mj} t^{nk} = t^{mn}, \quad m, n \in \{0, 1, \ldots, 2N-1\}. \tag{8.7}$$

If $\mathcal{T}$ is the matrix whose mnth entry is t^{mn} and $C = (c_{jk})$, then this equation becomes

$$\mathcal{T} C \mathcal{T} = \mathcal{T}$$

and hence we find that $C = \mathcal{T}^{-1}$. Since t is a primitive $2N$th root of unity, it follows after a few computations using (8.6) that

$$c_{jk} = \frac{1}{2N} t^{-jk}. \tag{8.8}$$

Hence, we arrive at the following formula for the universal R-matrix:

$$R = \frac{1}{2N} \sum_{j,k \in \mathbb{Z}_{2N}} t^{-jk} K^j \otimes K^k. \tag{8.9}$$

Note that this formula for R implies that the R-matrix is symmetric in the sense that $\sigma(R) = R$.

The proof that the quantum group is a quasitriangular Hopf algebra

We now prove that indeed the element R derived above has all the properties of a universal R-matrix for Hopf algebra $\mathcal{U}_t(\mathfrak{u}_{\mathbb{C}}(1))$.

Theorem 8.5. *$(\mathcal{U}_t(\mathfrak{u}_{\mathbb{C}}(1)), R)$ is a quasi-triangular Hopf algebra.*

Proof. We must check that R is invertible, and that it has the following properties

(qtha1) $\sigma \circ \Delta(a) = R\Delta(a)R^{-1}$, for all a;
(qtha2) $(\Delta \otimes I)(R) = R_{13}R_{23}$;
(qtha3) $(I \otimes \Delta)(R) = R_{13}R_{12}$.

Let us show that R is invertible by finding its inverse. By Lemma 8.1, the formula for R^{-1} should be

$$R^{-1} = (S \otimes I)(R) = \frac{1}{2N} \sum_{j,k \in \mathbb{Z}_{2N}} t^{-jk} K^{-j} \otimes K^{k}.$$

We rewrite this as

$$R^{-1} = \frac{1}{2N} \sum_{i,j \in \mathbb{Z}_{2N}} t^{jk} K^{-j} \otimes K^{-k}. \tag{8.10}$$

We may check that this element is inverse to R as follows:

$$\begin{aligned} RR^{-1} &= \frac{1}{4N^2} \sum_{j,j',k,k' \in \mathbb{Z}_{2N}} t^{j'k'-jk} K^{j-j'} \otimes K^{k-k'} \\ &= \frac{1}{4N^2} \sum_{m,n \in \mathbb{Z}_{2N}} \left(\sum_{j,k \in \mathbb{Z}_{2N}} t^{(j-m)(k-n)-jk} \right) K^{m} \otimes K^{n}. \end{aligned}$$

But

$$\sum_{j,k \in \mathbb{Z}_{2N}} t^{(j-m)(k-n)-jk} = t^{mn} \left(\sum_{j \in \mathbb{Z}_{2N}} t^{-jn} \right) \left(\sum_{k \in \mathbb{Z}_{2N}} t^{-km} \right).$$

By (8.6), this is zero unless $m = n = 0$, in which case it is equal to $4N^2$. Hence $RR^{-1} = 1 \otimes 1$. We conclude that R is invertible, with R^{-1} given by the above formula.

The identity (qtha1) follows trivially from the fact that $\mathcal{U}_t(\mathfrak{u}_{\mathbb{C}}(1)) = \mathbb{C}[\mathbb{Z}_{2N}]$ is both commutative and co-commutative.

Finally, let us establish (qtha2) and (qtha3). The two are similar and we only check (qtha3). Computing directly from the formula for R, we have

$$R_{13}R_{12} = \frac{1}{4N^2} \sum_{j,j',k,k' \in \mathbb{Z}_{2N}} t^{-jk-j'k'} K^{j+j'} \otimes K^{k'} \otimes K^k$$

$$= \frac{1}{4N^2} \sum_{m \in \mathbb{Z}_{2N}} \sum_{j,k,k' \in \mathbb{Z}_{2N}} t^{-jk-(m-j)k'} K^m \otimes K^{k'} \otimes K^k$$

$$= \frac{1}{4N^2} \sum_{m,k,k' \in \mathbb{Z}_{2N}} t^{-mk'} \left(\sum_{j \in \mathbb{Z}_{2N}} t^{j(k'-k)} \right) K^m \otimes K^{k'} \otimes K^k.$$

Now by (8.6), $\sum_{j \in \mathbb{Z}_{2N}} t^{j(k'-k)}$ is equal to zero, unless $k' = k$, in which case it is equal to $2N$. Hence,

$$R_{13}R_{12} = \frac{1}{2N} \sum_{m,k \in \mathbb{Z}_{2N}} t^{-mk} K^m \otimes K^k \otimes K^k = (I \otimes \Delta)(R)$$

which establishes (qtha3) and we are done. □

We derive the following formula for u (see §8.1.3):

$$u = \frac{1}{2N} \sum_{j,k \in \mathbb{Z}_{2N}} t^{-jk} K^{j-k}. \tag{8.11}$$

8.2.3 *The quantum group of theta functions is a ribbon Hopf algebra*

We want to prove the existence of a universal twist v that defines on $(\mathcal{U}_t(\mathfrak{u}_{\mathbb{C}}(1)), R)$ a ribbon Hopf algebra structure.

Since the algebra is commutative, the condition of v being in the center is superfluous. The maps

$$\phi^+ : V^k \to V^k, \quad \phi^+(x) = v^{-1}x$$

and

$$\phi^- : V^k \to V^k, \quad \phi^-(x) = vx$$

are intended to model a positive twist respectively a negative twist, as in Figure 8.17.

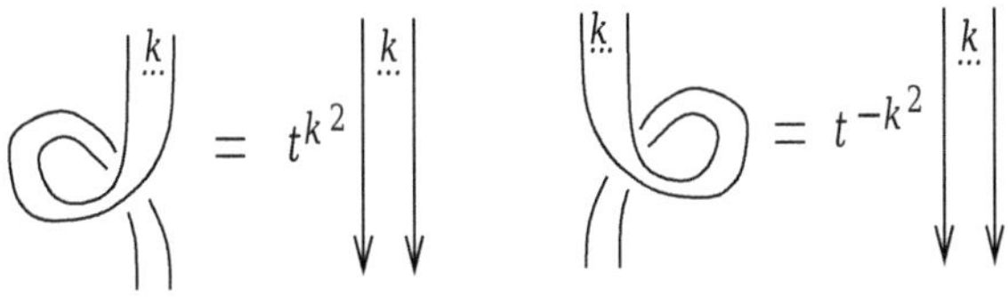

Fig. 8.17 The skein relation for a positive twist of k strands

The derivation of the formula for v

If we write $v = \sum_{j\in\mathbb{Z}_{2N}} c_j K^j$, then the above observation leads us to the system of equations

$$\sum_{j\in\mathbb{Z}_{2N}} c_j t^{jk} = t^{-k^2}, \quad k \in \mathbb{Z}_{2N} \tag{8.12}$$

for the coefficients c_j. From the formula (8.8) for the matrix $\mathcal{T}^{-1}$ derived in §8.2.2, we find

$$c_j = \frac{1}{2N} \sum_{k\in\mathbb{Z}_{2N}} t^{-jk} t^{-k^2},$$

which yields

$$v = \frac{1}{2N} \sum_{j,k\in\mathbb{Z}_{2N}} t^{-k(k+j)} K^j. \tag{8.13}$$

Before proceeding further, we determine a simpler expression for v. Note that

$$\sum_{k\in\mathbb{Z}_{2N}} t^{-(k+j)k} = \sum_{k=0}^{N-1} \left(t^{-(k+j)k} + t^{-(k+N+j)(k+N)} \right)$$
$$= \sum_{k=0}^{N-1} t^{-(k+j)k} (1 + t^{-2kN-jN-N^2}) = (1 + (-1)^{j+N}) \sum_{k=0}^{N-1} t^{-(k+j)k}.$$

Hence

$$v = \frac{1}{2N} \sum_{j\in\mathbb{Z}_{2N}} \left[(1 + (-1)^{j+N}) \sum_{k=0}^{N-1} t^{-(k+j)k} K^j \right]$$
$$= \frac{1}{N} \sum_{j=0}^{N-1} \sum_{k=0}^{N-1} t^{-(k+2j+N)k} K^{2j+N} = \frac{1}{N} \sum_{j,k\in\mathbb{Z}_N} (-1)^k t^{-(k+2j)k} K^{2j+N}$$
$$= \frac{1}{N} \sum_{j,k\in\mathbb{Z}_N} (-1)^{k-j} t^{-(k+j)(k-j)} K^{2j+N}$$

yielding the formula

$$v = \frac{1}{N}\left(\sum_{k\in\mathbb{Z}_N}(-1)^k t^{-k^2}\right)\left(\sum_{j\in\mathbb{Z}_N}(-1)^j t^{j^2}K^{2j+N}\right). \tag{8.14}$$

A similar computation starting with the negative twist yields

$$v^{-1} = \frac{1}{N}\left(\sum_{k\in\mathbb{Z}_N}(-1)^k t^{k^2}\right)\left(\sum_{j\in\mathbb{Z}_N}(-1)^j t^{-j^2}K^{2j+N}\right).$$

Let us check that

$$v^{-1}v = vv^{-1} = 1.$$

Firstly

$$\left(\sum_{k\in\mathbb{Z}_N}(-1)^k t^{k^2}\right)\left(\sum_{k'\in\mathbb{Z}_N}(-1)^{k'} t^{-k'^2}\right) = \sum_{k,k'\in\mathbb{Z}_N}(-1)^{k+k'}t^{(k-k')(k+k')}$$

$$= \sum_{k,k'\in\mathbb{Z}_N}(-1)^k t^{k(k+2k')} = \sum_{k\in\mathbb{Z}_N}\left[(-1)^k t^{k^2}\sum_{k'\in\mathbb{Z}_N}t^{2kk'}\right] = N$$

where the last equality follows from (8.6). Hence

$$\begin{aligned} v^{-1}v &= \frac{1}{N}\left(\sum_{j\in\mathbb{Z}_N}(-1)^j t^{-j^2}K^{2j+N}\right)\left(\sum_{j'\in\mathbb{Z}_N}(-1)^{j'} t^{j'^2}K^{2j'+N}\right)\\ &= \frac{1}{N}\sum_{j,j'\in\mathbb{Z}_N}(-1)^{j+j'}t^{j'^2-j^2}K^{2j+2j'}\\ &= \frac{1}{N}\sum_{m\in\mathbb{Z}_N}\sum_{j\in\mathbb{Z}_N}(-1)^m t^{(m-j)^2-j^2}K^{2m}\\ &= \frac{1}{N}\sum_{m\in\mathbb{Z}_N}(-1)^m t^{m^2}\left(\sum_{j\in\mathbb{Z}_N}t^{-2mj}\right)K^{2m} = 1, \end{aligned}$$

where the last line follows from (8.6), proving that v and v^{-1} are the inverse of each other.

The proof that the quantum group is a ribbon Hopf algebra

Theorem 8.6. *$(\mathcal{U}_t(\mathfrak{u}_{\mathbb{C}}(1)), R, v)$ is a ribbon Hopf algebra.*

Proof. We must check the following identities

(rha1) $v^2 = S(u)u$, where u is given by (8.11);
(rha2) $\Delta(v) = (\sigma(R)R)^{-1}(v \otimes v)$;
(rha3) $S(v) = v$;
(rha4) $\epsilon(v) = 1$.

Using (8.13) we have

$$v^2 = \frac{1}{4N^2} \sum_{j,k,l,m \in \mathbb{Z}_{2N}} t^{-k(j+k)-l(m+l)} K^j K^m$$

which after the change $j \mapsto j-k$, $m \mapsto m-l$ becomes

$$v^2 = \frac{1}{4N^2} \sum_{j,k,l,m \in \mathbb{Z}_{2N}} t^{-kj-lm} K^{j-k+l-m}.$$

On the other hand, using (8.11) we have

$$\begin{aligned} uS(u) &= \frac{1}{2N} \sum_{j,k \in \mathbb{Z}_{2N}} t^{-jk} K^{j-k} \frac{1}{2N} \sum_{m,l \in \mathbb{Z}_{2N}} t^{-ml} S(K^{m-l}) \\ &= \frac{1}{4N^2} \sum_{j,k,l,m \in \mathbb{Z}_{2N}} t^{-jk} t^{-ml} K^{j-k} K^{l-m} \\ &= \frac{1}{4N^2} \sum_{j,k,l,m \in \mathbb{Z}_{2N}} t^{-jk-ml} K^{j-k+l-m}. \end{aligned}$$

This proves (rha1).

Because $\sigma(R) = R$, $(\sigma(R)R)^{-1} = R^{-2}$. To show (rha2) we need to compute R^{-2}. From (8.9) we calculate

$$\begin{aligned} R^{-2} &= \frac{1}{4N^2} \sum_{j,j',k,k' \in \mathbb{Z}_{2N}} t^{jk+j'k'} K^{-j-j'} \otimes K^{-k-k'} \\ &= \frac{1}{4N^2} \sum_{m,n \in \mathbb{Z}_{2N}} \sum_{j,k \in \mathbb{Z}_{2N}} t^{jk+(m-j)(n-k)} K^{-m} \otimes K^{-n} \\ &= \frac{1}{4N^2} \sum_{m,n,k \in \mathbb{Z}_{2N}} t^{m(n-k)} \left(\sum_{j \in \mathbb{Z}_{2N}} t^{(2k-n)j} \right) K^{-m} \otimes K^{-n} \end{aligned}$$

$$= \frac{1}{2N} \sum_{m,k \in \mathbb{Z}_{2N}} t^{mk} K^{-m} \otimes K^{-2k} = \frac{1}{2N} \sum_{m,k \in \mathbb{Z}_{2N}} t^{mk} K^{m} \otimes K^{2k} =$$
$$= \frac{1}{2N} \sum_{m \in \mathbb{Z}_{2N}} \sum_{k=0}^{N-1} (t^{mk} + t^{m(k+N)}) K^m \otimes K^{2k}$$
$$= \frac{1}{2N} \sum_{m \in \mathbb{Z}_{2N}} \left[(1 + (-1)^m) \sum_{k=0}^{N-1} t^{mk} K^m \otimes K^{2k} \right]$$
$$= \frac{1}{N} \sum_{m=0}^{N-1} \sum_{k=0}^{N-1} t^{2mk} K^{2m} \otimes K^{2k},$$

where line 4 follows from (8.6). This yields

$$R^{-2} = \frac{1}{N} \sum_{j,k \in \mathbb{Z}_N} t^{2jk} K^{2j} \otimes K^{2k}. \tag{8.15}$$

Next, using (8.14), we may write

$$v \otimes v = \frac{1}{N^2} \Big(\sum_{r \in \mathbb{Z}_N} (-1)^r t^{-r^2} \Big)^2 \Big(\sum_{j,k \in \mathbb{Z}_N} (-1)^{j+k} t^{j^2+k^2} K^{2j+N} \otimes K^{2k+N} \Big).$$

From this and (8.15), we obtain that $R^{-2}(v \otimes v)$ is equal to

$$\frac{1}{N^3} \Big(\sum_{r \in \mathbb{Z}_N} (-1)^r t^{-r^2} \Big)^2 \sum_{m,n,j,k \in \mathbb{Z}_N} (-1)^{j+k} t^{2mn+j^2+k^2} K^{2m+2j+N} \otimes K^{2n+2k+N}$$
$$= \frac{1}{N^3} \Big(\sum_{r \in \mathbb{Z}_N} (-1)^r t^{-r^2} \Big)^2 \sum_{s,s' \in \mathbb{Z}_N} \Big(\sum_{j,k \in \mathbb{Z}_N} (-1)^{j+k} t^{2(s-j)(s'-k)+j^2+k^2} \Big)$$
$$\times K^{2s+N} \otimes K^{2s'+N}$$
$$= \frac{1}{N^3} \Big(\sum_{r \in \mathbb{Z}_N} (-1)^r t^{-r^2} \Big)^2 \sum_{s,s' \in \mathbb{Z}_N} t^{2ss'} \Big(\sum_{j,k \in \mathbb{Z}_N} (-1)^{j+k} t^{(j+k)^2 - 2sk - 2s'j} \Big)$$
$$\times K^{2s+N} \otimes K^{2s'+N}.$$

Now consider the coefficient

$$\sum_{j,k \in \mathbb{Z}_N} (-1)^{j+k} t^{(j+k)^2 - 2sk - 2s'j} = \sum_{j,k \in \mathbb{Z}_N} (-1)^j t^{j^2 - 2sk - 2s'(j-k)}$$
$$= \left(\sum_{j \in \mathbb{Z}_N} (-1)^j t^{j^2 - 2s'j} \right) \left(\sum_{k \in \mathbb{Z}_N} t^{2(s'-s)k} \right).$$

By (8.6), the right-hand factor is zero, unless $s = s'$, in which case it is equal to N. The left-hand factor is

$$\sum_{j \in \mathbb{Z}_N} (-1)^j t^{j^2 - 2s'j} = t^{-s'^2} \sum_{j \in \mathbb{Z}_N} (-1)^j t^{(j-s')^2} = (-1)^{s'} t^{-s'^2} \sum_{j \in \mathbb{Z}_N} (-1)^j t^{j^2}.$$

Consequently, $R^{-2}(v \otimes v)$ equals

$$\frac{1}{N^2}\Big(\sum_{r\in\mathbb{Z}_N}(-1)^r t^{-r^2}\Big)^2\Big(\sum_{j\in\mathbb{Z}_N}(-1)^j t^{j^2}\Big)\Big(\sum_{s\in\mathbb{Z}_N}(-1)^s t^{s^2} K^{2s+N}\otimes K^{2s+N}\Big)$$
$$=\frac{1}{N}\Big(\sum_{r\in\mathbb{Z}_N}(-1)^r t^{-r^2}\Big)\Big(\sum_{s\in\mathbb{Z}_N}(-1)^s t^{s^2} K^{2s+N}\otimes K^{2s+N}\Big)=\Delta(v),$$

where on the last line we have used the fact that $\Delta(K^j) = K^j \otimes K^j$. This proves (rha2).

Using (8.13), we have

$$S(v)=\frac{1}{2N}\sum_{j,k\in\mathbb{Z}_{2N}} t^{-k(k+j)}S(K^j)=\frac{1}{2N}\sum_{j,k\in\mathbb{Z}_{2N}} t^{-k(k+j)}K^{-j}$$
$$=\frac{1}{2N}\sum_{j,k\in\mathbb{Z}_{2N}} t^{-(-k)(-k-j)}K^{j}=\frac{1}{2N}\sum_{j,k\in\mathbb{Z}_{2N}} t^{-k(k+j)}K^{j}=v.$$

This proves (rha3).

Finally, for (rha4), we have

$$\epsilon(v)=\frac{1}{2N}\sum_{j,k\in\mathbb{Z}_{2N}} t^{-k(k+j)}\epsilon(K^j)=\frac{1}{2N}\sum_{j,k\in\mathbb{Z}_{2N}} t^{-k(k+j)}$$
$$=\frac{1}{2N}\sum_{k\in\mathbb{Z}_{2N}} t^{-k^2}\sum_{j\in\mathbb{Z}_{2N}} t^{-kj}.$$

By (8.6) the second sum is 0 unless $k = 0$, in which case it is equal to $2N$. Hence the result of the computation is $\frac{1}{2N}\cdot 2N = 1$. This proves (rha4) and we are done. □

Remark 8.4. There is a more restrictive notion, of a modular Hopf algebra, that allows us to connect quantum groups to modular functors. The reader can find details about this construction in [Turaev (1994)]. Unfortunately the quantum group constructed here does not satisfy all requirements to be modular (see [Sterling (2011)]). This is the reason why the book does not discuss that.

8.3 Modeling theta functions using the quantum group

It is time to model the theory of theta functions using the Reshetikhin-Turaev invariant associated to the quantum group $\mathcal{U}_t(\mathfrak{u}_{\mathbb{C}}(1))$.

Notation: In this unit we let $A_t := \mathcal{U}_t(\mathfrak{u}_{\mathbb{C}}(1))$.

At the heart of the construction lie oriented framed links in the genus g handlebody H_g, and in the cylinder over the genus g surface $\Sigma_g \times [0,1]$, links which are colored by representations of A_t. To such colored links we will associate theta functions and linear operators acting on them using the Reshetikhin-Turaev invariant. This is done by establishing a correspondence with the skein theory of theta functions developed in Chapter 5.

8.3.1 *The relationship between the linking number and the quantum group*

In §8.2 we constructed the quantum group that corresponds to the skein theory of the linking number. We now make that correspondence explicit.

Let L be an oriented framed link in a compact, oriented, smooth, 3-dimensional manifold M with link components $L_1, L_2, \ldots, L_n$. Consider a coloring

$$\mathbf{V} : \{L_1, L_2, \cdots, L_n\} \to \mathrm{Rep}(A_t)$$

of L by representations of A_t. If necessary, we might consider the more general case of the coloring by elements of the representation ring, thus allowing differences of representations as well. As before, let $\mathbf{V}(L)$ be the information consisting of the link L and the coloring $\mathbf{V}$.

Definition 8.9. Given an oriented compact 3-dimensional manifold M, we define $\mathcal{V}_{A_t}(M)$ to be the vector space whose basis consists of isotopy classes of oriented framed links in the interior of M, whose components are colored by **irreducible** representations of A_t.

There is a map

$$\mathcal{V}_{A_t}(M) \to \mathcal{L}_N(M), \tag{8.16}$$

defined by the cabling formula which replaces any link component colored by the irreducible representation V^j with j parallel copies of that component.

If $L \subset S^3$, consider the Reshetikhin-Turaev invariant

$$<\mathbf{V}(L)>$$

defined in 8.1.4, which we recall, is a topological invariant of framed oriented links.

Furthermore, since this invariant is distributive with respect to direct sums of representations, its definition extends to links that are colored by elements of the representation ring of our quantum group A_t. We may describe this extension explicitly as follows. If L is a link whose components $L_1, \ldots, L_m$ are colored by elements of the representation ring of A_t,

$$\mathbf{V} : L_j \mapsto \sum_{k=0}^{2N-1} c_{jk} V^k, \quad 1 \leq j \leq m; \tag{8.17}$$

then

$$< \mathbf{V}(L) >= \sum_{k_1=0}^{2N-1} \sum_{k_2=0}^{2N-1} \cdots \sum_{k_m=0}^{2N-1} c_{1k_1} c_{2k_2} \cdots c_{mk_m} < \mathbf{V}_{k_1,k_2,\ldots,k_m}(L) >;$$

where $\mathbf{V}_{k_1,k_2,\ldots,k_m}$ is the coloring of L that decorates the component L_j with the color V^{k_j}.

Theorem 8.7. *The following diagram commutes*

$$\begin{array}{ccc} \mathcal{V}_{A_t}(S^3) & & \\ {\scriptstyle \text{cabling}}\downarrow & \searrow^{\text{invariant}} & \\ \mathcal{L}_N(S^3) & = & \mathbb{C} \end{array}$$

where the map on the right assigns to a colored link $\mathbf{V}(L)$ its Reshetikhin-Turaev invariant $<\mathbf{V}(L)>$, and that on the left is the cabling map (8.16).

Proof. This is a basic consequence of our construction of the quantum group $A_t := \mathcal{U}_t(\mathfrak{u}_{\mathbb{C}}(1))$ which we carried out in §8.2.

Present the colored link diagram as a composition of horizontal slices, each containing exactly one event from Figure 8.12. Using the canonical basis elements e_j of V^j, the representation associated to each horizontal line may be identified with $\mathbb{C}$. Consequently, the homomorphism assigned to each horizontal slice is just multiplication by some complex number.

For the first event in Figure 8.12 with the "over" strand decorated by V^m and the "under" strand decorated by V^n, we have the following possibilities:

- if both strands are oriented downwards, then the complex number is t^{mn},
- if both strands are oriented to the right, then the number is $t^{(2N-m)n} = t^{-mn}$,
- if both strands are oriented upwards, then the number is $t^{(2N-m)(2N-n)} = t^{mn}$,
- if both strands are oriented to the left, then the number is $t^{(2N-m)n} = t^{-mn}$.

This is precisely what we get when we unlink n strands crossed by m strands in the linking number situation, precisely because we constructed the R matrix to satisfy this condition. A similar analysis for the second event with the same orientations yields respectively the numbers $t^{-mn}, t^{mn}, t^{-mn}, t^{mn}$, in agreement with what the linking number skein relations yield.

For the third event, if the strand is oriented from left to right and decorated by V^l, then $E(e^l \otimes e_l) = e^l(e_l) = 1$, so if we identify $(V^l)^* \otimes V^l$ with $\mathbb{C} \otimes \mathbb{C} = \mathbb{C}$, then E is just multiplication by 1. We recall that here e_l is the basis element of V^l.

If the strand is oriented from right to left, then the operator is E_{op}, for which the situation is slightly more complicated. We first compute

$$ue_l = \frac{1}{2N} \sum_{j,k \in \mathbb{Z}_{2N}} t^{-jk} K^{j-k} e_l = \frac{1}{2N} \sum_{j,k \in \mathbb{Z}_{2N}} t^{-jk} t^{(j-k)l} e_l$$
$$= \frac{1}{2N} \sum_{k \in \mathbb{Z}_{2N}} t^{-kl} \sum_{j \in \mathbb{Z}_{2N}} (t^{l-k})^j e_l.$$

The inner sum is zero unless $k \equiv l$. So $ue_l = t^{-l^2} e_l$. But v was constructed so that $ve_l = t^{-l^2} e_l$ and $v^{-1} e_l = t^{l^2} e_l$. Hence

$$E_{op}(e_l \otimes e^l) = e^l(v^{-1} ue_l) = 1$$

Similar computations show that the complex numbers corresponding to N and N_{op} are 1. So the homomorphisms $\mathbb{C} \to \mathbb{C}$ associated to maxima and minima are trivial. In particular trivial knot components have the quantum invariant equal to 1, as required by the skein module picture. The theorem is proved. □

Corollary 8.4. *Let* $\mathbf{V}$ *be a coloring of a link* L *in* S^3 *by irreducible representations, and suppose that some link component of* L *is colored by* V^n*, with* $0 \leq n \leq N-1$*. If* $\mathbf{V}'$ *denotes the coloring of* L *obtained by replacing the color of that link component by* V^{n+N}*, then*

$$< \mathbf{V}(L) > = < \mathbf{V}'(L) > .$$

It follows that we may factor the representation ring by the ideal generated by the single relation $V^N = 1$, without any affect on the invariants $< \mathbf{V}(L) >$. Let

$$R(A_t) := \mathbb{C}[V^1] / \left((V^1)^N - 1\right)$$

denote this quotient. Note that in $R(A_t)$,

$$(V^k)^* = V^{N-k} \quad \text{and} \quad V^m \otimes V^n = V^{m+n (\text{mod} N)};$$

where these equalities should be interpreted formally (i.e. inside $R(A_t)$). Thus, we can think of the link invariant defined above as an invariant of links colored by elements of $R(A_t)$.

Remark 8.5. One should be careful, this works for links but not for tangles.

8.3.2 *Theta functions as colored oriented framed links in a handlebody*

Consider the Heegaard decomposition

$$H_g \cup H_g = S^3.$$

This decomposition gives rise to a bilinear pairing

$$\begin{array}{rcccc} [\cdot,\cdot]_{\text{qgr}} : \mathcal{V}_{A_t}(H_g) \otimes \mathcal{V}_{A_t}(H_g) & \to & \mathcal{V}_{A_t}(S^3) & \to & \mathbb{C} \\ \mathbf{V}(L) \otimes \mathbf{V}'(L') & \mapsto & \mathbf{V}(L) \cup \mathbf{V}'(L') & \mapsto & < \mathbf{V}(L) \cup \mathbf{V}'(L') > \end{array} \tag{8.18}$$

on $\mathcal{V}_{A_t}(H_g)$. Due to the obvious diffeomorphism of S^3 that swaps the handlebodies, this pairing is symmetric. However, it is far from being nondegenerate, which leads us to the following definition.

Definition 8.10. The vector space $\mathcal{L}_{A_t}(H_g)$ is defined to be the quotient of the vector space $\mathcal{V}_{A_t}(H_g)$ by the annihilator

$$\text{Ann}(\mathcal{V}_{A_t}(H_g)) := \{x \in \mathcal{V}_{A_t}(H_g) : [x,y]_{\text{qgr}} = 0, \text{ for all } y \in \mathcal{V}_{A_t}(H_g)\}$$

of the bilinear form (8.18).

Now the pairing induced on $\mathcal{L}_{A_t}(H_g)$ by (8.18) is nondegenerate.

Proposition 8.1. *Consider the cabling map from $\mathcal{V}_{A_t}(H_g)$ to $\mathcal{L}(H_g)$ described in Definition 8.9. This map factors to an isomorphism,*

$$\mathcal{L}_{A_t}(H_g) \cong \mathcal{L}_N(H_g).$$

Proof. By Theorem 8.7, the following diagram commutes;

$$\begin{array}{ccc} \mathcal{V}_{A_t}(H_g) \otimes \mathcal{V}_{A_t}(H_g) & \xrightarrow{\text{cabling}} & \mathcal{L}_N(H_g) \otimes \mathcal{L}_N(H_g) \\ \downarrow & & \downarrow \\ \mathcal{V}_{A_t}(S^3) & \xrightarrow{\text{cabling}} & \mathcal{L}_N(S^3) \\ & \searrow^{\text{invariant}} \quad \mathbb{C} \quad \nearrow\!\!\!= & \end{array}$$

where the vertical map on the right is the pairing defined in Theorem 7.7. Since the cabling map from $\mathcal{V}_{A_t}(H_g)$ to $\mathcal{L}_N(H_g)$ is surjective, the result follows from the fact that the pairing on the right is nondegenerate, which was proved in Theorem 7.7. □

As a corollary we obtain the following result.

Theorem 8.8. *The space $\mathcal{L}_{A_t}(H_g)$ is isomorphic to the space of theta functions $\boldsymbol{\Theta}_N^{\Pi}(\Sigma_g)$.*

Proof. This is a consequence of Theorem 5.7. □

Consequently, we may represent the theta series $\theta_k^{\Pi}(\mathbf{z})$ as colored links in the handlebody H_g. More precisely, the coloring represented by Figure 8.18 of $a_1, \ldots, a_g$ by $V^{k_1}, \ldots, V^{k_g}$, $k_1, k_2, \ldots, k_g \in \{0, 1, \ldots, N-1\}$, corresponds to the theta series $\theta_{k_1,\ldots,k_g}^{\Pi}(\mathbf{z})$.

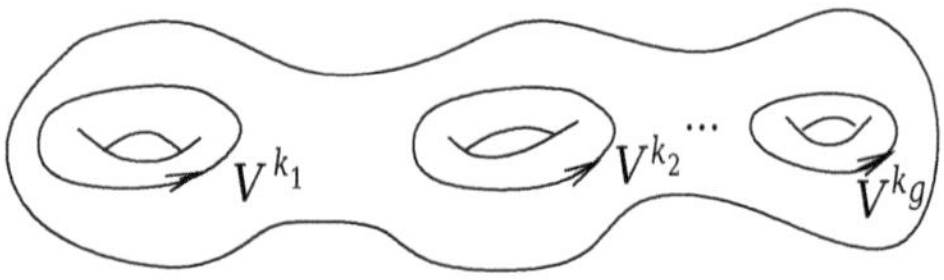

Fig. 8.18 The depiction of $\theta_{k_1,\ldots,k_g}^{\Pi}$ as a colored link in a handlebody

Remark 8.6. It is interesting to observe that theta series are modeled using only **half** of the irreducible representations of the quantum group.

8.3.3 *The Schrödinger representation and the action of the mapping class group via quantum group representations*

Let Σ_g be a genus g Riemann surface that bounds the standard handlebody H_g. Consider the spaces $\mathcal{V}_{A_t}(\Sigma_g \times [0,1])$ and $\mathcal{V}_{A_t}(H_g)$. Using the gluing identifications

$$\Sigma_g \times [0,1] \cup \Sigma_g \times [0,1] \approx \Sigma_g \times [0,1] \text{ and } \Sigma_g \times [0,1] \cup H_g \approx H_g$$

we can turn $\mathcal{V}_{A_t}(\Sigma_g \times [0,1])$ into an algebra and $\mathcal{V}_{A_t}(H_g)$ into a module over this algebra. Since the cabling maps to the corresponding reduced skein modules are equivariant with respect to this action, it follows from Proposition 8.1 that this action descends to $\mathcal{L}_{A_t}(H_g)$.

The Schrödinger representation

Endow Σ_g with a canonical basis $a_1, a_2, \ldots, a_g, b_1, b_2, \ldots, b_g$ of its first homology group, in such a way that after filling in the handlebody H_g, $a_1, a_2, \ldots, a_g$ becomes a canonical basis for the handlebody.

Using the canonical basis, associate to each element of the finite Heisenberg group $\mathbf{H}(\mathbb{Z}_N^g)$ of the form

$$(\mathbf{p}, \mathbf{q}, k), \quad \mathbf{p}, \mathbf{q} \in \{0, 1, \ldots, N-1\}, k \in \{0, 1, \ldots, 2N-1\}$$

an oriented framed multicurve $(\mathbf{p}, \mathbf{q})$ on Σ_g as explained in Chapter 3. Place this multicurve in $\Sigma_g \times [0,1]$ by identifying Σ with $\Sigma \times \{1/2\}$. We obtain an oriented framed link in $\Sigma_g \times [0,1]$. Add to it a trivial framed link component, with framing twisted k times. Color all link components by the representation V^1 of A_t. Denote the resulting colored link by $(\mathbf{p}, \mathbf{q}, k)_{\mathrm{qgr}}$. This is an element of $\mathcal{V}_{A_t}(\Sigma_g \times [0,1])$.

Theorem 8.9. *Let $\Phi_{qgr} : \mathbf{\Theta}_N^{\Pi}(\Sigma_g) \to \mathcal{L}_{A_t}(H_g)$ denote the isomorphism of Theorem 8.8. Then*

$$\Phi_{\mathrm{qgr}}[(\mathbf{p}, \mathbf{q}, k) \cdot \theta] = (\mathbf{p}, \mathbf{q}, k)_{\mathrm{qgr}} \cdot \Phi_{\mathrm{qgr}}[\theta],$$

for all $(\mathbf{p}, \mathbf{q}, k) \in \mathbf{H}(\mathbb{Z}_N^g)$ and $\theta \in \mathbf{\Theta}_N^{\Pi}(\Sigma_g)$.

Proof. Consider the commutative diagram

$$\begin{array}{ccccc}
\mathcal{V}_{A_t}(\Sigma_g \times [0,1]) \otimes \mathcal{L}_{A_t}(H_g) & \xrightarrow{\text{cabling}} & \mathcal{L}_N(\Sigma_g) \otimes \mathcal{L}_N(H_g) & \leftarrow & \mathbb{C}(\mathbf{H}(\mathbb{Z}_N^g)) \otimes \mathbf{\Theta}_N^{\Pi}(\Sigma_g) \\
\downarrow & & \downarrow & & \downarrow \\
\mathcal{L}_{A_t}(H_g) & \xrightarrow{\cong} & \mathcal{L}_N(H_g) & \xleftarrow{\cong} & \mathbf{\Theta}_N^{\Pi}(\Sigma_g)
\end{array}$$

where we have used Proposition 8.1 and Theorems 5.5 and 5.7. Using Theorem 8.3 we conclude that the image of $(\mathbf{p}, \mathbf{q}, k)$ in $\mathcal{L}_N(\Sigma_g)$ under the map Φ_N defined in Theorem 5.5 coincides with the image of $(\mathbf{p}, \mathbf{q}, k)_{\mathrm{qgr}}$ under the cabling map, and the theorem is proved. □

Given $\mathbf{p}, \mathbf{q} \in \{0, \ldots, N-1\}^g$, set

$$\Gamma(\mathbf{p}, \mathbf{q}) := V^{p_1}(a_1) V^{p_2}(a_2) \cdots V^{p_g}(a_g) V^{q_1}(b_1) V^{q_2}(b_2) \cdots V^{q_g}(b_g),$$

where $V^j(\gamma) \in \mathcal{V}_{A_t}(\Sigma_g \times [0,1])$ denotes the curve γ colored by the irreducible representation V^j.

Corollary 8.5. *We have*

$$\Phi_{\mathrm{qgr}}[(\mathbf{p}, \mathbf{q}, k) \cdot \theta] = t^{(k - \mathbf{p}^T \mathbf{q})} \Gamma(\mathbf{p}, \mathbf{q}) \cdot \Phi_{\mathrm{qgr}}[\theta],$$

for all $(\mathbf{p}, \mathbf{q}, k) \in \mathbf{H}(\mathbb{Z}_N^g)$ and $\theta \in \mathbf{\Theta}_N^{\Pi}(\Sigma_g)$.

A quantum group model for the finite Heisenberg group

We can describe the reduced linking number skein algebra $\mathcal{L}_N(\Sigma_g)$ as a quotient of $\mathcal{V}_{A_t}(\Sigma_g \times [0,1])$ and within it we can find a quantum group model of the finite Heisenberg group $\mathbf{H}(\mathbb{Z}_N^g)$.

Embed the cylinder $\Sigma_g \times [0,1]$ in the standard way in S^3, so that on each side lies one handlebody, H_g^1 inside and H_g^2 outside. We then have a decomposition of the 3-dimensional sphere as

$$S^3 = H_g^1 \cup (\Sigma_g \times [0,1]) \cup H_g^2. \tag{8.19}$$

Let L be an oriented framed link in $\Sigma_g \times [0,1]$ endowed with a coloring $\mathbf{V}$ by representations of A_t. Insert the colored link in (8.19). Then $\mathbf{V}(L)$ defines a bilinear pairing

$$\mathcal{V}_{A_t}(H_g^1) \otimes \mathcal{V}_{A_t}(H_g^2) \to \mathbb{C}$$

via the Reshetikhin-Turaev invariant in S^3. This pairing descends to a pairing

$$[\cdot,\cdot]_{\mathbf{V}(L)} : \mathcal{L}_{A_t}(H_g^1) \otimes \mathcal{L}_{A_t}(H_g^2) \to \mathbb{C}. \tag{8.20}$$

Because $[\cdot,\cdot]_{qgr}$ is nondegenerate, the bilinear map (8.20) defines a linear map

$$\mathrm{op}(\mathbf{V}(L)) : \mathcal{L}_{A_t}(H_g^1) \to \mathcal{L}_{A_t}(H_g^2),$$

by

$$[\mathrm{op}(\mathbf{V}(L))x, y]_{\mathrm{qgr}} = [x,y]_{\mathbf{V}(L)}.$$

Using the identifications of $\mathcal{L}_{A_t}(H_g^j)$, $j = 1,2$, with $\mathbf{\Theta}_N^{\Pi}(\Sigma_g)$, we deduce that $\mathrm{op}(\mathbf{V}(L)) \in L(\mathbf{\Theta}_N^{\Pi}(\Sigma_g))$.

Example 8.4. Let us consider the operator obtained by decorating the $(1,1)$ curve on the torus by the irreducible representation V^m. The theta series $\theta_j^{\tau}(z)$ is represented in the solid torus by the curve that is the core of the solid torus, decorated by V^j.

$$[\theta_j^{\tau}(z), \theta_k^{\tau}(z)]_{\mathrm{qgr}} = t^{2jk}.$$

The operator $\mathrm{op}(V^m(1,1))$ defined by coloring the $(1,1)$ curve by V^m is determined by requiring that for every $j,k = 0,1,2,\dots,N-1$,

$$[\mathrm{op}(V^m(1,1))\theta_j^{\tau}(z), \theta_k^{\tau}(z)]_{\mathrm{qgr}}$$

is equal to the Reshetikhin-Turaev invariant of the link from Figure 8.19.

This operator coincides with the one defined by the skein $<(m,m)>\in \mathcal{L}_N(\Sigma_g)$ acting on $\mathcal{L}_N(H_g)$.

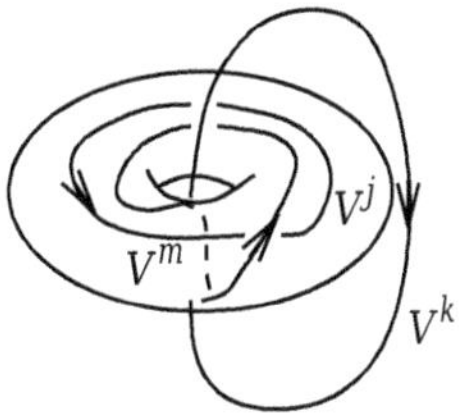

Fig. 8.19 The jk entry of op($\mathbf{V}(1,1)$)

The cabling map

$$\mathcal{V}_{A_t}(\Sigma_g \times [0,1]) \to \mathcal{L}_N(\Sigma_g),$$

is onto, by Theorem 8.9. Let $\mathcal{L}_{A_t}(\Sigma_g)$ be the quotient of $\mathcal{V}_{A_t}(\Sigma_g \times [0,1])$ by the kernel of this map. We thus obtain

Proposition 8.2. *$\mathcal{L}_{A_t}(\Sigma_g)$ is an algebra isomorphic to $\mathcal{L}_N(\Sigma_g)$, which is therefore isomorphic to the algebra of linear operators on the space of theta functions.*

Intuitively the algebra $L(\mathbf{\Theta}_N^{\Pi}(\Sigma_g))$ is an algebra of oriented simple closed curves on Σ_g decorated by irreducible representations of the quantum group A_t.

In particular, $\mathbf{H}(\mathbb{Z}_N^g)$ lies inside $\mathcal{L}_{A_t}(\Sigma_g)$, consisting of the equivalence classes of framed oriented links colored by irreducible representations. Its elements are $(\mathbf{p}, \mathbf{q}, k)_{\mathrm{qgr}}$, $\mathbf{p}, \mathbf{q} \in \{0, 1, \ldots, N-1\}^g$, $k \in \{0, 1, \ldots, 2N-1\}$. This is a consequence of Proposition 5.14.

Given the relationship between $\mathcal{L}_N(\Sigma_g) = L(\mathbf{\Theta}_N^{\Pi}(\Sigma_g))$ and the algebra $\widetilde{\mathbb{C}_t}[\mathbf{U}^{\pm 1}, \mathbf{V}^{\pm 1}]$ generated by the noncommutative exponentials of the quantum torus, we obtain a quantum group description of the quantum torus. We should however recall that $\widetilde{\mathbb{C}_t}[\mathbf{U}^{\pm 1}, \mathbf{V}^{\pm 1}]$ itself is not a quantum group.

In particular, the quantized exponentials of the quantum torus which is the deformation of $\mathcal{J}(\Sigma_g)$ are represented by multicurves on Σ_g colored by irreducible representations of A_t.

Example 8.5. The quantized exponential function

$$\mathrm{op}(e^{2\pi i(2x_1+x_2+2y_1+3y_2)})$$

on the Jacobian variety of Σ_2 is represented in Figure 8.20.

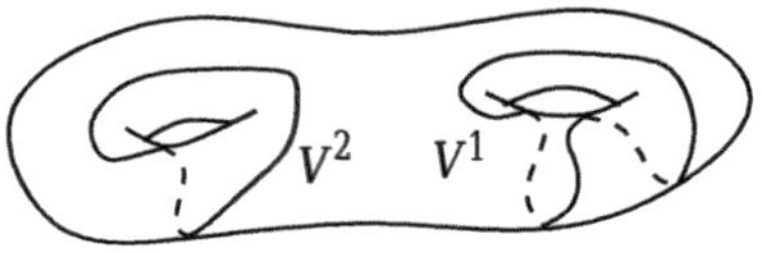

Fig. 8.20 Quantized exponential

The representation of the mapping class group

Let M be a compact oriented 3-dimensional manifold. Denote by $\mathcal{V}'_{A_t}(M)$ the vector space that is freely generated by isotopy classes of links colored by elements of the **representation ring** of A_t (thus we enlarge $\mathcal{V}_{A_t}(M)$ to allow colorings by general representations, not just irreducible, and even differences of representations). There is a multiplicative map from $\mathcal{V}'_{A_t}(M)$ onto $\mathcal{V}_{A_t}(M)$; if L is a link with m components whose coloring $\mathbf{V}$ by the representation ring is written as in (8.17), then this map is given by

$$\mathbf{V}(L) \mapsto \sum_{k_1,\ldots,k_m=0}^{2N-1} c_{1k_1}\cdots c_{mk_m}\mathbf{V}_{k_1,\ldots,k_m}(L);$$

where $\mathbf{V}_{k_1,\ldots,k_m}$ is the coloring of L that decorates the component L_j with the color V^{k_j}.

$\mathcal{V}'_{A_t}(\Sigma_g\times[0,1])$ is an algebra, and the above mentioned map to $\mathcal{V}_{A_t}(\Sigma_g\times[0,1])$ is an algebra homomorphism. Consequently, $\mathcal{L}_{A_t}(H_g)$ is a module over $\mathcal{V}'_{A_t}(\Sigma_g\times[0,1])$. In fact, in view of Corollary 8.4, we may assume that our links are colored by elements of the quotient ring $R(A_t)$.

Definition 8.11. Let

$$\Omega_{A_t} = N^{-\frac{1}{2}}\sum_{k=0}^{N-1} V^k.$$

If L is an oriented framed link in $\Sigma_g\times[0,1]$, we let the **coloring** of L by Ω_{A_t}, denoted by $\Omega_{A_t}(L)$, be the element of $\mathcal{V}'_{A_t}(\Sigma_g\times[0,1])$ obtained by coloring each component of L by Ω_{A_t}.

Theorem 8.10. *Let $h_L \in MCG(\Sigma_g)$ be a diffeomorphism that is represented by surgery on a framed link L. By Theorem 8.8 we may consider the discrete Fourier transform $\mathcal{F}(h_L)$ as an endomorphism of $\mathcal{L}_{A_t}(H_g)$. This endomorphism is given, up to multiplication by a constant of absolute value 1, by*

$$\mathcal{F}(h_L)(\beta) = \Omega_{A_t}(L)\cdot\beta, \quad \beta\in\mathcal{L}_{A_t}(H_g).$$

Proof. Since the image of $\Omega_{A_t}(L)$ in $\mathcal{L}_N(\Sigma_g \times [0,1])$ under the cabling map coincides with $\Omega(L)$, this is a consequence of Theorem 7.9 and the fact that the cabling map is equivariant. □

Here is an alternative way, for the case of the torus, of computing the discrete Fourier transform via quantum groups. The reader should recall Example 7.6. Let h be an element of the mapping class group of the torus, which we represent as

$$h = ST^{a_1}ST^{a_2}\cdots ST^{a_n}S.$$

Then up to multiplication by a constant the discrete Fourier transform $\mathcal{F}(h)$ has the matrix with j,k entry equal to the Reshetikhin-Turaev invariant of the link in Figure 7.27 with the left-most component colored by V^j, the right-most component colored by V^k and all other component colored by Ω_{A_t}.

For S and T there are even simpler expressions.

Example 8.6. The matrix of S is (up to multiplication by a constant) the matrix with the j,k entry equal to the Reshetikhin-Turaev invariant of the framed link in Figure 8.21.

Remark 8.7. Note that only half of the irreducible representations are used. If we were to use all irreducible representations, the matrix of S would become **singular**.

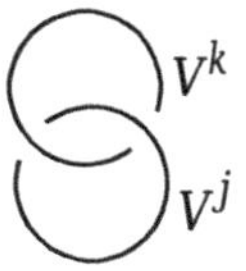

Fig. 8.21 Computation of the S-matrix using the quantum group

Example 8.7. The matrix of T is (up to multiplication by a constant) the diagonal matrix whose jth entry is the Reshetikhin-Turaev invariant of the framed link in Figure 8.22.

Fig. 8.22 Computation of the T-matrix using the quantum group

The Murakami-Ohtsuki-Okada as a Reshetikhin-Turaev manifold invariant

We conclude with the quantum group description of the Murakami-Ohtsuki-Okada invariant.

Set the quantum dimension $\dim_q V$ of a representation V to be the Reshetikhin-Turaev invariant of the trivial framed knot colored by that representation. In our situation,

$$\dim_q(V^j) = 1, \quad j = 0, 1, \ldots, 2N-1.$$

For a coloring $\mathbf{V}$ of a link L, let $\dim_q \mathbf{V}$ be the product of the dimensions of the colors of the link components. Then the Murakami-Ohtsuki-Okada link invariant of a closed, compact, oriented manifold M obtained by surgery on a framed link $L \in S^3$ is

$$Z_N(M) = \sum_{\mathbf{V}} \dim_q \mathbf{V} < \mathbf{V}(L) >,$$

where the sum is taken over all colorings of L by the irreducible representations $V^0, V^1, \ldots, V^{N-1}$. Again, we insist that **not** all irreducible representations of the quantum group are used.

We leave it to the reader to model the TQFT from Chapter 7 using quantum groups. The construction should parallel that from [Reshetikhin and Turaev (1991)] which was done for the quantum group of $SU(2)$.

Conclusions to the eighth chapter

In this chapter we constructed the **quantum group that models theta functions** by enforcing it correspond to the linking number via the Reshetikhin-Turaev link invariant.

For a closed Riemann surface, we represented its **theta series** as **curves in the handlebody** bounded by that surface **colored by irreducible representations** of the quantum group. We represented the **finite Heisenberg group** and the **discrete Fourier transform** associated to a homeomorphism as **curves in the cylinder over the surface colored by representations**, respectively by **linear combinations of representations**.

Chapter 9

An epilogue – Abelian Chern-Simons theory

In this concluding chapter we explain how the material from this book relates to Chern-Simons theory the way it is done from Witten's point of view. The discussion of concepts and results is sketchy, the reader can find more details about the algebraic geometry in [Griffiths and Harris (1994)] and [Farkas and Kra (1991)] and about the quantum field theory in [Tyurin (2003)] and [Manoliu (1998a)].

9.1 The Jacobian variety as a moduli space of connections

The Jacobian variety $\mathcal{J}(\Sigma_g)$ is also the moduli space $\mathcal{M}_g(U(1))$ of flat $\mathfrak{u}(1)$-connections on the surface Σ_g. To see why this is so, recall the notation from §4.1. Fix a point $p_0 \in \Sigma_g$ and consider the **Abel map** $\mathsf{u} : \Sigma_g \to \mathcal{J}(\Sigma_g)$,

$$\mathsf{u}(p) = \left(\int_{p_0}^{p} \zeta_1, \int_{p_0}^{p} \zeta_2, \dots \int_{p_0}^{p} \zeta_g \right) (\text{mod } \mathbf{\Lambda}(I, \Pi)).$$

Now recall from algebraic geometry that a **divisor** on Σ_g is a formal finite sum of points in Σ_g with integer coefficients

$$D = \sum_p n_p p.$$

The **degree** of the divisor is the sum of its coefficients. The set of divisors endowed with the addition $\sum_p n_p p + \sum_p n'_p p = \sum_p (n_p + n'_p) p$ is a abelian group, denoted $\mathrm{Div}(\Sigma_g)$.

Example 9.1. If $f : \Sigma_g \to \mathbb{C}$ is a meromorphic function then it has a naturally associated divisor

$$(f) = \sum_p (ord_p(f)) p,$$

where the sum is taken over all poles and zeros, and $ord_p(f)$ is the order of the zero or the negative of the order of the pole.[1] The divisors associated to meromorphic functions are called **principal**. Because a meromorphic function has the same number of zeros as poles, with multiplicity counted, the degree of a principal divisor is zero. We denote the set of principal divisors by $\mathrm{Prin}(\Sigma_g)$; it is a subgroup of $\mathrm{Div}^0(\Sigma_g)$, where the latter denotes the degree zero divisors.

The Abel map can be extended to divisors by linearity,

$$\mathsf{u}\left(\sum_p n_p p\right) = \sum_p n_p \mathsf{u}(p) (\mathrm{mod}\ \mathbf{\Lambda}(I, \Pi)).$$

One has the following two results, whose proofs the reader can find in [Farkas and Kra (1991)].

Theorem 9.1. *(Abel) A divisor D of degree zero is principal if and only if $\mathsf{u}(D) = \mathbf{0}$ in $\mathcal{J}(\Sigma_g)$.*

Theorem 9.2. *(Jacobi's Inversion Theorem) Every point in $\mathcal{J}(\Sigma_g)$ is the image of a degree 0 divisor.*

Consequently, the Jacobian variety of the surface Σ_g parametrizes the set of divisors of degree zero modulo principal divisors. Moreover, the map

$$\mathsf{u} : \mathrm{Div}^0(\Sigma_g)/\mathrm{Prin}(\Sigma_g) \to \mathcal{J}(\Sigma_g)$$

is a group isomorphism, where we recall from §4.1 that $\mathcal{J}(\Sigma_g)$ is a product of circles and therefore has a group structure.

To each divisor one associates a holomorphic line bundle. This is done by representing the divisor as what is called a Cartier divisor. For this choose a covering of Σ_g by simply connected charts $(U_j)_{j\in J}$, which we identify with complex disks in the plane. Then, for a divisor $D = \sum_p n_p p$, in each chart U_j choose a rational function f_j whose poles and zeros are precisely $p \in U_j$ with $n_p \neq 0$ and such that the order of f at such a point p is n_p.

Then for $j, k \in J$, the functions

$$c_{jk} = \frac{f_j}{f_k}$$

are holomorphic on $U_j \cap U_k$. They satisfy the conditions

$$c_{jk} = (c_{kj})^{-1}, \quad c_{jk}c_{kl}c_{lj} = 1,$$

[1] So we make a convention that zeros have positive order and poles have negative order.

so they define a line bundle, usually denoted by $[D]$. The degree of the line bundle is by definition the degree of the divisor.

Two line bundles are equivalent if one can be obtained from the other by tensoring with a trivial line bundle. The set of equivalence classes of line bundles, endowed with the tensor product, is a group, the **Picard group** $\mathrm{Pic}(\Sigma_g)$. The map

$$[\quad]: \mathrm{Div}(\Sigma_g) \to \mathrm{Pic}(\Sigma_g)$$

is a group homomorphism, whose kernel is the set of principal divisors. Consequently the equivalence classes of degree zero divisors forms a group, denoted by $\mathrm{Pic}^0(\Sigma_g)$. The map

$$[\quad] \circ \mathsf{u}^{-1}: \mathcal{J}(\Sigma_g) \to \mathrm{Pic}^0(\Sigma_g)$$

is a group isomorphism.

One should point out that the degree is a complete invariant from the purely topological point of view. So the Jacobian parametrizes the moduli space of holomorphic topologically trivial line bundles.

Theorem 9.3. *(Poincaré) Every line bundle $L \in Pic^0(\Sigma_g)$ admits a holomorphic connection with curvature zero, which is given by a character $\chi : \pi_1(\Sigma_g) \to \mathbb{C}\backslash\{0\}$.*

Such a connection is called **flat**. Every character gives rise to a connection, and hence corresponds to a line bundle, but several characters can correspond to the same line bundle. In fact, two characters χ and χ' correspond to the same line bundle if $\chi/|\chi| = \chi'/|\chi'|$. And we have a smooth map

$$\mathcal{J}(\Sigma_g) \to \{\rho : \pi_1(\Sigma_g) \to U(1)\},$$

which identifies, as smooth manifolds, the Jacobian variety with the $U(1)$-representation variety of the fundamental group of Σ_g. The latter can also be interpreted as the moduli space $\mathcal{M}_g(U(1))$ of flat $\mathfrak{u}(1)$-connections on the surface Σ_g. In plain words this is the space of $\mathfrak{u}(1)$-valued 1-forms on Σ_g modulo the equivalence relation $A \sim A'$ if

$$A' = g^{-1}Ag + g^{-1}dg,$$

for some smooth function $g : \Sigma_g \to U(1)$.

The moduli space $\mathcal{M}_g(U(1))$ has a complex structure defined as follows (see for example [Hitchin (1990)]). The tangent space to $\mathcal{M}_g(U(1))$ at an arbitrary point is $H^1(\Sigma_g, \mathbb{R})$, which, by Hodge theory (see Chapter 4), can be identified with the space of real-valued harmonic 1-forms on Σ_g. The complex structure is given by

$$J\alpha = - * \alpha,$$

where α is a harmonic form. In local coordinates, if $\alpha = udx + vdy$, then $J(udx + vdy) = vdx - udy$.

If we identify the space of real-valued harmonic 1-forms with the space of holomorphic 1-forms $H^{(1,0)}(\Sigma_g)$ by the map Φ given in local coordinates by $\Phi(udx + vdy) = (u - iv)dz$, then the complex structure becomes multiplication by i in $H^{(1,0)}(\Sigma_g)$.

The moduli space is a torus obtained by exponentiation

$$\mathcal{M}_g(U(1)) = H^1(\Sigma_g, \mathbb{R})/\mathbb{Z}^{2g}.$$

If we choose a basis of the space of real-valued harmonic forms $\alpha_1, \alpha_2, \ldots, \alpha_g, \beta_1, \ldots, \beta_g$ such that

$$\int_{a_j} \alpha_k = \delta_{jk}, \quad \int_{b_j} \alpha_k = 0, \quad \int_{a_j} \beta_k = 0, \quad \int_{b_j} \beta_k = \delta_{jk}, \tag{9.1}$$

then the above $\mathbb{Z}^{2g}$ is the period lattice of this basis.

On the other hand, if $\zeta_1, \zeta_2, \ldots, \zeta_g$, are the holomorphic forms introduced in §4.1, and if $\alpha_j' = \Phi^{-1}(\zeta_j)$ and $\beta_j' = \Phi^{-1}(-i\zeta_j)$, $j = 1, 2, \ldots, g$, then one can compute that

$$\int_{a_j} \alpha_k' = \delta_{jk}, \int_{b_j} \alpha_k' = \operatorname{Re} \pi_{jk}, \int_{a_j} \beta_k' = 0, \int_{b_j} \beta_k' = \operatorname{Im} \pi_{jk}.$$

The basis $\alpha_1', \ldots, \alpha_g', \beta_1', \ldots, \beta_g'$ determines coordinates (X', Y') in the tangent space to $\mathcal{M}_g(U(1))$. If we consider the change of coordinates $X' + iY' = X + \Pi Y$, then the moduli space is the quotient of $\mathbb{C}^g$ by the integer lattice $\mathbb{Z}^{2g}$. This is exactly what has been done in Section 4.1 to obtain the Jacobian variety. This shows that the complex structure on the Jacobian variety coincides with the standard complex structure on the moduli space of flat $\mathfrak{u}(1)$-connections on the surface.

The moduli space $\mathcal{M}_g(U(1))$ has a symplectic structure defined by the **Atiyah-Bott form** [Atiyah and Bott (1982)]. This form is given by

$$\omega(\alpha, \beta) = -\int_{\Sigma_g} \alpha \wedge \beta,$$

where α, β are real valued harmonic 1-forms, i.e. vectors in the tangent space to $\mathcal{M}_g(U(1))$. If α_j, β_j, $j = 1, 2, \ldots, g$, are as in (9.1), then $\omega(\alpha_j, \alpha_k) = \omega(\beta_j, \beta_j) = 0$ and $\omega(\alpha_j, \beta_k) = \delta_{jk}$ (which can be seen by identifying the space of real-valued harmonic 1-forms with $H^1(\Sigma_g, \mathbb{R})$ and using the topological definition of the cup product). This shows that the Atiyah-Bott form coincides with the symplectic form introduced in Section 4.1.

We conclude the Jacobian variety is homeomorphic with the moduli space of flat $\mathfrak{u}(1)$-connections, under a homeomorphism that preserves the symplectic structure and the complex structure.

For a $\mathfrak{u}(1)$-connection A and curve γ on the surface, we denote by $\mathrm{hol}_\gamma(A)$ the holonomy of A along γ. Let V be a representation of $U(1)$ and let us consider the trace of holonomies in this representation. The map

$$A \mapsto \mathrm{trace}(\mathrm{hol}_\gamma(A))$$

induces a function $W_{\gamma,V}$ on the moduli space $\mathcal{M}_g(U(1))$ called **Wilson line**. Since we work with the group $U(1)$, the holonomy is just a complex number of absolute value 1, and the trace is the number itself. If

$$[\gamma] = (\mathbf{p}, \mathbf{q}) \in H_1(\Sigma_g, \mathbb{Z}),$$

then the Wilson line associated to γ and to the 1-dimensional representation V given by $\lambda \cdot z = \lambda z$ is the function

$$(x, y) \mapsto e^{2\pi i(\mathbf{p}^T\mathbf{x}+\mathbf{q}^T\mathbf{y})}.$$

By taking products, we can associate Wilson lines to multicurves. We obtain an isomorphism between $H_1(\Sigma_g, \mathbb{Z})$ and the group of exponentials on the Jacobian variety.

In §4.2.1 we gave one explanation why the group of exponential functions on the Jacobian variety is isomorphic to the first homology group of the surface. The Wilson line picture gives another, perhaps better, explanation of this fact.

Note that the Wilson line associated to the curve γ and the representation $V^{\otimes n}$ is

$$(x, y) \mapsto e^{2\pi i n(\mathbf{p}^T\mathbf{x}+\mathbf{q}^T\mathbf{y})},$$

where $V^{\otimes n}$ denotes V tensored with itself n times, $n \geq 1$.

9.2 Weyl quantization versus quantum group quantization of the Jacobian variety

The goal is to quantize the moduli space of flat $\mathfrak{u}(1)$-connections on the closed Riemann surface Σ_g endowed with the Atiyah-Bott symplectic form. One procedure has been outlined in Chapter 4; it is Weyl quantization on the $2g$-dimensional torus in the holomorphic polarization.

Another quantization procedure, which works for general gauge groups, has been introduced by Witten in [Witten (1989)] using Feynman path integrals. Witten's approach is based on the Chern-Simons Lagrangian,

$$L(A) = \frac{1}{4\pi}\int_{\Sigma_g\times[0,1]} \mathrm{tr}\left(A\wedge dA + \frac{2}{3}A\wedge A\wedge A\right),$$

where A is a $\mathfrak{u}(1)$ connection, or in the general case a $\mathfrak{g}$-connection where $\mathfrak{g}$ is the Lie algebra of a compact Lie group G. With Planck's constant being $h = 1/N$ with N an even integer, the quantization of a Wilson line is the "integral" operator $\mathrm{op}(W_{\gamma,V})$ with "kernel"

$$< A_1|\mathrm{op}(W_{\gamma,V})|A_2 >= \int_{\mathcal{M}_{A_1,A_2}} e^{iNL(A)}W_{\gamma,V}(A)\mathcal{D}A,$$

where A_1 and A_2 are conjugacy classes of flat connections on Σ_g, modulo the gauge group, and A is a conjugacy class of a connection on $\Sigma_g \times [0,1]$, modulo the gauge group, such that $A|\Sigma_g \times \{0\} = A_1$ and $A|\Sigma_g \times \{1\} = A_2$.

States are "linear functionals" defined as well by path integrals of the form

$$< A_1|\mathrm{state}(W_{\gamma,V}) >= \int_{\mathcal{M}_{A_1}} e^{iNL(A)}W_{\gamma,V}(A)\mathcal{D}A,$$

where A_1 is the conjugacy class of flat connections on $\Sigma_g = \partial H_g$, modulo the gauge group, and A is a conjugacy class of a connection on H_g, modulo the gauge group, such that $A|\partial H_g = A_1$.

The quotation marks are motivated by the fact that the path integrals do not have a rigorous mathematical formulation. They should be thought of as averages over the space of all (flat) connections, with the averaging weight defined by the exponential of the Chern-Simons Lagrangian. According to Witten, states and observables should be representable as skeins in the skein modules of the linking number.

Both quantization models, Weyl and Feynmann integral, are rich in symmetries. If we accept that the Feynmann integral model yields a combinatorial scheme for quantization based on skein models, then this book gives an explanation of why the two quantization schemes coincide.

The relationship between Weyl quantization and Chern-Simons theory was first noticed in [Frohman and Gelca (2000)] and [Gelca and Uribe (2010)] for the gauge group $SU(2)$. Later Jørgen E. Andersen [Andersen (2005)] noticed the same relationship for abelian Chern-Simons theory.

There is a general philosophy for quantizing moduli spaces of flat $\mathfrak{g}$-connections on surfaces using quantum groups (see [Alexeev and Schomerus (1996)], [Gelca and Uribe (2010)]).

The states are linear combinations of loops or graphs colored by representations of the quantum group associated to $\mathfrak{g}$, such as the ones from Figure 8.18 or Figure 9.1, with choices of morphisms of the form $V^j \otimes V^k \to V^l$ at each trivalent vertex with adjacent edges colored by representations V^j, V^k, V^l. A certain family of such colored loops/graphs, whose structure is dictated by the representation theory of the quantum group, defines an orthogonal basis of the Hilbert space of the quantization. For abelian Chern-Simons theory, this basis is shown in Figure 8.18.

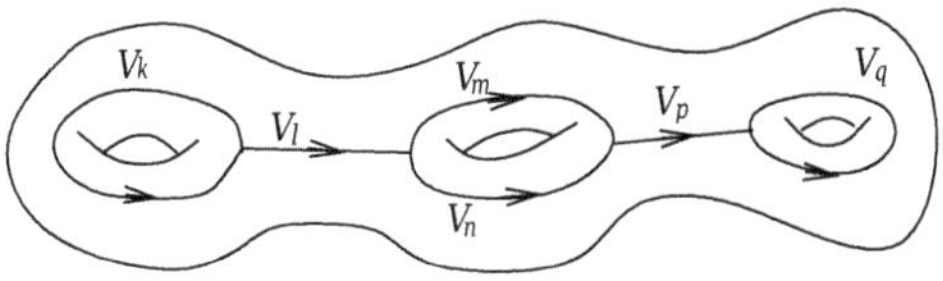

Fig. 9.1 States in the quantum group quantization of $\mathcal{M}_g(G)$

The quantum observable associated to the Wilson line $W_{\gamma,V}$ is the coloring of the curve $\gamma \in \Sigma_g$ by a representation of the quantum group that is canonically associated to V. In our situation, to the representation V of $U(1)$ introduced above one associates the representation V^1 of $\mathcal{U}_t(\mathfrak{u}_{\mathbb{C}}(1))$ defined in §8.2.1, and to $V^{\otimes n}$ one associates $(V^1)^n = V^1 \otimes V^1 \otimes \cdots \otimes V^1$.

The quantum observable is the linear operator defined by the bilinear form on the Hilbert space obtained by inserting the surface between the two handlebodies in the standard Heegaard decomposition of S^3, inserting basis elements in each handlebody, and then evaluating the quantum invariant of the resulting link/graph in S^3. This procedure was outlined in Chapter 8 for abelian Chern-Simons theory.

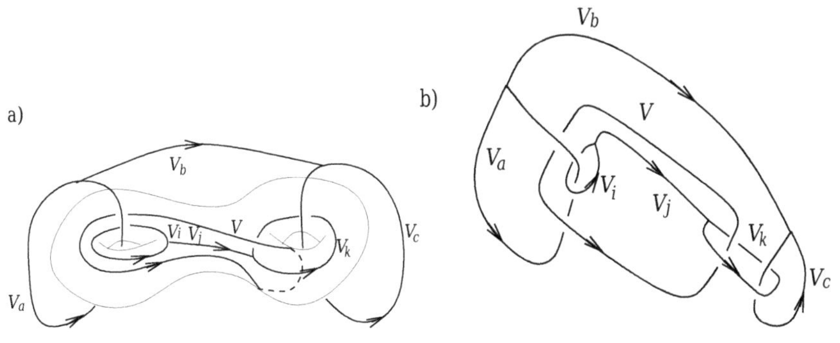

Fig. 9.2 Observables in the quantum group quantization of $\mathcal{M}_g(G)$

As a corollary of Theorem 8.9, we obtain

Theorem 9.4. *[Gelca and Hamilton (2012)] The Weyl quantization and the quantum group quantization of the moduli space of flat* $\mathfrak{u}(1)$*-connections on a closed surface are unitary equivalent.*

A similar result for the Chern-Simons theory with gauge group $SU(2)$, just on the **torus**, was obtained in [Gelca and Uribe (2010)].

Other points of view

We conclude our discussion by pointing out that there exists a different approach to abelian Chern-Simons theory in [Andersen (2005)], where everything is recovered from the **heat equation for theta functions** (the heat equation was introduced in §1.1). Also, for a different, more geometric construction of the topological quantum field theory from Chapter 7, albeit for 3-manifolds without embedded links, see [Manoliu (1998a)], [Manoliu (1998b)]. Finally, a very abstract construction in the framework of modular categories was given in [Sterling (2011)].

Bibliography

Alexeev, A.Yu. and Schomerus, V. (1996). Representation theory of Chern-Simons observables, *Duke Math. Journal* **85**, pp. 447–510.

Andersen, J.E. (2005). Deformation quantization and geometric quantization of abelian moduli spaces, *Commun. Math. Phys.* **255**, pp. 727–745.

Atiyah, M.F. (1988). Topological quantum field theory, *Publications Mathématiques de l'I.H.É.S.* **68**, pp. 175–186.

Atiyah, M.F. (1990). *The Geometry and Physics of Knots* (Cambridge Univ. Press).

Atiyah, M.F. and Bott, R. (1982). The Yang-Mills equations over a Riemann surface, *Phil. Trans. Royal. Soc. A* **308**, pp. 523–615.

Bargmann, V. (1961). On a Hilbert space of analytic functions and an associated integral transform, part 1, *Comm. Pure Appl. Math.* **14**, pp. 187–214.

Bayen, F., Flato, M., Fronsdal, C., Lichnerowicz, A. and Sternheimer, D. (1978a). Deformation theory and quantization I. Deformations of symplectic structures, *Annals of Physics* **111**, pp. 61–110.

Bayen, F., Flato, M., Fronsdal, C., Lichnerowicz, A. and Sternheimer, D. (1978b). Deformation theory and quantization II. Physical applications, *Annals of Physics* **111**, pp. 111–151.

Blanchet, C., Habegger, N., Masbaum, G. and Vogel, P. (1995). Topological quantum field theories derived from the Kauffman bracket, *Topology* **34**, pp. 883–927.

Bloch, A., Golse, F., Paul, T. and Uribe, A. (2003). Dispersionless Toda and Toeplitz operators, *Duke Math. J.* **117**, pp. 157–196.

Brendon, G.E. (1993). *Topology and Geometry* (Springer).

Cerf, J. (1970). La stratification naturelle des espaces de fonctions différentiable réelles et le théorème de la pseudoisotopie, *Publ. I.H.E.S* **39**, pp. 185–353.

Chern, S.S. and Simons, J. (1974). Characteristic forms and geometric invariants, *Ann. of Math.* **99(1)**, pp. 48–69.

Connes, A. (1994). *Noncommutative Geometry* (Academic Press Inc.).

Dehn, M. (1987). *Papers on Group Theory and Topology (Translated from German and with introductions and an appendix by J. Stillwell. With an appendix by O. Schreier)* (Springer).

Drinfeld, V.G. (1986a). Quantum groups, *Zap. Nauchn. Sem. LOMI* **155**, pp. 18–49.

Drinfeld, V.G. (1986b). Quantum groups, in *Proceedings of the International Congress of Mathematicians, Vol. 1* (Academic Press), pp. 798–820.

Egorov, Yu.V. (1969). The canonical transformations of pseudodifferential operators (Russian), *Uspehi Mat. Nauk* **24**, pp. 235–236, no 5 (149).

Faddeev, L.D. and Yakubovskii, O.A. (2009). *Lectures on Quantum Mechanics for Mathematics Students* (Amer. Math. Soc.).

Farb, B. and Margalit, D. (2011). *A Primer on Mapping Class Groups* (Princeton Univ. Press).

Farkas, H.M. and Kra, I. (1991). *Riemann Surfaces*, 2nd edn. (Springer).

Folland, G. (1989). *Harmonic Analysis in Phase Space* (Princeton University Press).

Frohman, Ch. and Gelca, R. (2000). Skein modules and the noncommutative torus, *Trans. Amer. Math. Soc.* **352**, pp. 4877–4888.

Fulton, W. and Harris, J. (1991). *Representation Theory* (Springer).

Gelca, R. and Hamilton, A. (2012). Classical theta functions in a quantum group perspective, ArXiv:1209.1135.

Gelca, R. and Hamilton, A. (2014). The topological quantum field theory associated to classical theta functions, `http://www.math.ttu.edu/~\rgelca/abeliancstqft.pdf`.

Gelca, R. and Uribe, A. (2003). The Weyl quantization and the quantum group quantization of the moduli space of flat $SU(2)$-connections on the torus are the same, *Commun. Math. Phys.* **233**, pp. 493–512.

Gelca, R. and Uribe, A. (2010). Quantum mechanics and nonabelian theta functions for the gauge group $SU(2)$, ArXiv:1007.2010.

Gelca, R. and Uribe, A. (2014). From classical theta functions to topological quantum field theory, in I. Gitler, E. Lupercio, H.G. Compéan and F. Turrubiates (eds.), *The influence of Solomon Lefschetz in geometry and topology: 50 years of Mathematics at Cinvestav* (Contemporary Math., Amer. Math. Soc.).

Gomez, C., Ruiz-Altaba, M. and Sierra, G. (1996). *Quantum Groups in Two-Dimensional Physics* (Cambridge Univ. Press).

Griffiths, Ph. (1994). *Introduction to Algebraic Curves* (John Wiley and Sons).

Griffiths, Ph. and Harris, J. (1994). *Principles of Algebraic Geometry* (John Wiley and Sons).

Hall, B.C. (2000). Holomorphic methods in analysis and mathematical physics , in S. Pérez-Esteva and C. Villegas-Blas (eds.), *First Summer School in Analysis and Mathematical Physics, Quantization, the Segal-Bargmann Transform and Semiclassical Analysis* (AMS), pp. 1–60.

Hall, B.C. (2013). *Quantum Theory for Mathematicians* (Springer).

Hamstrom, M.E. (1962). Some global properties of the space of homeomorphisms of a disk with holes, *Duke Math. J.* **29**, pp. 657–662.

Hatcher, A. and Thurston, W. (1980). A presentation of the mapping class group of a closed orientable surface, *Topology* **19**, pp. 221–237.

Hempel, J. (1976). *3-Manifolds* (Princeton Univ. Press).

Hirsch, M.W. (1976). *Differential Topology* (Springer).

Hitchin, N. (1990). Flat connections and geometric quantization, *Commun. Math. Phys.* **131**, pp. 347–380.

Houzel, C. (1978). Fonctions elliptiques et intégrales abéliennes, in J. Dieudonné (ed.), *Abregé d'Histoire des Mathématiques* (Hermann), pp. 1–114.

Humphries, S. (1977). Generators for the mapping class group, in *Proc. Second Sussex Conf, Lecture Notes in Math. 722* (Springer), pp. 44–47.

Jaco, W. (1997). *Lectures on Three-Manifold Topology* (Amer. Math. Soc.).

Jimbo, M. (1985). A q-difference analogue of $u_{\mathfrak{g}}$ and the Yang-Baxter equation, *Lett. Math. Phys.* **10**, pp. 63–69.

Jones, V.F.R. (1985). Polynomial invariants of knots via von Neumann algebras, *Bull. Amer. Math. Soc.* **12**, pp. 103–111.

Kashiwara, M. and Schapira, P. (1990). *Sheaves on Manifolds* (Springer).

Kassel, Ch. (1995). *Quantum Groups* (Springer).

Kač, V. and Peterson, D.H. (1984). Infinite-dimensional Lie algebras, theta functions and modular forms, *Adv. in Math.* **53**, pp. 125–264.

Kirby, R. (1978). A calculus for framed links in S^3, *Inventiones Math.* **45**, pp. 35–56.

Kirby, R. and Melvin, P. (1991). The 3-manifold invariants of Witten and Reshetikhin-Turaev for $sl(2)$, *Inventiones Math.* **105**, pp. 473–545.

Kirillov, A.N. and Reshetikhin, N.Yu. (1989). Representations of the algebra $u_q(sl_2)$, q-orthogonal polynomials and invariants of links, *Adv. Series in Math. Phys.* **7**, pp. 285–339.

Lang, S. (1994). *Algebraic Number Theory*, 2nd edn. (Springer).

Lickorish, W.B.R. (1962). A representation of orientable combinatorial 3-manifolds, *Ann. of Math.* **76**, pp. 531–540.

Lion, G. and Vergne, M. (1980). *The Weil Representation, Maslov Index, and Theta Series* (Birkhäuser).

Manin, Yu.I. (1990). Quantized theta functions, *Progress of Theor. Phys. Supplement* **102**.

Manin, Yu.I. (2001). Theta functions, quantum tori and Heisenberg groups, *Letters in Math. Phys.* **56:3**, pp. 295–320.

Manin, Yu.I. (2004). Functional equations for quantum theta functions, *Publ. Res. Inst. Math. Sci. Kyoto* **40:3**, pp. 605–624.

Manoliu, M. (1997). Quantization of symplectic tori in a real polarization, *J. Math. Phys.* **38**, pp. 2219–2254.

Manoliu, M. (1998a). Abelian Chern-Simons theory I, a topological quantum field theory, *J. Math. Phys.* **39**, pp. 170–206.

Manoliu, M. (1998b). Abelian Chern-Simons theory II, a functional integral approach, *J. Math. Phys.* **39**, pp. 207–217.

Melvin, P. (1984). 4-dimensional oriented bordism, *Contemporary Math.* **35**, pp. 399–405.

Moise, E. (1952). Affine structures in 3-manifolds. V. The triangulation theorem and Hauptvermutung, *Ann. of Math.* **56**, pp. 96–114.

Mumford, D. (1970). *Abelian Varieties* (Oxford Univ. Press).

Mumford, D. (1983). *Tata Lectures on Theta* (Birkhäuser).

Munkres, J. (1960). Obstructions to the smoothing of piecewise-differentiable homeomorphisms, *Ann. of Math.* **72**, pp. 521–554.

Munkres, J. (2000). *Topology*, 2nd edn. (Pearson).

Murakami, H., Ohtsuki, T. and Okada, M. (1992). Invariants of three-manifolds derived from linking matrices of framed links, *Osaka J. Math.* **29**, pp. 545–572.

Olds, C.D. (1963). *Continued Fractions* (L.W. Singer Co.).

Przytycki, J.H. (1991). Skein modules of 3-manifolds, *Bull. Pol. Acad. Sci.* **39**, pp. 91–100.

Przytycki, J.H. (1998). A q-analogue of the first homology group of a 3-manifold, in L. Coburn and M. Rieffel (eds.), *Contemporary Mathematics 214, Perspectives on Quantization (Proceedings of the joint AMS-IMS-SIAM Conference on Quantization, Mount Holyoke College, 1996)* (AMS), pp. 135–144.

Reidemeister, K. (1926). Elementare Begründung der Knotentheorie, *Abh. Math. Sem. Univ. Hamburg* **5**, pp. 24–32.

Reshetikhin, N.Yu. and Turaev, V.G. (1990). Ribbon graphs and their invariants derived from quantum groups, *Commun. Math. Phys.* **127**, pp. 1–26.

Reshetikhin, N.Yu. and Turaev, V.G. (1991). Invariants of 3-manifolds via link polynomials and quantum groups, *Inventiones Math.* **103**, pp. 547–597.

Rieffel, M. (1989). Deformation quantization of Heisenberg manifolds, *Commun. Math. Phys.* **122**, pp. 531–562.

Rokhlin, V.A. (1952). New results in the theory of 4-dimensional manifolds, *Dokl. Akad.Nauk. SSSR* **84**, pp. 221–224.

Rolfsen, D. (2003). *Knots and Links* (AMS Chelsea Publ.).

Schwarz, A. (1978). The partition function of degenerate quadratic functional and Ray-Singer invariants, *Lett. Math. Phys.* **2**, pp. 247–252.

Segal, I. (1962). Mathematical characterization of the physical vacuum for a linear Bose-Einstein field, *Illinois J. Math.* **6**, pp. 500–523.

Segal, I. (1963). Mathematical problems of relativistic physics, Chap. VI , in M. Kac (ed.), *Proceedings of the Summer Seminar, Boulder, Colorado, 1960, Vol. II* (Lectures in Applied Mathematics, Amer. Math. Soc.).

Serre, J.P. (1973). *A Course in Arithmetic* (Springer).

Sikora, A. (2000). Skein modules and TQFT , in C. M. Gordon (ed.), *Proc. of the Int. Conf. on Knot Theory* (World Scientific), pp. 436–439.

Smale, S. (1959). Diffeomorphisms of the 2-sphere, *Proc. Amer. Math. Soc.* **10**, pp. 621–626.

Śniatycki, J. (1980). *Geometric Quantization and Quantum Mechanics* (Springer).

Spanier, E. (1966). *Algebraic Topology* (McGraw-Hill).

Sterling, S. (2011). *Abelian Chern-Simons Theory with Toral Gauge Group, Modular Tensor Categories and Group Categories* (ProQuest, UMI Dissertation Publ.).

Tang, X., Weinstein, A. and Zhu, C. (2007). Hopfish algebras, *Pacific J. Math.* **231**, pp. 83–121.

Turaev, V.G. (1989). Algebras of loops on surfaces, algebras of knots, and quantization , in C.N. Yang and M.L. Ge (eds.), *Adv. Ser. in Math. Phys. 9*, pp.

59–95.
Turaev, V.G. (1994). *Quantum Invariants of Knots and 3-manifolds* (de Gruyter Studies in Math.).
Tyurin, A. (2003). *Quantization, Classical and Quantum Field Theory and Theta Functions* (Amer. Math. Soc.).
Walker, K. (1991). On Witten's 3-manifold invariants, `http://canyon23.net/math/1991TQFTNotes.pdf`.
Wall, C.T.C. (1969). Non-additivity of the signature, *Inventiones Math.* **7**, pp. 269–274.
Weil, A. (1958). *Variétés Kählériennes* (Hermann).
Weil, A. (1964). Sur certains groupes d'operateurs unitaires, *Acta Math.* **111**, pp. 143–211.
Whitehead, J.H.C. (1961). Manifolds with transverse fields in euclidean space, *Ann. of Math.* **73**, pp. 369–384.
Witten, E. (1989). Quantum field theory and the Jones polynomial, *Commun. Math. Phys.* **121**, pp. 351–399.
Woodhouse, N. (1980). *Geometric Quantization* (Clarendon Press Oxford).

Index

www.ingramcontent.com/pod-product-compliance
Ingram Content Group UK Ltd.
Pitfield, Milton Keynes, MK11 3LW, UK
UKHW022109170926
8624IPUK00024B/326

9 789814 520577